Lecture Notes in Artificial Intelligence   7414

Subseries of Lecture Notes in Computer Science

Lecture Notes in Artificial Intelligence   6214

Subseries of Lecture Notes in Computer Science

Tianrui Li   Hung Son Nguyen   Guoyin Wang
Jerzy Grzymala-Busse   Ryszard Janicki
Aboul Ella Hassanien   Hong Yu (Eds.)

Rough Sets and Knowledge Technology

7th International Conference, RSKT 2012
Chengdu, China, August 17-20, 2012
Proceedings

 Springer

Series Editors

Randy Goebel, University of Alberta, Edmonton, Canada
Jörg Siekmann, University of Saarland, Saarbrücken, Germany
Wolfgang Wahlster, DFKI and University of Saarland, Saarbrücken, Germany

Volume Editors

Tianrui Li
Southwest Jiaotong University, Chengdu, P.R. China
E-mail: trli@swjtu.edu.cn

Hung Son Nguyen
University of Warsaw, Poland
E-mail: son@mimuw.edu.pl

Guoyin Wang
Chongqing University of Posts and Telecommunications, P.R. China
E-mail: wanggy@cqupt.edu.cn

Jerzy Grzymala-Busse
University of Kansas, Lawrence, KS, USA
E-mail: jerzy@ku.edu

Ryszard Janicki
McMaster University, Hamilton, ON, Canada
E-mail: janicki@cas.mcmaster.ca

Aboul Ella Hassanien
Cairo University, Egypt
E-mail: aboitcairo@gmail.com

Hong Yu
Chongqing University of Posts and Telecommunications, P.R. China
E-mail: yuhong@cqupt.edu.cn

ISSN 0302-9743   e-ISSN 1611-3349
ISBN 978-3-642-31899-3   e-ISBN 978-3-642-31900-6
DOI 10.1007/978-3-642-31900-6
Springer Heidelberg Dordrecht London New York

Library of Congress Control Number: 2012941922

CR Subject Classification (1998): I.2.4, I.2.6, H.2.8, I.2, I.5, F.4.1,
I.4.8, G.1.2, G.1.6, F.4, H.4, H.3

LNCS Sublibrary: SL 7 – Artificial Intelligence

Typesetting: Camera-ready by author, data conversion by Scientific Publishing Services, Chennai, India
Printed on acid-free paper
Springer is part of Springer Science+Business Media (www.springer.com)

Preface

This volume contains the papers selected for presentation at RSKT 2012: The 7th International Conference on Rough Sets and Knowledge Technology (RSKT) held during August 17–20, 2012, in Chengdu, China, one of the co-located conferences of the 2012 Joint Rough Set Symposium (JRS 2012). JRS 2012 consisted of RSKT 2012 and the 8th International Conference on Rough Sets and Current Trends in Computing (RSCTC 2012).

RSKT has been held every year since 2006. It serves as a major forum that brings researchers and industry practitioners together to discuss and deliberate on fundamental issues of knowledge processing and management and knowledge-intensive practical solutions in the current knowledge age. Experts from around the world meet to present state-of-the-art scientific results, to nurture academic and industrial interaction, and to promote collaborative research in rough sets and knowledge technology. The first RSKT was held in 2006 in Chongqing, China, followed by RSKT 2007 in Toronto, Canada, RSKT 2008 in Chengdu, China, RSKT 2009 in Gold Coast, Australia, RSKT 2010 in Beijing, China, and RSKT 2011 in Banff, Canada.

JRS 2012 received 292 papers and competition submissions from 56 countries and regions, including Afghanistan, Antarctica, Antigua and Barbuda, Argentina, Australia, Austria, Bangladesh, Belgium, Brazil, Canada, China, Colombia, Croatia, Cuba, Denmark, Egypt, Finland, France, Germany, Ghana, Greece, Hong Kong, Hungary, India, Indonesia, Iran, Israel, Italy, Japan, Jordan, Korea, Lebanon, Mexico, The Netherlands, New Zealand, Pakistan, Poland, Portugal, Reunion, Romania, Russian Federation, Rwanda, Saudi Arabia, Serbia and Montenegro, Singapore, Slovenia, Spain, Sweden, Switzerland, Taiwan, Tunisia, Ukraine, UK, USA, Venezuela, and Vietnam.

Following the tradition of the previous RSCTC and RSKT conferences, all submissions underwent a very rigorous reviewing process. Every submission was reviewed by at least two Program Committee (PC) members and at least one external domain expert. On average, each submission received 3.6 reviews. About ten papers received more than six reviews each. Finally, the PC selected 63 papers (including 42 regular papers and 21 short papers), based on their originality, significance, correctness, relevance, and clarity of presentation, to be included in this volume of the proceedings. Revised camera-ready submissions were further reviewed by PC Chairs. Some authors were requested to make additional revisions. We would like to thank all the authors for submitting their papers for consideration for presentation at the conference. We also wish to congratulate those authors whose papers were selected for presentation and publication in the proceedings. Their contribution was crucial for the quality of this conference.

The JRS 2012 program was further enriched by four keynote speeches. We are grateful to RSKT keynote speakers, Andrzej Skowron and Zhi-Hua Zhou, as well as RSCTC keynote speakers, Yiyu Yao and Bo Zhang, for their inspiring talks on rough sets, knowledge technology, and current trends in computing.

The JRS 2012 program included one workshop, Advances in Granular Computing 2012, and five Special Sessions, Decision-Theoretic Rough Set Model and Applications, Intelligent Decision-Making and Granular Computing, Mining Complex Data with Granular Computing, Formal Concept Analysis and Granular Computing, and Rough Set Foundations. In addition, we selected papers written by the winners of the JRS 2012 Data Mining Competition: Topical Classification of Biomedical Research Papers.

This data mining competition was a special event associated with the JRS 2012 conference. It was organized by a research team from the University of Warsaw and co-funded by organizers of JRS 2012, Southwest Jiaotong University, and the SYNAT project. A task in this challenge was related to the problem of predicting topical classification of scientific publications in the field of biomedicine. It was an interactive on-line competition, hosted on the TunedIT platform (http://tunedit.org). The JRS 2012 Data Mining Competition attracted participants from 50 different countries across six continents. There were 126 active teams who submitted at least one solution to the leaderboard. Apart from submitting solutions, participants were asked to deliver short descriptions of their approaches. The most interesting of these reports were extended to conference papers and included in the RSCTC proceedings.

JRS 2012 would not have been successful without the support of many people and organizations. We wish to thank the members of the Steering Committee for their invaluable suggestions and support throughout the organization process. We are indebted to the PC members and external reviewers for their effort and engagement in providing a rich and rigorous scientific program. We express our gratitude to the Special Session Chairs (Mihir Kr. Chakraborty, Degang Chen, Davide Ciucci, Qinghua Hu, Andrzej Janusz, Xiuyi Jia, Adam Krasuski, Huaxiong Li, Jiye Liang, Tsau Young Lin, Dun Liu, Xiaodong Liu, Fan Min, Hung Son Nguyen, Jianjun Qi, Dominik Slezak, Sebastian Stawicki, Lidong Wang, Xizhao Wang, Ling Wei, JingTao Yao, Yiyu Yao, and Hong Yu) for selecting and coordinating the sessions on very interesting topics. Thanks also go to the Tutorial Chairs (Chris Cornelis and Qinghua Hu), Special Session/Workshop Chairs (Davide Ciucci and Wei-Zhi Wu), Publicity Chairs (Jianchao Han, Pawan Lingras, Dun Liu, Duoqian Miao, Mikhail Moshkov, Shusaku Tsumoto), and Organizing Chairs (Hongmei Chen, Yan Yang and Qinghua Zhang).

We are also grateful to Anping Zeng, Chuan Luo, Shaoyong Li, Jie Hu, Shengjiu Liu, and Junbo Zhang from Southwest Jiaotong University, whose great effort ensured the success of the conference. We greatly appreciate the co-operation, support, and sponsorship of various institutions, companies, and organizations, including Southwest Jiaotong University, the University of Regina, the University of Warsaw, the International Rough Set Society, the Rough Sets and Soft Computation Society, the Chinese Association for Artificial Intelligence, Infobright, the Chongqing Institute of Green and Intelligent Technology, the Chinese Academy of Sciences, Section of Intelligent Decision Support Systems and

Granular Computing of the Computer Science Committee of the Polish Academy of Sciences. Moreover, we would like to give special thanks for the support of the National Science Foundation of China (Funding Numbers: 61175047, 61170111, 61100117 and 61073146).
We acknowledge the use of the EasyChair conference system for paper submission, review, and editing of the proceedings. Its new feature of editing LNCS volumes is especially useful. We are thankful to Alfred Hofmann and the excellent LNCS team at Springer for their support and cooperation in publishing the proceedings as a volume of the *Lecture Notes in Computer Science*.

May 2012

Tianrui Li
Hung Son Nguyen
Guoyin Wang
Jerzy Grzyma-Busse
Ryszard Janicki
Aboul Ella Hassanien
Hong Yu

Organization

JRS 2012 Conference Committee

Honorary Chairs	Lotfi A. Zadeh, Bo Zhang
Conference Chairs	Roman Slowinski, Guoyin Wang
Program Chairs	Tianrui Li, Hung Son Nguyen, JingTao Yao
RSKT PC Co-chairs	Jerzy Grzymała-Busse, Ryszard Janicki, Aboul Ella Hassanien, Hong Yu
RSCTC PC Co-chairs	Salvatore Greco, Huaxiong Li, Sushmita Mitra, Lech Polkowski
Tutorial Chairs	Qinghua Hu, Chris Cornelis
Special Session/Workshop Chairs	Davide Ciucci, Wei-Zhi Wu
Publicity Chairs	Jianchao Han, Pawan Lingras, Dun Liu, Duoqian Miao, Mikhail Moshkov, Shusaku Tsumoto
Organizing Chairs	Hongmei Chen, Yan Yang, Qinghua Zhang
Steering Committee Chairs	Jiye Liang, Andrzej Skowron, Dominik Slezak, Yiyu Yao
Secretary-General	Anping Zeng
Secretaries	Jie Hu, Shaoyong Li, Shengjiu Liu, Chuan Luo, Junbo Zhang

Steering Committee

James F. Peters	Yuefeng Li	Lech Polkowski
Malcolm Beynon	Tsau Young Lin	Wladyslaw Skarbek
Hans-Dieter Burkhard	Jiming Liu	Roman Slowinski
Gianpiero Cattaneo	Qing Liu	Zbigniew Suraj
Nicholas Cercone	Jie Lu	Shusaku Tsumoto
Mihir K. Chakraborty	Stan Matwin	Julio V. Valdes
Juan-Carlos Cubero	Ernestina	Guoyin Wang
Didier Dubois	Menasalvas-Ruiz	Hui Wang
Ivo Duentsch	Duoqian Miao	S.K. Michael Wong
Salvatore Greco	Sadaaki Miyamoto	Bo Zhang
Jerzy Grzymala-Busse	Masoud Nikravesh	Wen-Xiu Zhang
Aboul E. Hassanien	Ewa Orlowska	Ning Zhong
Masahiro Inuiguchi	Sankar K. Pal	Wojciech Ziarko
Etienne Kerre	Witold Pedrycz	

Program Committee

Aijun An
Qiusheng An
Mohua Banerjee
Jan Bazan
Theresa Beaubouef
Jerzy Blaszczynski
Zbigniew Bonikowski
Maciej Borkowski
Cory Butz
Gianpiero Cattaneo
Nick Cercone
Mihir K. Chakraborty
Chien-Chung Chan
Hongmei Chen
Jiaxing Cheng
Davide Ciucci
Chris Cornelis
Krzysztof Cyran
Andrzej Czyzewski
Jianhua Dai
Martine De Cock
Dayong Deng
Ivo Düntsch
Lin Feng
Yang Gao
Anna Gomolinska
Xun Gong
Salvatore Greco
Jerzy Grzymała-Busse
Jianchao Han
Aboul Ella Hassanien
Jun He
Christopher Henry
Daryl Hepting
Joseph Herbert
Shoji Hirano
Jie Hu
Qinghua Hu
Xiaohua Hu
Masahiro Inuiguchi
Lakhmi Jain
Ryszard Janicki
Andrzej Janusz

Jouni Jarvinen
Richard Jensen
Xiuyi Jia
Chaozhe Jiang
Janusz Kacprzyk
John A. Keane
C. Maria Keet
Jan Komorowski
Jacek Koronacki
Bozena Kostek
Abd El-Monem Kozae
Krzysztof Krawiec
Marzena Kryszkiewicz
Yasuo Kudo
Henry Leung
Daoguo Li
Deyu Li
Fanchang Li
Huaxiong Li
Jinjin Li
Longshu Li
Tianrui Li
Yuefeng Li
Jiye Liang
Tsau Young Lin
Pawan Lingras
Dun Liu
Qing Liu
Qun Liu
Xiaodong Liu
Amjad Mahmood
Pradipta Maji
Benedetto Matarazzo
Lawrence Mazlack
Ernestina Menasalvas
Jusheng Mi
Duoqian Miao
Pabitra Mitra
Sushmita Mitra
Sadaaki Miyamoto
Mikhail Moshkov
Murice Mulvenna
Som Naimpally

Michinori Nakata
Sinh Hoa Nguyen
Ewa Orlowska
Hala Own
Sankar K. Pal
Krzysztof Pancerz
Neil Mac Parthalain
Puntip Pattaraintakorn
Witold Pedrycz
Bo Peng
Alberto Guillen Perales
Georg Peters
James Peters
Lech Polkowski
Yuhua Qian
Keyun Qin
Guofang Qiu
Anna Maria
    Radzikowska
Vijay Raghavan
Sheela Ramanna
C. Raghavendra Rao
Zbigniew Ras
Henryk Rybinski
Hiroshi Sakai
Lin Shang
Qiang Shen
Kaiquan Shi
Arul Siromoney
Władysław Skarbek
Andrzej Skowron
Dominik Slezak
Roman Slowinski
Hung Son Nguyen
Urszula Stanczyk
John Stell
Jaroslaw Stepaniuk
Zbigniew Suraj
Marcin Szczuka
Marcin Szellag
Fei Teng
Li-Shiang Tsay
Shusaku Tsumoto

Muhammad Zia
  Ur Rehman
Julio Valdes
Aida Vitoria
Alicja Wakulicz-Deja
Krzysztof Walczak
Guoyin Wang
Hongjun Wang
Hui Wang
Xin Wang
Anita Wasilewska
Piotr Wasilewski
Richard Weber
Ling Wei

Paul Wen
Szymon Wilk
Marcin Wolski
Tao Wu
Wei-Zhi Wu
Xiaohong Wu
Jiucheng Xu
Ronald Yager
Yan Yang
Yingjie Yang
Yong Yang
JingTao Yao
Yiyu Yao
Dongyi Ye

Hong Yu
Jian Yu
Slawomir Zadrozny
Xianhua Zeng
Bo Zhang
Ling Zhang
Qinghua Zhang
Yanping Zhang
Shu Zhao
Ning Zhong
Shuigeng Zhou
William Zhu
Wojciech Ziarko

Additional Reviewers

Stefano Aguzzoli
Piotr Artiemjew
Nouman Azam
Paweł Betliński
Bingzhen Sun
Long Chen
Xiaofei Deng
Martin Dimkovski
Yasunori Endo
Joanna Golinska-Pilarek
Dianxuan Gong
Przemysław Górecki
Feng He

Qiang He
Feng Hu
Md. Aquil Khan
Beata Konikowska
Adam Krasuski
Qingguo Li
Wen Li
Lihe Guan
Fan Min
Amit Mitra
Piero Pagliani
Wangren Qiu
Ying Sai

Mingwen Shao
Sebastian Stawicki
Lin Sun
Nele Verbiest
Jin Wang
Junhong Wang
Lidong Wang
Nan Zhang
Yan Zhang
Yan Zhao
Bing Zhou
Wojciech Świeboda

Table of Contents

Part III: Knowledge Technology

Part IV: Workshop: Advances in Granular Computing, 2012 (AGC2012)

Part V: Special Session: Decision-Theoretic Rough Set Model and Applications

Part VI: Special Session: Intelligent Decision-Making and Granular Computing

Part VII: Special Session: Rough Set Foundations

A Characterization of Rough Separability

Piotr Wasilewski[1] and Andrzej Skowron[2]

[1] Institute of Informatics
[2] Institute of Mathematics,
The University of Warsaw
Banacha 2, 02-097 Warsaw, Poland
{piotr,skowron}@mimuw.edu.pl

Abstract. Rough separability in topology is discussed by its connections with pseudometric spaces and rough sets. Pseudometric spaces are presented from the point of view of their connections with approximation spaces. A special way of determining equivalence relations by pseudometric spaces is considered and open sets in pseudometric spaces are studied. Investigations focus on the class of pseudometric spaces which are lower bounded in each point since open sets in these spaces coincide with definable sets of some prescribed approximation spaces. It is also shown that all equivalence and non transitive tolerance relations can be determined by pseudometric spaces in specified ways.

Keywords: rough sets, pseudometric spaces, topological spaces, approximation spaces, information systems, indiscernibility relations, informational representability, rough separability, clo-open topology.

1 Introduction

We study pseudometric spaces from a perspective of their connections with approximation spaces from the Rough Set theory [9–11] (see also e.g. [12]). A fundamental connection is based on the fact that every pseudometric space determines an equivalence relation identifying elements such that their distance with respect to a given pseudometric is equal to zero. We also show that every equivalence relation can be determined in that way. We call such relations atomizing, since they divides pseudometric spaces into atoms: a given pseudometric does not differentiate between elements of equivalence classes of its atomizing relation. This is closely related to the Pawlak's idea of information atoms [9–11]. We consider the form of open sets in pseudometric spaces and show that every open set in arbitrary pseudometric space is a union of equivalence classes of its atomizing relation, thus it is a definable set in an approximation space determined by this atomizing relation.

We study some properties of pseudometric spaces including a property of being lower bounded in each point characterizing pseudometric spaces where every atom is an open set. This implies that open sets in lower bounded in each point pseudometric spaces coincide with definable sets of approximation

T. Li et al. (Eds.): RSKT 2012, LNAI 7414, pp. 1–10, 2012.

spaces determined by corresponding atomizing relations of these spaces. This makes this property interesting from the Rough Set perspective applied in this paper so it is studied in sections 4–5. In Section 3 we introduce the notion of rough separability and shows that lower bounded pseudometric spaces are rough separable. We discuss pseudometrics determined by families of sets, they are examples of pseudometric space which are lower bounded in each point and we show that any equivalence relation determines a pseudometric space such that it is atomized by this relation (Section 4). We finish our presentation by showing that every lower bounded in each point pseudometric space is equivalent to double bounded pseudometric spaces determined by some partition of the space (Section 5).

Such study of pseudometric spaces is motivated by the Rough Set theory and possible applications in analysis of incomplete information. However, in this paper we focus on theoretical aspects only, leaving discussion of applications for future work.

2   Rough Sets and Indiscernibility Relations

Rough sets were introduced by Zdzislaw Pawlak as a tool for analyzing information systems – formal counterparts of information tables, where rows are labeled by names of objects and columns – by names of attributes. An *information system* is a triple $S = \langle Ob, At, Val_a \rangle$ where Ob is a set of objects, At is a set of attributes, and each Val_a is a value domain of an attribute $a \in At$, where $a : Ob \longrightarrow \mathcal{P}(Val_a)$ ($\mathcal{P}(Val_a)$ is a power set of Val_a). If $a(x) \neq \emptyset$ for all $x \in Ob$ and $a \in At$, then S is total. If $card(a(x)) = 1$ for every $x \in Ob$ and $a \in At$, then S is *deterministic*. Otherwise S is *indeterministic*. Referring to deterministic information systems we will use simply *information systems*.

According to Zdzisław Pawlak, knowledge is based on ability to discern objects by means of attributes [8, 9, 7, 10, 11]. In information systems, this ability is presented by indiscernibility relation. This idea can lead also to introducing a notion of a distance based on distinguishability of objects where a distance between objects is, loosely speaking, equal to the number of attributes distinguishing them (Section 5). For analyzing indiscernibility relations determined by information systems Zdzisław Pawlak proposed approximation spaces and on their basis Pawlak introduced rough sets as tools for representing knowledge contained in information systems.

Let $S = \langle U, At, Val_a \rangle$ be an information system, $B \subseteq At$ and $x, y \in U$. *Indiscernibility relation* $ind(B)$ is a relation such that $(x, y) \in ind(B) \Leftrightarrow a(x) = a(y)$ for all $a \in B$. $ind(B)$ can be further analyzed from abstract perspective of approximation spaces [9, 10]. An *approximation space* is a pair (U, R) where U is a non-empty set and R is an equivalence relation on U. (the family of all equivalence relations on a set U we denote by $Eq(U)$). The equivalence classes of R are called *atoms* [10]. Subsets of U which are unions of atoms are called *definable* (or *composed*). Otherwise they are called *rough* [9–11]. $Def_R(U)$ denotes the family of all definable sets in (U, R). For (U, R) and $X \subseteq U$, *lower* and *upper approximations* of X in (U, R) are defined as follows

$$R_*(X) = \bigcup\{Y \in U_{/R} : Y \subseteq X\} \qquad R^*(X) = \bigcup\{Y \in U_{/R} : Y \cap X \neq \emptyset\}.$$

Approximation spaces can be investigated by means of set spaces (see [21–23] where they are called *general approximation spaces*). They appeared to be appropriate tools for general investigations into approximation spaces and indiscernibility relations and their connections with concept lattices [22, 23]. Set spaces can also serve as basic structures for in foundations of granular computing [20]. A pair $(U, \mathcal{C})$ is a *set space* if U is non-empty set and $\mathcal{C} \subseteq \mathcal{P}(U)$. For any $\mathcal{C} \subseteq \mathcal{P}(U)$, $Sg^c(\mathcal{C})$ denotes the least complete field of sets containing $\mathcal{C}$. Elements of $Sg^c(\mathcal{C})$ are called definable sets in the set space $(U, \mathcal{C})$. For any set space $(U, \mathcal{C})$ two operators can be defined:

$$\mathcal{C}_*(X) := \bigcup\{A \in Sg^c(\mathcal{C}) : A \subseteq X\} \qquad \mathcal{C}^*(X) := \bigcap\{A \in Sg^c(\mathcal{C}) : X \subseteq A\}.$$

Let $C \subseteq U$ then an *indiscernibility relation with respect to C* and *indiscernibility relation with respect to family $\mathcal{C}$* are defined respectively as follows:

$$x \approx_C y \Leftrightarrow_{def} x \in C \Leftrightarrow y \in C \qquad (x,y) \in \approx_{\mathcal{C}} \Leftrightarrow_{def} (x,y) \in \bigcap_{C \in \mathcal{C}} \approx_C .$$

Observe that $\approx_C, \approx_{\mathcal{C}} \in Eq(U)$. Note also that set spaces can represent information systems, i.e. for every information system we can construct a set space such that both determine the same indiscernibility relation. It can be shown also that for any equivalence relation $R \in Eq(U)$ there is a family $\mathcal{C} \subseteq \mathcal{P}(U)$ such that $R = \approx_{\mathcal{C}}$. To show this one can prove that $R = \approx_{U_{/R}}$. Moreover, approximation spaces can be adequately represented by set spaces since the following conditions are equivalent: $R = \approx_{\mathcal{C}}$ iff for any $X \subseteq U$ $R_*(X) = \mathcal{C}_*(X)$ and $R^*(X) = \mathcal{C}^*(X)$ iff $Def_R(U) = Sg^c(\mathcal{C})$ [22, 23]. If $\mathcal{A} \subseteq U$ is a field of sets on U, then $At(\mathcal{A})$ denotes the family of all atoms of $\mathcal{A}$. For any family $\mathcal{C} \subseteq \mathcal{P}(U)$ it can be shown that $At(Sg^c(\mathcal{C})) = U_{/\approx_{\mathcal{C}}}$, i.e. atoms of $Sg^c(\mathcal{C})$ are precisely equivalence classes of $\approx_{\mathcal{C}}$. Thus $B \in Sg^c(\mathcal{C})$ if and only if B is a union of classes from $U_{/\approx_{\mathcal{C}}}$ [22, 26].

3 Rough Separability and Pseudometric Spaces

Here we introduce the notion of *rough separability*

Definition 1. *Let $(U, \mathcal{O})$ be a topological space. Points $x, y \in$ are said to be rough separable iff the following conditions are satisfied:*

(1) $\exists V \in \mathcal{O} \ (x \in V \ \& \ y \notin V)$,
(2) $\forall v \in V (v \neq x \Rightarrow \forall S \subseteq V : x \approx_S v)$,
(3) $\forall W(V \subseteq W \Rightarrow \exists w \in W \exists Q \subseteq W(x \not\approx_Q w))$, *i.e. set V is maximal with respect to property (2).*

Topological space $(U, \mathcal{O})$ is rough separable iff for every $x \in U$, there is $y \in$ such that x and y are rough separable.

Let us recall now definition of pseudometric space [3]. $\mathbb{R}_+$ denotes the set of nonnegative reals, i.e. $\mathbb{R}_+ := [0, +\infty)$, $\mathbb{R}_+^*$ denotes the set of positive reals, i.e. $\mathbb{R}_+^* := (0, +\infty)$, $\mathbb{N}$ denotes the set of natural numbers. Let U be any nonempty set. A function $p : U^2 \to \mathbb{R}_+$ is *pseudometric* iff the following conditions hold:

(PM1) $\forall x \in U : p(x, x) = 0$
(PM2) $\forall x, y \in U : p(x, y) = p(y, x)$
(PM3) $\forall x, y, z \in U : p(x, z) \leq p(x, y) + p(y, z)$

A pair of the form (U, p) is called a *pseudometric space* while a function p is called a *pseudometric on the set* U. We refer to the condition (PM2) as the *symmetry condition* and to the condition (PM3) as the *triangle inequality condition*. Pseudometric space (U, p) is *trivial* iff for all $x, y \in U$, $p(x, y) = 0$.

Let us note that every metric is also pseudometric, so pseudometric which is not a metric we will call *proper pseudometrics*. Now we present some examples of pseudometrics ($|\cdot|$ denotes the absolute value of a real number).

Example 1. Let a function $p : \mathbb{R}^2 \to \mathbb{R}_+$ be given by formula: for any $x, y \in \mathbb{R}$

$$p(x, y) := |x^2 - y^2|.$$

Let us note that a function p pseudometric on the set $\mathbb{R}$. Notice that $p(-2, 2) = 0$ and since $2 \neq -2$, then p is a proper pseudometric on $\mathbb{R}$.

Notice that every pseudometric is determined uniquely by some equivalence relation. This is relation which identify elements such that a distance between them, according to a given pseudometric, is equal to zero. Equivalence classes of this relation consist of elements which are not distinguish by a given pseudometric. Thus these classes play role of the basic components of a pseudometric space. Particularly, open sets in a given pseudometric space do not separate elements from equivalence classes of such relation. So, this remark justify call such equivalence classes *atoms*. Let us introduce the next definition:

Definition 2. *Let (U, p) be a pseudometric space and let $x, y \in U$. Then we define a relation $\sim_p$ putting:*

$$x \sim_p y := p(x, y) = 0.$$

We say that a relation $\sim_p$ atomize the space (U, p) or more generally, since $\sim_p \in Eq(U)$, a relation $\theta \in Eq(U)$ (an approximation space (U, θ)) atomizes a space (U, p) iff $\theta = \sim_p$. Equivalence classes of $\sim_p$ are called atoms of a pseudometric space (U, p), or shortly atoms.

Above definition directly entails the following corollary:

Corollary 1. *If a pseudometric space (U, p) is a metric space, then (U, p) is atomized by the identity relation on the space U, i.e. $\sim_p = \Delta_U$.*

Thus, for example, the relation atomizing of every metric space on $\mathbb{R}^n$ is the identity relation on the set $\mathbb{R}^n$. For the next atomizing relations see [26].

Note that the concept of a pseudometric space is a natural generalization of the concept of a metric space. Thus for the class of pseudometric space one can naturally generalize the concepts defined for metric spaces such as the concepts of *open ball, open set, closed set, continuous function*. Let us recall that $top(U, p)$ denotes the topology determined by pseudometric space (U, p), i.e. $top(U, p) = \{A \subseteq U : A \text{ is open in the space } (U, p)\}$.

In the theory of metric spaces, the class of bounded spaces is distinguished. In such spaces there is the least upper bound of distances between its points called a diameter of a given space (see [6, 3]). For example the real plane with Euclidean metric is not a bounded space, but arbitrary circle with metric induced by Euclidean metric is a bounded metric space. A diameter of this space is ordinary diameter of this circle and this justify used name. Thus one can note that concepts of diameter and bounded space are some generalizations of concepts form a classical geometry. These concepts can be also defined for pseudometric spaces. Here we also consider concepts of a *lower diameter lower bonded spaces*.

Definition 3. *Let (U, p) be a pseudometric space.*
A space (U, p) is upper bounded iff there is a number $a_0 \in \mathbb{R}_+$ such that

$$a_0 = \sup\{p(x, y) : x, y \in U\}.$$

The number a_0 is called the upper diameter of the space (U, p).
A space (U, p) is lower bounded iff there is a number $a_0 \in \mathbb{R}_+$ such that

$$a_0 = \inf\{p(x, y) : x, y \in U \text{ and } p(x, y) \neq 0\}.$$

The number a_0 is called the lower diameter of the space (U, p).
If a space (U, p) is upper and lover bounded, then it is called double bounded.

Let us note also that the pseudometric space from Example 1 neither is upper nor lower bounded. The following proposition holds:

Proposition 1. *If a pseudometric space (U, p) is lower bounded, then every union of atoms of (U, p) is an open set in the space (U, p).*

Proof. Assume that pseudometric space (U, p) is lower bounded. Let $r_0 \in \mathbb{R}_+$ be the lower diameter of space (U, p). Therefore for every $x \in U$ it holds that $K(x, r_0) = x_{/\sim_p}$, then every atom is an open ball with the center in its arbitrary element and a radius r_0.

For metric spaces it holds that the family of open sets of any metric space (U, d) is a topology on U. This fact can be generalized for the class of pseudometric spaces.

Proposition 2. *If (U, p) is a pseudometric space, then $top(U, p)$ is a topology on U.*

Definition 4. *Pseudometric space (U, p) is rough separable if and only if the topological space $(U, top(U, p))$ is rough separable.*

Corollary 2. *If a pseudometric space (U, p) is lower bounded, then (U, p) is rough separable.*

Let us note that it does not hold for every pseudometric space that every atom is an open set in that space. Moreover, in each pseudometric space an atom is an open ball if and only if it is an open set in that space. In order to characterize pseudometric spaces in which every atom is an open set we shall introduce a new concept.

Definition 5. *Let (U, p) be a pseudometric space. A space (U, p) is lower bounded in the point $x \in U$ if and only if there is a number $r \in \mathbb{R}_+^*$, such that r is a lower bound of the set*

$$\{p(x, y) : y \in U \text{ and } p(x, y) \neq 0\}$$

with respect to the natural order in the set of real numbers. Such number $r \in \mathbb{R}_+^*$ is called the lower bound of the space (U, p) in the point $x \in U$.

Definition 3 and above definition directly imply the following corollary:

Corollary 3. *If a pseudometric space (U, p) is lower bounded, then it is lower bounded in each its point (in each point $x \in U$).*

One can prove the following theorem (see [26]):

Theorem 1. *Let (U, p) be a pseudometric space and let $A \subseteq U$. The following conditions are equivalent:*

(1) *A pseudometric space (U, p) is lower bounded in each its point.*
(2) $U_{/\sim_p} \subseteq top(U, p)$, *i.e. every atom of the space (U, p) is an open set in (U, p).*
(3) *Every union of atoms of the space (U, p) is an open set in (U, p).*
(4) *A is an open set in the space (U, p) if and only if A is a union of atoms of the space (U, p).*
(5) *i) $top(U, p)$ is a clo-open topology on U and*
    ii) $U_{/\sim_p} \subseteq top(U, p)$.
(6) *i) $top(U, p)$ is a complete field of sets and*
    ii) $At(top(U, p)) = U_{/\sim_p}$, i.e. equivalence classes of the relation $\sim_p$ are exactly atoms of $top(U, p)$ as a complete field of sets.
(7) $top(U, p) = Sg^c(U_{/\sim_p})$, *i.e a topology $top(U, p)$ as a complete field of sets is completely generated by the family $U_{/\sim_p}$.*
(8) *The family of atoms of the pseudometric space (U, p) is a base of topology $top(U, p)$.*

Applications of Theorem 1 in rough set theory and rough set interpretations of construction given in Theorem 1 are presented in Sections 5 and 6.

4 Pseudometrics Determined by Families of Sets

We have pointed out that every pseudometric space determines some approximation space: namely an approximation space such that its equivalence relation atomizes initial pseudometric space. So the natural question arises: whether every approximation space determines a pseudometric space which is atomized be an equivalence relation taken from that approximation space. In order to answer this question we define here some distance functions determined by some families of sets,

Definition 6. *Let U be arbitrary non-empty set and let $C \subseteq U$. A function $d_C : U^2 \to \mathbb{R}_+$ is defined as follows:*

$$d_C(x, y) =_{def} \begin{cases} 0 : x \approx_C y \\ 1 : x \not\approx_C y, \end{cases}$$

i.e. d_C is a characteristic function of the relation $\not\approx_C$. We say that a function d_C is determined by the set C or that the set C determines a function d_C.
   Let $\mathcal{C} = \{C_i\}_{i \in I} \subseteq \mathcal{P}(U)$. A function $d_\mathcal{C} : U^2 \to \mathbb{R}_+$ is defined as follows:

$$d_\mathcal{C}(x, y) := \sum_{i \in I} d_{C_i}(x, y).$$

We say that a function $d_\mathcal{C}$ is *determined by the family $\mathcal{C}$ or that the family $\mathcal{C}$ determines a function $d_\mathcal{C}$.*

Proposition 3. *Let U be arbitrary non-empty set and let $C \subseteq U$. Then the following conditions hold:*

(1) *A function d_C is a pseudometric on U.*
(2) *$\sim_{d_C} = \approx_C$, i.e a relation $\approx_C$ atomizes a pseudometric space (U, d_C).*

Let us note that not every family of sets determines a pseudometrics, i.e. it can be found such family $\mathcal{C}$ such that a function $d_\mathcal{C}$ is not a pseudometric, for the example of such space see [26]. Of course a series $\sum_{i \in I} d_{C_i}$ for any finite family of sets $\{C_i\}_{i \in I}$ is convergent, thus the following corollary holds:

Corollary 4. *Let U be any non-empty set, let $\mathcal{C} \subseteq \mathcal{P}(U)$ and $|\mathcal{C}| < \aleph_0$. Then a function $d_\mathcal{C}$ is a pseudometric on U.*

Let us note also that a reverse implication to the above corollary 4 does not hold, i.e it is not true that if a function $d_\mathcal{C}$ is a pseudometric on U, then a family $\mathcal{C}$ is finite. For the example of such space see [26].

Proposition 4. *Let U be a non-empty set, let also a family $\mathcal{C} \subseteq \mathcal{P}(U)$ determine a pseudometric on U and let $\approx_\mathcal{C} \neq \nabla_U$. Then:*

(1) *Pseudometrics $d_\mathcal{C}$ takes natural numbers as its values, i.e. $d_\mathcal{C}(U \times U) \subseteq \mathbb{N}$.*
(2) *A relation $\approx_\mathcal{C}$ atomizes a space $(U, d_\mathcal{C})$, i.e. $\sim_{d_\mathcal{C}} = \approx_\mathcal{C}$.*

(3) *A pseudometric space $(U, d_{\mathcal{C}})$ is lower bounded, so it is lower bounded in each point.*

(4) $top(U, d_{\mathcal{C}}) = Sg^c(U_{/\approx_{\mathcal{C}}})$, *i.e. open sets in the pseudometric space $(U, d_{\mathcal{C}})$ are exactly definable sets in an approximation space $(U, \approx_{\mathcal{C}})$.*

Proof. Point (1) follows directly from definition 6.

(2) From definition of function $d_{\mathcal{C}}$ we have that for arbitrary $x, y \in U$, $d_{\mathcal{C}}(x, y) = 0$ iff for every $C \in \mathcal{C}$, $x \approx_{\mathcal{C}} y$. Thus, by the definitions of the relations $\sim_{d_{\mathcal{C}}}$ and $\approx_{\mathcal{C}}$ we have that $(x, y) \in \sim_{d_{\mathcal{C}}} \Leftrightarrow (x, y) \in \approx_{\mathcal{C}}$. Thus $\sim_{d_{\mathcal{C}}} = \approx_{\mathcal{C}}$.

(3) Since $\approx_{\mathcal{C}} \neq \nabla_U$, then there are elements $x, y \in U$ such that $d_{\mathcal{C}}(x, y) > 0$. Note that number 1 is a lower diameter of space $(U, d_{\mathcal{C}})$, thus space $(U, d_{\mathcal{C}})$ is lower bounded, so by corollary 3 we get that space $(U, d_{\mathcal{C}})$ is lower bounded in every point.

(4) From point (3) of this proposition we know that pseudometric space $(U, d_{\mathcal{C}})$ is lower bounded in every point. By virtue of theorem 1 this is equivalent to $top(U, d_{\mathcal{C}}) = Sg^c(U_{/\sim_{d_{\mathcal{C}}}})$. From point (2) we know that $\sim_{d_{\mathcal{C}}} = \approx_{\mathcal{C}}$. Thus $top(U, d_{\mathcal{C}}) = Sg^c(U_{/\approx_{\mathcal{C}}})$.

Notice that not every pseudometric space determined by a family of sets is upper bounded. For the example of such space see [26].

5   Equivalence of Pseudometric Spaces

The next concept which can be naturally generalized from the class of metric space onto the class of pseudometric spaces is that of equivalence of metric spaces: Pseudometric spaces (X, p) i (X, q) are said to be *equivalent* iff $top(X, p) = top(X, q)$. In other words pseudometric spaces are equivalent iff they determined the same open sets. For an example see [26]. Now we can present next characterization of pseudometric spaces which are lower bounded in each point.

Theorem 2. *Every pseudometric space lower bounded in each point is equivalent to a double bounded pseudometric space.*

Proof. Let (U, p) be a pseudometric space lower bounded in every point. Thus from theorem 1 we know that $top(U, p) = Sg^c(U_{/\sim_p})$. From proposition 4.1 we know that function $d_{U_{/\sim_p}}$ is a pseudometric on U, whereas from proposition 4.3 we know that $\approx_{U_{/\sim_p}}$ atomizes space $(U, d_{U_{/\sim_p}})$, then $\sim_{d_{U_{/\sim_p}}} = \approx_{U_{/\sim_p}}$. From proposition 4.2 we get that space $(U, d_{U_{/\sim_p}})$ is lower bounded in every point, what by theorem 1 is equivalent to $top(U, d_{U_{/\sim_p}}) = Sg^c(U_{/\approx_{U_{/\sim_p}}})$. From the Abstraction Principle it follows that $U_{/\sim_p} = U_{\approx_{U_{/\sim_p}}}$, therefore $Sg^c(U_{/\sim_p}) = Sg^c(U_{/\approx_{U_{/\sim_p}}})$, so $top(U, p) = top(U, d_{U_{/\sim_p}})$. Thus pseudometric spaces (U, p) and $(U, d_{U_{/\sim_p}})$ are equivalent. From proposition 4.2 it follows that space $(U, d_{U_{/\sim_p}})$ is bounded. We have thus proved that space (U, p) is equivalent to a double bounded space determined by a partition of set U.

6 Conclusions

We discussed pseudometric spaces from the perspective of the Rough Set theory. This discussion is based on a fundamental connection between pseudometric spaces and approximation spaces: namely that every pseudometric space determines an approximation space which atomizes it. We investigated open sets in pseudometric spaces from that perspective. We specially focused on pseudometric spaces which are lower bounded in each point since their open sets coincide with definable sets in approximation spaces which atomize them. We have shown also that every equivalence relation atomizes some pseudometric space.

Results presented within this paper can be used for defining approximation operators based on pseudometric spaces: namely operators which depend both on atomizing relations and distances with respect to appropriate pseudometrics. They are essential in tolerance rough sets methods. They can be also used in a study on the notion of *nearness* and its connections with rough sets. An interesting thing is to investigate connections of pseudometric spaces with *near sets* [13, 14] as well as with tolerance relations and proximity spaces which are closely related to *nearness* [25].

Obtained results, by their connection with tolerance spaces, can be used for construction new, generalized attributes which are essential for interactive information systems [18, 19]. Tolerance and pseudometric spaces are promising in modeling complex vague concepts [16] both for hierarchical approximation of complex vague concepts from lower-level data (e.g. sensory data) [1] and for decision making using Wisdom technology [4], where intelligent autonomous agents make adaptively of correct judgments to a satisfactory degree in the face of real-life constraints (e.g. time constraints). Therefore rough set methods based on pseudometric as well as tolerance spaces are important part or rough-granular approach to interactive computing [18, 19] whereas they are indispensable in perception based computing [17].

Acknowledgements. Research reported in this work has been supported by the individual research project realized within Homing Plus programme, edition 3/2011, of Foundation for Polish Science, co-financed from European Union, Regional Develop. Fund, and by the grant 2011/01/D/ST6/06981 from Polish National Science Centre.

References

1. Bazan, J.: Hierarchical Classifiers for Complex Spatio-temporal Concepts. In: Peters, J.F., Skowron, A., Rybiński, H. (eds.) Transactions on Rough Sets IX. LNCS, vol. 5390, pp. 474–750. Springer, Heidelberg (2008)
2. Demri, S., Orłowska, E.: Incomplete Information: Structures, Inference, Complexity. Springer (2002)
3. Engelking, R.: General Topology. PWN (1977)
4. Jankowski, J., Skowron, A.: Wisdom Technology: A Rough-Granular Approach. In: Marciniak, M., Mykowiecka, A. (eds.) Bolc Festschrift. LNCS, vol. 5070, pp. 3–41. Springer, Heidelberg (2009)

5. Kopelberg, S.: General theory of Boolean algebras. In: Monk, J.D., Bonnet, R. (eds.) Handbook of Boolean Algebras. North Holland (1989)
6. Kuratowski, K.: Wstęp do teorii mnogości i topologii. PWN (1980)
7. Orłowska, E., Pawlak, Z.: Representation of nondeterministic information. Theoretical Computer Science 29, 27–39 (1984)
8. Pawlak, Z.: Information Systems – theoretical foundation. Information Systems 6, 205–218 (1981)
9. Pawlak, Z.: Rough sets. International Journal of Computing and Information Sciences 18, 341–356 (1982)
10. Pawlak, Z.: Rough sets. Theoretical Aspects of Reasoning About Data. Kluwer Academic Publishers (1991)
11. Pawlak, Z.: Some Issues on Rough Sets. In: Peters, J.F., Skowron, A., Grzymała-Busse, J.W., Kostek, B.z., Świniarski, R.W., Szczuka, M.S. (eds.) Transactions on Rough Sets I. LNCS, vol. 3100, pp. 1–58. Springer, Heidelberg (2004)
12. Pawlak, Z., Skowron, A.: Rudiments of rough sets. Information Science 177, 3–27 (2007)
13. Peters, J.F.: Near sets. Special theory about nearness of objects. Fundamenta Informaticae 73, 1–27 (2006)
14. Peters, J.F., Wasilewski, P.: Foundations of near sets. Elsevier Science 179(18), 3091–3109 (2009)
15. Polkowski, L.: Rough Sets: Mathematical Foundations. Physica-Verlag (2002)
16. Skowron, A.: Rough sets and vague concepts. Fundamenta Informaticae 64(1-4), 417–431 (2005)
17. Skowron, A., Wasilewski, P.: An Introduction to Perception Based Computing. In: Kim, T.-H., Lee, Y.-H., Kang, B.-H., Ślęzak, D. (eds.) FGIT 2010. LNCS, vol. 6485, pp. 12–25. Springer, Heidelberg (2010)
18. Skowron, A., Wasilewski, P.: Information systems in modeling interactive computations on granules. Theoretical Computer Science 412(42), 5939–5959 (2011)
19. Skowron, A., Wasilewski, P.: Toward interactive rough–granular computing. Control & Cybernetics 40(2), 1–23 (2011)
20. Ślęzak, D., Wasilewski, P.: Granular Sets – Foundations and Case Study of Tolerance Spaces. In: An, A., Stefanowski, J., Ramanna, S., Butz, C.J., Pedrycz, W., Wang, G. (eds.) RSFDGrC 2007. LNCS (LNAI), vol. 4482, pp. 435–442. Springer, Heidelberg (2007)
21. Wasilewski, P.: Dependency and supervenience. In: Proc. of the Concurrency, Specification and Programming, CS & P 2003. Warsaw Univerity (2003)
22. Wasilewski, P.: On selected similarity relations and their applications into cognitive science. Unpublished doctoral dissertation, Jagiellonian University: Department of Logic, Cracow, Poland (2004) (in Polish)
23. Wasilewski, P.: Concept Lattices vs. Approximation Spaces. In: Ślęzak, D., Wang, G., Szczuka, M.S., Düntsch, I., Yao, Y. (eds.) RSFDGrC 2005, Part I. LNCS (LNAI), vol. 3641, pp. 114–123. Springer, Heidelberg (2005)
26. Wasilewski, P.: Indiscernibility relations (in preparation)
25. Wasilewski, P., Peters, J., Ramanna, S.: Perceptual Tolerance Intersection. In: Peters, J.F., Skowron, A., Chan, C.-C., Grzymala-Busse, J.W., Ziarko, W.P. (eds.) Transactions on Rough Sets XIII. LNCS, vol. 6499, pp. 159–174. Springer, Heidelberg (2011)
26. Wasilewski, P.: Pseudometric Spaces from Rough Sets Perspective. In: Skowron, A., Suraj, Z. (eds.) Rough Sets and Intelligent Systems - Professor Zdzisław Pawlak in Memoriam. ISRL, vol. 43, pp. 581–604. Springer, Heidelberg (2012)

Data-Driven Valued Tolerance Relation

Guoyin Wang[1,2] and Lihe Guan[1,2,*]

[1] Chongqing Key Laboratory of Computational Intelligence,
Chongqing University of Posts and Telecommunications,
Chongqing 400065, China
[2] School of Information Science and Technology, Southwest Jiaotong University,
Chengdu 610031, China
guanlihe@cqjtu.edu.cn, wanggy@ieee.org

Abstract. The valued tolerance relation in incomplete information systems is an important extension model of the classical rough set theory. However, the general calculation method of tolerance degree needs to know the probability distribution of an information system in advance, and it is also difficult to select a suitable threshold. In this paper, a data-driven valued tolerance relation is proposed based on the idea of data-driven data mining. The new calculation method of tolerance degree and the auto-selection method of threshold do not require any prior domain knowledge except the data set. Experiment results show that the data-driven valued tolerance relation can get better and more stable classification results than the other extension models of the classical rough set theory.

Keywords: rough set, valued tolerance relation, data-driven.

1 Introduction

The classical rough set theory developed by Professor Pawlak in 1982 [1], based on the conventional indiscernibility relation, is not much useful for analyzing incomplete information systems (IIS) where some attribute values are unknown. But in practice, because of the errors of data measuring, the limitations of acquiring data, some human factors, etc, IIS often occur in knowledge acquisition. Therefore, many researchers have drew attention upon this issue in recent years and endeavored to find out solutions. There are usually two methods in rough set theory to deal with IIS: data reparation [2] and model extension [3-8]. Compared with data reparation, model extension does not change the original information of IIS and the knowledge system generated is more objective. Thus model extension is more suitable than data reparation for processing IIS [8]. At present, various extension models of the classical rough set theory have been proposed, in which the indiscernibility relation was extended to some non-equivalence relations (or called generalized indiscernibility relations), such as tolerance relation [3], non-symmetric similarity relation [4], limited tolerance relation [5], valued tolerance relation [6], and characteristic relation [7].

---

* Corresponding author.

T. Li et al. (Eds.): RSKT 2012, LNAI 7414, pp. 11–19, 2012.
© Springer-Verlag Berlin Heidelberg 2012

However, these extension models have their own limitations. In tolerance relation, two objects that have no one known and same value on all attributes may be considered as indiscernible, and classified in the same class. In non-symmetric similarity relation, two objects that have a lot of known and same values on all attributes may be separated, and classified in the different classes. The limited tolerance relation inherits the merit of tolerance relation and non-symmetric similarity relation and avoids their limitations. But, in limited tolerance relation, two objects may be considered as similar if only they have one known and same value on all attributes. This condition is too lenient for the IIS with a large number of attributes. The characteristic relation, which is a generalization of both tolerance relation and non-symmetric similarity relation, can not avoid their limitations.

Compared with the other extension models, the valued tolerance relation proposed by Stefanowski and Tsoukis [6] has stronger adaptability. On the one hand, it is based on the tolerance degree which is measured using a comparison rule, and then the similar degree between objects can be accurately measured. On the other hand, it is based on a pre-given threshold to determine whether the objects are indiscernible, and then it can dynamically adjust the threshold to adapt to the different IIS. However, in valued tolerance relation, we must know the probability distribution of the information system in advance. Unfortunately, this is very difficult for a new IIS with some missing attribute values. So, it is usually supposed that there exists a uniform probability distribution among the possible values on each attribute. Obviously, this is too subjective and unreasonable. In addition, it is also difficult to select a suitable threshold for the different IIS. Therefore, there are two key problems in valued tolerance relation, which have not yet been fully resolved. The first is how to define a reasonable tolerance degree. The second is how to select a reasonable threshold according to the different IIS.

In [9], based on the idea that data mining is a process of knowledge transformation, a data-driven data mining model was proposed. In fact, the data-driven data mining is that any prior knowledge except data set is not required in the data mining process, i.e., the process of knowledge acquisition is completed independently by the data set. In resent years, this method has been successfully applied in artificial intelligence. Some data-driven data mining algorithms were also proposed, e.g., data-driven default rule generation algorithm [10], data-driven decision tree pre-pruning algorithm [11] and data-driven knowledge acquisition from concept lattice [12].

In this paper, based on the idea of data-driven data mining [9], we develop a new calculation method of tolerance degree and an auto-selection method of threshold. On this basis, a data-driven valued tolerance relation is proposed, and it is objective and reasonable for processing IIS.

In section 2, we review the valued tolerance relation. In section 3, we present a new calculating method of tolerance degree and an auto-selection method of threshold based on the data-driven. In section 4, some simulation experiment results are discussed. Conclusions are given in the last section.

2   Valued Tolerance Relation

In this section, the valued tolerance relation will be recalled briefly. The definitions of tolerance relation, non-symmetric similarity relation, limited tolerance relation, and characteristic relation, can be found in Refs.[3,4,5,7].

Formally, an information system (IS) is $< U, A >$, where U is a non-empty finite set of objects, called the universe, A is a set of attributes. If some attribute values are unknown, the IS is IIS, otherwise it is complete information system (CIS). In IIS, the unknown attribute value is denoted by "*".

Given an IIS $< U, A >$, a subset of attributes $B \subseteq A$. Let $I_U = \{< x, x > | x \in U\}$ be the identity relation on U, $P_B(x, y)$ be the tolerance degree between objects x and y with reference to B, and then for a given threshold $\lambda \in [0, 1]$ the valued tolerance relation VT_B^λ is defined as

$$VT_B^\lambda = \{< x, y >\in U^2 | P_B(x, y) \geq \lambda\} \cup I_U. \tag{1}$$

Obviously, VT_B^λ is a reflexive and symmetric relation, but not necessarily transitive. Further on, the valued tolerance class $VT_B^\lambda(x)$ of object x is defined as

$$VT_B^\lambda(x) = \{y \in U | < x, y >\in VT_B^\lambda\}.$$

It is noteworthy that we can define different types of $P_B(x, y)$ using different comparison rules, and then different types of valued tolerance relation will be got. In the general calculation method of tolerance degree, it is usually assumed that there exists a uniform probability distribution among the possible values on each attribute. Then, given any two objects $x, y \in U$ and an attribute $b \in B$, the probability $P_{\{b\}}(x, y)$ that x is similar to y on b can be calculated as

$$P_{\{b\}}(x, y) = \begin{cases} 1 & b(x) = b(y) \wedge b(x) \neq * \wedge b(y) \neq * \\ 0 & b(x) \neq b(y) \wedge b(x) \neq * \wedge b(y) \neq * \\ 1/|V_b| & b(x) = * \vee b(y) = * \end{cases} \tag{2}$$

where V_b denotes the domain of the attribute b. On this basis, the probability that two objects are similar on the whole set of attributes is calculated as the joint probability that their attribute values are same on all the attributes:

$$P_B(x, y) = \prod_{b \in B} P_{\{b\}}(x, y). \tag{3}$$

However, for any IIS, the hypothesis that there exists a uniform probability distribution among the attribute values is too subjective and unreasonable. In addition, the threshold λ is usually selected subjectively. Thus, a reasonable and objective calculation method of tolerance degree and an objective selecting method of threshold for different IIS are two key problems in valued tolerance relation.

3   Data-Driven Valued Tolerance Relation

In this section, based on the idea of data-driven data mining [9], we will present a new calculation method of tolerance degree and an auto-selection method of threshold, and get a data-driven valued tolerance relation.

3.1   Calculation of Tolerance Degree

Consider an attribute b of an IIS $< U, A >$ and the set $V_b' = \{k_b^1 b_1, k_b^2 b_2, \cdots, k_b^m b_m\}$, where $b_1, b_2, \cdots, b_m$ is all possible known values of b, and k_b^i denotes the cardinality of the set $\{x \in U | b(x) = b_i\}$. For any $x \in U$, the probability that $b(x) = b_i$ is $k_b^i/(k_b^1 + k_b^2 + \cdots + k_b^m)$. Thus, the formula (2) can be improved as

$$P_{\{b\}}(x, y) = \begin{cases} 1 & b(x) = b(y) \wedge b(x) \neq * \wedge b(y) \neq * \\ 0 & b(x) \neq b(y) \wedge b(x) \neq * \wedge b(y) \neq * \\ k_b^i / \sum\limits_{j=1}^{m} k_b^j & (b(x) = b_i \wedge b(y) = *) \vee (b(x) = * \wedge b(y) = b_i) \\ \sum\limits_{i=1}^{m} (k_b^i / \sum\limits_{j=1}^{m} k_b^j)^2 & b(x) = * \wedge b(y) = * \end{cases}$$

$$(4)$$

In formula (4), the frequency of attribute value is approximately taken as its probability of appearance, and then the approximate probability distribution among the attribute values is got. This calculation method of tolerance degree does not require any prior knowledge except data sets, and is objective.

In addition, the formula (3) considers only the possibility of two objects taking the same attribute values on all the attributes, but not the effect of the number of known and same attribute values between them. Thus it is too rough. On the contrary, the limited tolerance relation considers only the effect of the number of known and same attribute values between objects, and thus it is not accurate. Therefore the formula (3) can be improved as

$$P_B(x, y) = \prod\nolimits_{b \in B} P_{\{b\}}(x, y) \times N_B(x, y). \qquad (5)$$

where $N_B(x, y)$ is the weight factor of objects x and y taking the known and same attribute values on B, i.e.,

$$N_B(x, y) = \frac{|\{b \in B | b(x) = b(y) \neq *\}|}{|B|}.$$

Obviously, the formula (5) not only considers the possibility of two objects taking the same attribute values on the all attributes, but also considers the effect of the number of known and same attribute values between them.

Example 1. Suppose an IIS is given in Table 1, where $x_1, x_2, \cdots, x_{12}$ are the available objects, c_1, c_2, c_3, c_4 are four attributes.

Using the formulas (3) and (4), we have that

$$P_A(x_1, x_2) = \frac{4}{10} \times \frac{3}{6} \times \frac{4}{8} \times 1 = \frac{1}{10}$$

and

$$P_A(x_3, x_4) = \frac{1}{10} \times 1 \times 1 \times 1 = \frac{1}{10}.$$

Obviously, there is $P_A(x_1, x_2) = P_A(x_3, x_4)$.

Table 1. An IIS $< U, A >$

A	\multicolumn{12}{c}{U}
	x_1	x_2	x_3	x_4	x_5	x_6	x_7	x_8	x_9	x_{10}	x_{11}	x_{12}
c_1	*	1	*	2	3	4	5	6	1	1	1	6
c_2	2	*	3	3	*	2	*	2	*	*	3	*
c_3	*	2	1	1	2	*	1	*	2	2	*	1
c_4	1	1	0	0	3	1	3	*	2	*	0	2

Using the formulas (4) and (5), we have that

$$P_A(x_1, x_2) = (\frac{4}{10} \times \frac{3}{6} \times \frac{4}{8} \times 1) \times \frac{1}{4} = \frac{1}{40}$$

and

$$P_A(x_3, x_4) = (\frac{1}{10} \times 1 \times 1 \times 1) \times \frac{3}{4} = \frac{3}{40}.$$

At this time, there is $P_A(x_1, x_2) < P_A(x_3, x_4)$.

From table 1, it is clear that there is only one known and same attribute value between x_1 and x_2 on A, while there are three ones between x_3 and x_4. Obviously, the similar degree between x_3 and x_4 should be greater than between x_1 and x_2. Thus the formula (5) is more reasonable than the formula (3).

3.2   Auto-selection of Threshold

Consider an IIS $< U, A >$ and a tolerance relation T_B on $B \subseteq A$. For any x in U, the tolerance class $T_B(x)$ can be calculated. According to the calculation method of tolerance degree, for any y in $T_B(x)$, the tolerance degree $P_B(x, y)$ can be calculated. Let $\lambda_{\max}^x$ be the maximum, and $\lambda_{\min}^x$ be the minimum, i.e.,

$$\lambda_{\max}^x = \max\{P_B(x, y) | y \in T_B(x)\} \qquad (6)$$

and

$$\lambda_{\min}^x = \min\{P_B(x, y) | y \in T_B(x)\}. \qquad (7)$$

For any x in U, there are $\lambda_{\max}^x$ and $\lambda_{\min}^x$. Thus we may assume that

$$\lambda_{\max} = \min\{\lambda_{\max}^x | x \in U\} \qquad (8)$$

and

$$\lambda_{\min} = \max\{\lambda_{\min}^x | x \in U\}. \qquad (9)$$

Obviously, the relationship between $\lambda_{\max}$ and $\lambda_{\min}$ is uncertain. In valued tolerance relation, if $\lambda = \lambda_{\max}$, then the valued tolerance classes of some objects may be too small for the too large threshold when $\lambda_{\max} > \lambda_{\min}$. If $\lambda = \lambda_{\min}$, then the valued tolerance classes of some objects may be empty sets for the too large threshold when $\lambda_{\max} < \lambda_{\min}$. So, either $\lambda = \lambda_{\max}$ or $\lambda = \lambda_{\min}$, is unsuitable in some cases. But, if $\lambda = \min\{\lambda_{\max}, \lambda_{\min}\}$, then it is ok.

Table 2. $\lambda_{\max}^x$ and $\lambda_{\min}^x$ of each object x in U

U	Case1		Case2	
	$\lambda_{\max}^{x_i}$	$\lambda_{\min}^{x_i}$	$\lambda_{\max}^{x_i}$	$\lambda_{\min}^{x_i}$
x_1	1	1	1	0.4
x_2	0.9	0.5	0.7	0.5
x_3	0.8	0.6	0.7	0.3
x_4	0.5	0.2	0.7	0.2
x_5	0.8	0.4	0.7	0.4

Example 2. Suppose $U = \{x_1, x_2, x_3, x_4, x_5\}$, and the $\lambda_{\max}^x$ and $\lambda_{\min}^x$ of each object x in U are shown in table 2.

In case 1, there are $\lambda_{\max} = 0.5$ and $\lambda_{\min} = 1$. If $\lambda = 1$, the valued tolerance classes of x_2, x_3, x_4, x_5 will be empty sets. At this time, the threshold $\lambda = 1$ is obviously too great. However, if $\lambda = 0.5$, it is good.

In case 2, there are $\lambda_{\max} = 0.7$ and $\lambda_{\min} = 0.5$. If $\lambda = 0.7$, the valued tolerance classes of x_2, x_3, x_4, x_5 will be too small. If $\lambda = 0.5$, it is good.

Therefore, the threshold in valued tolerance relation may be calculated as

$$\lambda = \min\{\min_{x\in U}\{\max_{y\in T_B(x)}\{P_B(x,y)\}\}, \max_{x\in U}\{\min_{y\in T_B(x)}\{P_B(x,y)\}\}\}. \qquad (10)$$

In this way, we develop an auto-selection method of threshold.

Algorithm 1. Auto-selection Method of Threshold
Input: An IIS $< U, A >$, the attribute subset $B \subseteq A$.
Output: The threshold λ.
Step1: For any x in U, calculate its tolerance class $T_B(x)$.
Step2: For any x in U, calculate $\lambda_{\max}^x$ and $\lambda_{\min}^x$ by the formulas (6) and (7).
Step3: Calculate $\lambda_{\max}$ and $\lambda_{\min}$ by the formulas (8) and (9).
Step4: $\lambda = \min\{\lambda_{\max}, \lambda_{\min}\}$.
Step5: Return λ.

Especially, the valued tolerance relation, where the tolerance degree is defined by the formulas (4) and (5), and the threshold is calculated by the formula (10), is called a data-driven valued tolerance relation.

4   Experiment Results

In order to test the effectiveness of the data-driven valued tolerance relation, we compare its classification accuracy with that of tolerance relation, non-symmetric similarity relation, limited tolerance relation, uniform valued tolerance relation (defined by the formulas (1) to (3)), statistical valued tolerance relation(defined by the formulas (1),(3) and (4)).

Let E be an indiscernibility relation on the universe U, $[x]_E$ be the equivalent class of x in U, and R be a generalized indiscernibility relation on U. If R is symmetric then the generalized indiscernibility class of object x is denoted by

$R(x)$, otherwise the predecessor and successor sets of x are respectively denoted by $K_R(x)$ and $K_R^{-1}(x)$. Thus the classification accuracy of R is μ_R, i.e.,

$$\mu_R = \begin{cases} \dfrac{\sum\limits_{x \in U} \frac{|[x]_E \cap R(x)|}{|[x]_E \cup R(x)|}}{|U|} & R \text{ is symmetric} \\[3ex] \dfrac{\sum\limits_{x \in U} \frac{|[x]_E \cap (K_R(x) \cup K_R^{-1}(x))|}{|[x]_E \cup (K_R(x) \cup K_R^{-1}(x))|}}{|U|} & R \text{ is non-symmetric} \end{cases} \tag{11}$$

where $|X|$ denotes the cardinality of the set X. Obviously, there is $\mu_R \in [0,1]$, and the greater μ_R is, the better the classification of R is.

In our experiments, three complete data sets (Balance, Tic-Tac-Toe and Chess) in UCI are used. Each complete data set is modified by introducing 5%, 10% and 30% randomly chosen unknown attribute values, and then 9 incomplete data sets (Balance-5%, Balance-10%, Balance-30%, Tic-Tac-Toe-5%, Tic-Tac-Toe-10%, Tic-Tac-Toe-30%, Chess-5%, Chess-10%, Chess-30%) are got.

Table 3. Three complete data sets in UCI

Data sets	No. of objects	No. of Condition attributes	No. of Decision attributes
Balance	625	4	1
Tic-Tac-Toe	958	9	1
Chess	3196	36	1

The experiment results are shown in Table 4, where T, S, L, UVT^λ, SVT^λ, and DVT^λ denote respectively tolerance relation, non-symmetric similarity relation, limited tolerance relation, uniform valued tolerance relation, statistical valued tolerance relation, and data-driven valued tolerance relation.

Table 4. Classification accuracies of six generalized indiscernibility relations

Data sets	Generalized indiscernibility relations					
	T	S	L	UVT^λ	SVT^λ	DVT^λ
Balance-5%	0.711	0.799	0.711	0.761	0.795	0.809
Balance-10%	0.285	0.310	0.289	0.288	0.486	0.513
Balance-30%	0.021	0.042	0.023	0.021	0.329	0.346
Tic-Tac-Toe-5%	0.921	0.929	0.921	0.926	0.940	0.945
Tic-Tac-Toe-10%	0.666	0.679	0.666	0.709	0.766	0.764
Tic-Tac-Toe-30%	0.256	0.354	0.257	0.262	0.402	0.433
Chess-5%	0.797	0.758	0.799	0.800	0.836	0.858
Chess-10%	0.574	0.570	0.574	0.605	0.718	0.789
Chess-30%	0.283	0.298	0.285	0.300	0.518	0.528

From Table 4, we can see that the classification accuracy of DVT^λ is higher than that of T, S and L. The classification accuracy of SVT^λ is generally higher than that of UVT^λ, and slightly less than that of DVT^λ. In addition, the classification accuracies of T, S and L decrease quickly with the increasing of the incomplete degree of data set, and reduce to the minimum (below 5%) for the Balance-30%. The classification accuracy of DVT^λ is relatively stable, and meets the minimum for the Balance-30%, but more than 30%.

To further test the effectiveness of auto-selection method of threshold (ie., Algorithm 1), the classification accuracies of DVT^λ for different thresholds are calculated. Results are summarized in Table 5, where AVG denotes the average method (ie., the arithmetic mean of tolerance degree of any two objects satisfying tolerance relation is taken as the threshold).

Table 5. Classification accuracies of DVT^λ for different thresholds

Data sets	Thresholds						
	0.1	0.3	0.5	0.7	0.9	AVG	Algorithm1
Balance-5%	0.630	0.817	0.817	0.817	0.817	0.817	0.809
Balance-10%	0.476	0.565	0.565	0.565	0.565	0.476	0.513
Balance-30%	0.333	0.236	0.236	0.236	0.236	0.184	0.346
Tic-Tac-Toe-5%	0.942	0.936	0.632	0.632	0.632	0.632	0.945
Tic-Tac-Toe-10%	0.785	0.759	0.376	0.376	0.376	0.759	0.764
Tic-Tac-Toe-30%	0.297	0.198	0.038	0.038	0.038	0.455	0.433
Chess-5%	0.864	0.787	0.589	0.365	0.238	0.658	0.858
Chess-10%	0.767	0.478	0.228	0.103	0.039	0.686	0.789
Chess-30%	0.189	0.172	0.103	0.029	0.004	0.529	0.528

It can be found from Table 5 that the classification accuracy of DVT^λ is closely related to its threshold. In some cases, the classification accuracy is even slightly higher using the subjectively selected method and average method than Algorithm 1. However, the average method is less stable, and the subjectively selected method is uncertain. In comparison, the classification accuracy using Algorithm 1 can close or up to the best, and its change is more stable.

5   Conclusions

In this paper, a data-driven valued tolerance relation is developed based on the idea of data-driven data mining. The objective calculation method of tolerance degree is not only founded on the basis of the statistical characteristics of attribute values in IIS, but also considers the effect of the number of known and same attribute values between objects. In addition, an auto-selection method of threshold is too proposed. Experiment results show that the data-driven valued tolerance relation can get better and more stable classification results than the other extension models of the classical rough set theory.

Acknowledgments. This research has been supported by the National Natural Science Foundation of P. R. China (NSFC) under grant No. 61073146, and the Open Foundation of Key Laboratory of Computer Network and Communication Technology of Chongqing of P.R. China under grant No. CY-CNCL-2010-04.

References

1. Pawlak, Z.: Rough sets. International Journal of Computer and Information Sciences 11, 341–356 (1982)
2. Grzymała-Busse, J.W., Hu, M.: A Comparison of Several Approaches to Missing Attribute Values in Data Mining. In: Ziarko, W.P., Yao, Y. (eds.) RSCTC 2000. LNCS (LNAI), vol. 2005, pp. 378–385. Springer, Heidelberg (2001)
3. Kryszkiewicz, M.: Rough set approach to incomplete information systems. Information Sciences 112, 39–49 (1998)
4. Slowinski, R., Vanderpooten, D.: A generalized definition of rough approximations based on similarity. IEEE Transactions on Knowledge and Data Engineering 12(2), 331–336 (2000)
5. Wang, G.Y.: Extension of rough set under incomplete information systems. Journal of Computer Research and Development 39(10), 1238–1243 (2002)
6. Stefanowski, J., Tsoukiàs, A.: On the Extension of Rough Sets under Incomplete Information. In: Zhong, N., Skowron, A., Ohsuga, S. (eds.) RSFDGrC 1999. LNCS (LNAI), vol. 1711, pp. 73–82. Springer, Heidelberg (1999)
7. Grzymala-Busse, J.W.: Characteristic Relations for Incomplete Data: A Generalization of the Indiscernibility Relation. In: Peters, J.F., Skowron, A. (eds.) Transactions on Rough Sets IV. LNCS, vol. 3700, pp. 58–68. Springer, Heidelberg (2005)
8. Wang, G.Y., Guan, L.H., Hu, F.: Rough set extensions in incomplete information systems. Frontiers of Electrical and Electronic Engineering in China 3(4), 399–405 (2008)
9. Wang, G.Y., Wang, Y.: 3DM: Domain-oriented Data-driven Data Mining. Fundamenta Informaticae 90, 395–426 (2009)
10. Wang, G.Y., He, X.: A self-Learning Model under Uncertain Conditions. Journal of Software 14(6), 1096–1102 (2003) (in Chinese)
11. Wang, Y., Shen, Y.X., Tao, C.M.: Domain-oriented data-driven knowledge acquisition model and its implementation. Journal of Chongqing University of Posts and Telecommunications 21(4), 502–506 (2008) (in Chinese)
12. Wand, Y., Wand, G.Y., Deng, W.B.: Concept Lattice Based Data-Driven Uncertain Knowledge Acquisition. Pattern Recognition and Artificial Intelligence 20(5), 626–642 (2007) (in Chinese)

Optimistic Multi-Granulation Fuzzy Rough Set Model Based on Triangular Norm

Weihua Xu*, Wenxin Sun, and Yufeng Liu

School of Mathematics and Statistics, Chongqing University of Technology,
Chongqing, 400054, P.R. China
chxuwh@gmail.com, sunxuxin520@163.com, liuyufeng@cqut.edu.cn

Abstract. With granular computing point of view, the generalized T-fuzzy rough set model is based on a single fuzzy granulation in a T-fuzzy approximation space. This paper is devoted to the construction and study of the multi-granulation rough set based on triangular norm by defining the optimistic multi-granulation T-fuzzy lower and upper approximation operators in the generalized T-fuzzy approximation space. It is obvious that the generalize T-fuzzy lower and upper approximation operators defined on (U, R) are obtained as a special case of these operators. The main properties of the T-fuzzy lower and upper approximation operators are also studied.

Keywords: Rough set, Multi-granulation, T-fuzzy similarity relation, Triangular norm, Residual implication.

1 Introduction

The theory of rough sets, proposed by Pawlak [2], is a powerful mathematical approach to deal with inexact, uncertain or vague knowledge. And the fuzzy set theory also offers a wide variety of techniques for analyzing imprecise data. It seems quite natural to extend the Pawlak rough set by combining methods developed within both theories to construct hybrid structures. Such structures, called fuzzy rough sets and rough fuzzy sets, have been proposed in the literature[3,4].

On the other hand, the majority of studies on rough sets have been concerned on the point view of granular computing. Zadeh firstly proposed the concept of granular computing and discussed issues of fuzzy information granulation in 1979 [5]. In the point view of granular computing, the classical Pawlak rough set is based on a single granulation which can be regarded as an equivalence relation on the universe induced from an indiscernibility relation. However, when the rough set is based on many granulations induced from several relations, we can have some cases as follow:

Case 1. There exists a granulation at least such that the elements surely belong to the concept.

---

* W.H. Xu is a Ph. D and a Prof. of Chongqing University of Technology. His main research fields are rough set, fuzzy set and artificial intelligence.

T. Li et al. (Eds.): RSKT 2012, LNAI 7414, pp. 20–27, 2012.
© Springer-Verlag Berlin Heidelberg 2012

Case 2. There are some granulations such that the elements surely belong to the concept.

Case 3. All of the granulations such that the elements surely belong to the concept.

Case 4. There exists a granulation at least such that the elements possibly belong to the concept.

Case 5. There are some granulations such that the elements possibly belong to the concept.

Case 6. All of the granulations such that the elements possibly belong to the concept.

For the need of some practical issues, Qian and Xu extended the Pawlak rough set to multi-granulation rough set models where the approximation operators are defined by multiple equivalence relations on the universe [6,7,8]. On the basic, many researchers have been extended the multi-granulation rough set to the generalized multi-granulation rough sets[1,9,10].

Moreover, more generalizations of fuzzy rough sets were defined by using a residual implication and a triangular norm on $[0, 1]$ to define the lower and upper approximation operators. Several authors also have proposed a kind of implication[11], weak fuzzy partitions on the universe. In this paper, we intent to generalize the multi-granulation rough sets theory by using the concepts of a residual implication and a triangular norm on $[0, 1]$. In the following section, we recall some concepts and lemmas to be used in this paper. In Section 3, we proposed the definitions for the optimistic multi-granulation T-fuzzy lower and upper approximation operators and basic properties are studied. Section 4 concludes this article.

2   The Rough Set Based on Triangular Norm

Let U be a nonempty and finite set. The Cartesian product of U with U is denoted by $U \times U$. The classes of all crisp (fuzzy, respectively) subsets of U denoted by $P(U)$ ($F(U)$, respectively). Following[11], a binary operator T on the unit interval $I = [0, 1]$ is said to be a triangular norm, if $\forall a, b, c, d \in I$, we have

$(1)T(a, b) = T(b, a),$ $(2)T(a, 1) = a,$
$(3)a \leq c, b \leq d \Rightarrow T(a, b) \leq T(c, d),$ $(4)T(T(a, b), c) = T(a, T(b, c)).$

A fuzzy relation R from U to U is a fuzzy subset of $U \times U$, i.e., $R \in F(U \times U)$, and $R(x, y)$ is called the degree of relation between x and y. R is said to be reflexive on U, iff $\forall x \in U, R(x, x) = 1$; R is said to be symmetric on U, iff $\forall x \in U, R(x, y) = R(y, x)$; R is said to be T transitive on U, iff $\forall x, y, z \in U$, $R(x, x) \geq T(R(x, y), R(y, z))$. If R is reflexive, symmetric and T transitive on U, we then say that R is a T-fuzzy similarity relation on U.

Now, we define the following binary operator on I:

$$\theta(a, b) = \sup\{c \in I | T(a, c) \leq b\},$$

θ is called the residual implication based on a triangular norm T.

Lemma 2.1. Let T is a lower semi-continuous triangular norm, $\forall a, b, c \in I$ then the residual implication based on a triangular norm T satisfies the following properties.

$(\theta 1)\theta(a, 1) = 1, \quad \theta(1, a) = a$

$(\theta 2)a \leq b \Rightarrow \theta(c, a) \leq \theta(c, b)$

$(\theta 3)a \leq b \Rightarrow \theta(a, c) \geq \theta(b, c)$

$(\theta 4)T(\theta(a, c), \theta(c, b)) \leq \theta(a, b)$

$(\theta 5)\theta(a \vee b, c) = \theta(a, c) \wedge \theta(b, c)$

$(\theta 6)\theta(a, b \wedge c) = \theta(a, b) \wedge \theta(a, c)$

$(\theta 7)a \leq b \Leftrightarrow \theta(a, b) = 1$

$(\theta 8)\theta(a, \theta(b, c)) = \theta(b, \theta(a, c))$

$(\theta 9)\theta(T(a, b), c) = \theta(a, \theta(b, c))$

$(\theta 10)T(\theta(T(a, b), c), a) \leq \theta(b, c)$

$(\theta 11) \bigwedge_{a \in I} \theta(T(b, \theta(c, a)), a) = \theta(b, c)$

$(\theta 12)\theta(\theta(a, b), b) \geq a$

$(\theta 13) \bigwedge_{b \in I} \theta(\theta(a, b), b) = a$

$(\theta 14)T(\theta(a, b), c) \leq \theta(a, T(b, c))$

$(\theta 15) \bigwedge_{b \in I} \theta(\theta(a, b), \theta(c, b)) = \theta(c, a)$

$(\theta 16)\theta(a, b) \leq \theta(T(a, c), T(b, c))$

$(\theta 17)\theta(a, b \vee c) = \theta(a, b) \vee \theta(a, c)$

$(\theta 18)a \leq \theta(b, T(a, b))$

$(\theta 19)\theta(a \wedge b, c) = \theta(a, c) \vee \theta(b, c)$

$(\theta 20)\theta(a \wedge b, c) \geq \theta(a, c) \wedge \theta(b, c)$

Definition 2.1. Let U be a finite and nonempty sets called the universe, and R be a T-fuzzy similarity relation from U to U. The pair (U, R) is called a generalized T-fuzzy approximation space. For any $A \in F(U)$, we define two fuzzy set-theoretic operators from $F(U)$ to $F(U)$:

$$\underline{R}(A)(x) = \bigwedge_{y \in U} \theta(R(x, y), A(y)), \quad \overline{R}(A)(x) = \bigvee_{y \in U} T(R(x, y), A(y)), \quad x \in U.$$

Where $\underline{R}$ and $\overline{R}$ are referred to as the generalized T-fuzzy lower and upper approximation operators. The pair $(\underline{R}(A), \overline{R}(A))$ is called the generalized T-fuzzy rough set of A.

The following proposition reflects the relationships between $\underline{R}$ and $\overline{R}$.

Proposition 2.1. Let (U, R) be a fuzzy approximation space, for $\forall A, B \in F(U), (x, y) \in U \times U$, then

(1) $\underline{R}(A) \subseteq A \subseteq \overline{R}(A)$.

(2) $\underline{R}(A \cap B) = \underline{R}(A) \cap \underline{R}(B), \quad \overline{R}((A) \cup B) = \overline{R}(A) \cup \overline{R}(B)$.

(3) $\underline{R}(A \cup B) \supseteq \underline{R}(A) \cup \underline{R}(B), \quad \underline{R}(A \cap B) \subseteq \underline{R}(A) \cap \overline{R}(B)$.

(4) $A \subseteq B \Rightarrow \underline{R}(A) \subseteq \underline{R}(B), \overline{R}(A) \subseteq \overline{R}(B)$.

(5) $\underline{R}(\underline{R}(A)) = \underline{R}(A), \overline{R}(\overline{R}(A)) = \overline{R}(A)$.

(6) $\overline{R}(\underline{R}(A)) = \underline{R}(A), \underline{R}(\overline{R}(A)) = \overline{R}(A)$.

(7) $\overline{R}(A) = A \Leftrightarrow \overline{R}(A) = A$.

3   Optimistic Multi-Granulation Fuzzy Rough Set Based on Triangular Norm

In this section, we will study the optimistic multi-granulation fuzzy rough set based on triangular norm which is on the rough approximation problem in a generalized T-fuzzy approximation space.

Assumed that $(U, R_{A_1}, R_{A_2}, \cdots, R_{A_n})$ is a generalized T-fuzzy approximation space, if U is a finite and nonempty universe, and $\forall R_{A_i}$ is a T-fuzzy similarity relation from U to U.

In the following, we will give the definition of the optimistic multi-granulation rough set based on triangular norm.

Definition 3.1. Let $(U, R_{A_1}, R_{A_2}, \cdots, R_{A_n})$ be a generalized T-fuzzy approximation space. For any $X \in F(U)$, we can define the optimistic multi-granulation T-fuzzy lower and upper approximation of X as follows

$$OM_{\sum_{i=1}^{n} A_i}(X)(x) = \bigvee_{i=1}^{n} (\bigwedge_{u \in U} \theta(R_{A_i}(u, x), X(u))),$$

$$\overline{OM_{\sum_{i=1}^{n} A_i}}(X)(x) = \bigwedge_{i=1}^{n} (\bigvee_{u \in U} T(R_{A_i}(u, x), X(u))),$$

where "$\bigvee$" means "max", "$\bigwedge$" means "min", θ and T are defined in Section 2. $OM_{\sum_{i=1}^{n} A_i}$ and $\overline{OM_{\sum_{i=1}^{n} A_i}}$ are referred to as the generalized optimistic multi-granulation T-fuzzy lower and T upper approximation operators. The pair $(OM_{\sum_{i=1}^{n} A_i}(X), \overline{OM_{\sum_{i=1}^{n} A_i}}(X))$ is called the generalized optimistic multi-granulation T-fuzzy rough set of X.

In the following, we employ an example to illustrate the above concepts.

Example 3.1. Let (U, R_A, R_B) be a generalized T-fuzzy approximation space, where $U = \{x_1, x_2, x_3, x_4, x_5\}$,

$$R_A = \begin{pmatrix} 1 & 0.4 & 0.8 & 0.5 & 0.5 \\ 0.4 & 1 & 0.4 & 0.4 & 0.4 \\ 0.8 & 0.4 & 1 & 0.5 & 0.5 \\ 0.5 & 0.4 & 0.5 & 1 & 0.6 \\ 0.5 & 0.4 & 0.5 & 0.6 & 1 \end{pmatrix}, \quad R_B = \begin{pmatrix} 1 & 0.8 & 0.8 & 0.2 & 0.8 \\ 0.8 & 1 & 0.85 & 0.2 & 0.85 \\ 0.8 & 0.85 & 1 & 0.2 & 0.9 \\ 0.2 & 0.2 & 0.2 & 1 & 0.2 \\ 0.8 & 0.85 & 0.9 & 0.2 & 1 \end{pmatrix}.$$

Taking $T(x, y) = \min(x, y)$, $X = (0.5, 0.3, 0.3, 0.6, 0.5)$.

It is not difficult to verify the fuzzy relation R_A and R_B are both T-fuzzy similar relations. So we can obtain the generalized optimistic multi-granulation T-fuzzy lower and upper approximation of X as follow

$$OM_{A+B}(X) = (0.3, 0.3, 0.3, 0.6, 0.3), \quad \overline{OM_{A+B}}(X) = (0.5, 0.4, 0.5, 0.6, 0.5).$$

From the definitions of the optimistic multi-granulation T-fuzzy lower and upper approximation operators, it is possible to deduce the following properties by use of mathematical induction.

Proposition 3.1. Let $(U, R_{A_1}, R_{A_2}, \cdots, R_{A_n})$ be a generalized T-fuzzy approximation space, R_{A_i}, $i \in \{1, 2, 3..., n\}$ be the different T-fuzzy similarity relations, $\forall X, Y \in F(U)$. Then the optimistic multi-granulation T-fuzzy lower approximation operator has the following properties.

(1) $OM_{\sum\limits_{i=1}^{n} A_i}(X) \subseteq X.$

(2) $\overline{OM}_{\sum\limits_{i=1}^{n} A_i}(OM_{\sum\limits_{i=1}^{n} A_i}(X)) = OM_{\sum\limits_{i=1}^{n} A_i}(X).$

(3) $\overline{OM}_{\sum\limits_{i=1}^{n} A_i}(X \cap Y) \subseteq OM_{\sum\limits_{i=1}^{n} A_i}(X) \cap \overline{OM}_{\sum\limits_{i=1}^{n} A_i}(Y).$

(4) $\overline{X \subseteq Y} \Rightarrow OM_{\sum\limits_{i=1}^{n} A_i}(X) \subseteq OM_{\sum\limits_{i=1}^{n} A_i}(Y).$

(5) $OM_{\sum\limits_{i=1}^{n} A_i}(\overline{X \cup Y}) \supseteq \overline{OM}_{\sum\limits_{i=1}^{n} A_i}(X) \cup \overline{OM}_{\sum\limits_{i=1}^{n} A_i}(Y).$

Proof. Since the number of the granulations is finite, we only prove the results are true in a generalized T-fuzzy approximation space (U, R_A, R_B) for convenience. It is obvious that all terms hold when $R_A = R_B$. When $R_A \neq R_B$, the proposition can be proved as follows.

(1) For any $x \in U$, we have

$$\underline{OM}_{A+B}(X)(x) = \bigwedge_{u \in U} \theta(R_A(u,x), X(u)) \vee \bigwedge_{u \in U} \theta(R_B(u,x), X(u))$$
$$\leq \theta(R_A(x,x), X(x)) \vee \theta(R_B(x,x), X(x))$$
$$= \theta(1, X(x)) \vee \theta(1, X(x))$$
$$= X(x)$$

(2) For any $x \in U$, we can obtain

$$\underline{OM}_{A+B}(\underline{OM}_{A+B}(X))(x)$$
$$= \bigwedge_{u \in U} \theta(R_A(u,x), \underline{OM}_{A+B}(X)(u)) \vee \bigwedge_{u \in U} \theta(R_B(u,x), \underline{OM}_{A+B}(X)(u))$$
$$\geq \bigwedge_{u,v \in U} \theta(R_A(u,x), \theta(R_A(v,u), X(v))) \vee \bigwedge_{u,v \in U} \theta(R_B(u,x), \theta(R_B(v,u), X(v)))$$
$$\geq \bigwedge_{u,v \in U} \theta(T(R_A(u,x), R_A(v,u)), X(v)) \vee \bigwedge_{u,v \in U} \theta(T(R_B(u,x), R_B(v,u)), X(v))$$
$$\geq \bigwedge_{v \in U} \theta(R_A(v,x), X(v)) \vee \bigwedge_{v \in U} \theta(R_B(v,x), X(v))$$
$$= \underline{OM}_{A+B}(X)(x)$$

So, $\underline{OM}_{A+B}(\underline{OM}_{A+B}(X)) \supseteq \underline{OM}_{A+B}(X)$

On the other hand, we can obtain $\underline{OM}_{A+B}(\underline{OM}_{A+B}(X)) \subseteq \underline{OM}_{A+B}(X)$ by (1). Therefore, (2) have been proved.

(3) For any $x \in U$, we have

$$\underline{OM_{A+B}}(X \cap Y)(x)$$
$$= \bigwedge_{u \in U} \theta(R_A(u,x),(X \cap Y)(u)) \vee \bigwedge_{u \in U} \theta(R_B(u,x),(X \cap Y)(u))$$
$$\leq [(\bigwedge_{u \in U} \theta(R_A(u,x),X(u))) \vee (\bigwedge_{u \in U} \theta(R_B(u,x),X(u)))] \wedge$$
$$[(\bigwedge_{u \in U} \theta(R_A(u,x),Y(u))) \vee (\bigwedge_{u \in U} \theta(R_B(u,x),Y(u)))]$$
$$= \underline{OM_{A+B}}(X)(x) \wedge \underline{OM_{A+B}}(Y)(x)$$
$$= \underline{OM_{A+B}}(X \cap Y)(x)$$

i.e., $\underline{OM_{A+B}}(X \cap Y) \subseteq \underline{OM_{A+B}}(X) \cap \underline{OM_{A+B}}(Y)$.

(4) Since $X \subseteq Y$, then for any $x \in U$, we have $X(x) \leq Y(x)$. Therefore,

$$\underline{OM_{A+B}}(X)(x) = \bigwedge_{u \in U} \theta(R_A(u,x),X(u)) \vee \bigwedge_{u \in U} \theta(R_B(u,x),X(u))$$
$$\leq \bigwedge_{u \in U} \theta(R_A(u,x),Y(u)) \vee \bigwedge_{u \in U} \theta(R_B(u,x),Y(u))$$
$$= \underline{OM_{A+B}}(Y)(x)$$

(5) According to the proposition (4), this item can be proved easily.

Proposition 3.2. Let $(U, R_{A_1}, R_{A_2}, \cdots, R_{A_n})$ be a generalized T-fuzzy approximation space, R_{A_i}, $i \in \{1,2,3...,n\}$ be the different T-fuzzy similarity relations. For $\forall X, Y \in F(U)$, the optimistic multi-granulation T-fuzzy upper approximation operator has the following properties.

(1) $X \subseteq \overline{OM_{\sum_{i=1}^{n} A_i}}(X)$.

(2) $\overline{OM_{\sum_{i=1}^{n} A_i}}(\overline{OM_{\sum_{i=1}^{n} A_i}}(X)) = \overline{OM_{\sum_{i=1}^{n} A_i}}(X)$.

(3) $\overline{OM_{\sum_{i=1}^{n} A_i}}(X \cup Y) \supseteq \overline{OM_{\sum_{i=1}^{n} A_i}}(X) \cup \overline{OM_{\sum_{i=1}^{n} A_i}}(Y)$.

(4) $X \subseteq Y \Rightarrow \overline{OM_{\sum_{i=1}^{n} A_i}}(X) \subseteq \overline{OM_{\sum_{i=1}^{n} A_i}}(Y)$.

(5) $\overline{OM_{\sum_{i=1}^{n} A_i}}(X \cap Y) \subseteq \overline{OM_{\sum_{i=1}^{n} A_i}}(X) \cap \overline{OM_{\sum_{i=1}^{n} A_i}}(Y)$.

Proof. Since the number of the granulations is finite, we only prove the results are true in a generalized T-fuzzy approximation space (U, R_A, R_B) for convenience. When $R_A \neq R_B$, the proposition can be proved as follows.

(1) For any $x \in U$,

$$\overline{OM_{A+B}}(X)(x) = \bigvee_{u \in U} T(R_A(u,x),X(u)) \wedge \bigvee_{u \in U} T(R_B(u,x),X(u))$$
$$\geq T(R_A(x,x),X(x)) \wedge T(R_B(x,x),X(x))$$
$$= X(x)$$

(2) For any $x \in U$,

$$\overline{OM_{A+B}}(\overline{OM_{A+B}}(X))(x)$$

$$= \bigvee_{u \in U} T(R_A(u,x), \overline{OM_{A+B}}(X)(u)) \wedge \bigvee_{u \in U} T(R_B(u,x), \overline{OM_{A+B}}(X)(u))$$

$$\leq \bigvee_{u,v \in U} T(R_A(u,x), T(R_A(v,u), X(v))) \wedge \bigvee_{u,v \in U} T(R_B(u,x), T(R_B(v,u), X(v)))$$

$$= \bigvee_{u,v \in U} T(T(R_A(u,x), R_A(v,u)), X(v)) \wedge \bigvee_{u,v \in U} T(T(R_B(u,x), R_B(v,u)), X(v))$$

$$\leq \bigvee_{v \in U} T(R_A(v,x), X(v)) \wedge \bigvee_{v \in U} T(R_B(v,x), X(v)) = \overline{OM_{A+B}}(X)(x)$$

Moreover, we have know $\overline{OM_{A+B}}(X) \subseteq \overline{OM_{A+B}}(\overline{OM_{A+B}}(X))$ by the proposition (1).

(3) For any $x \in U$,

$$\overline{OM_{A+B}}(X \cup Y)(x)$$

$$= \bigvee_{u \in U} T(R_A(u,x), X(u) \vee Y(u)) \wedge \bigvee_{u \in U} T(R_B(u,x), X(u) \vee Y(u))$$

$$= [\bigvee_{u \in U} T(R_A(u,x), X(u)) \vee \bigvee_{u \in U} T(R_A(u,x), Y(u))] \wedge$$

$$[\bigvee_{u \in U} T(R_B(u,x), X(u)) \vee \bigvee_{u \in U} T(R_B(u,x), Y(u))]$$

$$\geq [\bigvee_{u \in U} T(R_A(u,x), X(u)) \wedge \bigvee_{u \in U} T(R_B(u,x), X(u))] \vee$$

$$[\bigvee_{u \in U} T(R_A(u,x), Y(u)) \wedge \bigvee_{u \in U} T(R_B(u,x), Y(u))]$$

$$= \overline{OM_{A+B}}(X \cup Y)(x)$$

(4) Since $X \subseteq Y$, then for any $x \in U$, we can have $X(x) \leq Y(x)$. So

$$\overline{OM_{A+B}}(X)(x) = \bigvee_{u \in U} T(R_A(u,x), X(u)) \wedge \bigvee_{u \in U} T(R_B(u,x), X(u))$$

$$\leq \bigvee_{u \in U} T(R_A(u,x), Y(u)) \wedge \bigvee_{u \in U} T(R_B(u,x), Y(u))$$

$$= \overline{OM_{A+B}}(Y)(x)$$

(5) This item can be proved by (4).

By the definitions of the optimistic multi-granulation T-fuzzy lower and upper approximation operators based on triangular norm, for $\forall X \in F(U)$, the relationships of the optimistic multi-granulation T-fuzzy lower approximation and upper approximation operators as follows:

$$OM_{\sum\limits_{i=1}^{n} A_i}(X) \subseteq X \subseteq \overline{OM_{\sum\limits_{i=1}^{n} A_i}}(X).$$

4   Conclusions

In this paper, the generalized T-fuzzy rough set model based on triangular norm has been significantly extended. In this extension, the approximations of sets were defined by using multiply T-fuzzy similarity relations on the universe. It is obvious that the generalize T-fuzzy lower and upper approximation operators defined on (U, R) were obtained as a special case of these operators. More properties of the optimistic multi-granulation T-fuzzy rough set based on triangular norm were discussed. And we investigated the relationships between the approximation operators. The construction of the optimistic multi-granulation fuzzy rough set model over T-fuzzy similarity relations on the universe is meaningful in terms of the generalization of rough set theory.

Acknowledgments. This work is supported by National Natural Science Foundation of China (No. 61105041, 71071124, and 11001227), Postdoctoral Science Foundation of China (No. 20100481331), Natural Science Foundation Project of CQ CSTC (No. cstc2011jjA40037), Graduate Innovation Foundation of Chongqing University of Technology (No. YCX2011312).

References

1. Khan, M. A., Ma, M.: A Modal Logic for Multiple-Source Tolerance Approximation Spaces. In: Banerjee, M., Seth, A. (eds.) Logic and Its Applications. LNCS (LNAI), vol. 6521, pp. 124–136. Springer, Heidelberg (2011)
2. Pawlak, Z.: Rough sets. International Journal of Computer and Information Sciences 11(5), 341–356 (1982)
3. Dubois, D., Prade, H.: Rough fuzzy sets and fuzzy rough sets. International Journal of General Systems 17, 191–208 (1990)
4. Dubois, D., Prade, H.: Putting rough sets and fuzzy sets together (1992)
5. Zadeh, L.A.: Fuzzy Sets and Information Granularity, Advances in fuzzy set theory and application. North Holland Publishing, Amstandam (1979)
6. Qian, Y.H., Liang, J.Y., Dang, C.Y.: Knowledge structure, knowledge granulation and knowledge distance in a knowledge base. International Journal of Approximate Reasoning 50, 174–188 (2009)
7. Qian, Y.H., Liang, J.Y., Yao, Y.Y., Dang, C.H.: MGRS: A multi-granulation rough set. Information Sciences 180, 949–970 (2010)
8. Xu, W.H., Zhang, X.Y., Zhang, W.X.: Two new types of multiple granulation rough set. Information Sciences (submitted)
9. Xu, W.H., Zhang, X.T., Wang, Q.R.: A Generalized Multi-granulation Rough Set Approach. In: Huang, D.-S., Gan, Y., Premaratne, P., Han, K. (eds.) ICIC 2011. LNCS, vol. 6840, pp. 681–689. Springer, Heidelberg (2012)
10. Xu, W.H., Wang, Q.R., Zhang, X.T.: Multi-granulation Fuzzy Rough Set Model on Tolerance Relations. In: Fourth International Workshop on Advanced Computational Intelligence, pp. 359-366 (2011)
11. Zhang, W.X., Wu, W.Z., Liang, J.Y., Li, D.Y.: Theory and method of rough sets (2001)

Rough Set Model Based on Hybrid Tolerance Relation

Junyi Zhou[1] and Xibei Yang[2,3,*]

[1] School of Economics and Management,
Jiangsu University of Science and Technology, Zhenjiang, 212003, China
zhoujy@just.edu.cn
[2] School of Computer Science and Engineering,
Jiangsu University of Science and Technology, Zhenjiang, 212003, P.R. China
[3] School of Computer Science and Technology,
Nanjing University of Science and Technology, Nanjing, 210094, China
yangxibei@hotmail.com

Abstract. In this paper, three different tolerance relations are deeply explored in incomplete information system. The strong tolerance relation is corresponding to the semantic explanation of conjunction while the weak tolerance relation is corresponding to the semantic explanation of disjunction. Based on such two explanations, to derive decision rules with both "AND" and "OR" connectives from an incomplete decision system, the concept of the hybrid tolerance relation is proposed through the fusion of strong and weak tolerance relations. Two different rough sets are then constructed by employing hybrid tolerance relation and hybrid tolerance relation based maximal consistent blocks. The relationships between such two different rough sets are also explored. Finally, based on the maximal consistent blocks technique, the generation of decision rules from incomplete decision system is addressed.

Keywords: incomplete system, strong tolerance relation, weak tolerance relation, hybrid tolerance relation, decision rule.

1 Introduction

Presently, with the rapid development of rough set theory[1,2], the rough set method has been successfully applied to the fields of knowledge discovery, decision supporting, pattern recognition, and so on. With the concepts of lower and upper approximation, knowledge hidden in information system may be expressed in the form of decision rules. Pawlak's rough set model was proposed through an indiscernibility relation and can only be used to deal with the complete information system. However, incomplete information systems[3] with unknown values can be seen everywhere in many real-world applications. At present, many types of binary relations have been proposed to solve such problems. In most of these binary relations, the used logical connective is conjunction, i.e. "AND".

---

* Corresponding author.

T. Li et al. (Eds.): RSKT 2012, LNAI 7414, pp. 28–33, 2012.

Nevertheless, In [4], a new binary relation has been proposed, which called weak indiscernibility relation, in which the used logical connective is disjunction, i.e. "OR". Correspondingly, Pawlak's indiscernibility is referred to as strong indiscernibility relation. Through weak indiscernibility relation, the "OR" decision rules can be generated from decision system.

In this paper, the concept of weak indiscernibility relation will be introduced into incomplete information system and then the weak tolerance relation is presented. Similar to the indiscernibility relation case, Kryszkiewicz's tolerance relation is called the strong tolerance relation. Based on weak and strong tolerance relation, the concept of hybrid tolerance relation is then proposed. Such hybrid tolerance relation is not only a fusion of weak and strong tolerance relations, but also can be used to derive decision rules with "AND " and "OR" logical connectives from the incomplete decision systems.

2   Preliminaries

In this section we review some basic notions of strong indiscernibility relation, weak indiscernibility relation, incomplete information system and tolerance relation.

2.1   Strong and Weak Indiscernibility Relations

A decision system is an information system $I = <U, AT \cup D>$, in which U is a non-empty finite set of objects called the universe of discourse and AT is a non-empty finite set of the condition attributes, D is the set of the decision attributes and $AT \cap D = \emptyset$.

Given a subset of attributes $A \subseteq AT$, the strong indiscernibility relation $IND(AT)$ and the weak indiscernibility relation $WIND(AT)$ are defined by

$$IND(AT) = \{(x, y) \in U^2 : \forall a \in AT, a(x) = a(y)\}. \tag{1}$$

$$WIND(AT) = \{(x, y) \in U^2 : \exists a \in AT, a(x) = a(y)\}. \tag{2}$$

Definition 1. *Let I be an information system in which $A \subseteq AT$, $\forall X \subseteq U$, the lower and upper approximations of X in terms of the strong indiscernibility relation $IND(AT)$ are denoted by $\underline{AT}_S(X)$ and $\overline{AT}_S(X)$, respectively,*

$$\underline{AT}_S(X) = \{x \in U : [x]_{AT} \subseteq X\}, \ \overline{AT}_S(X) = \{x \in U : [x]_{AT} \cap X \neq \emptyset\}.$$

$[x]_{AT}$ is the equivalence class based on indiscernibility relation $IND(AT)$ and is denoted as $[x]_{AT} = \{y \in U : (x, y) \in IND(AT)\}$.

Definition 2. *Let I be an information system in which $A \subseteq AT$, $\forall X \subseteq U$, the lower and upper approximations of X in terms of the strong indiscernibility relation $WIND(AT)$ are denoted by $\underline{AT}_W(X)$ and $\overline{AT}_W(X)$, respectively,*

$$\underline{AT}_W(X) = \{x \in U : [x]_{AT}^W \subseteq X\}, \ \overline{AT}_W(X) = \{x \in U : [x]_{AT}^W \cap X \neq \emptyset\}.$$

$[x]_{AT}^W$ is the equivalence class based on indiscernibility relation $WIND(AT)$ and is denoted as $[x]_{AT}^W = \{y \in U : (x, y) \in WIND(AT)\}$.

2.2   Incomplete Information System and Tolerance Relations

An incomplete information system indicates that the precise attributes' values for some objects are unknown, in this paper, we just consider the unknown value is "missing", but it does exist, it is considered as to be comparable with any value in the domain of the corresponding attribute, the unknown value is indicated as "*". From this point of view, Kryszkiewicz[5] proposed her tolerance relation as Eq. (3) shows. Given a subset of attributes $A \subseteq AT$, the tolerance relation T_A is defined by

$$T_A = \{(x, y) \in U^2 : \forall a \in A, a(x) = a(y) \lor a(x) = * \lor a(y) = *\}. \tag{3}$$

Similar to the strong indiscernibility relation, we call this tolerance relation as strong tolerance relation ST_A. Obviously, relation T_A is reflexive, symmetric.

Definition 3. *Let I be an incomplete information system in which $A \subseteq AT$, $\forall X \subseteq U$, the lower and upper approximations of X in terms of the strong tolerance relation ST_A are denoted by $\underline{AT}_W(X)$ and $\overline{AT}_W(X)$, respectively,*

$$\underline{SA}_T(X) = \{x \in U : T_A(x) \subseteq X\}, \quad \overline{SA}_T(X) = \{x \in U : T_A(x) \cap X \neq \emptyset\}.$$

$T_A(x)$ *is the equivalence class based on indiscernibility relation T_A and is denoted as $T_A(x) = \{y \in U : (x, y) \in T_A\}$.*

2.3   Weak Tolerance Relations

In strong tolerance relation, two objects are tolerated if and only if they are tolerant on all of the considered attributes. Following the basic idea of weak indiscernibility relation, it is natural to define a weak tolerance relation. In weak tolerance relation, two objects are tolerated when they are tolerant on at least only one of the considered attributes.

Definition 4. *Let I be an incomplete information system in which $A \subseteq AT$, the weak tolerance relation in terms of A is defined by WT_A*

$$WT_A = \{(x, y) \in U^2 : \exists a \in A, a(x) = a(y) \lor a(x) = * \lor a(y) = *\}.$$

Definition 5. *Let I be an incomplete information system in which $A \subseteq AT$, $\forall X \subseteq U$, the lower and upper approximations of X in terms of the tolerance relation WT_A are denoted by $\underline{WA}_T(X)$ and $\overline{WA}_T(X)$, respectively,*

$$\underline{WA}_T(X) = \{x \in U : WT_A(x) \subseteq X\}, \quad \overline{WA}_T(X) = \{x \in U : WT_A(x) \cap X \neq \emptyset\}.$$

$WT_A(x)$ *is the equivalence class based on indiscernibility relation WT_A and is denoted as $WT_A(x) = \{y \in U : (x, y) \in WT_A\}$.*

3   Hybrid Tolerance Relation and Decision Rule

Through strong tolerance relation, we can derive decision rules with "AND" logical connective from an incomplete information system, while through weak tolerance relation, we can derive decision rules with "OR" logical connective. In order to obtain a decision rule, which including both "AND" and "OR", new binary relation has become a necessity. This is what will be discussed in the following.

3.1   Hybrid Tolerance Relation

Definition 6. *Let I be an incomplete information system in which $A = A_1 \cup A_2 \subseteq AT$, the hybrid tolerance relation in terms of A is defined by HT_A*

$$HT_A = \{(x, y) \in U^2 : (\forall a \in A_1, a(x) = a(y) \vee a(x) = * \vee a(y) = *) \wedge$$
$$(\exists a \in A_2 a(x) = a(y) \vee a(x) = * \vee a(y) = *)\}.$$

Proposition 1. *Let I be an incomplete information system in which $A = A_1 \cup A_2 \subseteq AT$, we have $HT_A = T_{A_1} \cap WT_{A_2}$.*

Proof. $\forall(x, y) \in U^2$, by Definition 6, we have

$$(x, y) \in HT_A \Leftrightarrow (\forall a \in A_1, a(x) = a(y) \vee a(x) = * \vee a(y) = *) \wedge$$
$$(\exists a \in A_2 a(x) = a(y) \vee a(x) = * \vee a(y) = *)$$
$$\Leftrightarrow (x, y) \in T_{A_1} \wedge (x, y) \in WT_{A_2} \Leftrightarrow (x, y) \in T_{A_1} \cap WT_{A_2}$$

By Definition 6, hybrid tolerance relation HT_A is actually the fusion of strong tolerance relation T_{A_1} and weak tolerance relation WT_{A_2}. If there is only one attribute in A_2, then the hybrid tolerance relation will degenerated to be strong tolerance relation, if there is no attribute in A_1, then the hybrid tolerance relation will degenerate to be weak tolerance relation.

Definition 7. *Let I be an incomplete information system in which $A = A_1 \cup A_2 \subseteq AT$, $\forall X \subseteq U$, the lower and upper approximations of X in terms of the hybrid tolerance relation HT_A are denoted by $\underline{HA}_T(X)$ and $\overline{HA}_T(X)$, respectively,*

$$\underline{HA}_T(X) = \{x \in U : HT_A(x) \subseteq X\},$$
$$\overline{HA}_T(X) = \{x \in U : HT_A(x) \cap X \neq \emptyset\}.$$

$HT_A(x)$ *is the tolerance class based on hybrid tolerance relation HT_A and is denoted as $HT_A(x) = \{y \in U : (x, y) \in HT_A\}$.*

Obviously, hybrid tolerance relation HT_A is reflexive, symmetric, then similar to what have been discussed in [6,7], a family of the maximal consistent blocks can be obtained based on such relation.

Definition 8. *Let I be an incomplete information system in which $A = A_1 \cup A_2 \subseteq AT$, the complete covering in terms of the hybrid tolerance relation HT_A is denoted by $\mathcal{R}(A) = \{Y \subseteq U : Y^2 \subseteq HT_A \wedge (\forall x \notin Y, (Y \cup \{x\})^2) \not\subseteq HT_A\}$.*

Definition 9. *Let I be an incomplete information system in which $A = A_1 \cup A_2 \subseteq AT$, $\forall X \subseteq U$, the lower and upper approximations of X in terms of the complete covering $\mathcal{R}(A)$ are denoted by $\underline{AT}_{\mathcal{R}}(X)$ and $\overline{AT}_{\mathcal{R}}(X)$, respectively,*

$$\underline{AT}_{\mathcal{R}}(X) = \cup\{Y \in \mathcal{R}(A) : Y \subseteq X\}, \quad \overline{AT}_{\mathcal{R}}(X) = \cup\{Y \in \mathcal{R}(A) : Y \cap X \neq \emptyset\}.$$

Proposition 2. *Let I be an incomplete information system in which $A = A_1 \cup A_2 \subseteq AT$, $\forall X \subseteq U$, we have $HT_A(x) = \cup\{Y \in \mathcal{R}(A) : x \in Y\}$.*

Proof. "$\Rightarrow$" $\forall y \in HT_A(x)$, by Definition 6 we know that $(x, y) \in HT_A$, there must be $Y \in \mathcal{R}(A)$ such that $\{x, y\} \subseteq Y$, so it can be proved that $y \in \cup\{Y \in \mathcal{R}(A) : x \in Y\}$, thus $HT_A(x) \subseteq \cup\{Y \in \mathcal{R}(A) : x \in Y\}$.

"$\Leftarrow$" $\forall y \in \mathcal{R}(A) : x \in Y$, there must be $Y \in \mathcal{R}(A)$ such that $x \in Y$ and $y \in Y$, so we can get $\{x, y\} \subseteq Y$. By Definition 6, x and y satisfy HT_A such that $y \in HT_A(x)$, thus $HT_A(x) \supseteq \cup\{Y \in \mathcal{R}(A) : x \in Y\}$.

Proposition 3. *Let I be an incomplete information system in which $A = A_1 \cup A_2 \subseteq AT$, $\forall X \subseteq U$, we have $\underline{HA}_T(X) \subseteq \underline{AT}_{\mathcal{R}}(X)$, $\overline{HA}_T(X) = \underline{AT}_{\mathcal{R}}(X)$.*

Proof. $\forall x \in \underline{HA}_T(X)$, by Definition 7 we know that $HT_A(x) \subseteq X$. By Proposition 2, we know that $\cup\{Y \in \mathcal{R}(A) : x \in Y\} \subseteq X$, $\forall Y \in \mathcal{R}(A)$ and $x \in Y$ such that $Y \subseteq X$, by Definition 9 such that $x \in AT_{\wp}(X)$.

Similarity, $\overline{HA}_T(X) \subseteq \overline{AT}_{\mathcal{R}}(X)$. $\forall x \in \overline{AT}_{\mathcal{R}}(X)$, by Definition 9, there must be $Y \in \mathcal{R}(A)$ and $x \in Y$ such that $Y \cap X \neq \emptyset$. By Proposition 2, we know that $Y \subseteq HT_A(x) \Rightarrow HT_A(x) \cap X \neq \emptyset \Rightarrow x \in \overline{HA}_T(X) \Rightarrow \overline{HA}_T(X) \supseteq \underline{AT}_{\mathcal{R}}(X)$.

By Proposition 3, base on hybrid tolerance relation, we can generate bigger lower approximation in terms of maximal consistent block, so Definition 8 is prior to Definition 6.

3.2 Decision Rule

In terms of the maximal consistent block, the decision rule can be generated such that

$$r_x : a_1(x) = v_1 \wedge \cdots \wedge a_m(x) = v_m \wedge (a_{m+1}(x) = v_{m+1} \vee \cdots \vee a_n(x) = v_n)$$
$$\rightarrow d(x) = v_d$$

where $A1 = \{a_1, \ldots, a_m\}$, $A_2 = \{a_{m+1}, \ldots, a_n\}$, $(v_1, \ldots, v_n) \in V_{a_1} \times \ldots \times V_{a_n}$, d is the decision attribute, $v_d \in V_d$.

The decision rule r_x is actually the fusion of "AND" decision rule as $a_1(x) = v_1 \wedge \cdots \wedge a_m(x) = v_m \wedge a_{m+1}(x) = v_{m+1} \rightarrow d(x) = v_d$,

$$\vdots$$

$a_1(x) = v_1 \wedge \cdots \wedge a_m(x) = v_m \wedge a_n(x) = v_n \rightarrow d(x) = v_d$.

With the decision rule r_x, the certainty factor of the rule can be denoted as $Cer(r_x) = \max\{\frac{|Y \cap [x]_d|}{|Y|} : Y \in \mathcal{R}(A) \wedge x \in Y\}$.

Proposition 4. *Let I be a decision system in which $A = A_1 \cup A_2 \subseteq AT$, $\forall X \subseteq U$, we have:*

1. $x \in \underline{AT}_{\mathcal{R}}([x]_d) \Leftrightarrow Cer(r_x) = 1;$
2. $x \in \overline{AT}_{\mathcal{R}}([x]_d) \Leftrightarrow 0 < Cer(r_x) \leq 1.$

Proof. $\forall X \subseteq U$, by Definition 8,

$$x \in \underline{AT}_{\mathcal{R}}([x]_d)) \Leftrightarrow \exists Y \in \mathcal{R}(A) \wedge x \in Y \Rightarrow Y \subseteq [x]_d$$

$$\Leftrightarrow \exists Y \in \mathcal{R}(A) \wedge x \in Y \Rightarrow \frac{\mid Y \cap [x]_d \mid}{\mid Y \mid} = 1 \Leftrightarrow Cer(r_x) = 1.$$

The proof of 2 is similar to the proof of 1.

4 Conclusions

Tolerance relation plays a crucial role in the dealing with incomplete information system through rough set technique. By combining with strong and weak tolerance relations, the concept of hybrid tolerance relation is proposed. Two different rough set models are then constructed. These two rough sets are based on the hybrid tolerance relation and maximal consistent blocks induced by hybrid tolerance relation, respectively. Through the new constructed rough set, the decision rule including both "AND" and "OR" logical connectives can be generated from decision system.

Acknowledgment. This work is supported by Natural Science Foundation of China (No. 61100116), Natural Science Foundation of Jiangsu Province of China (No. BK2011492), Natural Science Foundation of Jiangsu Higher Education Institutions of China (No. 11KJB520004), Postdoctoral Science Foundation of China (No. 20100481149), Postdoctoral Science Foundation of Jiangsu Province of China (No. 1101137C).

References

1. Pawlak, Z.: Rough sets–theoretical aspects of reasoning about data. Kluwer Academic Publishers (1992)
2. Pawlak, Z., Skowron, A.: Rough sets: some extensions. Information Sciences 177, 28–40 (2007)
3. Yang, X.B., Yang, J.Y.: Incomplete information system and rough set theory: models and attribute reductions. Science Press Beijing & Springer (2012)
4. Zhao, Y., Yao, Y.Y., Luo, F.: Data analysis based on discernibility and indiscernibility. Information Sciences 177, 4959–4976 (2007)
5. Kryszkiewicz, M.: Rough set approach to incomplete information systems. Information Sciences 112, 39–49 (1998)
6. Leung, Y., Li, D.Y.: Maximal consistent block technique for rule acquisition in incomplete information systems. Information Sciences 153, 85–116 (2003)
7. Qian, Y.H., Liang, J.Y., Li, D.Y., et al.: Approximation reduction in inconsistent incomplete decision tables. Knowledge-Based Systems 23, 427–433 (2010)

Scalable Improved Quick Reduct: Sample Based

P.S.V.S. Sai Prasad and C. Raghavendra Rao

University of Hyderabad, Hyderabad
{saics,crrcs}@uohyd.ernet.in

Abstract. This paper develops an iterative sample based Improved Quick Reduct algorithm with Information Gain heuristic approach for recommending a quality reduct for large decision tables. The Methodology and its performance have been demonstrated by considering large datasets. It is recommended to use roughly 5 to 10% data size for obtaining an apt reduct.

Keywords: Quick Reduct, IQuick Reduct, Variable Precision, ξ-approximate Reduct.

1 Introduction

Given a decision system, feature selection is a process of selecting a subset of the conditional attributes that are relevant to the decision attribute and can be helpful in making robust learning models [1]. There are several feature selection approaches such as Regularized Trees, Minimum Redundancy Maximum Relevance (mRMR) etc [1]. The feature selection becomes complex in dealing with very large datasets. But it is mandatory to remove redundant attributes for building any classification or clustering model bearing in the impact of Curse of Dimensionality [2]. For large datasets sampling techniques helps in making the process of feature selection economical. Some of the works which uses sampling based approach for feature selection are [3,4]. The present paper develops a hybrid approach to avail the advantage of computational effort in using sampling and uses IQuickReduct strategy [9] to ascribe the desired accuracy for the given decision system.

Present paper confines to feature selection process driven by reduct based on Rough Set Theory [5]. Computation of all reducts of a decision table and finding shortest length reduct is proved to be NP hard [7]. Hence several heuristic based approaches are proposed which aim to find a single reduct of shorter length [6]. Quick Reduct algorithm (QRA) [8] is one such algorithm using additive method for finding a reduct. IQuick Reduct (IQRA) [9] algorithm is an improvement to QRA. Several variants of using IQRA are proposed in [10] using different heuristic functions and the variant using information gain heuristic IQRA_IG is used in this paper.

The extension of reduct computation to large decision tables has been addressed by several researchers in general by Divide and Conquer methods by using parallel and distributive approaches [11,15]. Sample based approaches to

T. Li et al. (Eds.): RSKT 2012, LNAI 7414, pp. 34–39, 2012.

reduct finding are done by [12,13]. Both the approaches use discernbility matrix approaches for reduct computation. As pointed out in [12] a sample based reduct computation results in general as a subset of reduct or an ξ -approximate reduct [14] and the suggested reduct may not qualify reduct properties for the original table. The additive method followed in QRA like algorithms is better suited for sample based approaches over deletion based approaches followed in [12] owing to computational efficiency. Hence this paper proposes a novel sample based reduct finding algorithm using IQRA_IG.

The paper is organized as follows. Section 2 describes the basics of IQRA_IG. Section 3 contains the proposed algorithm and its description. Section 4 gives the experiment results. The analysis of the results is done in Section 5 and the paper ends with conclusion and future work.

2   Overview of IQRA and IQRA_IG Algorithms

Basics of Rough Sets such as lower and upper approximations, POS region, Reduct and the concept of dependency measure kappa are described in [5]. This section gives brief overview of IQRA_IG which is used for the current work and the details of IQRA_IG are given in [10].

IQRA filters objects from the decision table from iteration to next iteration which are falling in Positive Region. The selection of an attribute in trivial ambiguity situations is achieved by VPRS heuristic [16]. In IQRA_IG the Information Gain heuristic is used to select attribute in ambiguity situations.

Taking the advantage of arriving at a reduct by using IQRA_IG principles for a reasonable size of decision table this paper proposes a novel scheme of sample based Reduct computation methodology for large decision table. The next section gives the proposed algorithm and its description.

3   Proposed SGIQRA_IG Algorithm

Let DT=(U, C $\cup\{d\}$) be the decision table as described before. Let 'N' denote the cardinality of U i.e. |U| and let M be |C|. If $N*M$ is very large the 'DT' is called a large decision table. Sample Guided IQRA_IG in short SGIQRA_IG algorithm is proposed for finding reduct in large datasets. The reduct set of attributes 'R' is initialized to null set. Let 's' be the given predefined manageable size. A sample of size 's' is selected from 'U' and a sample dataset 'DTS' with objects in the sample from 'DT' is constructed. Let 'IR' be the reduct obtained by using IQRA_IG on 'DTS'. Set R = R $\cup$ IR . Let 'U_p' be the set of objects of 'DT' in Positive region with respect to 'R' attributes i.e. $POS_R(\{d\})$. Let 'DTR' be the residual of DT by removing 'U_p' objects as well as attributes 'R' from C. If the resulting 'DTR' is empty then 'R' is the reduct or super reduct else the process is repeated with 'DTR'. The proposed algorithm is given below.

Algorithm Sample Guided IQRA_IG (SGIQRA_IG)

Input: 1) Decision Table DT(U, C $\bigcup$ {d}) where C is the
set of conditional attributes and 'd' is the decision attribute.
U is the set of objects.
2) Sample Size 's'
3) $\gamma_C(\{d\})$ Consistency measure of the dataset
Output: Approximate Super Reduct R.

1. R= Φ
2. DTR=DT
3. POSCOUNT=0
4. OBJECTCOUNT= |U |
5. While $\gamma_R(\{d\}) < \gamma_C(\{d\})$,
6. BEGIN
7. DTS=GenerateRandomSample(DTR,s).
8. IR=IQRA_IG(DTS)
9. R=R $\bigcup$ IR
10. Up= POS$_R(\{d\})$ //Set of obtained POS region objects in DTR
11. DTR=(U- Up,(C-IR) $\cup$ {d})
12. POSCOUNT=POSCOUNT+ | Up |
13. $\gamma_R(\{d\})$ = POSCOUNT/OBJECTCOUNT
14. ENDWHILE
15. Return R.

Note: In case the consistency measure $\gamma_C(\{d\})$ is not available, the stopping
criteria can be if $\gamma_R(\{d\})$ is 1 terminate else a persistence of trivial kappa gain
for L(predefined number) attribute insertions in R in sequence[15].

4   Experiments and Results

All experiments were conducted in Matlab environment. The datasets are from
a collection of large datasets for feature selection [17] maintained by Arizona
State University. Table 1 contains the details of the datasets along with the
results obtained using IQRA_IG algorithm. Table 1 also contains the summary
of the results with SGIQRA_IG for 100 trails. For understanding the impact of
sample sizes the SGIQRA_IG experiments are conducted on samples sizes of 5%,
10%, 15%, 20% on all datasets and also with 0.5%, 1% also on first two datasets.
For each sample size the experiment is repeated for 100 trails for all datasets.
The repeated trails help to understand the behavior of the algorithm.

The description of the terms given here is applicable to all tables. In reporting
SGIQRA_IG results Table 1 contains for each sample size, AL (Length of the rec-
ommended reduct giving minimum and maximum), IC (Iteration Count: Number
of iterations for a trial given are minimum and maximum), CT (Computational
Time in seconds, given are the mean(standard deviation)), EC(Excess Count:
Number of Redundant attributes found in the recommended reduct. Given as

minimum and maximum), RC (Reduct Count: Number of trails reduct realized). For IQRA_IG results the values for AL, EC, CT are provided. NOD denotes number of decision classes.

Table 1. IQRA_IG Results and Summary of SGIQRA_IG Results for 100 Trails

Dataset	IQRA_IG Results	Sample Size	AL	IC	CT	EC,RC
BASEHOCK	AL=47	0.5%	109,169	90,156	420.86(56.2)	(43,109),0
|U|=1993	EC=7	1%	59,95	28,61	270.41(33.59)	(10,41),0
|C|=4862	CT=369.97	5%	42,60	4,6	195.23(21.4)	(2,13),0
NOD=2		10%	37,60	2,3	213.46(30.2)	(1,13),0
		15%	35,56	2,3	261.6(38.9)	(0,16),1
		20%	40,59	2,2	275.01(49.2)	(1,17),0
PCMAC	AL=39	0.5%	96,178	88,162	282.13(45.06)	(40,117),0
|U|=1943	EC=2	1%	52,86	26,54	171.84(22.7)	(5,33),0
|C|=3289	CT=457.38	5%	41,58	4,6	134.33(13.6)	(1,11),0
NOD=2		10%	39,53	3,4	156.34(21.19)	(0,9),5
		15%	38,54	2,3	191.28(23.37)	(0,11),2
		20%	39,56	2,2	203.12(34.67)	(0,17),2
AR10P	AL=2	5%	3,3	3,3	3.35(0.06)	(0,0),100
|U|=130	EC=0	10%	2,3	2,3	4.78(0.57)	(0,0),100
|C|=2400	CT=10.33	15%	2,3	2,3	5.83(0.58)	(0,1),98
NOD=10		20%	2,3	2,3	6.62(0.61)	(0,1),89
ORL10P	AL=2	5%	3,3	3,3	10.13(0.23)	(0,0),100
|U|=100	EC=0	10%	3,3	3,3	16.24(1.82)	(0,0),100
|C|=10304	CT=33.55	15%	2,3	2,3	20.43(2.14)	(0,1),99
NOD=10		20%	2,3	2,3	24.78(1.76)	(0,1),95

A comparative analysis is done with results obtained by Hu, et al. [12] for sampling approximate reduct algorithm(SARA) in large datasets and with the results given by Wang, et al. [15] for scalable reducts generation RGonCRS algorithm. The datasets used for comparative analysis in Table 2 and Table 3 are from UCI Machine Learning Repository [18]. Table 2 gives the summary of the comparative results with the SARA algorithm.

Table 2. Comparitive Results with Sampling Approximate Reduct Algorithm

Dataset Details			SARA Results				SGIQRA_IG Results			
Dataset	|U|	Attr	Sample Size	Samples	AL	CT	Sample Size	IC	AL	CT
Satimage	4435	37	100	50	11	4.56	100	4	7	3.67
Shuttle	43500	10	100	450	4	6.49	100	3	4	1.67

RGonCRS algorithm returns all possible reducts for the dataset. To compare to our algorithm we have taken the time required to compute one reduct by RGonCRS as the ratio of time for computing reducts by number of reducts found by RGonCRS as an estimate called TSR (Time for Single Reduct). The summary of these results are given in Table 3.

Table 3. Comparitive Results with algorithm RGonCRS

Dataset Details			RGonCRS Results			SGIQRA_IG Results			
Dataset	|U|	Attr	Reducts	CT	T.S.R	Sample Size	IC	AL	CT
Dermatology	358	35	112708	136297.82	1.21	36	5	6	0.42
Breast Cancer	699	10	20	0.41	0.02	70	2	4	0.07

5   Analysis of Results

Observing the results given in Table 2 and Table 3 it can be seen that SGIQRA_IG has given comparable performance with SARA algorithm in getting recommended reduct using fewer samples and having comparable average computational time with RGonCRS. The idea behind these comparisons is not to conclude in absolute terms, to do so require working with these algorithms in a unified platform, but to relatively point out that SGIQRA_IG is computationally efficient having performance comparable to existing scalable algorithms.

SGIQRA_IG produces reduct or super reduct in contrast SARA algorithm [12] which gives only sub reduct. In this sampling method the recommended reduct iteration wise derived through sampling is getting validated on the original dataset. Thus SGIQRA_IG will be insensitive to the sample.

The following observations can be made from the results given in Table 1. For datasets AR10P, ORL10P SGIQRA_IG could result in reduct without any redundant attributes in almost all trails. Basehock and PCMAC datasets resulted in good computational gains. In Basehock dataset, for example, the SGIQRA_IG could get lesser EC than IQRA_IG for some trails. To sum up the results, the computational gain achieved over IQRA_IG is significant. It is also observed that too small samples, like 0.5%, 1%, may not be representative leading to inferior performance. The number of objects used in samples is small for 5% and 10% and have lesser computational times than 15% and 20%. As there is no significant difference concerning presence of excess attributes an empirically arrived conclusion is 5 to 10% sample sizes will be giving reasonable performance.

6   Conclusion

The SGIQRA_IG algorithm is proposed to find reduct for large datasets. Experiments on bench mark large datasets have shown significant computational gains over an efficient reduct computation algorithm IQRA_IG. It is also having comparable performance with Sampling approximation Reduct algorithm and RGonCRS algorithm. This algorithm is expected to give quality suggestive reduct by utilizing approximately 20 to 30% of the data (overall). It is proposed to develop a unified computational infrastructure for validating these scalable algorithms.

References

1. Feature Selection in wikipedia,
   http://en.wikipedia.org/wiki/Feature_selection
2. Bellman, R.E.: Adaptive Control Processes: a guided toor. Princeton University Press (1961)
3. Liu, H., Motoda, H., Yu, L.: Feature selection with selective sampling. In: Proceedings of the Nineteenth International Conference on Machine Learning, pp. 395–402 (2002)
4. Yang, J., Olafsson, S.: Optimization based feature selection using adaptive instance sampling. Computers & Operational Research 33(11), 3088–3106 (2006)
5. Pawlak, Z.: Rough Sets: Theoretical Aspects and Reasoning about Data. Kluwer Academic Publications (1991)
6. Jensen, R., Shen, Q.: Rough set based feature selection: A review. In: Hassanien, A.E., Suraj, Z., Slezak, D., Lingras, P. (eds.) Rough Computing: Theories, Technologies and Applications. IGI global (2007)
7. Nguyen, H.S., Skowron, A.: Boolean Reasoning for Feature Extraction Problems. In: Raś, Z.W., Skowron, A. (eds.) ISMIS 1997. LNCS, vol. 1325, pp. 117–126. Springer, Heidelberg (1997)
8. Chouchoulas, A., Shen, Q.: Rough Set-Aided Keyword Reduction for Text Categorization. Applied Artificial Intelligence 15(9), 843–873 (2001)
9. Sai Prasad, P.S.V.S., Raghavendra Rao, C.: IQuickReduct: An Improvement to Quick Reduct Algorithm. In: Sakai, H., Chakraborty, M.K., Hassanien, A.E., Ślęzak, D., Zhu, W. (eds.) RSFDGrC 2009. LNCS (LNAI), vol. 5908, pp. 152–159. Springer, Heidelberg (2009)
10. Sai Prasad, P.S.V.S., Raghavendra Rao, C.: Extensions to iQuickReduct. In: Sombattheera, C., Agarwal, A., Udgata, S.K., Lavangnananda, K. (eds.) MIWAI 2011. LNCS, vol. 7080, pp. 351–362. Springer, Heidelberg (2011)
11. Strąkowski, T., Rybiński, H.: A New Approach to Distributed Algorithms for Reduct Calculation. In: Peters, J.F., Skowron, A., Rybiński, H. (eds.) Transactions on Rough Sets IX. LNCS, vol. 5390, pp. 365–378. Springer, Heidelberg (2008)
12. Hu, K., Lu, Y., Shi, C.: Feature Ranking in Rough Sets. AI Communications 16, 41–50 (2003)
13. Boussouf, M., Quafafou, M.: Scalable Feature Selection Using Rough Set Theory. In: Ziarko, W., Yao, Y. (eds.) RSCTC 2000. LNCS (LNAI), vol. 2005, pp. 131–138. Springer, Heidelberg (2001)
14. Slezak, D.: Approximate Entropy Reducts. Fundamenta Informatica 53(3-4), 365–390 (2002)
15. Wang, P.: Highly Scalable Rough Set Reducts Generation. Journal of Information Science and Engineering 23, 1281–1298 (2007)
16. Mi, J.-S., Wu, W.-Z., Zhang, W.-X.: Approaches to knowledge reduction based on variable precision rough set model. Information Sciences 159, 255–272 (2004)
17. Datasets for feature selection, http://featureselection.asu.edu/datasets.php
18. UCI machine learning repository, http://archive.ics.uci.edu/ml/

Soft Rough Sets Based on Similarity Measures

Keyun Qin, Zhenming Song, and Yang Xu

College of Mathematics, Southwest Jiaotong University,
Chengdu, Sichuan 610031, China
keyunqin@263.net, {zhmsong,xuyang}@home.swjtu.edu.cn

Abstract. This paper is devoted to the discussion of the combinations of fuzzy set, rough set and soft set. Based on the analysis of the existing soft fuzzy rough approaches, the notions of similarity measures induced by soft set and soft fuzzy set are presented. Some new soft fuzzy rough set models are proposed by introducing confidence threshold values. The related lower and upper approximation operators are presented and their properties are surveyed.

Keywords: Fuzzy set, rough set, soft set, similarity measure, soft fuzzy rough set.

1 Introduction

To solve complicated problems in economics, engineering, environmental science and social science, methods in classical mathematics are not always successful because of various types of uncertainties presented in these problems. While probability theory, fuzzy set theory [18], rough set theory [9,10], and other mathematical tools are well-known and often useful approaches to describing uncertainty, each of these theories has its inherent difficulties as pointed out in [7,8]. Consequently, Molodtsov [7] proposed a completely new approach for modeling vagueness and uncertainty in 1999. This approach called soft set theory is free from the difficulties affecting existing methods.

Soft set theory, fuzzy set theory and rough set theory are all mathematical tools to deal with uncertainty. It has been found that soft set, fuzzy set and rough set are closely related concepts [3]. The combinations of soft set, fuzzy set and rough set have received much attention recently [4,13,15]. Specifically, Feng et al. [3] provided a framework to combine fuzzy sets, rough sets and soft sets all together, which gives rise to several interesting new concepts such as rough soft sets, soft rough sets and soft rough fuzzy sets. Meng et al. [6] introduced a more general model called soft fuzzy rough set. In this model, fuzzy soft set is employed to granulate the universe of discourse. In the present paper, we attempt to conduct a further study along this line. Based on the analysis of the existing soft fuzzy rough approaches, the notions of similarity measures induced by soft set and soft fuzzy set are introduced. These similarity measures are employed to granulate the universe of discourse and the related soft fuzzy rough set models are presented with their basic properties being discussed.

T. Li et al. (Eds.): RSKT 2012, LNAI 7414, pp. 40–48, 2012.

2   Preliminaries

This section presents a review of some fundamental notions of fuzzy sets, soft sets and fuzzy soft sets. We refer to [4,7,18] for details.

Fuzzy set theory initiated by Zadeh [18] provides an appropriate framework for representing and processing vague concepts by allowing partial memberships. Let U be a nonempty set, called universe. A fuzzy set μ on U is defined by a membership function $\mu : U \to [0,1]$. For $x \in U$, the membership value $\mu(x)$ essentially specifies the degree to which x belongs to the fuzzy set μ. The operations on fuzzy sets can be defined componentwise [18]. In what follows, the family of all subsets of U and the family of all fuzzy sets on U are denoted by $P(U)$ and $F(U)$ respectively.

In 1999, Molodtsov [7] introduced the concept of soft sets. Let U be the universe set and E the set of all possible parameters under consideration with respect to U. Usually, parameters are attributes, characteristics, or properties of objects in U. Molodtsov defined the notion of a soft set in the following way:

Definition 1. *[7] A pair (F, A) is called a soft set over U, where $A \subseteq E$ and F is a mapping given by $F : A \to P(U)$.*

In other words, a soft set over U is a parameterized family of subsets of U. For $e \in A$, $F(e)$ may be considered as the set of $e-$approximate elements of the soft set (F, A). For illustration, we consider the following example of soft sets.

Example 1. Suppose that there are six houses in the universe U given by $U = \{h_1, h_2, h_3, h_4, h_5\}$ and $E = \{e_1, e_2, e_3, e_4, e_5\}$ is the set of parameters. Where e_1 stands for the parameter 'expensive', e_2 stands for the parameter 'beautiful', e_3 stands for the parameter 'wooden', e_4 stands for the parameter 'cheap' and e_5 stands for the parameter 'in the green surroundings'.

In this case, to define a soft set means to point out expensive houses, beautiful houses, and so on. The soft set (F, E) may describe the 'attractiveness of the houses' which Mr.X is going to buy. Suppose that $F(e_1) = \{h_2, h_4\}$, $F(e_2) = \{h_3\}$, $F(e_3) = \{h_3, h_4, h_5\}$, $F(e_4) = \{h_1, h_3, h_4, h_5\}$, $F(e_5) = \{h_2, h_3\}$. Then the soft set (F, E) is a parameterized family $\{F(e_i); 1 \le i \le 5\}$ of subsets of U and give us a collection of approximate descriptions of an object. $F(e_1) = \{h_2, h_4\}$ means 'houses h_2 and h_4' are 'expensive'.

Maji et al. [4] initiated the study on hybrid structures involving both fuzzy sets and soft sets. The notion of fuzzy soft sets was introduced as a fuzzy generalization of soft sets.

Definition 2. *[4] Let U be the universe set and E the set of all possible parameters under consideration with respect to U. A pair (F, A) is called a fuzzy soft set over U, where $A \subseteq E$ and F is a mapping given by $F : A \to F(U)$.*

In the definition of fuzzy soft set, fuzzy sets in the universe U are used as substitutes for the crisp subsets of U. Hence, every soft set may be considered as a fuzzy soft set. Based on operations on soft sets presented by Maji ea al. [5] and Ali et al. [1], Qin et al. [11] defined several operations on fuzzy soft sets and established the lattice structure of fuzzy soft sets.

3 Analysis of Soft Rough Approximations

In this section, we make a theoretical analysis of the existing soft rough approaches and point out some interesting connections between soft rough approximations and generalized rough approximations.

Rough set theory was proposed by Pawlak [9,10]. Let U be a universe of discourse, and R an equivalence relation on U. The pair (U, R) is called a Pawlak approximation space. R will generate a partition $U/R = \{[x]_R; x \in U\}$ on U, where $[x]_R$ is the equivalence class with respect to R containing x. These equivalence classes are building blocks (concepts) of rough approximations. For each $X \subseteq U$, the upper approximation $\overline{R}(X)$ and lower approximation $\underline{R}(X)$ of X with respect to (U, R) are defined as [9]:

$$\overline{R}(X) = \{x \in U; [x]_R \cap X \neq \emptyset\} \tag{1}$$

$$\underline{R}(X) = \{x \in U; [x]_R \subseteq X\} \tag{2}$$

X is called definable in (U, R) if $\underline{R}(X) = \overline{R}(X)$; otherwise X is called a rough set. Thus, in rough set theory, a rough concept is characterized by a couple of exact concepts, namely, its lower approximation and upper approximation.

By replacing the equivalence relation by an arbitrary relation or fuzzy relation, different kinds of generalizations of Pawlak rough set model were obtained [2,12,14,16,17]. Dubois and Prade [2] initiated the study of fuzzy rough sets.

Definition 3. *[2] Let U be a universe of discourse, and R a fuzzy relation on U. The pair (U, R) is called a fuzzy approximation space. The lower and upper rough approximations of $\mu \in F(U)$ in (U, R) are denoted by $\underline{R}(\mu)$ and $\overline{R}(\mu)$, respectively, which are fuzzy subsets in U defined by:*

$$\underline{R}(\mu)(x) = \wedge_{y \in U}((1 - R(x, y)) \vee \mu(y)), \tag{3}$$

$$\overline{R}(\mu)(x) = \vee_{y \in U}(R(x, y) \wedge \mu(y)), \tag{4}$$

for all $x \in U$.

Clearly, fuzzy rough sets are natural extensions of rough sets.

In [3], Feng introduced the notions of soft rough sets and soft rough fuzzy sets. The key point is that a soft set instead of an equivalence relation is used to granulate the universe of discourse.

Definition 4. *[3] Let $\mathfrak{S} = (f, A)$ be a soft set over U. The pair $S = (U, \mathfrak{S})$ is called a soft approximation space. Based on S, we define the following two operations:*

$$\underline{apr}_S(X) = \{u \in U; \exists a \in A(u \in f(a) \subseteq X)\}, \tag{5}$$

$$\overline{apr}_S(X) = \{u \in U; \exists a \in A(u \in f(a), f(a) \cap X \neq \emptyset)\}, \tag{6}$$

assigning to every subset $X \subseteq U$ two sets $\underline{apr}_S(X)$ and $\overline{apr}_S(X)$ called the lower and upper soft rough approximations of X in S, respectively.

Definition 5. *[3] Let $\mathfrak{S} = (f, A)$ be a full soft set over U and $S = (U, \mathfrak{S})$ be a soft approximation space. For a fuzzy set $\mu \in F(U)$, the lower and upper soft rough approximations of μ with respect to S are denoted by $\underline{sap}_S(\mu)$ and $\overline{sap}_S(\mu)$, respectively, which are fuzzy sets in U given by:*

$$\underline{sap}_S(\mu)(x) = \wedge\{\mu(y); \exists a \in A(\{x, y\} \subseteq f(a))\}, \tag{7}$$

$$\overline{sap}_S(\mu)(x) = \vee\{\mu(y); \exists a \in A(\{x, y\} \subseteq f(a))\}, \tag{8}$$

for all $x \in U$. The operators $\underline{sap}_S$ and $\overline{sap}_S$ are called the lower and upper soft rough approximation operators on fuzzy sets.

Note that $\underline{sap}_S$ and $\overline{sap}_S$ are dual to each other, i.e., $\overline{sap}_S(\mu^c) = (\underline{sap}_S(\mu))^c$ for every $\mu \in F(U)$. Moreover, $\overline{sap}_S$ is a generalization of $\overline{apr}_S$ [6], i.e., $\overline{sap}_S(\mu) = \overline{apr}_S(\mu)$ if $\mu \in P(U)$.

Meng et al. presented a new soft rough fuzzy set model in [6].

Definition 6. *[6] Let $\mathfrak{S} = (f, A)$ be a full soft set over U and $S = (U, \mathfrak{S})$ be a soft approximation space. For a fuzzy set $\mu \in F(U)$, the lower soft rough approximation $\underline{sap}'_S(\mu)$ and upper soft rough approximation $\overline{sap}'_S(\mu)$ of μ are fuzzy sets in U given by:*

$$\underline{sap}'_S(\mu)(x) = \vee_{x \in f(a)} \wedge_{y \in f(a)} \mu(y), \tag{9}$$

$$\overline{sap}'_S(\mu)(x) = \wedge_{x \in f(a)} \vee_{y \in f(a)} \mu(y), \tag{10}$$

for all $x \in U$.

It is proved that [6] $\underline{sap}'_S$ and $\overline{sap}'_S$ are dual to each other, and $\underline{sap}'_S$ is a generalization of $\underline{apr}_S$.

We consider the soft set $\mathfrak{S} = (F, E)$ given in Example 1. $S = (U, \mathfrak{S})$ is a soft approximation space. By Definition 4, we have

$$\overline{apr}_S(\{h_1\}) = \{h_1, h_2, h_3, h_4\}.$$

For the five parameters we considered, h_1 and h_3 are similar with respect to only one parameter e_4. But $h_3 \in \overline{apr}_S(\{h_1\})$. Theoretically, by (6), we have

$$\overline{apr}_S(\{x\}) = \cup_{a \in A}\{f(a); f(a) \cap \{x\} \neq \emptyset\} = \cup\{f(a); a \in A, x \in f(a)\},$$

and

$$y \in \overline{apr}_S(\{x\}) \Leftrightarrow \exists a \in A(x \in f(a), y \in f(a)).$$

That is, $y \in \overline{apr}_S(\{x\})$ prescribe a minimum condition for roughness in the sense that there exists at least one parameter $a \in A$ such that x and y are similar with respect to a. In practice, however, it seems that considering the proportion of parameters that x and y are similar with respect to them may lead to a better utilization of soft rough approach. Formally, Let $\mathfrak{S} = (f, A)$ be a soft set over U

and $S = (U, \mathfrak{S})$ a soft approximation space. Based on S, we define the similarity degree $D_{\mathfrak{S}}(x, y)$ between $x, y \in U$ by:

$$D_{\mathfrak{S}}(x, y) = \frac{|\{a \in A; x \in f(a)\} \cap \{a \in A; y \in f(a)\}|}{|\{a \in A; x \in f(a)\} \cup \{a \in A; y \in f(a)\}|}. \tag{11}$$

Clearly, we have $D_{\mathfrak{S}}(x, x) = 1$ and $D_{\mathfrak{S}}(x, y) = D_{\mathfrak{S}}(y, x)$ for all $x, y \in U$.

Let us note that the existing rough approximations and the measure $D_{\mathfrak{S}}$ are related:

$$\underline{sap}_S(\mu)(x) = \wedge\{\mu(y); D_{\mathfrak{S}}(x, y) > 0\}, \tag{12}$$

$$\overline{sap}_S(\mu)(x) = \vee\{\mu(y); D_{\mathfrak{S}}(x, y) > 0\}. \tag{13}$$

Meng et al. [6] proposed a more general model called soft fuzzy rough set where fuzzy soft set is employed to granulate the universe of discourse.

Definition 7. *[6] Let $\mathfrak{F} = (f, A)$ be a fuzzy soft set over U. The pair $SF = (U, \mathfrak{F})$ is called a soft fuzzy approximation space. For a fuzzy set $\mu \in F(U)$, the lower and upper soft fuzzy rough approximations of μ with respect to SF are denoted by $\underline{Apr}_{SF}(\mu)$ and $\overline{Apr}_{SF}(\mu)$, respectively, which are fuzzy sets in U given by:*

$$\underline{Apr}_{SF}(\mu)(x) = \wedge_{a \in A}((1 - f(a)(x)) \vee (\wedge_{y \in U}((1 - f(a)(y)) \vee \mu(y)))), \tag{14}$$

$$\overline{Apr}_{SF}(\mu)(x) = \vee_{a \in A}(f(a)(x) \wedge (\vee_{y \in U}(f(a)(y) \wedge \mu(y)))), \tag{15}$$

for all $x \in U$. The operators $\underline{Apr}_{SF}$ and $\overline{Apr}_{SF}$ are called the lower and upper soft fuzzy rough approximation operators on fuzzy sets.

It is proved [6] that operators $\underline{Apr}_{SF}$ and $\overline{Apr}_{SF}$ are extensions of operators $\underline{sap}_{SF}$ and $\overline{sap}_{SF}$ respectively.

Theorem 1. *Let $\mathfrak{F} = (f, A)$ be a fuzzy soft set over U, $SF = (U, \mathfrak{F})$, $\mu \in F(U)$ and $R_{\mathfrak{F}}$ the fuzzy relation on U given by: $R_{\mathfrak{F}}(x, y) = \vee_{a \in A}(f(a)(x) \wedge f(a)(y))$.*
(1) $\underline{Apr}_{SF}(\mu) = \underline{R_{\mathfrak{F}}}(\mu)$,
(2) $\overline{Apr}_{SF}(\mu) = \overline{R_{\mathfrak{F}}}(\mu)$.

Proof. (1) For each $x \in U$, we have

$$\begin{aligned}
\underline{Apr}_{SF}(\mu)(x) &= \wedge_{a \in A}((1 - f(a)(x)) \vee (\wedge_{y \in U}((1 - f(a)(y)) \vee \mu(y)))) \\
&= \wedge_{a \in A} \wedge_{y \in U}((1 - f(a)(x)) \vee ((1 - f(a)(y)) \vee \mu(y)) \\
&= \wedge_{y \in U}(\wedge_{a \in A}((1 - f(a)(x)) \vee (1 - f(a)(y))) \vee \mu(y)) \\
&= \wedge_{y \in U}(1 - \vee_{a \in A}(f(a)(x) \wedge f(a)(y))) \vee \mu(y) \\
&= \wedge_{y \in U}(1 - R_{\mathfrak{F}}(x, y)) \vee \mu(y) = \underline{R_{\mathfrak{F}}}(\mu)(x).
\end{aligned}$$

(2) can be proved similarly.

By this theorem, the soft fuzzy rough approximation presented in Definition 7 is a kind of fuzzy rough approximation in the sense of Definition 3. Note that $R_{\mathfrak{F}}(x, y)$ describes a kind of similarity between x and y, and $R_{\mathfrak{F}}$ is symmetric but $R_{\mathfrak{F}}(x, x) \neq 1$ in general.

4 Soft Fuzzy Rough Sets Based on Similarity Measures

In this section, we present a general soft fuzzy rough set approach based on similarity measures.

Definition 8. *Let* $\mathfrak{S} = (f, A)$ *be a full soft set over* U, $S = (U, \mathfrak{S})$ *a soft approximation space and* $\alpha \in [0, 1)$. *For a fuzzy set* $\mu \in F(U)$, *the* $\alpha-$*lower and* $\alpha-$*upper soft rough approximations of* μ *with respect to* S *are denoted by* $\underline{S^{\alpha}}(\mu)$ *and* $\overline{S^{\alpha}}(\mu)$, *respectively, which are fuzzy sets in* U *given by:*

$$\underline{S^{\alpha}}(\mu)(x) = \wedge\{\mu(y); D_{\mathfrak{S}}(x, y) > \alpha\}, \tag{16}$$

$$\overline{S^{\alpha}}(\mu)(x) = \vee\{\mu(y); D_{\mathfrak{S}}(x, y) > \alpha\}, \tag{17}$$

for all $x \in U$.

The following results are easily obtained from the definition of $\alpha-$soft rough approximations.

Theorem 2. *Let* $\mathfrak{S} = (f, A)$ *be a full soft set over* U, $S = (U, \mathfrak{S})$, $\alpha, \beta \in [0, 1)$ *and* $\mu, \nu \in F(U)$.
  (1) $\underline{S^{\alpha}}(\emptyset) = \emptyset = \overline{S^{\alpha}}(\emptyset)$, $\underline{S^{\alpha}}(U) = U = \overline{S^{\alpha}}(U)$.
  (2) $\underline{S^{\alpha}}(\mu) \subseteq \mu \subseteq \overline{S^{\alpha}}(\mu)$.
  (3) $\mu \subseteq \nu \Rightarrow \underline{S^{\alpha}}(\mu) \subseteq \underline{S^{\alpha}}(\nu), \overline{S^{\alpha}}(\mu) \subseteq \overline{S^{\alpha}}(\nu)$.
  (4) $\underline{S^{\alpha}}(\mu \cap \nu) = \underline{S^{\alpha}}(\mu) \cap \underline{S^{\alpha}}(\nu), \underline{S^{\alpha}}(\mu \cup \nu) \supseteq \underline{S^{\alpha}}(\mu) \cup \underline{S^{\alpha}}(\nu)$.
  (5) $\overline{S^{\alpha}}(\mu \cap \nu) \subseteq \overline{S^{\alpha}}(\mu) \cap \overline{S^{\alpha}}(\nu), \overline{S^{\alpha}}(\mu \cup \nu) = \overline{S^{\alpha}}(\mu) \cup \overline{S^{\alpha}}(\nu)$.
  (6) $\alpha \leq \beta \Rightarrow \underline{S^{\alpha}}(\mu) \subseteq \underline{S^{\beta}}(\mu), \overline{S^{\beta}}(\mu) \subseteq \overline{S^{\alpha}}(\mu)$.
  (7) $\underline{S^{0}}(\mu) = \underline{sap}_S(\mu), \overline{S^{0}}(\mu) = \overline{sap}_S(\mu)$.

From Theorem 2(7), we know that $\alpha-$soft rough approximation operators are the generalizations of soft rough approximation operators $\underline{sap}_S$ and $\overline{sap}_S$.

By taking $D_{\mathfrak{S}}$ as the elementary knowledge of the universe, we propose another soft rough fuzzy set model in the following definition.

Definition 9. *Let* $\mathfrak{S} = (f, A)$ *be a full soft set over* U, $S = (U, \mathfrak{S})$ *a soft approximation space. For a fuzzy set* $\mu \in F(U)$, *the lower and upper soft rough approximations of* μ *with respect to* S *are denoted by* $\underline{S}(\mu)$ *and* $\overline{S}(\mu)$, *respectively, which are fuzzy sets in* U *given by:*

$$\underline{S}(\mu)(x) = \wedge_{y \in U}((1 - D_{\mathfrak{S}}(x, y)) \vee \mu(y)), \tag{18}$$

$$\overline{S}(\mu)(x) = \vee_{y \in U}(D_{\mathfrak{S}}(x, y) \wedge \mu(y)), \tag{19}$$

for all $x \in U$.

The following theorem shows some connections between operators $\overline{S^{\alpha}}$, $\underline{S^{\alpha}}$, $\overline{S}$ and $\underline{S}$.

Theorem 3. *Let* $\mathfrak{S} = (f, A)$ *be a full soft set over* U, $S = (U, \mathfrak{S})$, *and* $\mu \in F(U)$.

(1) *If* $\vee_{x \in U}(1 - \mu(x)) \leq \alpha$, *then* $\underline{S}(\mu) \subseteq \underline{S}^\alpha(\mu)$.
(2) *If* $\vee_{x \in U}\mu(x) \leq \alpha$, *then* $\overline{S}^\alpha(\mu) \subseteq \overline{S}(\mu)$.
(3) $\overline{S}(\mu) \subseteq \overline{S}^0(\mu)$, $\underline{S}^0(\mu) \subseteq \underline{S}(\mu)$.

Proof. (1) Assume that $\vee_{x \in U}(1 - \mu(x)) \leq \alpha$. It follows that $1 - \wedge_{x \in U}\mu(x) \leq \alpha$ and hence $1 - \alpha \leq \wedge_{x \in U}\mu(x)$. For each $x \in U$, we have $1 - \alpha \leq \wedge_{y \in U}\mu(y) \leq \wedge_{y \in U, D_\mathfrak{S}(x,y) > \alpha}\mu(y)$ and consequently

$$
\begin{aligned}
\underline{S}(\mu)(x) &= \wedge_{y \in U}((1 - D_\mathfrak{S}(x, y)) \vee \mu(y)) \\
&\leq \wedge_{y \in U, D_\mathfrak{S}(x,y) > \alpha}((1 - D_\mathfrak{S}(x, y)) \vee \mu(y)) \\
&\leq \wedge_{y \in U, D_\mathfrak{S}(x,y) > \alpha}((1 - \alpha) \vee \mu(y)) = (1 - \alpha) \vee (\wedge_{y \in U, D_\mathfrak{S}(x,y) > \alpha}\mu(y)) \\
&= \wedge_{y \in U, D_\mathfrak{S}(x,y) > \alpha}\mu(y) = \underline{S}^\alpha(\mu)(x).
\end{aligned}
$$

(2) and (3) can be proved similarly.

Given below is a list of some basic properties of operators $\underline{S}$ and $\overline{S}$.

Theorem 4. *Let* $\mathfrak{S} = (f, A)$ *be a full soft set over* U, $S = (U, \mathfrak{S})$, *and* $\mu, \nu \in F(U)$.

(1) $\underline{S}(\emptyset) = \emptyset = \overline{S}(\emptyset)$, $\underline{S}(U) = U = \overline{S}(U)$.
(2) $\underline{S}(\mu) \subseteq \mu \subseteq \overline{S}(\mu)$.
(3) $\underline{S}(\mu^c) = (\overline{S}(\mu))^c$, $\overline{S}(\mu^c) = (\underline{S}(\mu))^c$.
(4) $\mu \subseteq \nu \Rightarrow \underline{S}(\mu) \subseteq \underline{S}(\nu)$, $\overline{S}(\mu) \subseteq \overline{S}(\nu)$.
(5) $\underline{S}(\mu \cap \nu) = \underline{S}(\mu) \cap \underline{S}(\nu)$, $\underline{S}(\mu \cup \nu) \supseteq \underline{S}(\mu) \cup \underline{S}(\nu)$.
(6) $\overline{S}(\mu \cap \nu) \subseteq \overline{S}(\mu) \cap \overline{S}(\nu)$, $\overline{S}(\mu \cup \nu) = \overline{S}(\mu) \cup \overline{S}(\nu)$.

The notion of similarity measure induced by a soft set can be naturally extended to fuzzy soft set. Let $\mathfrak{F} = (f, A)$ be a fuzzy soft set over U. We define the similarity degree $FD_\mathfrak{F}(x, y)$ between $x, y \in U$ by:

$$
FD_\mathfrak{F}(x, y) = \frac{\sum_{a \in A} f(a)(x) \wedge f(a)(y)}{\sum_{a \in A} f(a)(x) \vee f(a)(y)}. \tag{20}
$$

Clearly, we have $FD_\mathfrak{F}(x, x) = 1$ and $FD_\mathfrak{F}(x, y) = FD_\mathfrak{F}(y, x)$ for all $x, y \in U$. That is, $FD_\mathfrak{F}$ is a reflexive and symmetric fuzzy relation on U.

Definition 10. *Let* $\mathfrak{F} = (f, A)$ *be a fuzzy soft set over* U, $SF = (U, \mathfrak{F})$ *a soft fuzzy approximation space and* $\alpha \in [0, 1)$. *For a fuzzy set* $\mu \in F(U)$, *the* $\alpha-$*lower and* $\alpha-$*upper soft rough approximations of* μ *with respect to* SF *are denoted by* $\underline{SF}^\alpha(\mu)$ *and* $\overline{SF}^\alpha(\mu)$, *respectively, which are fuzzy sets in* U *given by:*

$$
\underline{SF}^\alpha(\mu)(x) = \wedge\{\mu(y); FD_\mathfrak{F}(x, y) > \alpha\}, \tag{21}
$$

$$
\overline{SF}^\alpha(\mu)(x) = \vee\{\mu(y); FD_\mathfrak{F}(x, y) > \alpha\}, \tag{22}
$$

for all $x \in U$.

Assume that $\mathfrak{F} = (f, A)$ is a soft set over U. It follows that $f(a)(x) = 0$ or $f(a)(x) = 1$ for each $a \in A$ and $x \in U$, and consequently

$$FD_{\mathfrak{F}}(x, y) = \frac{|\{a \in A; x \in f(a) \wedge y \in f(a)\}|}{|\{a \in A; x \in f(a) \vee y \in f(a)\}|} = D_{\mathfrak{F}}(x, y).$$

Thus, $FD_{\mathfrak{F}}$ is a generalization of $D_{\mathfrak{G}}$. Consequently, operators $\underline{SF^\alpha}$ and $\overline{SF^\alpha}$ are generalizations of $\underline{S^\alpha}$ and $\overline{S^\alpha}$ respectively.

Given below is a list of some basic properties of operators $\underline{SF^\alpha}$ and $\overline{SF^\alpha}$.

Theorem 5. *Let $\mathfrak{F} = (f, A)$ be a fuzzy soft set over U, $SF = (U, \mathfrak{F})$, $\alpha, \beta \in [0, 1)$ and $\mu, \nu \in F(U)$.*
(1) $\underline{SF^\alpha}(\emptyset) = \emptyset = \overline{SF^\alpha}(\emptyset)$, $\underline{SF^\alpha}(U) = U = \overline{SF^\alpha}(U)$.
(2) $\underline{SF^\alpha}(\mu) \subseteq \mu \subseteq \overline{SF^\alpha}(\mu)$.
(3) $\mu \subseteq \nu \Rightarrow \underline{SF^\alpha}(\mu) \subseteq \underline{SF^\alpha}(\nu), \overline{SF^\alpha}(\mu) \subseteq \overline{SF^\alpha}(\nu)$.
(4) $\underline{SF^\alpha}(\mu \cap \nu) = \underline{SF^\alpha}(\mu) \cap \underline{SF^\alpha}(\nu)$, $\underline{SF^\alpha}(\mu \cup \nu) \supseteq \underline{SF^\alpha}(\mu) \cup \underline{SF^\alpha}(\nu)$.
(5) $\overline{SF^\alpha}(\mu \cap \nu) \subseteq \overline{SF^\alpha}(\mu) \cap \overline{SF^\alpha}(\nu)$, $\overline{SF^\alpha}(\mu \cup \nu) = \overline{SF^\alpha}(\mu) \cup \overline{SF^\alpha}(\nu)$.
(6) $\alpha \leq \beta \Rightarrow \underline{SF^\alpha}(\mu) \subseteq \underline{SF^\beta}(\mu), \overline{SF^\beta}(\mu) \subseteq \overline{SF^\alpha}(\mu)$.

5 Concluding Remarks

Fuzzy set theory, soft set theory and rough set theory are all mathematical tools for dealing with uncertainties. This paper is devoted to the discussion of the combinations of fuzzy set, rough set and soft set. The notions of similarity measures induced by soft set and soft fuzzy set are presented. Based on these similarity measures, some new soft fuzzy rough set models are proposed and their properties are surveyed.

In further research, the axiomatization of the approximation operators is an important and interesting issue to be addressed.

Acknowledgements. This work has been supported by the National Natural Science Foundation of China (Grant No. 61175055, 61175044) and Sichuan Key Technology Research and Development Program (Grant No.2011FZ0051).

References

1. Ali, M.I., Feng, F., Liu, X., Min, W.K., Shabir, M.: On some new operations in soft set theory. Computers and Mathematics with Applications 57, 1547–1553 (2009)
2. Dubois, D., Prade, H.: Rough fuzzy set and fuzzy rough sets. International Journal of General Systems 17, 191–209 (1990)
3. Feng, F., Li, C.X., Davvaz, B., Ali, M.I.: Soft sets combined with fuzzy sets and rough sets: a tentative approach. Soft Computing 14, 899–911 (2010)
4. Maji, P.K., Biswas, R., Roy, A.R.: Fuzzy soft sets. The Journal of Fuzzy Mathematics 9, 589–602 (2001)
5. Maji, P.K., Biswas, R., Roy, A.R.: Soft set theory. Computers and Mathematics with Applications 45, 555–562 (2003)

6. Meng, D., Zhang, X.H., Qin, K.Y.: Soft rough fuzzy sets and soft fuzzy rough sets. Computers and Mathematics with Applications 62, 4635–4645 (2011)
7. Molodtsov, D.: Soft set theory-First results. Computers and Mathematics with Applications 37, 19–31 (1999)
8. Molodtsov, D.: The theory of soft sets. URSS Publishers, Moscow (2004) (in Russian)
9. Pawlak, Z.: Rough sets. International Journal of Computer and Information Science 11, 341–356 (1982)
10. Pawlak, Z.: Rough sets: Theoretical Aspects of Reasoning About Data. Kluwer Academic Publishers, Boston (1991)
11. Qin, K., Zhao, H.: Lattice Structures of Fuzzy Soft Sets. In: Huang, D.-S., Zhao, Z., Bevilacqua, V., Figueroa, J.C. (eds.) ICIC 2010. LNCS, vol. 6215, pp. 126–133. Springer, Heidelberg (2010)
12. Radzikowska, A.M., Kerre, E.E.: A comparative study of fuzzy rough sets. Fuzzy Sets and Systems 126, 137–155 (2002)
13. Roy, A.R., Maji, P.K.: A fuzzy soft set theoretic approach to decision making problems. Journal of Computational and Applied Mathematics 203, 412–418 (2007)
14. Wu, W.Z., Mi, J.S., Zhang, W.X.: Generalized fuzzy rough sets. Information Sciences 151, 263–282 (2003)
15. Yang, X.B., Lin, T.Y., Yang, J.Y., Li, Y., Yu, D.J.: Combination of interval-valued fuzzy set and soft set. Computers and Mathematics with Applications 58, 521–527 (2009)
16. Yao, Y.Y.: A comparative study of fuzzy sets and rough sets. Information Sciences 109, 227–242 (1998)
17. Yao, Y.Y., Wong, S.K.M.: Generalization of rough sets using relationships between attribute values. In: Proceedings of the Second Annual Joint Conference Information Sciences, pp. 30–33 (1995)
18. Zadeh, L.A.: Fuzzy sets. Information and Control 8, 338–353 (1965)

Theory of Truth Degrees
in Three Valued Formed System RSL

Huan Liu, Yingcang Ma*, and Xuezhen Dai

School of Science, Xi'an Polytechnic University
Xi'an 710048, China
mayingcang@126.com

Abstract. By means of the function induced by a logical formula A, the concept of truth of the logical formula A is introduced in three valued formed system RSL, and some reasoning rules of truth degree based on the logic system are given. Moreover, the similarity degree and pseudo-distance between two formulas are defined by using the truth degree concept, and their properties are discussed. This offers a theory framework for approximate reasoning in three valued formed system RSL.

Keywords: Truth degree, Similarity degree, Pseudo-metric, Formed system RSL.

1 Introduction

Since Pawlak proposed Rough Sets [1], many scholars have tried to study rough logical theory to implement approximate reasoning and problem solving in artificial intelligence with it. Pawlak proposed Rough Logic and Decision Logic in 1987 and in 1991 respectively [2,3]. The former set up five truth values: true, false, roughly true, roughly false and roughly inconsistent; the latter is based on the information table, which is essentially a special case in two-valued logic. Further studies have done in [4-8]. In [8], Zhang make rough logic and fuzzy logic together and proposed a rough logic system RSL, and prove which is a extension of logical system IMTL.

In recent years, there is an research method to make logical reasoning and numerical computation together to study, for example, the quantitative logic proposed by Wang in [9]. The theory of truth degrees of propositions in 2-valued propositional logic was proposed in [10]. In the following, the theory of truth degrees of Lukasiewicz system and other fuzzy logical system are studied [11−12]. In the present paper, according to the method of quantitative logic, we introduce the theory of truth degrees of three valued formed system RSL and provide a type of metric approximate reasoning theory in RSL.

The paper is organized as follows. After this introduction, Section 2 will introduce the basic content of rough set and formed system RSL. In Section 3, the

---

* Corresponding author.

T. Li et al. (Eds.): RSKT 2012, LNAI 7414, pp. 49–54, 2012.

theory of truth degree in three valued formed system RSL is built through introducing a truth degree's definition, and some properties are studied. In section 4, a type of metric approximate reasoning theory in three valued formed system RSL based on the proposed pseudo-metrics are established. The final section offers the conclusion.

2   Basic of Rough Set and Formed System RSL

Let (U, R) be an approximation space, if $A \subseteq U$, the lower approximation $\underline{R}A$ of A is the union of equivalence classes contained in A, while its upper approximation $\overline{R}A$ is the union of equivalence classes intersecting A, i.e.

$\underline{R}A = \cup \{[x] \, | [x] \subseteq A\} = \{x \, | x \in U, [x] \subseteq A\}$

$\overline{R}A = \cup \{[x] \, | [x] \cap X \neq \phi\} = \{x \, | x \in U, [x] \cap A \neq \phi\}$

where $[x]$ denotes the equivalence class containing x. We call $\underline{R}A(\overline{R}A)$ lower (upper) approximation with respect to R of A.

The language of formed system RSL consists of atomic formula $S = \{p, q, r, \cdots\}$, logical symbols $(*, +, \wedge, \rightarrow)$ and parentheses. The formula set of formed system RSL is generated by the following two rules in finite times:

(i) If A is a atomic formula, then A is a formula;

(ii) If A and B are formulas, then $A^*, A^+, A \wedge B, A \rightarrow B$ are formulas;

The set of all formulas in formed system RSL is denoted by $F(S)$. Further connectives are defined as following, for any wffs A, B of formed system RSL. $A \vee B = \neg(\neg A \wedge \neg B)$, $\neg A = A \rightarrow \bot$, $\top = \bot \rightarrow \bot$, $A\&B = \neg(A \rightarrow \neg B)$.

Definition 1.[8] *Let $S = \{p, q, r, \cdots\}$ be a set of propositional variables. A free algebra $F(S)$ generated by S is a structure $(*, +, \wedge, \rightarrow)$, $\bot$ is a propositional constant. The axioms in the formed system RSL are consisted of the following formulas:*

(Ax1) $(A \rightarrow B) \rightarrow ((B \rightarrow C) \rightarrow (A \rightarrow C))$;

(Ax2) $A \wedge B \rightarrow C$;

(Ax3) $A \wedge B \rightarrow B \wedge A$;

(Ax4) $A \rightarrow ((A \rightarrow B) \rightarrow (A \wedge B))$;

(Ax5) $((A \rightarrow B) \rightarrow C) \rightarrow (((B \rightarrow A) \rightarrow C) \rightarrow C)$;

(Ax6) $(\neg A \rightarrow \neg B) \rightarrow (B \rightarrow A)$;

(Ax7) $A \wedge (A \wedge B)^* \rightarrow A \wedge B^*$, $A \wedge B^* \rightarrow A \wedge (A \wedge B)^*$;

(Ax8) $A \vee (A \vee B)^+ \rightarrow A \vee B^+$, $A \vee B^+ \rightarrow A \vee (A \vee B)^+$;

(Ax9) $A \rightarrow \bot^*$, $\top^+ \rightarrow A$;

(Ax10) $(A \rightarrow B) \rightarrow (A^+ \vee B) \wedge (A^+ \vee B^{*+}) \wedge (A \vee A^+ \vee B^{*+})$, $(A^+ \vee B) \wedge (A^+ \vee B^{*+}) \wedge (A \vee A^+ \vee B^{*+}) \rightarrow (A \rightarrow B)$.

Deduction rules of URL is Modus Ponens(MP): from $A, A \rightarrow B$ infer B.

Definition 2.[8] *Algebra structure $(L, \vee, \wedge, *, +, \otimes, \rightarrow, 0, 1)$ is RSL-algebra if $(L, \vee, \wedge, *, +, 0, 1)$ is regular double Stone algebra and $(L, \vee, \wedge, \otimes, \rightarrow, 0, 1)$ is ITML-algebra and $A \otimes B = (A^{++} \wedge B) \vee (A \wedge B^{++})$, $A \rightarrow B = (A^+ \vee B) \wedge (A^+ \vee B^{*+}) \wedge (A \vee A^* \vee B^{*+})$.*

Definition 3.[8] *A valuation v in RSL is a map from the set of rough formulas* $F(S)$ *to any RSL-algebra* $(L, \vee, \wedge, *, +, \otimes, \rightarrow, 0, 1)$ *satisfying* $\forall A, B \in F(S)$,

$$v(A \wedge B) = v(A) \wedge v(B), \quad v(A \vee B) = v(A) \vee v(B), v(A^*) = v(A)^*,$$
$$v(A^+) = v(A)^+, v(A \otimes B) = v(A) \otimes v(B), \quad v(A \rightarrow B) = v(A) \rightarrow v(B).$$

i.e., v is $(\vee, \wedge, *, +, \otimes, \rightarrow)$ *type homomorphism.*

Remark 1. From [8], RSL is sound and complete relative to the class of all RSL- algebras. i.e. $\forall A \in F(S), \Gamma \vdash A$ if and only if $\Gamma \models A$, where Γ is a theory in RSL logic.

3   The Theory of Truth Degree

Definition 4. *Let* $T = \{0, \frac{1}{2}, 1\}$, *define logical operations on* T *as follows:*

$$\forall x, y \in T, \neg 0 = 1, \neg\frac{1}{2} = \frac{1}{2}, \neg 1 = 0, \ 0^* = 1, \frac{1}{2}^* = 0, 1^* = 0;$$
$$0^+ = 1, \frac{1}{2}^+ = 1, 1^+ = 0, \ x \wedge y = \min(x, y), x \vee y = \max(x, y),$$
$$x \rightarrow y = \begin{cases} \neg x \vee y, & (x, y) \neq (\frac{1}{2}, \frac{1}{2}) \\ 1, & (x, y) = (\frac{1}{2}, \frac{1}{2}) \end{cases}, \quad x \otimes y = \begin{cases} x \wedge y, & (x, y) \neq (\frac{1}{2}, \frac{1}{2}) \\ 0, & (x, y) = (\frac{1}{2}, \frac{1}{2}) \end{cases}.$$

Then T *is an algebra of type* $(\vee, \wedge, *, +, \otimes, \rightarrow)$, *which is called three valued formed system RSL, denoted by* T.

A valuation of $v : F(S) \rightarrow T$ *is a homomorphism of type* $(\vee, \wedge, *, +, \otimes, \rightarrow)$, $v(A)$ *is the valuation of A with respect to v. The set consisting of all valuations of* $F(S)$ *is denoted by* Ω_3.

Suppose that logic formula $A(p_1, p_2, \cdots, p_n)$ contains n atomic formulas in $F(S)$, then A can induce a function $f_A : \{0, \frac{1}{2}, 1\}^m \rightarrow \{0, \frac{1}{2}, 1\}$ naturally. $f_A(x_1, x_2, \cdots, x_m)$ is making $x_1, x_2, \cdots, x_m$ composed together by operators $(*, +, \wedge, \rightarrow)$, which is similar with the formula $A(p_1, p_2, \cdots, p_m)$ makes $p_1, p_2, \cdots, p_m$ composed together by connective $(*, +, \wedge, \rightarrow)$.

Definition 5. *Let* $A(p_1, p_2, \cdots, p_n) \in F(S)$ *contains* n *atomic formulas in RSL logic system,* $f_A(x_1, x_2, \cdots, x_n)$ *is the induced function of A, then* $\tau(A)$ *which is the truth degree of A is defined as*

$$\tau(A) = \frac{1}{3^n} \times |f_A^{-1}(1)| \tag{1}$$

where, $|f_A^{-1}(1)|$ *is the number of valuation satisfying* $f_A(v(p_1), \cdots, v(p_n)) = 1$.

Remark 2. From the Definition 5, we can see that $\tau(A)$ is the portion of the valuation v's satisfying $v(A) = 1$ with respect to Ω_3, and therefore it is reasonable to call $\tau(A)$ the truth degree of A. Moreover, logically equivalent formulas have the same truth degree.

Example: Calculate the rough truth degrees of the following formulas, where p, q are difference atom formulas.

1) $\tau(\neg p)$, $\tau(p^*)$; 2) $\tau(p^* \wedge q^+)$, $\tau(p \vee q)$; 3) $\tau(p \rightarrow q)$.

Solution: From Definition 5, we have:

1) $\tau(\neg p) = \frac{|f_{\neg p}^{-1}(1)|}{3} = \frac{|f_p^{-1}(0)|}{3} = \frac{1}{3}$, $\tau(p^*) = \frac{|v(p^*)=1|}{3} = \frac{|v(p)=0|}{3} = \frac{1}{3}$.

2) $\tau(p^* \wedge q^+) = \frac{|v(p^*)=1 \ and \ v(q^+)=1|}{3^2} = \frac{|(0,0),(0,\frac{1}{2})|}{9} = \frac{2}{9}$,

$$\tau(p \vee q) = \frac{|v(p \vee q)=1|}{3^2} = \frac{|v(p)=1| \times 3 + |v(q)=1| \times 3 - |v(p)=1 \text{ and } v(q)=1|}{3^2} = \frac{3+3-1}{3^2} = \frac{5}{9}.$$

$$3)\tau(p \rightarrow q) = \frac{|v(p \rightarrow q)=1|}{3^2} = \frac{|v(\neg p \vee q)=1| + |v(p)=\frac{1}{2} \text{ and } v(q)=\frac{1}{2}|}{3^2}$$

$$= \frac{|v(\neg p)=1| \times 3 + |v(q)=1| \times 3 - |v(\neg p)=1 \text{ and } v(q)=1| + 1}{3^2} = \frac{3+3-1+1}{3^2} = \frac{6}{9}.$$

Proposition 1. *Let τ be truth degree. Then*

1) $0 \leq \tau(A) \leq 1$;

2) $\tau(A) = 1$ if and only if A is a theorem in formed system RSL;

3) $\tau(A^*) = \tau(\neg A) \leq 1 - \tau(A) = \tau(A^+)$;

4) If $\vdash A \rightarrow B$, then $\tau(A) \leq \tau(B)$;

5) $\tau(A \vee B) = \tau(A) + \tau(B) - \tau(A \wedge B)$

Proof: 1) and 2) can be get immediately from the Definition 5 and Remark 1.

3) Being $\tau(A^*) = \frac{|v(A^*)=1|}{3^n} = \frac{|v(A)=0|}{3^n} = \frac{|v(\neg A)=1|}{3^n} = \tau(\neg A)$; $\tau(\neg A) = \frac{|v(\neg A)=1|}{3^n} = \frac{|f_{\neg A}^{-1}(1)|}{3^n} = 1 - \frac{|f_A^{-1}(\frac{1}{2})|}{3^n} - \frac{|f_A^{-1}(1)|}{3^n} \leq 1 - \tau(A)$; $\tau(A^+) = \frac{|v(A^+)=1|}{3^n} = \frac{|v(A)=0 \text{ or } v(A)=\frac{1}{2}|}{3^n} = \frac{3^n - |v(A)=1|}{3^n} = 1 - \tau(A)$; So, $\tau(A^*) = \tau(\neg A) \leq 1 - \tau(A) = \tau(A^+)$.

4) From [8], being $A \rightarrow B = < X_u' \cup Y_d \cup (Y_u \cap X_d'), X_d' \cup Y_u >$, from $\vdash A \rightarrow B$, $X_u' \cup Y_d \cup Y_u = U$ and $X_u' \cup Y_d \cup X_d' = U$ and $X_d' \cup Y_u = U$, then $X_u' \cup Y_u = U$ and $X_u' \cup Y_d = U$ and $X_d' \cup Y_u = U$, So $X_u \subseteq Y_u$ and $X_d \subseteq Y_d$, we have $v(A) \leq v(B)$, hence $\tau(A) \leq \tau(B)$.

5) In general, we assume that both A and B have the same atomic formulas, So we can get $|f_{A \vee B}^{-1}(1)| = |f_A^{-1}(1)| + |f_B^{-1}(1)| - |f_{A \wedge B}^{-1}(1)|$. From Definition 5 we can get $\tau(A \vee B) = \tau(A) + \tau(B) - \tau(A \wedge B)$.

Theorem 1. *Suppose that A is a tautology in logical system RSL, then $\tau(A \rightarrow B) = \tau(B)$, $\tau(B \rightarrow A) = 1$.*

Proof: From Definition 5

$\tau(A \rightarrow B) = \frac{1}{3^n}|v(A \rightarrow B) = 1| = \frac{1}{3^n}|(v(A) \rightarrow v(B)) = 1| = \frac{1}{3^n}|(1 \rightarrow v(B)) = 1| = \frac{1}{3^n}|v(B) = 1| = \tau(B)$.

$\tau(B \rightarrow A) = \frac{1}{3^n}|v(B \rightarrow A) = 1| = \frac{1}{3^n}|(v(B) \rightarrow v(A)) = 1| = \frac{1}{3^n}|(v(B) \rightarrow 1) = 1| = 1$.

Theorem 2. *Suppose that $A, B, C \in F(S)$, if $\tau(A) \geq s, \tau(A \rightarrow B) \geq t$, then $\tau(B) \geq s + t - 1$.*

Proof: From $\tau(A) = \frac{1}{3^n} \times |f_A^{-1}(1)| \geq s$, then we have $|f_A^{-1}(1)| \geq 3^n \cdot s$, From $\tau(A \rightarrow B) = \frac{1}{3^n} \times |f_{A \rightarrow B}^{-1}(1)| \geq t$, we have $|f_{A \rightarrow B}^{-1}(1)| \geq 3^n \cdot t$.

Let $X = A, Y = A \rightarrow B$, Being $|f_{X \wedge Y}^{-1}(1)| = |f_X^{-1}(1)| + |f_Y^{-1}(1)| - |f_{X \vee Y}^{-1}(1)| \geq 3^n \cdot s + 3^n \cdot t - 3^n$, so we need to prove $|f_{X \wedge Y}^{-1}(1)| \leq |f_B^{-1}(1)|$.

Being $X \wedge Y = A \wedge (A \rightarrow B) = (A^* \vee B) \wedge A \wedge B^{*+}$, then $v(X \wedge Y) = 1$ holds iff $v(A) = 1$ and $v(B) = 1$. So if $v(A \wedge B) = 1$, then $v(X \wedge Y) = 1$, that is $|f_{X \wedge Y}^{-1}(1)| = |f_{A \wedge B}^{-1}(1)|$, we know that $A \wedge B \rightarrow B$ is a theorem in RSL, So we can get $f_{A \wedge B}^{-1}(1) \subseteq f_B^{-1}(1)$.

So we can get $|f_{X \wedge Y}^{-1}(1)| \leq |f_B^{-1}(1)|$, that is $|f_B^{-1}(1)| \geq 3^n \cdot s + 3^n \cdot t - 3^n$, hence $\tau(B) = \frac{|f_B^{-1}(1)|}{3^n} \geq s + t - 1$ holds.

Theorem 3. *Suppose that* $A, B, C \in F(S)$, *if* $\tau(A \to B) \geq s, \tau(B \to C) \geq t$, *then* $\tau(A \to C) \geq s + t - 1$.

Proof: Being $(A \to B) \to ((B \to C) \to (A \to C))$ is theorem in RSL, and from $\vdash A \to B$, we can get $\tau(A) \leq \tau(B)$, so $\tau(A \to B) \leq \tau((B \to C) \to (A \to C))$, moreover we can get $\tau((B \to C) \to (A \to C)) \geq s$. Being $\tau(B \to C) \geq t$, then from Theorem 2 we can get $\tau(A \to B) \geq s + t - 1$.

Theorem 4. *Suppose that* $A, B, C \in F(S)$, *if* $\tau(A \to B) \geq s, \tau(B \to C) \geq t$, *then* $\tau(A \to B \wedge C) \geq s + t - 1$.

Proof: Let $E = \{v \in X | v \in \Omega_3, v(A \to B) = 1\}$, $F = \{v \in X | v \in \Omega_3, v(B \to C) = 1\}$, $G = \{v \in X | v \in \Omega_3, v(A \to B \wedge C) = 1\}$.

Being $\tau(A \vee B) = \tau(A) + \tau(B) - \tau(A \wedge B)$, So $|E \cap F| = |E| + |F| - |E \cup F| \geq 3^n(s + t - 1)$. So we need to prove $|G| \geq |E \cap F|$, that is $G \supseteq E \cap F$. If $v \in E \cap F$, then $v(A \to B) = 1$ and $v(A \to C) = 1$, and being $((A \to B) \wedge (A \to C)) \to (A \to B \wedge C)$ is a theorem in RSL. So $v(A \to B \wedge C) = v((A \to B) \wedge (A \to C)) = v(A \to B) \wedge v(A \to C)$. Then we have $v \in X$, that is $E \cap F \subseteq G$.

Hence, $\tau(A \to B \wedge C) = \frac{1}{3^n}|G| \geq \frac{1}{3^n}|E \cap F| \geq \alpha + \beta - 1$ holds.

4 Similarity Degree and a Pseudo-metric among Formulas

Definition 6. *Suppose that* $A, B \in F(S)$, *Let* $\xi(A, B) = \tau((A \to B) \wedge (B \to A))$, *then* $\xi(A, B)$ *is called the the similarity degree between* A *and* B.

Theorem 5. *Suppose that* $A, B \in F(S)$, *then*
1) $\xi(A, B) = 1$ *iff* $\vdash A \leftrightarrow B$; 2) $\xi(A, B) + \xi(B, C) \leq \xi(A, C) + 1$.

Proof: 1) can be get immediately from the definition 6.

2) In general, assume that A, B and C have the same atomic formulas $p_1, \cdots, p_m$, From RSL, we can get that $(A \to B) \vee (B \to A)$ is a tautology. So $\xi(A, B) + \xi(B, C) = \tau((A \to B) \wedge (B \to A)) + \tau((B \to C) \wedge (C \to B)) = (\tau(A \to B) + \tau(B \to A)) - \tau((A \to B) \vee (B \to A)) + (\tau(B \to C) + \tau(C \to B)) - \tau((B \to C) \vee (C \to B)) = (\tau(A \to B) + \tau(B \to C)) - 1 + (\tau(C \to B) + \tau(B \to A)) - 1 = \tau((A \to B) \wedge (B \to C)) + \tau((A \to B) \vee (B \to C)) - 1 + \tau((C \to B) \wedge (B \to A)) + \tau((C \to B) \vee (B \to A)) - 1 \leq \tau((A \to B) \wedge (B \to C)) + \tau((C \to B) \wedge (B \to A)) \leq \tau(A \to C) + \tau(C \to A) = \tau((A \to C) \vee (C \to A)) + \tau((A \to C) \wedge (C \to A)) = \tau((A \to C) \wedge (C \to A)) + 1$,

Hence $\xi(A, B) + \xi(B, C) \leq \xi(A, C) + 1$ holds.

Definition 7. *Define non-negative functions* $\rho : F(S) \times F(S) \to [0, 1]$, *which satisfy* $\forall A, B \in F(S)$, $\rho(A, B) = 1 - \xi(A, B)$, *then* ρ *is called the pseudo-metrics on the set of formulas in RSL.*

Theorem 6. *Suppose that* ρ *is the pseudo-metrics on the set of formulas in RSL, then* $\forall A, B \in F(S)$:
1) $\rho(A, A)) = 0$; 2) $\rho(A, B) = \rho(B, A)$; 3) $\rho(A, C) \leq \rho(A, B) + \rho(B, C)$.

Proof: 1) and 2) can be get immediately from Definition 7.

(3) From Definition 7 and Theorem 5, we can get $\rho(A,B) + \rho(B,C) = 1 - \xi(A,B) + 1 - \xi(B,C) = 2 - (\xi(B,C) + \xi(A,B)) \geq 2 - \xi(A,C) - 1 = 1 - \xi(A,C) = \rho(A,C)$, Hence $\rho(A,C) \leq \rho(A,B) + \rho(B,C)$ holds.

5 Conclusion

In this paper we give the theory of truth degrees of three valued rough logic system RSL, and some properties and reasoning rules of truth degrees are studied, moreover, the similarity degree and pseudo-distance between two formulas are defined, and their properties are discussed. This offers a theory framework for approximate reasoning in three valued formed system RSL.

Acknowledgement. This work is supported by Scientific Research Program Funded by Shaanxi Provincial Education Department (Program No.2010JK567) and Doctor Scientific Research Foundation Program of Xi'an Polytechnic University.

References

1. Pawlak, Z.: Rough sets. International Journal of Computer and Information Sciences 11(5), 341–356 (1982)
2. Pawlak, Z.: Rough logic. Bull. Polish Acad. Sc. (Tech. Sc.) 35, 253–258 (1987)
3. Pawlak, Z.: Rough Sets-Theoretical aspects of reasoning about data. Kluwer Academic Publishers, Dordrecht (1991)
4. Banerjee, M., Chakraborty, M.K.: Rough Sets Through Algebraic Logic. Fundamenta Informaticae 28, 211–221 (1996)
5. Duntsch, I.: A logic for rough sets. Theoretical Computer Science 179, 427–436 (1997)
6. Banerjee, M.: Logic for rough truth. Fundamenta Informaticae 71, 139–151 (2006)
7. Vakarelov, D.: A modal logic for similarity relations in Pawlak knowledge representaion systems. Fundamenta Informaticae 15, 61–79 (1991)
8. Zhang, X., Zhu, W.: Rough Logic System RSL and Fuzzy Logic System Luk. Journal of University of Electronic Science and Technology of China 40(2), 296–302 (2011)
9. Wang, G.J., Zhou, H.J.: Quantitative logic. Information Sciences 179, 226–247 (2009)
10. She, Y.H., He, X.L., Wang, G.J.: Rough truth degrees of formulas and approximate reasoning in rough logic. Fundamenta Informaticae 107, 1–17 (2011)
11. Li, J., Wang, G.J.: Theory of truth degrees of propositions in the logic system L_n^*. Science in China (Series F: Information Sciences) 49(4), 471–483 (2006)
12. Hui, X.J.: Generalization of Fundamental Theorem of Probability Logic in Multivalued Propositional Logic. Acta Mathematicae Applicatae Sinica 34(2), 217–228 (2011)

Upper Approximation Reduction Based on Intuitionistic Fuzzy $\mathcal{T}$ Equivalence Information Systems

Weihua Xu*, Yufeng Liu, and Wenxin Sun

School of Mathematics and Statistics,
Chongqing University of Technology, Chongqing, 400054, P.R. China
chxuwh@gmail.com, liuyufeng@cqut.edu.cn, sunxuxin520@163.com

Abstract. Attribute reduction, as one research problem, has played an important role in rough set theory. In this paper, the concept of upper approximation reduction is introduced in intuitionistic fuzzy $\mathcal{T}$ equivalence information systems. Moreover, rough set approach to upper approximation reduction is presented, and the judgement theorems and discernibility matrices are obtained in intuitionistic fuzzy $\mathcal{T}$ equivalence information systems. An example illustrates the validity of the approach, and shows that it is an efficient tool for knowledge discovery in intuitionistic fuzzy $\mathcal{T}$ equivalence information systems.

Keywords: Discernability matrix, Intuitionistic fuzzy relation, Intuitionistic fuzzy rough sets, Upper approximation reduction.

1 Introduction

The theory of rough sets, proposed by Pawlak [9,10], is a powerful mathematical approach to deal with inexact, uncertain or vague knowledge. It has been successfully applied to various fields of artificial intelligence such as pattern recognition, machine learning, and automated knowledge acquisition.

One important application of rough sets theory is attribute reduction in databases. For a data set with discrete attribute values, this can be done by reducing the number of redundant attributes and find a subset of the original attribute set that contains the same information as the original one. Then, people have been attempting to find all reducts. Much study on this area has been reported and many useful results were obtained [2,5,11,7,12].

In 1986, Atanassov [1] introduced the concept of intuitionistic fuzzy (IF) set. Combining IF set theory and rough set theory may result in a new hybrid mathematical structure for the requirement of knowledge-handling systems. The existing researches on IF rough sets are mainly concentrated on the approximation of IF sets. For example, according to fuzzy rough sets in the sense of Ntheda and Majumda [8], Jena and Ghosh [6] independently proposed the concept of

---

* W.H. Xu is a Ph. D and a Prof. of Chongqing University of Technology. His main research fields are rough set, fuzzy set and artificial intelligence.

T. Li et al. (Eds.): RSKT 2012, LNAI 7414, pp. 55–62, 2012.
© Springer-Verlag Berlin Heidelberg 2012

the IF rough set in which the lower and upper approximations are both IF sets. Zhou and Wu [13] explored a general framework for the study of various relation-based IF rough approximation operators when the IF triangular norm $\mathcal{T} = \min$. Though Zhou and Wu [14] present a general framework for the study of relation based $(\mathcal{I}, \mathcal{T})$-IF rough sets by using constructive and axiomatic approachs, the reduction of IF rough sets based on IF $\mathcal{T}$ equivalence relation has not been considered. In this paper, we introduce formal concepts of upper approximation reductions with IF rough sets. The method using discernibility matrix to compute all the attribute reductions is developed.

The rest of this paper is structured as follows: to facilitate our discussion, some preliminary concepts are briefly recalled in Section 2. In Section 3, upper approximation reduction is proposed for the IF information systems. Moreover, the judgement theorems and discernibility matrices are obtained, from which we can provide an approach to attribute reductions in IF $\mathcal{T}$ equivalence information systems. In Section 4, an example illustrates the validity of this method, which shows that the method is effective in complicated information systems.

2   Intuitionistic Fuzzy Rough Sets and Intuitionistic Fuzzy $\mathcal{T}$ Equivalence Information Systems

In this section we mainly review the basic contents of IF information systems and IF rough sets based on IF $\mathcal{T}$ equivalence relation.

Definition 2.1.[4] Let $L^* = \{(\alpha_1, \alpha_2) \in I^2 | \alpha_1 + \alpha_2 \leq 1\}$. We define a relation $\leq_{L^*}$ on L^* as follows: for all $(\alpha_1, \alpha_2), (\beta_1, \beta_2) \in L^*$, $(\alpha_1, \alpha_2) \leq_{L^*} (\beta_1, \beta_2) \Leftrightarrow \alpha_1 \leq \beta_1$ and $\alpha_2 \geq \beta_2$.

Then the relation $\leq_{L^*}$ is a partial ordering on L^* and the pair $(L^*, \leq_{L^*})$ is a complete lattice with the smallest element $0_{L^*} = (0, 1)$ and the greatest element $1_{L^*} = (1, 0)$. The meet operator $\wedge$, join operator $\vee$ and complement operator $\sim$ on $(L^*, \leq_{L^*})$ which are linked to the ordering $\leq_{L^*}$ are, respectively, defined as follows: for all $(\alpha_1, \alpha_2), (\beta_1, \beta_2) \in L^*$, $(\alpha_1, \alpha_2) \wedge (\beta_1, \beta_2) = (\min(\alpha_1, \beta_1), \max(\alpha_2, \beta_2))$, $(\alpha_1, \alpha_2) \vee (\beta_1, \beta_2) = (\max(\alpha_1, \beta_1), \min(\alpha_2, \beta_2))$. $\sim (\alpha_1, \alpha_2) = (\alpha_2, \alpha_1)$.

Definition 2.2.[4] An IF t-norm on L^* is an increasing, commutative, associative mapping $\mathcal{T} : L^* \times L^* \to L^*$ satisfying $\mathcal{T}(1_{L^*}, \alpha) = \alpha$ for all $\alpha \in L^*$. An IF t-conorm on L^* is an increasing, commutative, associative mapping $\mathcal{S} : L^* \times L^* \to L^*$ satisfying $\mathcal{T}(0_{L^*}, \alpha) = \alpha$ for all $\alpha \in L^*$.

Definition 2.3.[4] An IF negator on L^* is a decreasing mapping $\mathcal{N} : L^* \to L^*$ satisfying $\mathcal{N}(0_{L^*}) = 1_{L^*}$ and $\mathcal{N}(1_{L^*}) = 0_{L^*}$. If $\mathcal{N}(\mathcal{N}(\alpha)) = \alpha$ for all $\alpha \in L^*$, then $\mathcal{N}$ is called an involutive IF negator.

The mapping $\mathcal{N}_s$, defined as $\mathcal{N}_s(\alpha_1, \alpha_2) = (\alpha_2, \alpha_1)$, $\forall (\alpha_1, \alpha_2) \in L^*$, is called the standard IF negator.

An IF t-norm $\mathcal{T}$ and an IF t-conorm $\mathcal{S}$ on L^* are said to be dual with respect to an IF negator $\mathcal{N}$ if

$$\mathcal{T}(\mathcal{N}(\alpha),\mathcal{N}(\beta)) = \mathcal{N}(\mathcal{S}(\alpha,\beta)), \quad \mathcal{S}(\mathcal{N}(\alpha),\mathcal{N}(\beta)) = \mathcal{N}(\mathcal{T}(\alpha,\beta)) \quad \forall \alpha, \beta \in L^*.$$

Definition 2.4.[1] Let a set U be fixed. An IF set $\widetilde{A}$ in U is an object having the form $\widetilde{A} = \{\langle x, \mu_{\widetilde{A}}(x), \nu_{\widetilde{A}}(x)\rangle | x \in U\}$, where $\mu_{\widetilde{A}} : U \to I$ and $\nu_{\widetilde{A}} : U \to I$ satisfy $0 \le \mu_{\widetilde{A}}(x) + \nu_{\widetilde{A}}(x) \le 1$ for all $x \in U$, $\mu_{\widetilde{A}}(x)$ and $\nu_{\widetilde{A}}(x)$ are called the degree of membership and the degree of non-membership of the element $x \in U$ to $\widetilde{A}$, respectively. The family of all IF subsets of U is denoted by $IF(U)$. The complement of an IF set $\widetilde{A}$ is defined by $\sim \widetilde{A} = \{\langle x, \nu_{\widetilde{A}}(x), \mu_{\widetilde{A}}(x)\rangle | x \in U\}$. The IF universe set is $\widetilde{1}_U = \widetilde{(1,0)} = \widetilde{1}_{L^*} = \{\langle x, 1, 0\rangle | x \in U\}$ and the IF empty set is $\widetilde{1}_{\emptyset} = \widetilde{(0,1)} = \widetilde{0}_{L^*} = \{\langle x, 0, 1\rangle | x \in U\}$.

Definition 2.5.[3] An IF binary relation $\widetilde{R}$ on U is an IF subset of $U \times U$, namely, $\widetilde{R}$ is given by $\widetilde{R} = \{\langle (x,y), \mu_{\widetilde{R}}(x,y), \nu_{\widetilde{R}}(x,y)\rangle | (x,y) \in U \times U\}$, where $\mu_{\widetilde{R}} : U \times U \to I$ and $\nu_{\widetilde{R}} : U \times U \to I$, $0 \le \mu_{\widetilde{R}}(x,y) + \nu_{\widetilde{R}}(x,y) \le 1$ for all $(x,y) \in U \times U$. $IFR(U \times U)$ will be used to denote the family of all IF relations on U.

An IF $\mathcal{T}$ equivalence relation $\widetilde{R}$ is an IF relation on U which is reflexive ($\widetilde{R}(x,x) = 1$), symmetric ($\widetilde{R}(x,y) = \widetilde{R}(y,x)$) and $\mathcal{T}$ transitive ($\widetilde{R}(x,z) \ge_{L^*} \mathcal{T}(\widetilde{R}(x,y), \widetilde{R}(y,z))$), for every $x, y, z \in U$.

An IF information system is an ordered quadruple $I = (U, AT, V, f)$, where $U = \{x_1, x_2, \cdots, x_n\}$ is a non-empty finite set of objects, $AT = \{a_1, a_2, \cdots, a_p\}$ is a non-empty finite set of attributes, $V = \bigcup_{a \in AT} V_a$ and V_a is a domain of attribute a, $f : U \times AT \to V$ is a function such that $f(a,x) \in V_a$, for each $a \in AT$, $x \in U$, called an information function, where V_a is an IF set of universe U. That is $f(a,x) = (\mu_a(x), \nu_a(x))$, for all $a \in AT$, where $\mu_a : U \to [0,1]$ and $\nu_a : U \to [0,1]$ satisfy $0 \le \mu_a(x) + \nu_a(x) \le 1$, for all $x \in U$. μ_a and ν_a are, respectively, called the degree of membership and the degree of non-membership of the element $x \in U$ to attribute a. We denote $\widetilde{a}(x) = (\mu_a(x), \nu_a(x))$, then it is clear that $\widetilde{a}$ is an IF set of U.

Definition 2.6. An IF $\mathcal{T}$ equivalence information system is an ordered quintuple $\widetilde{\mathcal{I}} = (U, AT, V, f, F)$, where (U, AT, V, f) is an IF information system, F is the mapping from power set AT into the family set $\mathbf{\widetilde{R}}$ of IF $\mathcal{T}$ equivalence relation. Let $\widetilde{\mathcal{I}} = (U, AT, V, f, F)$ be an IF $\mathcal{T}$ equivalence information system, for any $A \subseteq AT$, $a \in A$, $\widetilde{R}_a \in \mathbf{\widetilde{R}}$ be an IF $\mathcal{T}$ equivalence relation with respect to attribute a, denoted as $\widetilde{R}_A = \bigcap_{a \in A} \widetilde{R}_a$.

Definition 2.7.[14] Let $\widetilde{\mathcal{I}} = (U, AT, V, f, F)$ be an IF $\mathcal{T}$ equivalence information system. $\widetilde{X} \in IF(U)$ and $A \subseteq AT$, the $\mathcal{T}$-upper and $\mathcal{S}$-lower approximations of $\widetilde{X}$ with respect to IF $\mathcal{T}$ equivalence relation $\widetilde{R}_A$ are respectively defined by

$$\overline{\widetilde{R}_A}(\widetilde{X})(x) = \bigvee_{y \in U} \mathcal{T}(\widetilde{R}_A(x,y), \widetilde{X}(y)), \quad \forall x \in U,$$

$$\underline{\widetilde{R}_A}(\widetilde{X})(x) = \bigwedge_{y \in U} \mathcal{S}(\sim \widetilde{R}_A(x,y), \widetilde{X}(y)), \quad \forall x \in U.$$

From the definition of IF rough approximation, the following important properties in IF $\mathcal{T}$ equivalence information systems have been proved.

Theorem 2.1.[14] Let $\widetilde{\mathcal{I}} = (U, AT, V, f, F)$ be an IF $\mathcal{T}$ equivalence information system. $\widetilde{X}, \widetilde{Y} \in IF(U)$ and $A \subseteq AT$ then the $\mathcal{T}$-upper and $\mathcal{S}$-lower approximations satisfy the following properties.

(1) $\underline{\widetilde{R}_A}(\sim \widetilde{X}) = \sim\overline{\widetilde{R}_A}(\widetilde{X}), \quad \overline{\widetilde{R}_A}(\sim \widetilde{X}) = \sim \underline{\widetilde{R}_A}(\widetilde{X})$.

(2) $\underline{\widetilde{R}_A}(\widetilde{X}) \subseteq \widetilde{X} \subseteq \overline{\widetilde{R}_A}(\widetilde{X})$.

(3) $\underline{\widetilde{R}_A}(\widetilde{X} \cap \widetilde{Y}) = \underline{\widetilde{R}_A}(\widetilde{X}) \cap \underline{\widetilde{R}_A}(\widetilde{Y}), \quad \overline{\widetilde{R}_A}(\widetilde{X} \cup \widetilde{Y}) = \overline{\widetilde{R}_A}(\widetilde{X}) \cup \overline{\widetilde{R}_A}(\widetilde{Y})$.

(4) $\widetilde{X} \subseteq \widetilde{Y} \Rightarrow \underline{\widetilde{R}_A}(\widetilde{X}) \subseteq \underline{\widetilde{R}_A}(\widetilde{Y})$ and $\overline{\widetilde{R}_A}(\widetilde{X}) \subseteq \overline{\widetilde{R}_A}(\widetilde{Y})$.

(5) $\underline{\widetilde{R}_A}(\widetilde{X} \cup \widetilde{Y}) \supseteq \underline{\widetilde{R}_A}(\widetilde{X}) \cup \underline{\widetilde{R}_A}(\widetilde{Y}), \quad \overline{\widetilde{R}_A}(\widetilde{X} \cap \widetilde{Y}) \subseteq \overline{\widetilde{R}_A}(\widetilde{X}) \cup \overline{\widetilde{R}_A}(\widetilde{Y})$.

(6) $\underline{\widetilde{R}_A}(\widetilde{\alpha}) = \widetilde{\alpha}, \quad \overline{\widetilde{R}_A}(\widetilde{\alpha}) = \widetilde{\alpha}$, for $\alpha = (\alpha_1, \alpha_2) \in L^*$.

   In particular, $\underline{\widetilde{R}_A}(\widetilde{1}_\emptyset) = \overline{\widetilde{R}_A}(\widetilde{1}_\emptyset) = \widetilde{1}_\emptyset, \quad \underline{\widetilde{R}_A}(\widetilde{1}_U) = \overline{\widetilde{R}_A}(\widetilde{1}_U) = \widetilde{1}_U$.

(7) $\underline{\widetilde{R}_A}(\underline{\widetilde{R}_A}(\widetilde{X})) = \underline{\widetilde{R}_A}(\widetilde{X}), \quad \overline{\widetilde{R}_A}(\overline{\widetilde{R}_A}(\widetilde{X})) = \overline{\widetilde{R}_A}(\widetilde{X})$.

3   Upper Approximation Reduction in IF $\mathcal{T}$ Equivalence Information Systems with Decision

In this section we define upper approximation reduction with respect to single IF decision class; we also develop the method based on discernibility matrix to compute all the upper approximation reductions.

An IF $\mathcal{T}$ equivalence information system with decision, is a special case of an IF $\mathcal{T}$ equivalence information system $\widetilde{\mathcal{I}} = (U, AT \cup D, V, f, F)$, where $\widetilde{D} = \{\widetilde{D}_k | k = 1, 2, \cdots, n\}$, $\widetilde{D}_k$ is an IF set of U called IF decision class.

Definition 3.1. Let $\widetilde{\mathcal{I}} = (U, AT \cup D, V, f, F)$ be an IF $\mathcal{T}$ equivalence information system with decision, $\widetilde{D}_k \in \widetilde{D}$ be the IF decision class, and $B \subseteq AT$. If $\overline{\widetilde{R}_B}(\widetilde{D}_k)(x) = \overline{\widetilde{R}_{AT}}(\widetilde{D}_k)(x)$ for any $x \in U$, we say that B is an upper consistent set of AT relative to $\widetilde{D}_k$. Moreover, if any proper subset of B is not the upper approximation set, then B is called one upper approximation reduction of this IF information system.

Theorem 3.1. Let $\widetilde{\mathcal{I}} = (U, AT \cup D, V, f, F)$ be an IF $\mathcal{T}$ equivalence information system with decision, $\widetilde{D}_k \in \widetilde{D}$ be the IF decision class, $B \subseteq AT$. Attribute set B is an upper approximation consistent set if and only if for any $x_i, x_j \in U$, there must exist $a_r \in B$ such that $\mathcal{T}(\widetilde{R}_{a_r}(x_i, x_j), \widetilde{D}_k(x_j)) \leq \overline{\widetilde{R}_{AT}}(D_k)(x_i)$.

Proof. On the one hand, if $B \subseteq AT$, then $\widetilde{R}_B \supseteq \widetilde{R}_{AT}$, we can easily show that $\overline{\widetilde{R}_B}(\widetilde{D}_k)(x_i) \geq \overline{\widetilde{R}_{AT}}(\widetilde{D}_k)(x_i)$, for any $x_i \in U$.

On the other hand,

$$\overline{\widetilde{R}_B}(\widetilde{D}_k)(x_i) \leq \overline{\widetilde{R}_{AT}}(\widetilde{D}_k)(x_i)$$

$$\Leftrightarrow \bigvee_{x_j \in U} \mathcal{T}(\widetilde{R}_B(x_i, x_j), D_k(x_j)) \leq \overline{\widetilde{R}_{AT}}(\widetilde{D}_k)(x_i)$$

$$\Leftrightarrow \bigvee_{x_j \in U} \mathcal{T}(\bigwedge_{a_k \in B} \widetilde{R}_{a_k}(x_i, x_j), D_k(x_j)) \leq \overline{\widetilde{R}_{AT}}(\widetilde{D}_k)(x_i)$$

$$\Leftrightarrow \bigvee_{x_j \in U} \bigwedge_{a_k \in B} \mathcal{T}(\widetilde{R}_{a_k}(x_i, x_j), D_k(x_j)) \leq \overline{\widetilde{R}_{AT}}(\widetilde{D}_k)(x_i)$$

$$\Leftrightarrow \forall x_j \in U, \exists a_r \in B, \text{such that}, \mathcal{T}(\widetilde{R}_{a_r}(x_i, x_j), D_k(x_j)) \leq \overline{\widetilde{R}_{AT}}(\widetilde{D}_k)(x_i)$$

The theorem is proved.

Definition 3.2. Let $\widetilde{\mathcal{I}} = (U, AT \cup D, V, f, F)$ be an IF $\mathcal{T}$ equivalence information system with decision, $\widetilde{D}_k \in \widetilde{D}$ be the IF decision class, $B \subseteq AT$. For any $x_i, x_j \in U$, we denote

$\text{UDis}(x_i, x_j) = \{a_r \in AT | \mathcal{T}(\widetilde{R}_{a_r}(x_i, x_j), \widetilde{D}_k(x_j)) \leq \overline{\widetilde{R}_{AT}}(D_k)(x_i)\}$,

$\text{UM} = (u_{ij})_{n \times n}$,

where $u_{ij} = \text{UDis}(x_i, x_j)$, then $\text{UDis}(x_i, x_j)$ is said to be upper approximation discernibility attribute set between objects x_i and x_j. And matrix UM is referred to as upper approximation discernibility matrix of the IF $\mathcal{T}$ equivalence information system with decision.

Theorem 3.2. Let $\widetilde{\mathcal{I}} = (U, AT \cup D, V, f, F)$ be an IF $\mathcal{T}$ equivalence information system with decision, $\widetilde{D}_k \in \widetilde{D}$ be the IF decision class, $B \subseteq AT$. Attribute set B is an upper approximation consistent set if and only if $B \cap \text{UDis}(x_i, x_j) \neq \emptyset$ for all $x_i, x_j \in U$

Proof. It can be obtained from Theorem 3.1 and Definition 3.2.

Definition 3.3. Let $\widetilde{\mathcal{I}} = (U, AT \cup D, V, f, F)$ be an IF $\mathcal{T}$ equivalence information system with decision, $\widetilde{D}_k \in \widetilde{D}$ be the IF decision class, $B \subseteq AT$. UM be the upper approximation discernibility matrix of the IF $\mathcal{T}$ equivalence information system with decision $\widetilde{\mathcal{I}}$. Let

$$UF = \wedge\{\vee\{a | a \in \text{UDis}(x_i, x_j)\} | x_i, x_j \in U\}.$$

Then UF is called discernibility formulas of upper approximation in IF $\mathcal{T}$ equivalence information system with decision $\widetilde{\mathcal{I}}$.

Theorem 3.3. Let $\widetilde{\mathcal{I}} = (U, AT \cup D, V, f, F)$ be an IF $\mathcal{T}$ equivalence information system with decision, $\widetilde{D}_k \in \widetilde{D}$ be the IF decision class, $B \subseteq AT$. The minimal disjunctive normal form of discernibility formula of upper approximation is

$$UF = \bigvee_{k=1}^{p} (\bigwedge_{s=1}^{q_k} a_s).$$

Let $UB_k = \{a_s | s = 1, 2, \cdots, q_k\}$. Then $\{UB_k | k = 1, 2, \cdots, p\}$ is just set of all upper approximation reductions in IF $\mathcal{T}$ equivalence information system with decision $\mathcal{I}$.

Proof. For any $x_i, x_j \in U$, by the definition of minimum disjunctive normal form, we have that UB_k is upper approximation consistent set. If one element of UB_k is reduced in $UF = \bigvee_{k=1}^{p} (UB_k)$, without loss of generality and the result denoted by UB_k', then there exist $x_{i_0}, x_{j_0} \in U$ such that $UB_k' \cap \mathrm{UDis}(x_{i_0}, x_{j_0}) = \emptyset$. So, UB_k' is no an upper approximation consistent set. So UB_k is an upper approximation reduction in IF $\mathcal{T}$ equivalence information system with decision.

On the other hand, we known that the discernibility formula of upper approximation includes all $\mathrm{UDis}(x_i, x_j)$. Thus, there is not other upper approximation reduction except UB_k.

4   An Illustrated Example

In this section, we employ an example to illustrate our approach in this paper.

Example 4.1. Table 1 shows an IF information system, where $U = \{x_1, x_2, x_3, x_4, x_5, x_6, x_7, x_8, x_9, x_{10}\}$, $AT = \{a_1, a_2, a_3, a_4, a_5\}$. The membership degree and non-membership degree of every object are given in the following table.

Table 1. An IF information system

U	a_1	a_2	a_3	a_4	a_5
x_1	$(0.3, 0.5)$	$(0.6, 0.4)$	$(0.5, 0.2)$	$(0.7, 0.1)$	$(0.5, 0.4)$
x_2	$(0.2, 0.7)$	$(0.1, 0.8)$	$(0.4, 0.5)$	$(0.1, 0.8)$	$(0.2, 0.8)$
x_3	$(0.2, 0.7)$	$(0.1, 0.8)$	$(0.4, 0.5)$	$(0.7, 0.1)$	$(0.2, 0.8)$
x_4	$(0.1, 0.8)$	$(0.1, 0.8)$	$(0.2, 0.7)$	$(0.1, 0.8)$	$(0.2, 0.8)$
x_5	$(0.9, 0.1)$	$(0.8, 0.1)$	$(0.8, 0.1)$	$(0.9, 0.0)$	$(0.7, 0.1)$
x_6	$(0.4, 0.6)$	$(0.8, 0.1)$	$(0.6, 0.3)$	$(0.9, 0.0)$	$(0.7, 0.1)$
x_7	$(0.3, 0.5)$	$(0.7, 0.3)$	$(0.5, 0.1)$	$(0.7, 0.1)$	$(0.6, 0.3)$
x_8	$(0.5, 0.3)$	$(0.8, 0.1)$	$(0.7, 0.1)$	$(1.0, 0.0)$	$(0.7, 0.1)$
x_9	$(0.6, 0.3)$	$(0.9, 0.0)$	$(0.7, 0.1)$	$(0.8, 0.2)$	$(0.8, 0.0)$
x_{10}	$(0.9, 0.1)$	$(0.9, 0.0)$	$(0.8, 0.1)$	$(0.6, 0.3)$	$(1.0, 0.0)$

Every IF attribute a_k can define an IF $\mathcal{T}$ equivalence relation $\widetilde{R}_{a_k}$ as:

$$\widetilde{R}_{a_k}(x_i, x_j) = (\mu_{\widetilde{R}_{a_k}}(x_i, x_j), \nu_{\widetilde{R}_{a_k}}(x_i, x_j)),$$

where, $\mu_{\widetilde{R}_{a_k}}(x_i, x_j) = 1 - \max\{|\mu_{a_k}(x_i) - \mu_{a_k}(x_j)|, |\nu_{a_k}(x_i) - \nu_{a_k}(x_j)|\}$ and $\nu_{R_{a_k}}(x_i, x_j) = \frac{1}{2}(|\mu_{a_k}(x_i) - \mu_{a_k}(x_j)| + |\nu_{a_k}(x_i) - \nu_{a_k}(x_j)|)$. Consider the IF t-norm $\mathcal{T}$: $\mathcal{T}(\alpha, \beta) = (\max\{0, \alpha_1 + \beta_1 - 1\}, \min\{1, \alpha_2 + \beta_2\})$ for $\alpha = (\alpha_1, \alpha_2), \beta = (\beta_1, \beta_2) \in L^*$. Let $U = \{x_1, x_2, x_3, x_4, x_5, x_6, x_7, x_8, x_9, x_{10}\}$, $A = \{(0.6, 0.3), (0.3, 0.5), (0.9, 0.1), (0.6, 0.3), (0.7, 0.1), (0.2, 0.7), (0.4, 0.5), (0.5, 0.2), (0.7, 0.2), (1.0, 0)\}$,

$\widetilde{R}_{AT}$ is computed as follows:

$$
\begin{pmatrix}
(1.0,0) & (0.3,0.65) & (0.5,0.45) & (0.3,0.65) & (0.4,0.5) & (0.7,0.25) & (0.9,0.1) & (0.7,0.25) & (0.6,0.35) & (0.4,0.5) \\
(0.3,0.65) & (1.0,0) & (0.3,0.65) & (0.8,0.2) & (0.2,0.8) & (0.2,0.8) & (0.3,0.65) & (0.1,0.85) & (0.2,0.8) & (0.2,0.8) \\
(0.5,0.45) & (0.3,0.65) & (1.0,0) & (0.3,0.65) & (0.3,0.7) & (0.3,0.7) & (0.4,0.55) & (0.3,0.7) & (0.2,0.8) & (0.2,0.8) \\
(0.3,0.65) & (0.8,0.2) & (0.3,0.65) & (1.0,0) & (0.2,0.8) & (0.2,0.8) & (0.3,0.65) & (0.1,0.85) & (0.2,0.8) & (0.2,0.8) \\
(0.4,0.5) & (0.2,0.8) & (0.3,0.7) & (0.2,0.8) & (1.0,0.5) & (0.5,0.5) & (0.4,0.5) & (0.6,0.3) & (0.7,0.25) & (0.7,0.3) \\
(0.7,0.25) & (0.2,0.8) & (0.3,0.7) & (0.2,0.8) & (0.5,0) & (1.0,0) & (0.8,0.15) & (0.7,0.2) & (0.7,0.25) & (0.5,0.5) \\
(0.9,0.1) & (0.3,0.65) & (0.4,0.55) & (0.3,0.65) & (0.4,0.15) & (0.8,0.15) & (1.0,0) & (0.7,0.2) & (0.7,0.25) & (0.4,0.5) \\
(0.7,0.25) & (0.1,0.85) & (0.3,0.7) & (0.1,0.85) & (0.6,0.2) & (0.7,0.2) & (0.7,0.2) & (1.0,0) & (0.8,0.2) & (0.6,0.35) \\
(0.6,0.35) & (0.2,0.8) & (0.2,0.8) & (0.2,0.8) & (0.7,0.25) & (0.7,0.25) & (0.7,0.25) & (0.8,0.2) & (1.0,0) & (0.7,0.25) \\
(0.4,0.5) & (0.2,0.8) & (0.2,0.8) & (0.2,0.8) & (0.7,0.3) & (0.5,0.5) & (0.4,0.5) & (0.6,0.35) & (0.7,0.25) & (1.0,0)
\end{pmatrix}.
$$

Suppose an IF decision is $\widetilde{D} = \{(1,0),(0.6,0.4),(0.5,0.4),(0.7,0.1),(0.4,0.3),$
$(1,0),(0,1),(0.5,0.5),(0.3,0.5),(0.7,0.2)\}$, then $\overline{\widetilde{R}}_{AT}(\widetilde{D}) = \{(1.0,0),(0.6,0.3),$
$(0.5,0.4),(0.7,0.1),(0.5,0.3),(1.0,0),(0.9,0.1),(0.7,0.2),(0.7,0.25),(0.7,0.2)\}$ and
the upper approximation discernibility matrix of UM is as follows:

$$
UM = \begin{pmatrix}
AT & AT & AT & AT & AT & AT & AT & AT & AT & AT \\
\{2,4,5\} & AT & AT & \{3\} & AT & \{2,4,5\} & AT & AT & AT & AT \\
\{2\} & \{4\} & AT & \{4\} & AT & \{2,5\} & AT & AT & AT & \{1,2,3,5\} \\
AT & AT & AT & AT & AT & AT & AT & AT & AT & AT \\
\{1\} & AT & AT & AT & AT & \{1\} & AT & AT & AT & \{4,5\} \\
AT & AT & AT & AT & AT & AT & AT & AT & AT & AT \\
\{2,5\} & AT & AT & AT & AT & AT & AT & AT & AT & AT \\
\{2,4,5\} & AT & AT & AT & AT & \{1\} & AT & AT & AT & AT \\
\{1,2,5\} & AT & AT & AT & AT & \{1\} & AT & AT & AT & \{1,3,4,5\} \\
\{1,2,3,5\} & AT & AT & AT & AT & \{1,4,5\} & AT & AT & AT & AT
\end{pmatrix}.
$$

Where 1,2,3,4,5 means a_1, a_2, a_3, a_4, a_5. We can get that $\{a_1, a_2, a_3, a_4\}$ is the
only reduction of AT.

5   Conclusions

Intuitionistic fuzzy rough sets are the extension of fuzzy rough sets to deal with
both fuzziness and vagueness in data. It is more material and concise than
fuzzy rough sets to describe the essence of fuzziness. The existing researches
on intuitionistic fuzzy rough sets are mainly concentrated on the construction
of approximation operators. Less effort has been put on the attributes reduction of databases with intuitionistic fuzzy rough sets. In this paper, the concept
of upper approximation reduction has been constructed in intuitionistic fuzzy
$\mathcal{T}$ equivalence information systems. Moreover, rough set approach to upper approximation reductions has been presented and the judgement theorems and
discernibility matrices have been obtained in intuitionistic fuzzy $\mathcal{T}$ equivalence
information systems. The effectiveness of the approach to attribute reduction
has been demonstrated by an example.

Acknowledgments. This work is supported by National Natural Science Foundation of China (No.61105041,71071124 and 11001227), Postdoctoral Science

Foundation of China (No.20100481331) and National Natural Science Foundation of CQ CSTC (No. cstc2011jjA40037) and Graduate Innovation Foundation of Chongqing University of Technology (No.YCX2011312).

References

1. Atanassov, K.: Intuitionistic fuzzy sets. Fuzzy Sets and Systems 20, 87–96 (1986)
2. Beynon, M.: Reducts within the variable precision rough sets model: a further investigation. European Journal of Operational Research 134, 592–605 (2001)
3. Bustince, H., Burillo, P.: Structures on Intuitionistic Fuzzy Relations. Fuzzy Sets and Systems 78, 293–303 (1996)
4. Cornelis, C., Deschrijver, G., Kerre, E.E.: Implication in intuitionistic fuzzy and interval-valued fuzzy set theory: construction, classification, application. International Journal of Approximate Reasoning 35, 55–95 (2004)
5. He, Q., Wu, C.X., Chen, D.G., Zhao, S.Y.: Fuzzy rough set based attribute reduction for information systems with fuzzy decisions. Knowledge-Based Systems 24, 689–696 (2011)
6. Jena, S.P., Ghosh, S.K.: Intuitionistic fuzzy rough sets. Notes on Intuitionistic Fuzzy Sets 8, 1–18 (2002)
7. Mi, J.S., Wu, W.Z., Zhang, W.X.: Comparative studies of knowledge reductions in inconsistent systems. Fuzzy Systems and Mathematics 17, 54–60 (2003)
8. Nanda, S., Majumda, S.: Fuzzy rough sets. Fuzzy Sets and Systems 45, 157–160 (1992)
9. Pawlak, Z.: Rough sets. International Journal of Computer and Information Sciences 11, 341–356 (1982)
10. Pawlak, Z., Skowron, A.: Rudiments of rough sets. Information Sciences 177, 3–27 (2007)
11. Qian, Y.H., Liang, J.Y., Li, D.Y., Wang, F., Ma, N.N.: Approximation reduction in inconsistent incomplete decision tables. Knowledge-Based Systems 23, 427–433 (2010)
12. Xu, W.H., Zhang, X.Y., Zhong, J.M., Zhang, W.X.: Attribute reduction in ordered information systems based on evidence theory. Knowledge and Information Systems 25, 169–184 (2010)
13. Zhou, L., Wu, W.Z.: On generalized intuitionistic fuzzy approximation operators. Information Sciences 178, 2448–2465 (2008)
14. Zhou, L., Wu, W.Z., Zhang, W.X.: On characterization of intuitionistic fuzzy rough sets based on intuitionistic fuzzy implicators. Information Sciences 179, 883–898 (2008)

A Fuzzy-Rough Sets Based Compact Rule Induction Method for Classifying Hybrid Data

Yang Liu[1,2], Qinglei Zhou[1], Elisabeth Rakus-Andersson[3], and Guohua Bai[2]

[1] School of Information Engineering,
Zhengzhou University, Zhengzhou, Henan 450001, China
[2] School of Engineering, Blekinge Institute of Technology, Karlskrona,
Blekinge 371 79, Sweden
[3] School of Applied Mathematics, Blekinge Institute of Technology,
Karlskrona, Blekinge 371 79, Sweden

Abstract. Rule induction plays an important role in knowledge discovery process. Rough set based rule induction algorithms are characterized by excellent accuracy, but they lack the abilities to deal with hybrid attributes such as numeric or fuzzy attributes. In real-world applications, data usually exists with hybrid formats, and thus a unified rule induction algorithm for hybrid data learning is desirable. We firstly model different types of attributes in equivalence relationship, and define the key concepts of block, minimal complex and local covering based on fuzzy rough sets model, then propose a rule induction algorithm for hybrid data learning. Furthermore, in order to estimate performance of the proposed method, we compare it with state-of-the-art methods for hybrid data learning. Comparative studies indicate that rule sets extracted by this method can not only achieve comparable accuracy, but also get more compact rule sets. It is therefore concluded that the proposed method is effective for hybrid data learning.

Keywords: Knowledge discovery, classification, rough sets, rule induction, hybrid data.

1 Introduction

Knowledge acquisition system has been successful in many application areas [1]. There are two most commonly used methods to generate expressive and human readable knowledge, i.e., decision tree and rule induction methods. Decision tree methods have drawn significant attention over the last several years, and incorporate various advanced speed, memory, and pruning optimization techniques [2]. However, rule induction algorithms also exhibit a lot of desirable properties. As reported by some problems, rule learner methods were found to outperform decision tree methods since the production of rule sets in expert system appears to be more human-comprehensible than decision trees [3]. In addition, rule set can be post-processed and analyzed in modality, which is very important when a decision maker needs to understand and validate the generated results.

T. Li et al. (Eds.): RSKT 2012, LNAI 7414, pp. 63–70, 2012.
© Springer-Verlag Berlin Heidelberg 2012

The main concern about rule induction method is the problem of inconsistencies and uncertainty in datasets. There have been designed well-known rule induction methods inspired by rough sets theory, and LERS system (Learning from examples based on rough sets) is one of the most widely used systems for real-world applications [4]. As we know, Pawlak's rough sets model works in case that only symbolic attributes exist in information system [5]. However, complex application problems are likely to present a number of hybrid attributes [6]. Numerical attributes must be discretized by a front-end discretization tool. The discretization procedure may change the discernibility power of original training data and decrease the classification accuracy of testing data that have values near the boundary of cut points [7]. Algorithm MLEM2 is proposed as an extension of LEM2, which computes cut points by averaging any two consecutive values of a sorted value list [8]. However, the selection of crisp cut points is also crucial to the performance of subsequent rule learning.

Some generalizations of rough sets model were proposed to deal with this problem. Dubois et al. firstly proposed fuzzy rough sets and rough fuzzy sets [9]. Hu et al. examined uncertainty model of fuzzy rough sets for feature selection and classification in hybrid data [6]. Jensen et al. proposed a fuzzy rough sets based rule induction method for fuzzy information system [10]. The fuzzy inference engine can take either fuzzy measurements or crisp measurements as inputs from the real world to suggest a classification or decision that is in crisp form or fuzzy form. However, these methods mainly focused on obtaining better classification accuracy, and do not consider extracting compact rule sets. In reality, compact knowledge can be easily comprehended by people and may avoid over-fitting training data. In this study, we focus on using fuzzy equivalence relations to model different types of attributes, and studying rule induction method that can extract small size of rule sets on hybrid data.

2   Preliminary Notion of Rough Sets Based Rule Induction Method

LERS is a rough sets based rule induction system. For algorithm LEM2 in LERS, the Pawlak's lower and upper approximation sets of a concept are used as input sets, which resolves the conflicting problems in training data. Algorithm LEM2 explores search space of attribute-value pairs from an input set, and consequently computes a local covering that can be converted into a rule set. We firstly quote a few definitions in algorithm LEM2.

The main notion of algorithm LEM2 is a block of an object-attribute pair. For an object-attribute pair $t = (x, c), x \in U, c \in C$, a *block* of t, denoted by $[t]$, is a set of examples from U such that $[t] = \{y \mid f(y, c) = f(x, c), y \in U\}$. Let B be a nonempty lower or upper approximation set of a concept and T be a set of object-attribute pairs. Set B depends on T if and only if $\emptyset \neq [T] = \cap_{t \in T}[t] \subseteq B$. Set T is called as a *minimal complex* of B if and only if B *depends on* T and no proper subset T' of T exists such that B depends on T'. Let $\mathbb{T}$ be a nonempty sets of minimal complex, $\mathbb{T}$ is a *local covering* of B if and only if:

(1). $\cup_{T \in \mathbb{T}}[T] = B$,

(2). $\mathbb{T}$ is minimal, i.e., there exists no proper subset $\mathbb{T}'$ of $\mathbb{T}$ satisfy condition (1).

When selecting an appropriate object-attribute pair t, algorithm LEM2 uses a priority-based strategy. It searches for an object-attribute pair t with the highest priority, and then adds it to partial minimal complex. When a local covering is found, it is converted to a crisp rule set for classification of symbolic data.

3   Fuzzy Rough Sets Based Compact Rule Learner

Pawlak's rough sets model can only deal with data containing symbolic attributes. As we know, real-world applications usually contain hybrid attributes such as real-valued or fuzzy attributes. Let us denote that HDT $= \langle U, C^s \bigcup C^n \bigcup C^f \bigcup \{d\}, V, f \rangle$ is a hybrid decision table, where C^s is a set of symbolic attributes, C^n is a set of numeric attributes, C^f is a set of fuzzy attributes, and d is a decision attribute.

In fuzzy rough sets, it assumes that each attribute generate a fuzzy equivalence relation. If c is a symbolic attribute, the crisp equivalence relation can be viewed as a fuzzy one. If c is a numeric attribute, the values will be firstly normalized into the interval $[0, 1]$. A fuzzy similarity relation can be used:

$$Sim^c(x_i, x_j) = \begin{cases} 1 - d/\delta, & d < \delta, \\ 0, & otherwise. \end{cases} \tag{1}$$

where $d = |f(x_i, c) - f(x_j, c)|$, $\delta \in [0, 0.5]$. In experimental section, we set $\delta = 0.25$ for all experiments.

A fuzzy equivalence relation can be computed from fuzzy similarity relation with max-min transitivity operation. If c is a fuzzy attribute, let m_c be the number of fuzzy sets $F_1^c, \dots F_{m_c}^c$ in the attribute c. There are a great many candidate similarity measures:

(1). Hamming similarity measure:

$$Sim^c(x_i, x_j) = \frac{1}{m_c} \sum_{k=1}^{m_c} (1 - |\mu_{F_k^c}(x_i) - \mu_{F_k^c}(x_j)|);$$

(2). Max-Min similarity measure:

$$Sim^c(x_i, x_j) = \frac{1}{m_c} \sum_{k=1}^{m_c} \frac{\min(\mu_{F_k^c}(x_i), \mu_{F_k^c}(x_j))}{\max(\mu_{F_k^c}(x_i), \mu_{F_k^c}(x_j))}.$$

Definition 1. Fuzzy block of an object-attribute pair. Let $R = (r_{ij})_{n \times n}$ be a fuzzy equivalence relation on universe U induced by an attribute $c \in C$, $t = (x_i, c)$ be an object-attribute pair. The *fuzzy block* of t is defined as

$$[t] = \int_{x_j \in U} r_{ij}/x_j, \tag{2}$$

Obviously, $[t]$ is the fuzzy equivalence class generated by x_i on fuzzy equivalence relation R. Therefore, $[t]$ is a fuzzy set.

Some problems encountered by fuzzy lower and upper approximation set are proposed [11]. The computation problems are solved by using a crisp lower and upper approximation set. Let R be a fuzzy equivalence relations and X be a crisp set of objects, the crisp definition of lower or upper approximation set is defined as:

$$\underline{R}(X) = \{x_i \mid [x_i]_R \subseteq X, x_i \in U\}, \tag{3}$$

$$\overline{R}(X) = \{x_i \mid [x_i]_R \cap X \neq \emptyset, x_i \in U\}, \tag{4}$$

where $A \subseteq B$ means $\forall x \in U, \mu_A(x) \leqslant \mu_B(x)$.

Definition 2. α-depends on. Let X be a crisp set, T be a set of object-attribute pairs. Set X fuzzily α-*depends on* set T if and only if

$$\emptyset \neq [T] = \cap\{[t] \mid t \in T\} \subseteq_\alpha X, \tag{5}$$

where $A \cap B = \int_{x \in U} \min(\mu_A(x), \mu_B(x))/x$, $A \subseteq_\alpha B$ if and only if $\forall x \in U, \max(1 - \mu_A(x), \mu_B(x)) \geqslant \alpha$.

Let B be a nonempty lower or upper approximation set of a concept, T be a set of object-attribute pairs. Set T is a α-*minimal complex* of B if and only B fuzzily α-depends on T and no proper subset T' of T exists such that B fuzzily α-depends on T'.

Definition 3. (α, k)-local covering. Let $\mathbb{T}$ be a collection of α-minimal complex and B be a nonempty crisp set. $\mathbb{T}$ is (α, k)-*local covering* of B if and only if the following conditions are satisfied:

(1). $I(\cup_{T \in \mathbb{T}}[T], B) \leqslant k$, and
(2). $\mathbb{T}$ is minimal, i.e., $\mathbb{T}$ has the smallest possible number of members,

where similarity function of fuzzy sets $I(A, B) = (|\overline{A} \cap B| + |A \cap \overline{B}|)/|A \cup B|$, $|\bullet| = \sum_{x \in U} \mu_\bullet(x)$, $\overline{A} = \int_{x \in U}(1 - \mu_A(x))/x$, $I(A, B) \in [0, 1]$.

It is obvious that α and k provide a new stopping condition for rule induction method. When $\alpha = 0, k = 0$, and attributes are symbolic, the stopping condition in new algorithm is degraded into the same condition used in algorithm LEM2.

Definition 4. Score of an object-attribute pair. Let $t = (x, c)$ be an object-attribute pair and G be a fuzzy set. The *score* of t related to G is defined as:

$$Score(t, G) = |[t] \cap G|. \tag{6}$$

Let us denote B is a crisp lower or upper approximation of a concept set. Normal fuzzy set operations in the algorithm are based on Zadeh's definition. The followings present the pseudocode of proposed rule induction algorithm on hybrid data.

Algorithm. FRLEM2 Rule Induction Method on Hybrid Data
Input: A non-empty lower or upper approximation set of a concept B, and thresholds α and k.
Output: A (α, k)-local covering $\mathbb{T}$.

```
 1: G ⇐ B;// G is a fuzzy set of objects
 2: T ⇐ ∅; //T is a crisp set of minimal complex
 3: while I(∪_{T∈T}[T], B) > k do
 4:     T ⇐ ∅;//T is a crisp set of of object-attribute pairs
 5:     while (T = ∅) or (not ([T] ⊆_α B)) do
 6:         t ⇐ arg max_{t∉T} Score(t, G);
 7:         T ⇐ T ∪ {t};
 8:         G ⇐ [t] ∩ G;
 9:     end while
10:     for ∀t ∈ T do
11:         if [T \ {t}] ⊆_α B and T \ {t} ≠ ∅ then
12:             T ⇐ T \ {t};
13:         end if
14:     end for
15:     T ⇐ T ∪ {T};
16:     G ⇐ B \ ∪_{T∈T}[T];
17: end while
18: for ∀T ∈ T do
19:     if I(∪_{S∈T\{T}}[S], B) ⩽ k then
20:         T ⇐ T \ {T};
21:     end if
22: end for
23: return T;
```

Theorem 1. *When $k = \alpha = 0$, and attributes are symbolic, the rule learner FRLEM2 is degraded to LEM2.*

Proof. If $k = \alpha = 0$, and attributes are symbolic, the following properties in FRLEM2 learner hold:

(1). Fuzzy equivalence relation degenerates to classical equivalence relation.
(2). The definition of α-depends on is equivalent with classical definition.
(3). The definition of (α, k)-local covering is equivalent with classical definition.
(4). The fuzzy set operation is degraded to crisp set operation.

From (1), (2), (3) and (4), the key concepts and termination criteria are equivalent between FRLEM2 and LEM2. Therefore, the theorem holds. ∎

Theorem 1 means that algorithm LEM2 is a special case of FRLEM2 when $k = 0, \alpha = 0$, and attributes are symbolic. The proposed learner uses two threshold values to reinforce the stopping condition. These two threshold values are set to be near to zero, unless a user specifies alternative value. The pruning threshold α is used to prune very specific rules. A minimal complex generation process is terminated if input set α-depends on the partial minimal complex. Generalization threshold k is used to generate a small amount of rules. This threshold is used to relax the requirement of LINE 3 of FRLEM2 learner, by which it allows the local covering does not cover a small portion of input set. Such a mechanism is especially valuable in case of data containing overlapping classes, i.e., data having inconsistent cases in training data.

4   Experiments and Discussion

We tested FRLEM2 on some benchmark problems with the same parameters and conditions. Here, threshold parameters α and k are set to be 0.02 and 0.025 respectively. In order to evaluate the performance of FRLEM2 for hybrid data learning, systematic comparative experiments are conducted. The methods employed for the comparison comprise a decision tree based method C4.5 [2], and two rule learners LEM2 [8] and DataSqueezer [12], both of which use CAIM as the front-end discretization tool [13].

Table 1. Description of datasets

abbr.	Dataset	cases	class	$|C|$	$|C^s|$	$|C^n|$	test data
tae	TA Evaluation	151	3	5	1	4	10F-CV
hea	StatLog heart disease	270	2	13	7	6	10F-CV
cle	Cleve database	303	2	14	6	8	10F-CV
bos	Boston housing	506	3	13	12	1	10F-CV
cra	Credit approval	690	2	15	6	9	10F-CV
aca	Australian credit approval	690	2	14	6	8	10F-CV
gec	StatLog German credit data	1000	2	20	7	13	10F-CV
hyp	Hypothyroid disease	3163	2	25	7	18	10F-CV
adu	Adult	48842	2	14	6	8	16281

A detailed description of datasets is presented in Table 1. The datasets were obtained from the UCI ML repository [14]. We note that *10F-CV* denotes ten-fold cross-validation experiment on a dataset. We focus on reporting accuracy and rule complexity of each method on a variety of realistic datasets.

Table 2 reports results of classification accuracy of each method. On average, C4.5 obtains the highest average accuracy, with FRLEM2 being the second best. Closer observation reveals that there is no universally best learner in this comparison group. In Table 3, we present our experimental measurements of rule complexity. In this table, "r" represents the number of rules extracted by each method, and "l/r" means the average length per rule. The smallest average number of rules is generated by FRLEM2, with DataSqueezer being the second best. The number of rules generated by FRLEM2 is close to the average number of rules generated by DataSqueezer. We observe that FRLEM2 generates rule

Table 2. Comparison results of classification accuracy

Set	C4.5	LEM2	DataSqueezer	FRLEM2
tae	60 ±12.8	54 ±12.5	55 ±7.2	58 ±10.3
hea	77 ±8.0	79 ±6.7	79 ±6.5	83 ±7.6
cle	55 ±10.3	56 ±13.2	46 ±23.5	58 ±13.7
bos	95 ±6.8	93 ±8.4	90 ±6.3	93 ±9.5
cra	86 ±3.5	75 ±1.9	73 ±5.2	81 ±6.2
aca	86 ±6.3	72 ±3.4	62 ±6.8	74 ±5.7
gec	71 ±3.6	82 ±4.3	80 ±7.9	83 ±3.5
hyp	99 ±3.9	83 ±8.6	95 ±3.0	94 ±3.8
adu	79.4	72.6	71.3	76.2
Mean	79.4	74.8	73.6	78.4

sets that are on average two-third of the size obtained by LEM2, which shows a significant improvement for LEM2.

Table 3. Comparison results of rule sets complexity

Set	C4.5		LEM2		DataSqueezer		FRLEM2	
	r	l/r	r	l/r	r	l/r	r	l/r
tae	17	6.8	43	2.4	21	2.5	19	2.9
hea	27	3.7	36	2.3	7	3.2	17	2.4
cle	21	2.8	38	3.5	16	3.0	25	2.6
bos	6	3.3	49	3.8	22	5.2	23	3.9
cra	30	4.2	52	4.1	27	3.6	25	4.7
aca	31	4.4	57	5.3	20	2.5	20	3.7
gec	103	6.1	160	4.3	134	3.3	63	4.7
hyp	7	3.9	46	4.1	15	4.2	14	4.5
adu	56	9.1	303	3.3	360	6.7	359	3.2
Mean	89.6	4.9	87.1	3.7	69.1	3.8	62.8	3.6

To summarize, the FRLEM2 is characterized by good accuracy that is comparable with state-of-the-art methods. It can generate more compact rule sets than other methods, and reduce one-third size of rule sets generated by LEM2. The good accuracies and compact rule sets are observed in the comparison experiments.

5   Conclusion

The classical rough sets based rule induction methods just work in nominal domain. The discretization methods have great influence on the performance of rule induction algorithms. In this paper, we propose a fuzzy-rough sets based rule induction method that can exhibits on hybrid data. The method overcomes limitations of the classical algorithms that only support symbolic attribute-type and have stringently stopping conditions. The proposed method is proven to be a powerful extension of algorithm LEM2, in which the performance of the method can be tuned in a flexible way even if types of attributes are symbolic. Experiments show that proposed method can extract compact rule sets on hybrid data that maintain comparable classification accuracies as that of other methods.

References

1. Gebus, S., Leiviska, K.: Knowledge acquisition for decision support systems on an electronic assembly line. Expert Systems with Applications 36(1), 93–101 (2009)
2. Quinlan, J.: Induction of decision trees. Machine Learning 1, 81–106 (1986)
3. Garcia, S., Fernandez, A., Herrera, F.: Enhancing the effectiveness and interpretability of decision tree and rule induction classifiers with evolutionary training set selection over imbalanced problems. Applied Soft Computing 9(4), 1304–1314 (2009)
4. Grzymala-Busse, J.W., Marepally, S.R., Yao, Y.: An Empirical Comparison of Rule Sets Induced by LERS and Probabilistic Rough Classification. In: Szczuka, M., Kryszkiewicz, M., Ramanna, S., Jensen, R., Hu, Q. (eds.) RSCTC 2010. LNCS, vol. 6086, pp. 590–599. Springer, Heidelberg (2010)
5. Pawlak, Z., Skowron, A.: Rudiments of rough sets. Information Science 177(1), 3–27 (2007)
6. Hu, Q., Xie, Z., Yu, D.: Hybrid attribute reduction based on a novel fuzzy-rough model and information granulation. Pattern Recognition 40(12), 3509–3521 (2007)
7. Blajdo, P., Hippe, Z., Mroczek, T., Grzymala-Busse, J., Knap, M., Piatek, L.: An Extended Comparison of Six Approaches to Discretization–A Rough Set Approach. Fundamenta Informaticae 94(2), 121–131 (2009)
8. Grzymala-Busse, J.W.: Mining Numerical Data – A Rough Set Approach. In: Peters, J.F., Skowron, A. (eds.) Transactions on Rough Sets XI. LNCS, vol. 5946, pp. 1–13. Springer, Heidelberg (2010)
9. Dubois, D., Prade, H.: Rough fuzzy sets and fuzzy rough sets. International Journal of General Systems 17(2-3), 191–209 (1990)
10. Jensen, R., Cornelis, C., Shen, Q.: Hybrid fuzzy-rough rule induction and feature selection. In: IEEE International Conference on Fuzzy Systems, pp. 1151–1156 (2009)
11. Bhatt, R.B., Gopal, M.: On the compact computational domain of fuzzy-rough sets. Pattern Recognition Letters 26(11), 1632–1640 (2005)
12. Kurgan, L.A., Cios, K.J., Dick, S.: Highly scalable and robust rule learner: Performance evaluation and comparison. IEEE Transactions on Systems Man and Cybernetics 36, 32–53 (2006)
13. Kurgan, L.A., Cios, K.J.: CAIM discretization algorithm. IEEE Transactions on Data and Knowldge Engineering 16(2), 145–153 (2004)
14. Frank, A., Asuncion, A.: UCI machine learning repository (2010), http://archive.ics.uci.edu/ml

A New Intuitionistic Fuzzy Rough Set Approach for Decision Support

Junyi Chai, James N.K. Liu, and Anming Li

Department of Computing,
The Hong Kong Polytechnic University,
Hung Hom, Kowloon, Hong Kong SAR
{csjchai,csnkliu}@comp.polyu.edu.hk, lianming03@gmail.com

Abstract. The rough set theory was proved of its effectiveness in dealing with the imprecise and ambiguous information. Dominance-based Rough Set Approach (DRSA), as one of the extensions, is effective and fundamentally important for Multiple Criteria Decision Analysis (MCDA). However, most of existing DRSA models cannot directly examine uncertain information within rough boundary regions, which might miss the significant knowledge for decision support. In this paper, we propose a new believe factor in terms of an intuitionistic fuzzy value as foundation, further to induce a kind of new uncertain rule, called believable rules, for better performance in decision-making. We provide an example to demonstrate the effectiveness of the proposed approach in multicriteria sorting and also a comparison with existing representative DRSA models.

Keywords: Multicriteria decision analysis, Rough set, Intuitionistic fuzzy set, Rule-based approach, Sorting.

1 Introduction

Rough set methodology is an effective mathematical tool for Multicriteria Decision Analysis (MCDA) because of its strength in data analysis and knowledge discovery from imprecise and ambiguous data. The classical Pawlak's rough set had been successfully applied in medical diagnosis [13], supplier selection [5], etc. However, it cannot deal with the preference-ordered data. With substitution of indiscernibility relations by dominance relations, Classical Dominance-based Rough Set Approach (C-DRSA) was firstly generated by Greco et al. [8]. Compared with Pawlak's Rough Set, the key idea of C-DRSA is mainly in two aspects: (1) the knowledge granules generated from multiple criteria are dominance cones rather than the concept of indiscernibility; (2) the objective sets of rough approximations are the upward and downward unions of preference-ordered classes, rather than the binary-relation-based non-preference classes. Such properties let C-DRSA be a suitable means for decision supports, particularly with respect to multicriteria ranking, sorting and choice.

C-DRSA is the core procedure for calculation of rough approximations, in which consistency data are assigned to lower approximations and inconsistency

T. Li et al. (Eds.): RSKT 2012, LNAI 7414, pp. 71–80, 2012.

data are put into the rough boundary regions. The purpose of applying DRSA models is to induce decision rules and then employ them for providing assignments to pre-defined decision classes. Various extensions of DRSA models also appeared. Variable-Precision DRSA (VP-DRSA) [9] defined a threshold called the precision to control the membership of inconsistent objects into the lower approximations. Quasi-DRSA [6] hybridized Pawlak's rough set and C-DRSA for lower error rates in natural selection. Chai and Liu [3] provided a class-based rough approximation model and studied the reducts preserving the singleton class rather than the traditional class unions.

However, most of previous DRSA models aim to generate a minimal rule set, which might neglect valuable uncertain information within rough boundary regions [8]. Even though such possible rules and approximate rules as uncertain rules are able to extract uncertain information, they rarely can be employed in real world. A significant extension of C-DRSA is Variable-Consistency DRSA (VC-DRSA) [7] that relaxes the strict dominance principle and hence admits several inconsistent objects to the lower approximations. This approach indeed enhances the opportunity of discovering the strong rule patterns, and is particularly useful for large datasets. Yet, it is still far from satisfactory.

In this paper, we develop a new DRSA model through inducing a new kind of uncertain rule called believable rule, in order to better extract valuable uncertain information. To this end, we introduce a new believe factor in terms of the concept of intuitionistic fuzzy value [4], [11]. Three related measurements are generated for exploring rough boundary region. Finally, aided by the proposed believe factor, we define a new kind of uncertain rule, called believable rule, for better examination of uncertain information within rough boundary regions. Through comparing with previous representative DRSA models, an example is provided to verify the capability of the proposed model in solving sorting problems.

The rest of this paper is organized as follows. Section 2 provides the preliminaries, including the principles of DRSA methodology and intuitionistic fuzzy theory. Section 3 presents believable rule induction aided by believe factor. In section 4, we demonstrate the capability of the proposed model via an illustrative example with a comparison. Finally, we draw the conclusion and outline the future work in Section 5.

2   Preliminaries

2.1   Dominance-Based Rough Set Approach

An information table can be transferred to a *decision table* via distinguishing condition criteria and decision criteria. Formally, a decision table is the 4-tuple $S = \langle U, Q, V, f \rangle$, which includes (1) a finite set of objects denoted by U, $x \in U = \{x_1, ..., x_m\}$; (2) a finite set of criteria denoted by $Q = C \bigcup D$, where condition criteria set $C \neq \emptyset$, decision criteria set $D \neq \emptyset$ (usually the singleton set $D = \{d\}$), and $q \in Q = \{q_1 ..., q_n\}$; (3) the scale of criterion q denoted by V_q, where $V = \bigcup_{q \in Q} V_q$; (4) information function denoted by $f_q(x) : U \times Q \to V$, where $f_q(x) \in V_q$ for each $q \in Q$, $x \in U$. In addition, each object x from U is described by a vector called *decision description* in terms of the decision information on the criteria, denoted by $Des_Q(x) = [f_{q_1}(x), ..., f_{q_n}(x)]$. As such, information function $f_q(x)$ also can be called *decision values* in MCDA.

The objective sets of dominance-based rough approximations are the upward or downward unions of predefined decision classes. Suppose the decision criterion d makes a partition of U into a finite number of classes $CL = \{Cl_t, t = 1, ..., l\}$. We assume that Cl_{t+1} is superior to Cl_t according to DM's preference. Each object x from U belongs to *one and only one* class Cl_t. The upward and downward unions of classes are represented respectively as: $Cl_t^{\geq} = \bigcup_{s \geq t} Cl_s$, $Cl_t^{\leq} = \bigcup_{s \leq t} Cl_s$, where $t = 1, ..., l$.

Then, the following operational laws are valid: $Cl_1^{\leq} = Cl_1$; $Cl_l^{\geq} = Cl_l$; $Cl_t^{\geq} = U - Cl_{t-1}^{\leq}$; $Cl_t^{\leq} = U - Cl_{t+1}^{\geq}$; $Cl_1^{\geq} = Cl_l^{\leq} = CL$; $Cl_0^{\geq} = Cl_{l+1}^{\leq} = \emptyset$.

The granules of knowledge in DRSA theory are *dominance cones* with respect to values space of the considered criteria. If two decision values are with the dominance relation like $f_q(x) \geq f_q(y)$ for every considered criterion $q \in P \subseteq C$, we say x *dominates* y, denoted by xD_py. The dominance relation is reflexive and transitive. With this in mind, the *dominance cone* can be represented by: P-dominating set $D_P^+(x) = \{y \in U : yD_px\}$; P-dominated set $D_P^-(x) = \{y \in U : xD_py\}$.

The key concept in DRSA theory is the *Dominance Principle*: if the decision value of object x is no worse than that of object y on all considered condition criteria (saying x is dominating y on $P \subseteq C$), object x should also be assigned to a decision class no worse than that of object y (saying x is dominating y on D). Founded on such dominance principle, the definitions of rough approximations are given in the following.

P-lower approximations denoted as $\underline{P}(Cl_t^{\geq})$ and $\underline{P}(Cl_t^{\leq})$, are represented as: $\underline{P}(Cl_t^{\geq}) = \{x \in U : D_P^+(x) \subseteq Cl_t^{\geq}\}$; $\underline{P}(Cl_t^{\leq}) = \{x \in U : D_P^-(x) \subseteq Cl_t^{\leq}\}$.

P-upper approximations denoted as $\overline{P}(Cl_t^{\geq})$ and $\overline{P}(Cl_t^{\leq})$, are represented as: $\overline{P}(Cl_t^{\geq}) = \{x \in U : D_P^-(x) \cap Cl_t^{\geq} \neq \emptyset\}$; $\overline{P}(Cl_t^{\leq}) = \{x \in U : D_P^+(x) \cap Cl_t^{\leq} \neq \emptyset\}$.

VC-DRSA model accepts a limited number of inconsistent objects which are controlled by the predefined threshold called *consistency level*. For $P \subseteq C$, the P-lower approximations of VC-DRSA can be represented as: $\underline{P}^l(Cl_t^{\geq}) = \{x \in Cl_t^{\geq} : \frac{|D_P^+(x) \cap Cl_t^{\geq}|}{|D_P^+(x)|} \geq l\}$; $\underline{P}^l(Cl_t^{\leq}) = \{x \in Cl_t^{\leq} : \frac{|D_P^-(x) \cap Cl_t^{\leq}|}{|D_P^-(x)|} \geq l\}$, where consistency level l means that object x from U belongs to the class union $Cl_t^{\geq}$ (or $Cl_t^{\leq}$) with no ambiguity at level $l \in (0, 1]$.

2.2 Intuitionistic Fuzzy Theory

This section revisits the principles of intuitionistic fuzzy theory as one of our preliminaries. Atanassov [1] extended Zadeh's fuzzy set employed by a membership function, and defined the notion of intuitionistic fuzzy set (IFS) via further considering a non-membership function. An IFS A in a finite set X can be written as: $A = \{< x, \mu_A(x), \nu_A(x) > | x \in X\}$ s.t. $0 \leq \mu_A + \nu_A \leq 1, x \in X$; with

$\mu_A : X \to [0,1], x \in X \to \mu_A(x) \in [0,1]$; $\nu_A : X \to [0,1], x \in X \to \nu_A(x) \in [0,1]$. The hesitation degrees [10] can be defined as: $\pi_A = 1 - \mu_A - \nu_A$.

Xu [11] extracted the basic element from IFS as the Intuitionistic Fuzzy Value (IFV) denoted as $a = (\mu_a, \nu_a, \pi_a)$, where the membership degree $\mu_a \in [0,1]$, the non-membership degree $\nu_a \in [0,1]$, and the hesitation degree $\pi_a \in [0,1]$ with $\pi_a = 1 - \mu_a - \nu_a$. Let a_1 and a_2 be two IFVs. The related operations [12] are revisited in the following. Complement: $\bar{a} = (\nu_a, \mu_a)$; Addition: $a_1 \oplus a_2 = \{\mu_{a_1} + \mu_{a_2} - \mu_{a_1}\mu_{a_2}, \nu_{a_1}\nu_{a_2}\}$; Multiplication: $a_1 \otimes a_2 = \{\mu_{a_1}\mu_{a_2}, \nu_{a_1} + \nu_{a_2} - \nu_{a_1}\nu_{a_2}\}$; Multiple law: $\lambda a = (1 - (1 - \mu_a)^\lambda, \nu_a^\lambda)$, $\lambda > 0$; Exponent law: $a^\lambda = (\mu_a^\lambda, 1 - (1 - \nu_a)^\lambda)$, $\lambda > 0$; The Score Function: $s(a) = \mu_a - \nu_a$; The Accuracy Function: $h(a) = \mu_a + \nu_a$. The method for comparing two intuitionistic fuzzy values through using $s(a)$ and $h(a)$ is presented: If $s(a_1) < s(a_2)$, then $a_1 < a_2$. If $s(a_1) = s(a_2)$, then, 1) If $h(a_1) = h(a_2)$, then $a_1 = a_2$; 2) If $h(a_1) < h(a_2)$, $a_1 < a_2$; 3) If $h(a_1) > h(a_2)$, then $a_1 > a_2$.

3   Uncertain Rule Induction

3.1   Believe Factor

Considering the assignment of object $x \in U$, dominance cones $D_P^+(x)$ and $D_P^-(x)$ can be divided into three subsets, denoted as X_1, X_2 and X_3: (a) for $D_P^+(x)$, we have $X_1 \subseteq \underline{P}(Cl_t^\geq)$, $X_2 \subseteq Cl_t^\geq - \underline{P}(Cl_t^\geq)$, $X_3 \subseteq Cl_{t-1}^\leq$; (b) for $D_P^-(x)$, we have $X_1 \subseteq \underline{P}(Cl_t^\leq)$, $X_2 \subseteq Cl_t^\leq - \underline{P}(Cl_t^\leq)$, $X_3 \subseteq Cl_{t+1}^\geq$. With respect to the objects belonging to the class unions $Cl_t^\geq$ and $Cl_t^\leq$ but failing to be assigned to the corresponding lower approximations, the following assertions are valid: (1) For $t = 2, ..., l$, we have $Bn_P(Cl_t^\geq) = Bn_P(Cl_{t-1}^\leq) = (Cl_t^\geq - \underline{P}(Cl_t^\geq)) \bigcup (Cl_{t-1}^\leq - \underline{P}(Cl_{t-1}^\leq))$. (2) For $x \in Cl_t^\geq - \underline{P}(Cl_t^\geq)$, $t = 2, ..., l$, we have $D_P^+(x) = X_1 \bigcup X_2 \bigcup X_3$ subject to $X_1 \subseteq \underline{P}(Cl_t^\geq)$, $X_2 \subseteq Cl_t^\geq - \underline{P}(Cl_t^\geq)$, $X_3 \subseteq Cl_{t-1}^\leq$. (3) For $x \in Cl_t^\leq - \underline{P}(Cl_t^\leq)$, $t = 1, ..., l - 1$, we have $D_P^-(x) = X_1 \bigcup X_2 \bigcup X_3$ subject to $X_1 \subseteq \underline{P}(Cl_t^\leq)$, $X_2 \subseteq Cl_t^\leq - \underline{P}(Cl_t^\leq)$, $X_3 \subseteq Cl_{t+1}^\geq$.

Lemma 1. *For $x \in Bn_P(Cl_t^\geq)$ (or $x \in Bn_P(Cl_t^\leq)$), the following assertions are valid:*

$$(a) \ |X_1| \geq 0; \ (b) \ |X_2| \geq 1; \ (c) \ |X_3| \geq 1,$$

where the number of objects in a set is denoted by $| \bullet |$.

Proof. We take $x \in Cl_t^\geq - \underline{P}(Cl_t^\geq)$ as example. For (a), it is given by nature. For (b), assuming $|X_2| = 0$, we get $D_P^+(x) \bigcap (Cl_t^\geq - \underline{P}(Cl_t^\geq)) = \emptyset$. Since we held $x \in D_P^+(x)$, we then infer $x \notin Cl_t^\geq - \underline{P}(Cl_t^\geq)$, which is contradictory to our premises: $x \in Cl_t^\geq - \underline{P}(Cl_t^\geq)$. Therefore, the assumption $|X_2| = 0$ does not hold. Finally, we obtain $|X_2| \geq 1$. For (c), assuming $|X_3| = 0$, we get $D_P^+(x) \bigcap Cl_{t-1}^\leq = \emptyset$. Since we held $U - Cl_{t-1}^\leq = Cl_t^\geq$, we then get $D_P^+(x) \subseteq Cl_t^\geq$. According to the definition of $\underline{P}(Cl_t^\geq)$, we then hold $x \in \underline{P}(Cl_t^\geq)$, which is contradictory to our premises : $x \in Cl_t^{\geq} - \underline{P}(Cl_t^{\geq})$. Therefore, the assumption $|X_3| = 0$ does not hold. Finally, we hold $|X_3| \geq 1$. For $x \in Cl_t^{\leq} - \underline{P}(Cl_t^{\leq})$, the proof is in the similar processing.

Based on these observations, we propose a new coefficient, called Believe Factor of upward and downward unions (*Believe Factor* for short). The definition is given as follows.

Definition 1. *For $x \in Cl_t^{\geq} - \underline{P}(Cl_t^{\geq}), t = 2, ..., l$, we have the believe factor of upward union of decision classes (upward believe factor for short):*

$$\beta(x \to Cl_t^{\geq}) = (\mu_t^{\geq}(x), \nu_t^{\geq}(x), \pi_t^{\geq}(x)) \ s.t. \ \mu_t^{\geq}(x) = \frac{|D_P^+(x) \cap \underline{P}(Cl_t^{\geq})|}{|D_P^+(x)|},$$

$$\nu_t^{\geq}(x) = \frac{|D_P^+(x) \cap Cl_{t-1}^{\leq}|}{|D_P^+(x)|}, \ \pi_t^{\geq}(x) = \frac{|D_P^+(x) \cap (Cl_t^{\geq} - \underline{P}(Cl_t^{\geq}))|}{|D_P^+(x)|} \ .$$

Definition 2. *For $x \in Cl_t^{\leq} - \underline{P}(Cl_t^{\leq})$, $t = 1, ..., l-1$, we have the believe factor of downward union of decision classes (downward believe factor, for short):*

$$\beta(x \to Cl_t^{\leq}) = (\mu_t^{\leq}(x), \nu_t^{\leq}(x), \pi_t^{\leq}(x)) \ s.t. \ \mu_t^{\leq}(x) = \frac{|D_P^-(x) \cap \underline{P}(Cl_t^{\leq})|}{|D_P^-(x)|},$$

$$\nu_t^{\leq}(x) = \frac{|D_P^-(x) \cap Cl_{t+1}^{\geq}|}{|D_P^-(x)|}, \ \pi_t^{\leq}(x) = \frac{|D_P^-(x) \cap (Cl_t^{\leq} - \underline{P}(Cl_t^{\leq}))|}{|D_P^-(x)|} \ .$$

Remark that the symbol "$\to$" in $\beta(x \to Cl_t^{\geq})$ and $\beta(x \to Cl_t^{\leq})$ can be understood as "be assigned to" or "belongs to". For object $x \in U$, $\mu(x)$ (including $\mu_t^{\geq}(x)$ and $\mu_t^{\leq}(x)$) is called *positive score* ; $\nu(x)$ (including $\nu_t^{\geq}(x)$ and $\nu_t^{\leq}(x)$) is called *negative score*; $\pi(x)$ (including $\pi_t^{\geq}(x)$ and $\pi_t^{\leq}(x)$) is called *hesitancy score*. The forms of upward/downward believe factors can be regarded as intuitionistic fuzzy values [4], [11].

Lemma 2. *For object $x \in Cl_t, t = 1, ..., l$, the following assertions are valid:*

$$\mu_t^{\geq}(x) + \nu_t^{\geq}(x) + \pi_t^{\geq}(x) = 1; \ \mu_t^{\leq}(x) + \nu_t^{\leq}(x) + \pi_t^{\leq}(x) = 1 \ .$$

Proof. It can be easily proved according to definition 1 and definition 2.

Lemma 3. $\beta(x \to Cl_t^{\geq}) = (\mu_t^{\geq}(x), \nu_t^{\geq}(x), \pi_t^{\geq}(x)) = (1, 0, 0)$ *is valid for $x \in \underline{P}(Cl_t^{\geq})$. $\beta(x \to Cl_t^{\leq}) = (\mu_t^{\leq}(x), \nu_t^{\leq}(x), \pi_t^{\leq}(x)) = (1, 0, 0)$ is valid for $x \in \underline{P}(Cl_t^{\leq})$.*

Proof. It can be easily proved according to definition 1 and definition 2.

3.2 Measurements

We introduce three measurements related to believe factor for uncertain rule induction.

Definition 3. *(Confidence degree) For object $x \in U$, the confidence degree of believe factor, denoted by $L(x)$, is defined by: $L(x) = \mu(x) + \pi(x)$, where $\mu(x)$ is positive score and $\pi(x)$ is hesitancy score. Specifically, we hold:*

$$L(x \rightarrow Cl_t^{\leq}) = \mu_t^{\leq}(x) + \pi_t^{\leq}(x); L(x \rightarrow Cl_t^{\geq}) = \mu_t^{\geq}(x) + \pi_t^{\geq}(x).$$

Definition 4. *(Believe degree) For object $x \in U$, the believe degree of believe factor, denoted by $S(x)$, is defined by: $S(x) = \mu(x) - \nu(x)$, where $\mu(x)$ is positive score and $\nu(x)$ is negative score. Specifically, we hold:*

$$S(x \rightarrow Cl_t^{\leq}) = \mu_t^{\leq}(x) - \nu_t^{\leq}(x); S(x \rightarrow Cl_t^{\geq}) = \mu_t^{\geq}(x) - \nu_t^{\geq}(x).$$

Definition 5. *(Accuracy degree) For object $x \in U$, the accuracy degree of believe factor, denoted by $H(x)$, is defined by: $H(x) = \mu(x) + \nu(x)$, where $\mu(x)$ is positive score and $\nu(x)$ is negative score. Specifically, we hold:*

$$H(x \rightarrow Cl_t^{\leq}) = \mu_t^{\leq}(x) + \nu_t^{\leq}(x); H(x \rightarrow Cl_t^{\geq}) = \mu_t^{\geq}(x) + \nu_t^{\geq}(x).$$

3.3   Believable Rule Induction

Given a decision table, each object x from U has a *decision description* in terms of the evaluations on the considered criteria: $Des_P(x) = [f_{q_1}(x), ..., f_{q_n}(x)]$, where information function $f_q(x) \in V_q$, for $V = \bigcup_{q \in P} V_q$, $q \in P \subseteq C$. We say each $Des_P(x)$ is able to induce an uncertain rule based on *cumulated preferences*. Considering $Des_P(x)$ of boundary object x which is coming from $Bn_P(Cl_t^{\geq})$, there are two kinds of decision descriptions in the separated rough boundary regions as: $Des_P(x) = [r_{q_1}^{\geq}, r_{q_2}^{\geq}, ..., r_{q_n}^{\geq}]$, for $x \in Cl_t^{\geq} - \underline{P}(Cl_t^{\geq})$; $Des_P(x) = [r_{q_1}^{\leq}, r_{q_2}^{\leq}, ..., r_{q_n}^{\leq}]$, for $x \in Cl_{t-1}^{\leq} - \underline{P}(Cl_{t-1}^{\leq})$; where $(Cl_t^{\geq} - \underline{P}(Cl_t^{\geq})) + (Cl_{t-1}^{\leq} - \underline{P}(Cl_{t-1}^{\leq})) = Bn_P(Cl_t^{\geq}) = Bn_P(Cl_{t-1}^{\leq})$.

With this in mind, the boundary objects carry the *valuable* uncertain information for decision making on the following conditions: (1) Considering the believe factor of object $x \in Cl_t^{\geq} - \underline{P}(Cl_t^{\geq})$, if believe degree $S(x \rightarrow Cl_t^{\geq}) > 0$, we say object x carries the *believable* decision information as: Providing the assignment to class union $Cl_t^{\geq}$ in some degree. (2) Considering the believe factor of object $x \in Cl_{t-1}^{\leq} - \underline{P}(Cl_{t-1}^{\leq})$, if believe factor $S(x \rightarrow Cl_{t-1}^{\leq}) > 0$, we say object x carries the *believable* decision information as: Providing the assignment to class union $Cl_{t-1}^{\leq}$ in some degree.

The boundary objects satisfying the above conditions are called *valuable* objects. The induced uncertain rules on the basis of these *valuable* objects are called *believable rules*. In the following, the strategies are given in order to induce a set of believable rules.

Strategy I (Upward believable rule): Considering the object x_i from the separated boundary region $Cl_t^{\geq} - \underline{P}(Cl_t^{\geq})$, if $S(x_i \rightarrow Cl_t^{\geq}) = \mu_t^{\geq}(x_i) - \nu_t^{\geq}(x_i) > 0$ is satisfied, we then induce an upward believable rule $BR_t^{\geq}$ based on the decision description $Des_P(x_i) = [r_{q_1}^{\geq}, r_{q_2}^{\geq}, ..., r_{q_n}^{\geq}]$: If $f_{q_1}(x) \geq r_{q_1}^{\geq}$ and $f_{q_2}(x) \geq r_{q_2}^{\geq} ... f_{q_n}(x) \geq r_{q_n}^{\geq}$, then $x \in Cl_t^{\geq}$, which is with three measuring degrees: $L(x_i \rightarrow Cl_t^{\geq})$, $S(x_i \rightarrow Cl_t^{\geq})$ and $H(x_i \rightarrow Cl_t^{\geq})$.

Strategy II (Downward believable rule): Considering the object x_i from the separated boundary region $Cl_{t-1}^{\leq} - \underline{P}(Cl_{t-1}^{\leq})$, if $S(x_i \rightarrow Cl_{t-1}^{\leq}) = \mu_{t-1}^{\leq}(x_i) -$

$v_{t-1}^{\leq}(x_i) > 0$ is satisfied, we then induce a downward believable rule $BR_{t-1}^{\leq}$ based on the decision description $Des_P(x_i) = [r_{q_1}^{\leq}, r_{q_2}^{\leq}, ..., r_{q_n}^{\leq}]$: If $f_{q_1}(x) \leq r_{q_1}^{\leq}$ and $f_{q_2}(x) \leq r_{q_2}^{\leq}...f_{q_n}(x) \leq r_{q_n}^{\leq}$, then $x \in Cl_{t-1}^{\leq}$, which is with three measuring degrees: $L(x_i \to Cl_{t-1}^{\leq})$, $S(x_i \to Cl_{t-1}^{\leq})$ and $H(x_i \to Cl_{t-1}^{\leq})$.

4 Illustrative Example

4.1 Decision Table and Rough Approximations

In this section, we use an example to illustrate the application of believable rule for multicriteria sorting (also known as ordinal classification). We use synthetic data set as shown in Table 1. We consider that the decision table is monotonic, which means a better decision value on condition criteria tends to contribute a better assignment in decision class, rather than the worse one, or vice versa. The decision information is summarized below: object set $\{S1, S2,...,S50\}$; condition criterion set $\{A, B, C\}$; single decision criterion $\{D\}$; decision values scale $[1, 2, 3, 4, 5]$, where the larger number is superior to the smaller one according to DM's preference; decision class scale [III, II, I], where Class I is superior than Class II, and then Class III, denoted by Class I $=Cl_3$; Class II$=Cl_2$; Class III$=Cl_1$.

Table 1. Decision table

Object	A	B	C	D	Object	A	B	C	D	Object	A	B	C	D
S 1	3	4	3	I	S 18	3	4	3	I	S 35	2	3	3	III
S 2	4	3	3	I	S 19	5	2	4	II	S 36	1	2	3	III
S 3	5	3	4	I	S 20	3	4	2	II	S 37	2	3	3	III
S 4	5	3	4	I	S 21	4	2	3	II	S 38	2	2	3	III
S 5	5	4	3	I	S 22	5	2	4	II	S 39	1	3	2	III
S 6	3	4	3	I	S 23	5	2	4	II	S 40	1	2	3	III
S 7	5	3	3	I	S 24	1	4	2	II	S 41	2	3	2	III
S 8	1	3	3	I	S 25	1	4	2	II	S 42	1	3	2	III
S 9	4	3	4	I	S 26	2	4	3	II	S 43	1	3	2	III
S 10	4	3	4	I	S 27	3	4	2	II	S 44	2	3	2	III
S 11	4	4	3	I	S 28	1	4	2	II	S 45	3	2	3	III
S 12	1	3	3	I	S 29	1	4	2	II	S 46	4	2	3	III
S 13	3	4	3	I	S 30	2	1	3	II	S 47	3	2	3	III
S 14	4	3	3	I	S 31	3	2	4	II	S 48	5	3	3	III
S 15	5	3	4	I	S 32	3	2	4	II	S 49	3	2	3	III
S 16	5	3	4	I	S 33	5	2	4	II	S 50	3	2	3	III
S 17	5	4	3	I	S 34	1	3	2	III					

Each object belongs to *one and only one* decision class. The upward and downward unions of decision classes are given as: $Cl_1^{\leq} = Cl_1$; $Cl_2^{\leq} = Cl_1 \bigcup Cl_2$; $Cl_2^{\geq} = Cl_3 \bigcup Cl_2$; $Cl_3^{\geq} = Cl_3$; $Cl_1^{\geq} = Cl_1^{\leq} = Cl_1 \bigcup Cl_2 \bigcup Cl_3$.

According to the strict dominance principle, we can obtain the C-lower approximation. Then, we can further obtain the separated boundary regions as: $Cl_3^{\geq}-\underline{C}(Cl_3^{\geq})=\{S2; S7; S8; S12; S14\}; Cl_2^{\geq}-\underline{C}(Cl_2^{\geq})=\{S2; S7; S8; S12; S14; S30;$ $S21\}; Cl_1^{\leq}-\underline{C}(Cl_1^{\leq})=\{S35; S37; S48; S50; S49; S47; S46; S45; S38\}; Cl_2^{\leq}-\underline{C}(Cl_2^{\leq})$ $=\{S35; S37; S48; S26\}.$

4.2   Believable Rule Induction

We calculate the believe factor of each object which is from the rough boundary regions, as shown in Table 2. In this table, the believe degree of S46 is equal to zero rather than a positive value. Thus, S46 is not a valuable object, and it is unable to provide any assignment for decision-making. Excluding S46, other objects are all valuable objects and are able to induce believable rules. According to Strategy I and Strategy II in section 3.3, we generate the believable rules together with their measurements, as shown in Table 3.

Table 2. Believe factor of rough boundary objects

Regions	Boundary objects	Believe factors			Measurements		
		$\mu(x)$	$\pi(x)$	$\nu(x)$	$S(x)$	$H(x)$	$L(x)$
$Cl_3^{\geq} - \underline{P}(Cl_3^{\geq})$ S2; S14		9/13	3/13	1/13	8/13	10/13	12/13
S7		6/8	1/8	1/8	5/8	7/8	7/8
S8; S12		13/22	5/22	4/22	9/22	17/22	18/22
$Cl_2^{\geq} - \underline{P}(Cl_2^{\geq})$ S2; S14		9/13	3/13	1/13	8/13	10/13	12/13
S7		6/8	1/8	1/8	5/8	7/8	7/8
S8; S12		14/22	5/22	3/22	11/22	17/22	19/22
S21		13/19	4/19	2/19	11/19	15/19	17/19
S30		20/34	5/34	9/34	11/34	29/34	25/34
$Cl_1^{\leq} - \underline{P}(Cl_1^{\leq})$ S35; S37		8/14	3/14	3/14	5/14	11/14	11/14
S38		2/4	1/4	1/4	1/4	3/4	3/4
S45; S47; S49; S50		2/8	5/8	1/8	1/8	3/8	7/8
S46		**2/10**	**6/10**	**2/10**	**0/10**	**4/10**	**8/10**
S48		8/24	9/24	7/24	15/24	15/24	17/24
$Cl_2^{\leq} - \underline{P}(Cl_2^{\leq})$ S26		14/19	3/19	2/19	12/19	16/19	17/19
S35; S37		10/14	2/14	2/14	8/14	12/14	12/14
S48		16/24	3/24	5/24	11/24	21/24	19/24

4.3   Verification of Sorting Capability

This section aims to verify the sorting capability of our induced rules. We choose existing representative DRSA models as competitors, including C-DRSA model [8], VC-DRSA model [7], and the extended scheme [2] of DRSA models. Hereinto, C-DRSA can be regarded as consistency level L=1.0. VC-DRSA can be denoted as L<1.0, i.e. L=0.9, L=0.8, L=0.7, etc. The extended scheme is with the symbol ^, i.e. $\hat{L}= 1.0, \hat{L}= 0.9, \hat{L}= 0.8$, etc.

Table 3. Induction of believable decision rules

Believable rules	Conditional criteria			Assign-ments	Confidence degree	Accuracy degree	Base(s) of rules
	A	B	C				
[B1]	≥ 4	≥ 3	≥ 3	$\geq$I	0.9231	0.7692	S2; S14
[B2]	≥ 5	≥ 3	≥ 3	$\geq$I	0.8750	0.8750	S7
[B3]	≥ 1	≥ 3	≥ 3	$\geq$I	0.8182	0.7727	S8; S12
[B4]	≥ 4	≥ 3	≥ 3	$\geq$II	0.9231	0.7692	S2; S14
[B5]	≥ 5	≥ 3	≥ 3	$\geq$II	0.8750	0.8750	S7
[B6]	≥ 1	≥ 3	≥ 3	$\geq$II	0.8636	0.7727	S8; S12
[B7]	≥ 4	≥ 2	≥ 3	$\geq$II	0.8947	0.7895	S21
[B8]	≥ 2	≥ 1	≥ 3	$\geq$II	0.7353	0.8529	S30
[B9]	≤ 2	≤ 3	≤ 3	$\leq$III	0.7857	0.7857	S35; S37
[B10]	≤ 2	≤ 2	≤ 3	$\leq$III	0.7500	0.7500	S38
[B11]	≤ 3	≤ 2	≤ 3	$\leq$III	0.8750	0.3750	S45; S47; S49; S50
[B12]	≤ 5	≤ 3	≤ 3	$\leq$III	0.7083	0.6250	S48
[B13]	≤ 2	≤ 4	≤ 3	$\leq$II	0.8947	0.8421	S26
[B14]	≤ 2	≤ 3	≤ 3	$\leq$II	0.8571	0.8571	S35; S37
[B15]	≤ 5	≤ 3	≤ 3	$\leq$II	0.7917	0.8750	S48

Table 4 illustrates the statistical results of assignments in multicriteria sorting. Through generated certain rules, sorting rates are given in the first six rows of this table. The proposed believable rules together with the induced certain rules (via C-DRSA) are also tested for sorting. The comparison result indicates that our proposal provides the highest *correct sorting rate* (the ratio of the number of correctly classified objects over the total number of testing objects), which equals to 0.94.

Table 4. A comparison of correct sorting rate

Alternative Proposals	Correctly Sorted objects		Incorrectly Sorted objects		Unknown objects	
1. L=1.0	42	84%	0	0%	8	16%
2. L=0.9	45	90%	1	2%	4	8%
3. L=0.8	41	82%	2	4%	7	14%
4. Ĺ=1.0	36	72%	6	12%	8	16%
5. Ĺ=0.9	43	86%	5	10%	2	4%
6. Ĺ=0.8	42	84%	8	16%	0	0%
7. Our proposal	47	94%	3	6%	0	0%

5 Conclusion

This paper provided a new idea for extracting uncertain information within rough boundary regions. In terms of intuitionistic fuzzy values, we proposed a new coefficient, called believe factor, together with three measuring degrees. Aided by these measurements, we further provided the method for inducing believable rule as a new kind of uncertain rule for sorting problems. In the experimental testing, we illustrated the process of believable rule induction, and verified its sorting capability via comparison with other representative proposals. In the future work, we shall develop this approach for multicriteria ranking and expand its ability for prediction. We would have been investigating some real-world applications using the proposed approach like supplier chain management (i.e. supplier selection).

References

1. Atanassov, K.: Intuitionistic fuzzy sets. Fuzzy Sets and Systems 20, 87–96 (1986)
2. Blaszczynski, J., Greco, S., Slowinski, R.: Multi-criteria classification: a new scheme for application of dominance-based decision rules. European Journal of Operational Research 181, 1030–1044 (2007)
3. Chai, J.Y., Liu, J.N.K.: Class-based rough approximation with dominance principle. In: Proceedings of IEEE International Conference on Granular Computing (GrC), pp. 77–82 (2011)
4. Chai, J.Y., Liu, J.N.K., Xu, Z.S.: A new rule-based SIR approach to supplier selection under intuitionistic fuzzy environments. International Journal of Uncertainty, Fuzziness, and Knowledge-Based Systems 20(3) (2012)
5. Chang, B., Hung, H.F.: A study of using RST to create the supplier selection model and decision-making rules. Expert Systems with Applications 37, 8284–8295 (2010)
6. Cyran, K.A.: Quasi Dominance Rough Set Approach in Testing for Traces of Natural Selection at Molecular Level. In: Cyran, K.A., Kozielski, S., Peters, J.F., Stańczyk, U., Wakulicz-Deja, A. (eds.) Man-Machine Interactions. AISC, vol. 59, pp. 163–172. Springer, Heidelberg (2009)
7. Greco, S., Matarazzo, B., Słowiński, R., Stefanowski, J.: Variable Consistency Model of Dominance-Based Rough Sets Approach. In: Ziarko, W.P., Yao, Y. (eds.) RSCTC 2000. LNCS (LNAI), vol. 2005, pp. 170–181. Springer, Heidelberg (2001)
8. Greco, S., Matarazzo, B., Slowinski, R.: Decision rule approach. In: Figueira, J., Greco, S., Ehrgott, M. (eds.) Multiple Criteria Decision Analysis: State of the Art Surveys, pp. 507–561. Springer, Berlin (2005)
9. Inuiguchi, M., Yoshioka, Y., Kusunoki, Y.: Variable-precision dominance-based rough set approach and attribute reduction. International Journal of Approximate Reasoning 50, 1199–1214 (2009)
10. Szmidt, E., Kacprzyk, J.: Distances between intuitionistic fuzzy sets. Fuzzy Sets and Systems 114, 505–518 (2000)
11. Xu, Z.S.: Intuitionistic fuzzy aggregation operators. IEEE Transactions on Fuzzy Systems 15, 1179–1187 (2007)
12. Xu, Z.S., Cai, X.Q.: Recent advances in intuitionistic fuzzy information aggregation. Fuzzy Optimization and Decision Making 9(4), 359–381 (2010)
13. Zhang, Z.W., Shi, Y., Gao, G.X.: A rough set-based multicriteria criteria linear programming approach for the medical diagnosis and prognosis. Expert Systems with Applications 36(5), 8932–8937 (2009)

A New Rule Induction Method
from a Decision Table
Using a Statistical Test

Tsukasa Matsubayashi[1], Yuichi Kato[1], and Tetsuro Saeki[2]

[1] Shimane University,
1060 Nishikawatsu-cho, Matsue city, Shimane 690-8504, Japan
ykato@cis.shimane-u.ac.jp
[2] Yamaguchi University,
2-16-1 Tokiwadai, Ube city, Yamaguchi 755-8611, Japan
tsaeki@yamaguchi-u.ac.jp

Abstract. Rough Sets theory provides a method of estimating and/or inducing knowledge structure of if-then rules from various databases, using approximations of accuracy and coverage indices. Several recent studies have examined the confidence of these indices. In these studies their estimated rules were based on a sample data set obtained from a population, and the sampling affects the confidence of the estimation. However, these studies of the quality of the approximation evaluate the effects on rule estimation indirectly. In this paper, we propose a new rule induction method by statistical testing which directly contains the effect of sampling. The validity and usefulness of our method are confirmed by a computer simulation experiment and comparison of the results with those by other well-known methods.

1 Introduction

Data mining from databases aims at extracting implicit knowledge contained within the database. Rough Sets theory introduced by Pawlak[1] has been studied as an object of mathematical interest, and provides areas such as machine learning and data mining with many useful analytical methods. One of important tasks of rough sets is to compute the attribute reduct, which provides the construction by the minimum number of attributes, maintaining knowledge of the decision table obtained from the database of interest. Another task is to extract if-then rules from the decision table, by the use of approximations such as accuracy and coverage indices. This is useful for diagnosis systems of diseases, discrimination problems, and other aspects. The theoretical studies and practical algorithms for the above two tasks are given in reference [2–7].

On the other hand, data mining by rough sets is studied from the statistical viewpoint because the results of the two tasks are highly dependent on the decision table sampling the population, and the results from one sample will differ from those from another. Several considerations for this statistical problem have

T. Li et al. (Eds.): RSKT 2012, LNAI 7414, pp. 81–90, 2012.
© Springer-Verlag Berlin Heidelberg 2012

Table 1. An example of a decision table

U	$C(1)$	$C(2)$	$C(3)$	$C(4)$	$C(5)$	$C(6)$	D
1	1	1	4	3	6	1	1
2	4	1	1	5	4	4	2
3	6	4	6	1	6	1	1
4	2	2	4	3	1	1	2
...	...	...	...	...	...	...	...
$N-1$	2	2	1	1	6	4	2
N	4	2	3	6	3	5	1

been studied as a problem of evaluating the confidence coefficient of accuracy and coverage indices [8–10]. Jaworski[11] summarized these studies, identified the problems so far, and proposed a confidence coefficient with significance γ_n for those indices in an extended approximation which was introduced into the population of the decision table sampled.

However, previous discussions on the confidence of those indices were based on indirect effects on the rule estimation, and the degree of confidence by γ_n is much harder to understand. Consequently, we here propose a new rule estimation method by use of a statistical hypothesis test that directly extracts if-then rules using the sampled data. Specifically, we regard the rule estimation problem by use of the decision table as the problem of identifying a black box containing rules. The inputs and outputs of the box are the former part of an if-then rule and the latter part of the rule, respectively. We conduct a preliminary experiment in a white box which specifies if-then rules in advance, generates the input of the former part randomly, and decides the output by use of the specified rules. From this preliminary experiment, we found that the decision table basically contains two types of data set. The first is the data set controlled by the rules in the rule box, and the second is the data set not to be applied to the rules, and the output is obtained by chance. Accordingly, we propose the following rule estimation strategy: (1) We set up the null hypothesis that a trying rule as a test does not exist in the black box of the rule box. (2) We test the hypothesis using the data set of the decision table sampling the population, and decide whether the hypothesis is true or not. (3) We repeat procedures 1 and 2 changing the trying rule systematically and efficiently, and estimate the rules in the black box. The validity and usefulness is confirmed by comparing the results from our proposed method with those by other well-known methods in a simulation experiment.

2   Model of Data Generation

The decision table shown in Table 1 is first obtained. This table is conventionally denoted with $S = (U, A = C \cup D, V, \rho)$. Here, $U = \{u(i)|i = 1, ..., |U| = N\}$ is a sample set, A is an attribute set, $C = \{C(j)|j = 1, ..., |C|\}$ is a condition attribute set, $C(j)$ $(j = 1, ..., |C|)$ is the member of C and a condition attribute,

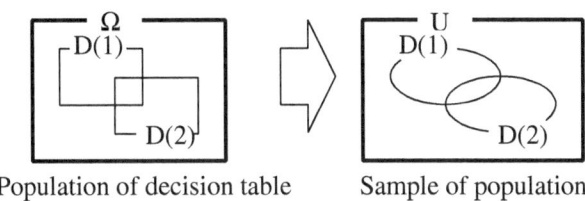

Population of decision table   Sample of population

Fig. 1. Relationship between population and sample data set: A decision table is obtained from the population Ω

Fig. 2. Relationship between the condition attribute value and the decision attribute value

and D is a decision attribute. V is a set of attribute values and denoted with $V = \cup_{a \in A} V_a$ and is characterized by an information function $\rho : U \times A \rightarrow V$. In this example, if $a = C(j) \in A$ ($j = 1, ..., |C|$) then $V_a = \{1, 2, ..., 6\}$ and if $a = D$ then $V_a = \{1, 2\}$. This decision table was obtained from a population Ω as a sample data set, as shown in Fig. 1.

On the other hand, this sampling of the population is regarded as the input-output system shown in Fig. 2. The input is $u^C(i)$, which is the condition attributes' value of $u(i)$; and the output is $u^D(i)$, which is decided by use of the rules in the rule box corresponding to the input. The rule box contains several if-then rules, such as:

if $R(k)$ then $D = d(k)$ ($\in V_D$), where $R(k) = \wedge_{j_k} (C(j_k) = c(l_k)$ ($\in V_{C(j_k)}$)) ($k = 1, 2, ...$)

The aim of this paper is to propose a method and an effective algorithm to directly estimate the above if-then rules using the data set $(u^C(i), u^D(i))$ ($i = 1, ..., N$); that is, the decision table observed.

The following three cases occur with regards to the input-output relationship:

Case 1) $u^C(i)$ can be applied to $R(k)$, and $u^D(i)$ is uniquely determined as $D = d(k)$.

Case 2) $u^C(i)$ can not be applied to any $R(k)$, and $u^D(i)$ can only be determined randomly.

Case 3) $u^C(i)$ can be applied to several $R(k)$ ($k = k1, k2, ...$), and their outputs of $u^C(i)$ conflict with each other. Accordingly, the output of $u^C(i)$ must be randomly determined from the conflicted outputs.

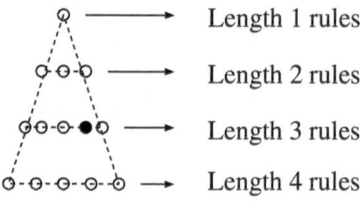

Fig. 3. Image of *Pyramid* and *rePyramid* (∘: $r(k) \in Pyramid(l)$, •: *rePyramid*)

Case 2) is an indifferent one for the system, and **case 3)** is an insistent one. In real-life programs, the above three specifications of generating $u(i)$ do not seem to be unusual. For example, let the following rules be specified:

R1: if $C(1) = 1 \wedge C(2) = 1 \vee C(3) = 1 \wedge C(4) = 1$ then $D = 1$
R2: if $C(1) = 2 \wedge C(2) = 2 \vee C(3) = 2 \wedge C(4) = 2$ then $D = 2$

Here, $\wedge$ is conjunction and $\vee$ is disjunction. The **case 1)** is $u(1)$ and $u(4)$, **2)** is $u(2)$, $u(3)$ and $u(N)$ and **3)** is $u(N-1)$ in Table 1.

3   Proposal of Rule Estimation Method by Statistical Test

We generated $u^C(i)$ by use of random numbers with the uniform distribution at each position of the attribute by 10000 and determined $u^D(i)$ by use of $R1$ and $R2$ of the above example rules and the above three cases for determining $u^D(i)$. We deleted the same $u(i)$, obtained the different ones by $N_{actual} = 9482$, and arranged them as in Table 1. We use this decision table to examine the way of estimating rules in the rule box, and to propose a new rule estimation method. For this table, we count the frequency of the samples $u^{D=1}(i)$ and $u^{D=2}(i)$ respectively which have $u^C(i) = r = \wedge_j(C(j) = V_{C(j)})$ and arrange them as $(|\{u^{D=1}(i)\}| = n1, |\{u^{D=2}(i)\}| = n2)$.

Table 2 shows some examples of r and its frequency $(n1, n2)$. For example, $r(1)$ in Table 2 means that if $R(k)(=r(1))$ is $C(1) = 1$ (hereafter we denote this $R(k)$ "100000" for the sake of convenience and call the rule length 1, where "0" denotes any value of at the position of the attribute), then it has $(n1, n2) = (920, 653)$; $r(11)$ means that if $R(k)(=r(11))$ is $C(1) = 1 \wedge C(2) = 1$ (= "110000" and the rule length 2) then it has $(n1, n2) = (265, 3)$; $r(17)$ means that if $R(k)(=r(17))$ is $C(1) = 1 \wedge C(2) = 1 \wedge C(6) = 4$ (= "110004" and the rule length 3) then it has $(n1, n2) = (61, 1)$, and so on. Table 2 shows that:

A1) The longer the rule length gets, the fewer the samples the rule can apply to.
A2) In this case, r containing the part of the rules existing in the rule box where $R1$ and $R2$ are specified causes a partiality of the distribution of $(n1, n2)$. For example, see $(n1, n2)$ of $r(1)$, $r(11)$ and $r(17)$, which result from **Case 1)** and **Case 3)**.

Table 2. An example of trying rule $r(k)$ and the corresponding frequency of their decision attribute values (z is a standard score and $(n1, n2) = (|\{u^{D=1}(i)\}|, |\{u^{D=2}(i)\}|)$ for the corresponding trying rule, and "0" denotes any value of $C(j)$)

trying rule $r(k)$	$C(1)$	$C(2)$	$C(3)$	$C(4)$	$C(5)$	$C(6)$	$(n1, n2)$	z
1	1	0	0	0	0	0	(920, 653)	6.757
2	2	0	0	0	0	0	(680, 949)	-6.640
3	3	0	0	0	0	0	(788, 783)	0.151
4	4	0	0	0	0	0	(821, 784)	0.949
5	5	0	0	0	0	0	(777, 805)	-0.679
6	6	0	0	0	0	0	(723, 799)	-1.922
7	0	1	0	0	0	0	(920, 683)	5.944
8	0	2	0	0	0	0	(639, 956)	-7.912
9	0	3	0	0	0	0	(773, 786)	-0.304
10	0	4	0	0	0	0	(802, 803)	0.000
11	1	1	0	0	0	0	(265, 3)	16.065
12	1	2	0	0	0	0	(131, 127)	0.311
13	1	3	0	0	0	0	(133, 127)	0.434
14	1	4	0	0	0	0	(123, 143)	-1.165
15	1	5	0	0	0	0	(133, 122)	0.751
16	1	6	0	0	0	0	(135, 131)	0.307
17	1	1	0	0	0	4	(61, 1)	7.747
18	1	0	0	0	4	5	(39, 21)	2.453
19	1	0	0	0	6	6	(33, 27)	0.904
20	2	2	0	3	0	0	(0, 61)	-7.682
21	2	4	0	0	0	4	(28, 32)	-0.387
22	2	6	0	0	6	0	(40, 22)	2.413
23	2	0	1	0	2	0	(32, 30)	0.381
24	2	0	5	0	0	2	(29, 31)	-0.129
25	2	0	6	2	0	0	(26, 34)	-0.904
26	2	0	0	6	0	4	(23, 41)	-2.125
27	3	0	4	0	0	2	(25, 39)	-1.625

A3) The partiality of the example in **A2)** has a related tendency since their rules are combined by the inclusion relationship; $100000 \supseteq 110000 \supseteq 110004$.

A4) $(n1, n2) = (256, 3)$ of 110000 and $(61, 1)$ of 110004 have the inconsistent samples of three and one, respectively (cf. **Case 3)**).

A5) An indifferent rule r which does not have any part of the rules existing in the rule box does not cause a severe partiality. See $(n1, n2)$ of $r(3)$, $r(27)$ and on the like which result from **Case2)**.

A6) To summarize **A2)–A5)**, the partiality of $(n1, n2)$ can be utilized to determine whether r contains the part of the rules in the rule box or not, regardless of the inconsistency of r.

The above experimental knowledge leads to the following strategies for estimating rules in the rule box by use of the decision table:

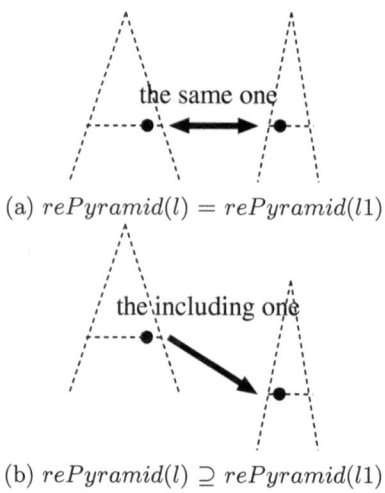

(a) $rePyramid(l) = rePyramid(l1)$

(b) $rePyramid(l) \supseteq rePyramid(l1)$

Fig. 4. Reduction of $rePyramid$

1) Specify the following null hypothesis $H0$ and alternative hypothesis $H1$ (cf. **A6**)):
H0: a trying rule r for test is not one of $R(k)$ $(k = 1, 2, ...)$.
H1: a trying rule r for test is one of $R(k)$ $(k = 1, 2, ...)$.
2) Make a trying rule r of the rule length 1 and count its frequency $(n1, n2)$.
3) Test $H0$ by use of $(n1, n2)$. Specifically, select a proper test statistic which depends on the problem under consideration. For the problem with two decision attribute values, the following z statistic can be adopted as one of methods:

$$z = \frac{(n1 + 0.5 - np)}{\sqrt{p(1 - p)}},$$

here $p = 0.5$ (cf. **Case 2**)), and $n1$ obeys a binominal distribution. If $n1 + n2 \geq N0 = 20$ then z obeys the standard normal distribution. Then specify the statistical rule that if $|z| \geq 3.0$ then $H0$ is rejected and $H1$ is adopted; this specifies the level of significance less than 1 [%]. The value of z also is shown in Table 2. If $z \geq 3.0$ then the trying rule r is the rule for $D = 1$, and if $z \leq -3.0$ then the trying rule r is the rule for $D = 2$.
4) Reserve the trying rules once when $H0$ is rejected (cf. **A6**)).
5) If procedures from **2)** to **4)** for all of r with rule length 1 have been completed, repeat the same procedures with the increased length rules until the condition $n1 + n2 \geq N0 = 20$ does not hold (cf. **A1**)).
6) Classify the reserved rules into several rule sets by use of the order relation $\supseteq$ (cf. **A3**)). Denote the class with $Pyramid(l)$ $(l = 1, 2, ...)$, and then the following features hold in this class:
$\forall r(k) \in Pyramid(l) \Rightarrow \exists r(k1) \in Pyramid(l): r(k) \subseteq r(k1)$ or $r(k) \supseteq r(k1)$.
Accordingly, $Pyramid(l)$ forms an upper semi-lattice.

```
int main(void) {
  int rule[|C|]={0,...,0}; //initialize trying rules
  int tail=-1; //initial vale set
  input data; // set decion table
  rule_check(tail,rule); // 1)-5) strategies
  make Pyramid(l) (l=1,2,...) so that every r(k) belongs to one Pyramid at least; // strategy 6)
  make rePyramid(l) (l=1,2,...); // strategy 7)
  reduce rePyramid; // strategy 8)
} // end of main

int rule_check(int tail,int rule[|C|]) {
  for (ci=tail+1; cj<|C|; ci++) {
    for (cj=1; cj<=|C[ci]|; cj++) {
      rule[ci]=cj; // a trying rule sets for test
      count frequency of the trying rule; // count n1 n2
      if (frequency>=NO) { //sufficient frequency ?
        if (|z|>3.0) { //sufficient evidence ?
          store necessary data such as rule, frequency of n1 and n2, and z
        } // end of if |z|
        rule_check(ci,rule);
      } // end of if frequency
    } // end of for cj
    rule[ci]=0; // trying rules reset
  } // end of for ci
} // end of rule_check
```

Fig. 5. An algorithm for STRIM (Statistical Test Rule Induction Method)

7) Find $r(k)$ in $Pyramid(l)$ which has the maximum of $|z|$ in $Pyramid(l)$ and denote it with $rePyramid(l)$ (representation of $Pyramid(l)$).

8) Let $rePyramid(l')$ with the maximum of $|z|$ represent $rePyramide(l)$ if the relationship $\supseteq$ exists between $rePyramid(l)$ ($l = 1, 2, ...$), and determine the final rule estimation result $rePyramid(l)$ ($l = 1, 2, ...$).

Figures 3 and 4 show the image of strategy **7)** and **8)** respectively. Figure 4 (a) shows an example of the same $rePyramid$, and Fig. 4 (b) shows an example of $rePyramids$ with relationship $\supseteq$.

4 Algorithm for Rule Induction

Figure 5 shows an algorithm for the STRIM (Statistical Test Rule Induction Method) described in C language style, based on the considerations in 3. Strategies from **1)** to **5)** are executed in a function `rule_check()` recursively. This function explores trying rules for tests satisfying the condition $n1 + n2 \geq N0$ without overlapping. This part of the example is shown in Table 2.

5 Experimental Studies

We developed the algorithm shown in Fig. 5 into a piece of software using C language, and implemented it in a personal computer to compare the execution time [sec] and the results of estimated rules with those of existing LEM2[12] and FDMM software[7] which quickly execute the decision matrix method [6], using three sets of the decision table generated by the procedures shown in 3.

Table 3. Three cases of comparisons of the time and the rule length on reducing rules between FDMM, LEM2 and STRIM (N_{actual}: actual data number)

Case No. Data Number (N_{actual})	Method	Reducing Time [sec]	Number of Rule Length						
			1	2	3	4	5	6	Total
Case 1	FDMM	100	0	0	48	631	3650	35	4364
10000	LEM2	7061	0	0	47	523	3819	35	4424
(9482)	STRIM	5	0	4	1	0	0	0	5
Case 2	FDMM	101	0	0	51	609	3751	37	4448
10000	LEM2	8504	0	0	47	527	3920	37	4531
(9417)	STRIM	5	0	4	5	0	0	0	9
Case 3	FDMM	101	0	0	45	602	3770	42	4459
10000	LEM2	7059	0	0	43	501	3945	42	4531
(9432)	STRIM	5	0	4	2	0	0	0	6

Table 4. Results of estimated rules for three cases by STRIM

Case No. Data Number (N_{actual})	estimated rule $R(i)$	$C(1)$	$C(2)$	$C(3)$	$C(4)$	$C(5)$	$C(6)$	D	$(n1, n2)$	$p - value(z)$	accuracy	coverage
	1	1	1	0	0	0	0	1	(265, 3)	0(16.07)	0.989	0.056
Case 1	2	2	2	0	0	0	0	2	(5, 292)	0(-16.60)	0.983	0.061
10000	3	0	0	1	1	0	0	1	(264, 6)	0(15.76)	0.978	0.056
(9482)	4	0	0	2	2	0	0	2	(5, 223)	0(-14.37)	0.978	0.047
	5	5	3	0	0	5	0	2	(9, 29)	1.03e-3(-3.08)	0.763	0.006
	1	1	1	0	0	0	0	1	(244, 2)	0(15.49)	0.992	0.052
	2	2	2	0	0	0	0	2	(5, 258)	0(-15.54)	0.981	0.055
	3	0	0	1	1	0	0	1	(252, 4)	0(15.56)	0.984	0.054
Case 2	4	0	0	2	2	0	0	2	(2, 235)	0(-15.07)	0.992	0.050
10000	5	4	6	0	5	0	0	1	(29, 11)	1.33e-3(3.00)	0.725	0.006
(9417)	6	6	0	0	3	0	6	2	(11, 32)	1.14e-3(-3.05)	0.744	0.007
	7	6	0	0	0	3	5	1	(29, 8)	1.49e-3(3.62)	0.784	0.006
	8	0	4	5	0	0	6	1	(35,13)	4.51e-4(3.32)	0.729	0.007
	9	0	0	4	5	0	6	1	(24, 7)	6.12e-4(3.23)	0.774	0.005
	1	1	1	0	0	0	0	1	(251, 3)	0(15.62)	0.988	0.054
Case 3	2	2	2	0	0	0	0	2	(3, 271)	0(-16.13)	0.989	0.057
10000	3	0	0	1	1	0	0	1	(212, 7)	0(13.92)	0.968	0.045
(9432)	4	0	0	2	2	0	0	2	(1, 251)	0(-15.69)	0.996	0.053
	5	6	5	0	4	0	0	1	(39, 16)	6.06e-4(3.24)	0.709	0.008
	6	0	0	6	0	2	1	2	(13, 35)	1.24e-3(-3.03)	0.729	0.007

Table 3 shows the comparison of the execution time of the rule induction between the three methods when using a PC with a Celeron(R) CPU with 2.67 [GHz] clock speed and 992 [MB] of RAM memory. The times shown in the table are truncated to the nearest second. LEM2 and FDMM use the lower approximation (that is, the accuracy index is one), while STRIM needs no such approximation. To compare the time of the rule induction in the three cases, FDMM is faster than LEM2 by one digit, and our proposed STRIM is faster than FDMM by one digit. That is, STRIM is much faster than LEM2 and FDMM. The reasons for these differences seem to be that LEM2 or FDMM needs the set or logical operations and implements them in the computer, whereas STRIM spends most of the time in calculating the frequency of the trying rule simply by accessing the main memory.

With regards to the results of the estimated rules, LEM2 and FDMM execute the operation of the set and/or the logic and reduce about 9400 of samples by only about 40 [%] at most. About 80 [%] of the estimated rules are those with length five, and it is evident that the results do not include rules specified in advance, that is $R1$ and $R2$. Simply put, they cannot extract rules. On the contrary, STRIM extracts the number of rules of one digit, and includes the true rules, as well as the errors as shown in Table 4. This table also shows the values of $(n1, n2)$, $p - value(z)$, accuracy and coverage indices as by-products corresponding to the estimated rules. Here, $p - value$ means the probability of errors of rejecting $H0$ and adopting $H1$. It is also clear that the values of $p - value(z)$, accuracy and coverage indices of the errors are worse than those of the true rules. The validity and effectiveness of our experimental studies has been confirmed.

6   Conclusions

Rough Sets theory extracts if-then rules using the decision table obtained from the database. The decision table is a sample set from its population. Accordingly, studies of the rule estimation problem by Rough Sets theory to date have investigated the effects of the sampling, and the effects for the estimated rules on the accuracy and coverage indices [8–11]. However, these studies are indirect ones to the effects. In this paper, we proposed a direct rule estimation method using the sample data and its algorithm by use of the experimental knowledge obtained from a preliminary experiment, developed the algorithm into a piece of software, implemented this software in a PC, and conducted a simulation experiment. The results showed:

1) The time [sec] taken by our proposed algorithm for extracting if-rules is much shorter (by one to three digits) than those by the algorithms in LEM2 and FDMM;

2) The extracted results improved the results by LEM2 and FDMM dramatically;

3) Our proposed algorithm statistically and directly extracted if-then rules without using the concept of approximation. As a result, the rules extracted have the confidence coefficient of the $p - value$, and also had accuracy and coverage indices as by-products.

Focus for future studies:

1) To expand the applicability of the algorithm to problems with more than three values of the decision attribute.

2) To examine how many samples are needed for extracting rules with high precision, depending on the rule extraction problems.

3) To study how many and which rules extracted in Table 4 should be adopted as a family of rules by use of a multiple comparison procedure.

4) Application to real-life data, such as that in the repository of the University of California at Irvine[13].

References

1. Pawlak, Z.: Rough sets. Internat. J. Inform. Comput. Sci. 11(5), 341–356 (1982)
2. Skowron, A., Rauser, C.M.: The Discernibility Matrix and Functions in Information Systems. In: Słowiński, R. (ed.) Intelligent Decision Support, Handbook of Application and Advances of Rough Set Theory, pp. 331–362. Kluwer Academic Publishers (1992)
3. Bao, Y.G., Du, X.Y., Deng, M.G., Ishii, N.: An Efficient Method for Computing All Reducts. Transactions of the Japanese Society for Artificial Intelligence 19(3), 166–173 (2004)
4. Grzymala-Busse, J.W.: LERS- A system for learning from examples based on rough sets. In: Słowiński, R. (ed.) Intelligent Decision Support. Handbook of Applications and Advances of the Rough Sets Theory, pp. 3–18. Kluwer Academic Publishers (1992)
5. Ziarko, W.: Variable precision rough set model. Journal of Computer and System Science 46, 39–59 (1993)
6. Shan, N., Ziarko, W.: Data-based acquisition and incremental modification of classification rules. Computational Intelligence 11(2), 357–370 (1995)
7. Nishimura, T., Kato, Y., Saeki, T.: Studies on an Effective Algorithm to Reduce the Decision Matrix. In: Kuznetsov, S.O., Ślęzak, D., Hepting, D.H., Mirkin, B.G. (eds.) RSFDGrC 2011. LNCS (LNAI), vol. 6743, pp. 240–243. Springer, Heidelberg (2011)
8. Gediga, G., Düntsch, I.: Statistical technique for rough set data analysis. In: Polkowski, L., et al. (eds.) Rough Set Methods and Applications: New Developments in Knowledge Discovery in Information System, pp. 545–565. Physica Verlag, Heidelberg (2000)
9. Tsumoto, S.: Accuracy and Coverage in Rough Set Rule Induction. In: Alpigini, J.J., Peters, J.F., Skowron, A., Zhong, N. (eds.) RSCTC 2002. LNCS (LNAI), vol. 2475, pp. 373–380. Springer, Heidelberg (2002)
10. Gillet, F., Hamilton, H. (eds.): Quality Measures in Data Mining. SCI, vol. 43. Springer, Heidelberg (2007)
11. Jaworski, W.: Rule Induction: Combining Rough Set and Statistical Approaches. In: Chan, C.-C., Grzymala-Busse, J.W., Ziarko, W.P. (eds.) RSCTC 2008. LNCS (LNAI), vol. 5306, pp. 170–180. Springer, Heidelberg (2008)
12. Laboratory of Intelligent Decision Support System (IDSS), http://idss.cs.put.poznan.pl/site/idss-en.html
13. Asuncion, A., Newman, D.J.: UCI Machine Learning Repository, University of California, School of Information and Computer Science, Irvine (2007), http://www.ics.uci.edu/~mlearn/MLRepository.html

A Rough Neurocomputing Approach for Illumination Invariant Face Recognition System

Singh Kavita[1], Zaveri Mukesh[2], and Raghuwanshi Mukesh[3]

[1] Computer Technology Department, Y.C.C.E., Nagpur, 441110, India
[2] Computer Engineering Department, S.V.N.I.T., Surat, 329507, India
[3] NYSS College of Engineering and Research, Nagpur, 441110, India
singhkavita19@yahoo.co.in, mzaveri@coed.svnit.ac.in,
m_raghwanshi@rediffmail.com

Abstract. In this paper we are presenting an illumination invariant face recognition system that will be able to identify the facial images under varying range of illumination of (azimuth, elevation) (+120-65 and -120+65). The main focus of this work is to address the problem of variations in illumination through the strength of the rough sets to recognize the faces under varying illumination. The proposed approach consists of three major parts, namely, illumination normalization, feature extraction and recognition procedure. Illumination normalization part utilizes the existing approaches for illumination normalization based on a mathematical model called rough membership function (rmf) illumination classifier. After normalizing the illumination, geometrical features are extracted and feature vector is formed using the geometrical relations between facial fiducial points. Finally, in recognition part feature vectors of training images are given as input to approximation-decider neuron network (ADNN). The efficiency and robustness of the proposed system are demonstrated on data set of significant size and are compared with state of the art classifier techniques. Our recognizer has achieved 93.56% accuracy for Yale database and 85% accuracy for CMU-PIE database in recognizing the facial images with different types of illumination.

Keywords: Rough Sets, Neurocomputing, Illumination Variation, Face Recognition.

1 Introduction

In past, several illumination invariant face recognition systems, with different illumination normalization techniques and recognizers have been proposed [1], [2] to address the illumination problem. Almost all proposed techniques used machine learning approach to face recognition. In addition, the various classifiers used in face recognition system were Bayesian classifier [4] and Probabilistic networks [5], [6], [7]. These classifiers classify training vectors into their correct classes by approximating their distribution or densities in a feature space. Another one is Nearest-neighbor classifier (NNC) [8] that has, consistently shown high performance, without a priori assumptions about the distributions from

T. Li et al. (Eds.): RSKT 2012, LNAI 7414, pp. 91–100, 2012.
© Springer-Verlag Berlin Heidelberg 2012

which the training images are drawn. One advantage of NNC is that, new samples can be added to the database at any time. Yet, another classifier, RBF network [9] is a form of neural network that have been used widely in face detection [10], [11] when detection rate is in focus. A neural network has the ability to adjust its weights according to the differences it encounters during training. As a result, it delivers high efficiency in the classification of linearly as well as nonlinearly separable classes, but it is of a different nature compared to probabilistic approach. However, neural network needs high computation between the layers of the neural networks and also problems in adjusting the topology of the network. Through literature survey it has been observed that two widely used classification approaches such as neural network approach and probabilistic approach are quite different in nature. However, the similarity between these two approaches is that, both of them need high computation to achieve high efficiency in the classification.

To overcome the above mentioned problem we are presenting a Rough Neurocomputing Recognition System (RNRS) for illumination invariant face recognition. It is worth emphasizing that (i) proposed RNRS inherits the ability of neural network to adjust its weights according to the differences it encounters during training by em-bedding rough membership function value and (ii) it is based on probabilistic approach that classifies training vectors into their correct face classes by approximating their distribution in a feature space. The efficiency of proposed system is demonstrated on data set of significant size and is compared with state-of-art classifiers.

The paper is organized as follows: Section 1 presents introduction to the proposed RNRS system. Section 2 briefly presents an overview of the rough set theory. Section 3 describes the different modules of RNRS system that is used for illumination invariant face recognition. Section 4 presents the detailed design of ADNN that acts as recognizer for the proposed system. Section 5 describes the experimentation for the proposed system and presents the results. Section 6 concludes the paper.

2   Rough Set Theory (RST)

In RST [12], information about the real world is given in the form of an information table. An information table can be represented as a pair $A = (U, A)$,where, U is a non-empty finite set of objects called the universe and A is a non-empty finite set of attributes such that information function $f_a : U \rightarrow V_a$,for every $a \in A$. The set V_a is called the value set of a. Furthermore, a decision system is any information system of the form $A = (U, A \cup d)$, where $d \notin A$ is a decision attribute. The main concept involves in rough set theory is an indiscernibility relation. For every set of attributes $B \subseteq A$, an indiscernibility relation $IND(B)$ is defined in the following way: two objects, x_i and x_j, are indiscernible by the set of attributes $B \subseteq A$, if $b(x_i) = b(x_j)$ for every $b \subseteq B$. The equivalence class of $IND(B)$ is called elementary set in B because it represents the smallest discernible groups of objects. For any element x_i of U, the equivalence class of x_i in relation $IND(B)$ is represented as$[x_i]_{IND(B)}$. The notation $[x]_B$ denotes equivalence classes of $IND_A(B)$. The partitions induced by an equivalence relation can be used to build new subsets of the universe. The construction of equivalence classes is the first step in classification with rough sets.

Rough sets can also be defined by rough membership function. A rough membership function (rmf) makes it possible to measure the degree that any specified object with given attribute values belongs to a given decision set X.

3   Proposed RNRS Architecture

The proposed system consists of three major modules viz: (1) illumination normalization (2) feature extraction and (3) recognizer.The complete flow of proposed system is shown in Fig 1.

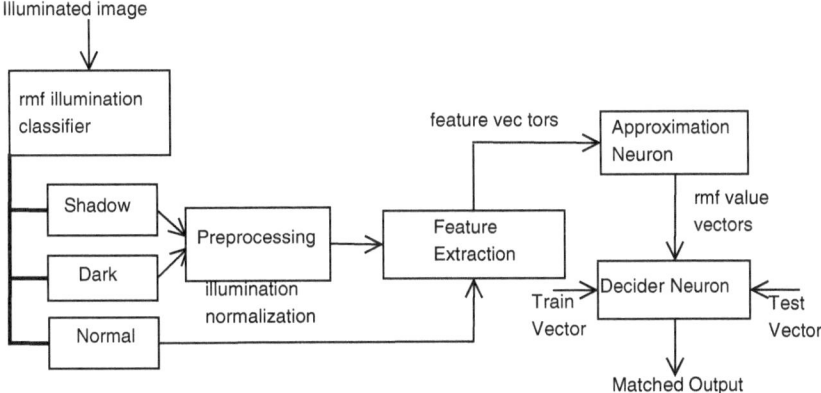

Fig. 1. Architecture of Proposed RNRS System

1. *Illumination Normalization*: In this module classified shadow and dark images must be normalized with Self Quotient Image (SQI) filter [13] and Homomorphic filter [19] respectively before feature extraction. Features are directly extracted from properly illuminated images classified as normal, without any illumination normalization. Readers can refer the details of *rmf* illumination classifier from [16].

2. *Feature Extraction*: There are many geometrical parameters one can extract from the facial geometry however; we concentrated only on five geometrical features as represented in [14]. It is to be noted that, if geometrical feature points such as nose tip, eye corners and mouth corners are measured, then it can be observed that the relationships between these features are not much dispersed from each other for different faces. Recognizing a face under such circumstances through a machine learning model becomes a difficult task. Such a problem can be very well tackled with rough set theory [12], as it maps the ambiguity and vagueness in data very efficiently.

3. *Recognizer*: Finally, in last module we use an approximation-decider neuron network (ADNN) as a recognizer for the proposed recognition system as illumination invariant face recognition.

4   Design of ADNN

Here we define two types of rough neurons that are approximation neuron and decider neuron for face recognition influenced from the work in [15].

4.1   Approximation Neuron

Approximation neurons compute the degree of overlap of test image with each of the subject samples in the trained database. The input to each approximation neuron is the set of feature vector $b_i, i = 1 \ldots d, b \subseteq B$, where B being the total number of attributes (features) under consideration and d being the input dimensionality. Here, we have considered five features as input to approximation neuron. The feature vectors are created for all sample facial images N and are normalized through discretization process that is defined for each feature. These discretized feature values are used to build decision table. Now, at this point, the indiscernibility relation is used to partition the set of sample faces in X into equivalence classes to compute the rough membership function value as discussed in [16].

4.2   Decider Neuron

The decider neuron is responsible for taking the decision of assigning a test facial image to its correct face class C_i. The input to decider neuron is the matrix of rmf values computed for each test facial image with respect to each approximation neuron. From all the rmf values given to decider neuron, it assigns the test facial image P_i to a specific subject face class C_i, $i = 1 \ldots K$ with highest rmf value. See (Step 1) in Algorithm 1, where, $K \subseteq N$ is the total no of subject classes. However, if the test facial image may actually belong to subject face class C_i, whereas, it has been approximated to more than one face classes since there holds a condition $max\{rmf_k\} = rmf_i = rmf_j = rmf_k$, $i, j, k \subseteq K$, provided $i \neq j \neq k$. That is, this decision of approximating the match of a test face image to more than one class can be resolved through the step 2 and step 3 given in Algorithm 1.

Algorithm1:Decider Neuron

```
Input: feature matrix of face samples with equal rmfs
Output: matched image, X
Step1: given the rough membership function values for each test
image
```

with respect to each train image
-if $rmf_i = max\{rmf_K\}, i \subseteq K$ assign P_i to subject class C_i
elseif
-count the occurrences of $max\{rmf_K\}$
-if the occurrences are greater than one go to step2
Step 2:given the reduced feature matrix for number of faces with
equal$rmfs$
- initialize the best match with random number
- for each feature vector of given reduced feature matrix
Step 3: compute Euclidean Distance w.r.t test face
- compare the difference with best match
- if the difference is less than best match return the
index of matched face

The Algorithm1 is used to declare a match or to assign a test image to correct class C_i, in cases where same $rmfs$ have been obtained. A feature matrix is formed from the feature values of the faces with equal $rmfs$. A minimum difference computed through Euclidian distance between the input feature matrix and test facial image tells that there is a match between experimental and ideal feature values.

4.3   Architecture of ADNN

The complete architecture of ADNN as shown in Fig. 2 is built around two layers: input layer consisting of approximation neuron and output layer with a single decider neuron. The number of approximation neurons in first layer is equal to the number of facial images in training database. Thus the architecture of an ADNN for face recognition is dependent on the number of faces available. Each face will have its own approximation neuron. Here, in our case hundred samples of twenty facial images have been used for training. The computation by a layer of approximation neurons constitutes a parallel process (i.e. for one test facial image each approximation neuron computes rough membership function value with each trained sample in database.) In our case, the relationships between the weight values and other parameters are expressed by rough membership function value. For each test facial image, the rmf values are computed in parallel by all approximation neurons and get modified, after every iteration. Since there is no hidden layer in the network any number of training images can be added at any point of time without disturbing the topology of the network. Only one has to add approximation neurons without any overhead for intermediate layer.

The vector of all $rmfs$ that is output of approximation neuron is the input to decider neuron. A corresponding match is found for test facial image with highest rmf value. Experimenting with many values, later rmf threshold is set at (0.5). That means a given test image is said to be matched with a train image for an rmf value of greater than or equal to 0.5.

The advantage of RNRS is that the correct face match is estimated at the approximation layer itself based on the highest rmf value. On the contrary, if

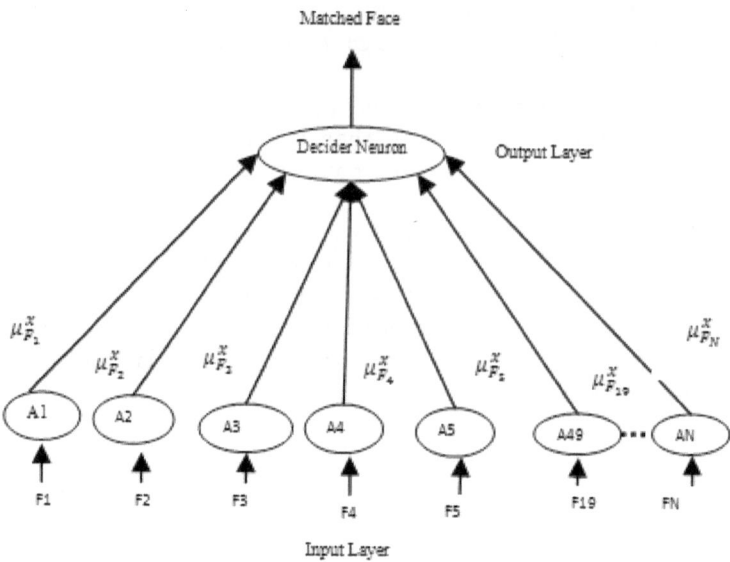

Fig. 2. ADNN architecture

it is not being done at this layer then decider neuron does this against reduced number of sample faces. Thus, recognizing time for each test image is greatly reduced. Moreover, the ambiguity present in feature values are very efficiently dealt with rough sets.

5　Experimental Results

In order to test the performance of the RNRS, all the experiments have been performed on the Extended YaleB face database [17] and CMU-PIE face database [18]. Only frontal images have been considered from both the databases with large variation in illumination.

For experimentation, dataset is distributed into two sets. One is used as training data set and the other is used as testing or validation data set. Training set consists of 100 images of 20 randomly selected different individuals with five face samples per individual. RNRS was trained only with normal images that have been classified as *normal* by *rmf* illumination classifier. Recognition performance of the proposed system was assessed on images with varying illumination range in the azimuth and elevation angle with 6 face samples per individual. Thus illuminated images were grouped into three separate testing subsets of 40 images each, set1, set2 and set3. Set1 is the set of classified *normal* images; set2 consists of classified *shadow* images and set3 is of classified *dark* images. Same experimentation setup was used for both the database. In first step of experimentation every image is first classified as *normal, dark* and *shadow* based on *rmf* illumination classifier. The classified *normal, dark* and *shadow* illuminated images were normalized through an appropriate filter as said in Section 3. The resultant normalized images are shown in Fig 3.

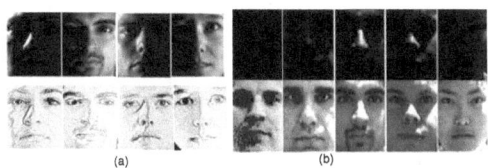

(a)   (b)

Fig. 3. Normalized Samples faces from YaleB database: (a) Shadow+SQI (b) Dark+ Homomorphic

The features for face recognition were extracted from the normalized images by applying the canny edge detection algorithm in order to locate the fiducial points. Locating the desired feature points manually, the geometrical spatial relations between the facial feature points are computed to represent a face. It is observed that more clear and relevant fiducial points were obtained from images that were pre-processed with an appropriate filter. This led to better feature extraction and hence better face recognition. Extracted feature values are then discretized to form a decision table. For this decision table; ADNN is used as a recognizer.

Selection of training images greatly affects the performance of face recognition system. Therefore, we performed the training with training sets consisting different number of samples per subject with S=1, S=3 and S=5. (S is number of training sample per subject). The respective recognition rate achieved for each training set is depicted in Table 1. Table 1 shows that recognition rate improves to a great extent if number of samples per subject increases. This is due to the fact that more feature values helps in defining more appropriate function values and thus more accurate training model. In spite, even if number of training samples per subject was increased more than five, no improvement in recognition rate was observed. Therefore, we continued the training of the proposed recognition system with S=5. In order to prove the effectiveness of

Table 1. Recognition Rate for Samples per subject

Illumination type+	Recognition Rate for Samples per subject		
normalization technique	$s = 1$	$s = 3$	$s = 5$
Normal	50.78	80.23	90.18
Shadow+SQL	45.00	76.02	92.51
Dark+homomorphic	43.00	75.10	90

proposed recognition system, we compared RNRS against state-of-art classifiers such as nearest-neighbor [8] as it has always shown high accuracy in face recognition and Naive Baye's [3] a probabilistic model which is based on conditional probability. All these classifiers were compared against the same experimentation setup. Face recognition rates for different classifiers on YaleB and CMU-PIE face databases are given in Table 2 and Table 3 respectively.

Table 2. Face Recognition Rates for different classifiers(YaleB face database)

Classifiers / Illumination type	Recognition Rates in percentage.		
	Nearest Neighbor [8]	Naive Baye's [3]	RNRS
Normal	98.18	67.27	98.18
Shadow+SQL	60	57.5	92.51
Dark+homomorphic	52.15	55	90

Here we list the former 10 targets and the final 10 targets for the different punctuation sets. The results are described as Table 3.(the training terms is expressed as A(B,C), in which A stands for Chinese word, B stands for the same word in English, C stands for the Hanyu Pinyin of the word.)

Although nearest-neighbor has shown good results when normal images were compared against *normal* images, the recognition rate degrades when the *shadow* and *dark* images are tested against the *normal* images. In contrast, RNRS have shown better results for all type of illumination. However, recognition rate falls to certain range for RNRS when *shadow* and *dark* images are compared against normal images also but still it maintains the accuracy comparatively.

Table 3. Face Recognition rates for different classifiers (CMU-PIE face database)

Classifiers / Illumination type	Recognition Rates in percentage.		
	Nearest Neighbor [8]	Naive Baye's [3]	RNRS
Normal	98	88	90
Shadow+SQL	68	64	80

For CMU-PIE database, proposed recognition system has shown less recognition result compared to YaleB database because the images in former database are not at equal scale. This affects the feature values as the feature extraction technique we have used is not a scale invariant one. In turn this affects the recognition rate. However, the limitation of our implemented system is that it is not invariant to large zooming in or zooming out as our training set did not have samples with scale changes. It can be seen that comparatively RNRS has outperformed against the other two classifiers. It shows an average accuracy of 93.56% for YaleB database and 85% accuracy for CMU-PIE database. The number of true positives and false positives are are plotted as shown in Fig. 4. Graph in Fig. 4 depicts that only one false positive is there for RNRS and nearest neighbor and so the curve exactly fit with the target face curve, where as for Naive Baye's classifier there are more number of false positives and so the output curve is not in line with the target face curve. From all the graphs shown in Fig. 4 it can be observed that RNRS has outperformed comparatively to both classifiers for both the databases.

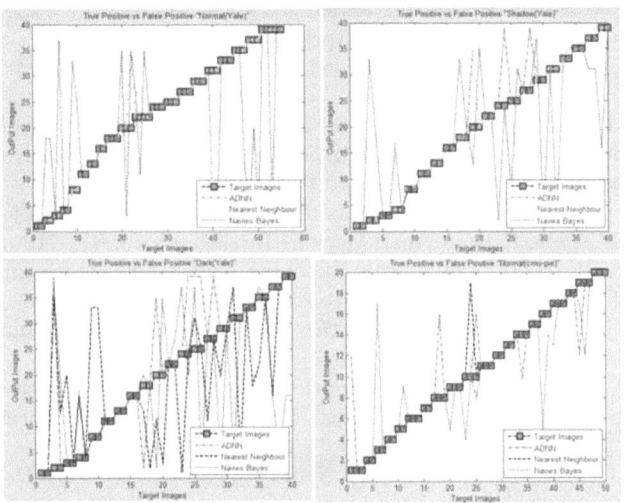

Fig. 4. Graphs showing the true positives and false positives for different classifiers

6   Conclusion

This paper has presented a face recognition system for 2D still images, which is robust to illuminated images. Robustness is achieved by using the concept of rough member-ship function from rough sets to cope with illumination variations that inhibit the performance of 2D face recognition. An "ADNN" is introduced as recognizer which is capable of recognizing the facial images in an efficient manner compared to other classifiers. Proposed RNRS is compared with few state-of-the-art classifiers reviewed in the literature survey. On an average, the system shows 93.56% accuracy for Yale database and 85% accuracy for CMU-PIE database.

There is a scope of several improvements to this recognition system. The first and the important one are by employing some automatic feature extraction technique with RNRS. Another possible enhancement is to do face recognition using ADNN recognizer under variations in poses.

Acknowledgement. This project is supported by AICTE of India (Grant No: 8023/RID/RPS-81/2010-11, Dated: March 31, 2011).

References

1. Turk, M., Pentland, A.: Eigenfaces for Recognition. Journal of Cognitive Neuroscience 3, 71–86 (1991)
2. Belhumeur, P.N., Hespanha, J.P., Kriegman, D.J.: Eigenfaces vs. Fisherfaces: recognition using class specific linear projection. IEEE Transactions on Pattern Analysis and Machine Intelligence 19, 711–720 (1997)
3. Xiaogang, W.S., Xiaoou, T.: Bayesian Face Recognition Using Gabor Features. In: Proceedings of the ACM SIGMM Workshop, pp. 70–73 (2003)
4. Alpaydin, E.: Introduction to Machine Learning. MIT Press (2004) ISBN-81-203-2791-8
5. Yiu, K., Mak, M., Li, C.: Gaussian Mixture Models and Probabilistic Decision-Based Neural Networks for Pattern Classification: A Comparative Study. Hong Kong Polytechnic University, HK (1999)
6. Moghaddam, B., Pentland, A.: Probablistic visual learning for object representation. IEEE Trans. Pattern Anal. Machine Intelligence 19, 696–710 (1997)
7. Liu, C., Wechsler, H.: Probablistic reasoning models for face recognition. In: Proceeding of Computer Vision and Pattern Recognition, pp. 827–832 (1998)
8. Sang-II, C., Chong-Ho, C.: An effective Face Recognition under Illumination and Pose Variation. IJCNN, 914–919 (2007)
9. Howell, A., Buxton, H.: Face recognition using radial basis function neural networks. In: Proceedings of the British Machine Vision Conference, pp. 455–464 (1996)
10. Rowley, H., Baluja, S., Kanade, T.: Neural Network-Based Face Detection. IEEE Trans. Pattern Analysis and Machine Intelligence 20, 23–38 (1998)
11. Garica, C., Delakis, M.: Convolutional Face Finder:A Neural Architecture of Fast and Robust Face Detection. IEEE Trans. Pattern Analysis and Machine Intelligence 26, 1408–1423 (2004)
12. Pawlak, Z.: Rough sets. Int. Journal of Computer and Information Sciences 11(5), 341–356 (1982)
13. Wang, H., Li, S.Z., Wang, Y.: Face recognition under varying lightening conditions using self quotient image. In: Proc. IEEE International Conference on Automatic Face and Gesture Recognition (FGR), pp. 819–824 (2004)
14. Kavita, S.R., Mukeshl, Z.A., Mukesh, R.M.: Extraction of Pose Invariant Facial Features. In: Das, V.V., Vijaykumar, R. (eds.) ICT 2010. CCIS, vol. 101, pp. 535–539. Springer, Heidelberg (2010)
15. Han, L., Peters, J.F.: Rough Neural Fault Classification of Power System Signals. In: Peters, J.F., Skowron, A. (eds.) Transactions on Rough Sets VIII. LNCS, vol. 5084, pp. 396–519. Springer, Heidelberg (2008)
16. Singh, K.R., Zaveri, M.A., Raghuwanshi, M.M.: A Robust Illumination Classifier using Rough Sets. In: IEEE CSAE, pp. 10–12 (2011)
17. www.cvc.yale.edu/projects/yalefaces/yalefaces.html
18. Sim, T., Baker, S., Bsat, M.: The CMU Pose, Illumination, and Expression Database. IEEE Trans Pattern Analysis and Machine Intelligence 25, 1615–1618 (2003)
19. de Solar, J.R., Quinteros, J.: Illumination Compensation and normalization in Eigen space based face Recognition: A comparative study of different pre-processing approaches. Computer, Vision Graphics and Image Processing 42, 342–350 (1999)

An Approximation Decision Entropy Based Decision Tree Algorithm and Its Application in Intrusion Detection

Hongbo Zhao[1], Feng Jiang[1,*], and Chunping Wang[2]

[1] College of Information Science and Technology,
Qingdao University of Science and Technology, Qingdao 266061, P.R. China
[2] College of Computer Science and Technology,
Zhejiang University of Technology, Hangzhou 310023, P.R. China
jiangkong@163.net

Abstract. In this paper, we propose a novel decision tree algorithm DTADE within the framework of rough set theory, and apply DTADE to intrusion detection. We define a new information entropy model — approximation decision entropy (ADE) in rough sets, which combines the concept of conditional entropy in Shannon's information theory and the concept of approximation accuracy in rough sets. In algorithm DTADE, ADE is adopted as the heuristic information for the selection of splitting attributes. Moreover, we present a method of decision tree pre-pruning based on the concept of knowledge entropy proposed by Düntsch and Gediga. Finally, the KDDCUP99 data set is used to verify the effectiveness of our algorithm in intrusion detection.

Keywords: Intrusion detection, decision tree, rough sets, approximation decision entropy, approximation accuracy, knowledge entropy.

1 Introduction

Intrusion detection systems (IDS) attract much attention in recent years [1-2]. Due to the large volumes of security audit data as well as complex and dynamic properties of intrusion behaviors, many artificial intelligence techniques have been utilized to IDS, where decision tree technology plays an important role [2].

Decision tree is an important technology of data mining to deal with classification [3-4]. When constructing a decision tree, the most important step is to select the splitting attributes [3]. Different standards for choosing splitting attributes have been proposed. For instance, Shannon's information entropy has been used to select splitting attributes [5]. Many decision tree algorithms such as ID3 and C4.5 have been proposed based on information entropy [3-4].

Rough set theory was first proposed by Pawlak in 1982 [6]. Recently, information entropy has been widely applied to rough set theory. Many different

---

* Corresponding author.

T. Li et al. (Eds.): RSKT 2012, LNAI 7414, pp. 101–106, 2012.
© Springer-Verlag Berlin Heidelberg 2012

information entropy models have been proposed in rough sets [7-10], among which the model proposed by Düntsch and Gediga is widely used [7].

In this paper, we propose a new information entropy model called *approximation decision entropy* (ADE) in rough sets, which combines the concept of conditional entropy in Shannon's information theory and the concept of approximation accuracy in rough sets [5-6]. By virtue of this model, we propose a novel decision tree algorithm DTADE and apply it to intrusion detection. In our algorithm, we use ADE to select splitting attributes during the process of constructing a decision tree. Moreover, we present a method of decision tree prepruning based on the concept of knowledge entropy [7], which can reduce the size of decision trees and avoid overfitting in building decision trees [11].

2   Preliminaries

In rough sets, a *decision table* is a 5-tuple $DT = (U, C, D, V, f)$, where U is a non-empty finite set of objects; C is a non-empty finite set of condition attributes; D is a non-empty finite set of decision attributes; $V = \bigcup\limits_{a \in C \cup D} V_a$ is the union of attribute domains; $f : U \times (C \cup D) \to V$ is a function such that for any $a \in C \cup D$ and $x \in U$, $f(x, a) \in V_a$ [6].

Given a decision table $DT = (U, C, D, V, f)$, for any $B \subseteq C \cup D$, we call binary relation $IND(B)$ an indiscernibility relation, which is defined as $IND(B) = \{(x, y) \in U \times U : \forall a \in B(f(x, a) = f(y, a))\}$. $IND(B)$ partitions U into disjoint equivalence classes, let $U/IND(B)$ denote the family of all equivalence classes of $IND(B)$ [6].

Definition 1. *Given a decision table $DT = (U, C, D, V, f)$, for any $B \subseteq C \cup D$ and $X \subseteq U(X \neq \emptyset)$, the approximation accuracy $\alpha_B(X)$ of set X with respect to relation $IND(B)$ is defined as follows [6].*

$$\alpha_B(X) = \frac{|\underline{X}_B|}{|\overline{X}_B|}, \tag{1}$$

where $\underline{X}_B$ and $\overline{X}_B$ respectively denote the B-lower and B-upper approximations of set X.

Definition 2. *Given a decision table $DT = (U, C, D, V, f)$, for any $B \subseteq C \cup D$, let $U/IND(B) = \{X_1, ..., X_n\}$ denote the partition of U induced by $IND(B)$. The knowledge entropy $I(B)$ of relation $IND(B)$ is defined as follows [7].*

$$I(B) = -\sum_{i=1}^{n} \frac{|X_i|}{|U|} \log_2 \frac{|X_i|}{|U|}. \tag{2}$$

3   Approximation Decision Entropy

Definition 3. [Approximation Decision Entropy]. *Given a decision table $DT = (U, C, D, V, f)$, for any $B \subseteq C$, let $U/IND(B) = \{X_1, ..., X_n\}$ and*

$U/IND(D) = \{Y_1, ..., Y_m\}$. The approximation decision entropy $ADE(D|B)$ of D with respect to B in DT is defined as follows.

$$ADE(D|B) = -\sum_{j=1}^{m} \log_2(2 - \alpha_B(Y_j)) \sum_{i=1}^{n} p(X_i)p(Y_j|X_i) \log_2(p(Y_j|X_i)), \quad (3)$$

where $p(X_i) = |X_i|/|U|$, $p(Y_j|X_i) = |X_i \cap Y_j|/|X_i|$, and $\alpha_B(Y_j)$ denotes the approximation accuracy of Y_j with respect to $IND(B)$, $1 \le i \le n$, $1 \le j \le m$.

Theorem 1. *Given a decision table $DT = (U, C, D, V, f)$, where $U = \{x_1, x_2, ..., x_n\}$. For any $B \subseteq C$, we have the following conclusions.*
(1) $0 \le ADE(D|B) \le \log_2 n$.
(2) *If $U/IND(B) \preceq U/IND(D)$, then $ADE(D|B)$ achieves the minimum value 0, where $\preceq$ is a partial order relation on the set of all partitions of U, which is defined as: $U/IND(B) \preceq U/IND(D) \Leftrightarrow \forall X \in U/IND(B), \exists Y \in U/IND(D)(X \subseteq Y)$.*
(3) *If $U/IND(D) = \{\{x_1\}, \{x_2\}, ..., \{x_n\}\}$ and $U/IND(B) = \{U\}$, then $ADE(D|B)$ achieves the maximum value $\log_2 n$.*

Theorem 2. *Given a decision table $DT = (U, C, D, V, f)$, where $U = \{x_1, x_2, ..., x_n\}$. For any $B \subset C$ and $a \in C - B$, $ADE(D|B) \ge ADE(D|B \cup \{a\})$.*

We omit the proofs of Theorem 1 and Theorem 2, due to limited space.

4   Decision Tree Algorithm DTADE

Algorithm 1. *DTADE.*

Input: decision table $T_1 = (U, C, D, V, f)$ and threshold value σ, where $|D| = 1$.
Output: set R of decision rules.

Function Main(T_1)
(1) *For each continuous attribute $c \in C$, discretize c by virtue of a discretization approach [12]. Let $T_2 = (U_2, C, D, V_2, f_2)$ denote the discretized dataset.*
(2) *Decision_Tree(T_2). //Call a function defined below to induce a decision tree.*
(3) *Generate a set R of rules from the above decision tree.*
(4) *Return R.*

Function Decision_Tree$(T_{current})$ // $T_{current} = (U', C', D, V', f')$ *is the*
                                                    decision table that is currently in use.
(1) *Create a node N.*
(2) *If all objects in U' belong to the same class d, then return N as a leaf node, labeled with the class d.*
(3) *If $C' = \emptyset$, then return N as a leaf node, labeled with the most common class in U'.*
(4) *Calculate the knowledge entropy $I(D)$ of relation $IND(D)$ in $T_{current}$.*

(5) *If $I(D) < \sigma$ then return N as a leaf node, labeled with the most common class in U'.*

(6) *For each attribute $a \in C'$, calculate the approximation decision entropy $ADE(D|\{a\})$ of D with respect to $\{a\}$ in $T_{current}$.*

(7) *Select attribute t from C' such that $ADE(D|\{t\})$ is the maximum value of set $\{ADE(D|\{a\}) : a \in C'\}$ (If there exist more than one attributes in C' satisfy such a condition, then arbitrarily select one from them). Let t be the current splitting attribute.*

(8) *Label node N with attribute t, and let $C' = C' - \{t\}$.*

(9) *For each value v_j of attribute t*
>   (9.1) *Generate one branch of N, labeled with $t = v_j$.*
>   (9.2) *Let $S_j = \{o \in U' : f'(o, t) = v_j\}$.*
>   (9.3) *If $S_j = \emptyset$, then add one leaf node L, labeled with the most common class in U';*
>   (9.4) *Else obtain a sub-table $T_{sub} = \{S_j, C', D, V_{sub}, f_{sub}\}$ of $T_{current}$, and call function Decision_Tree (T_{sub}) to construct a sub-tree of N.*

In function Decision_Tree of algorithm 1, before computing the partition $U'/IND(D)$, we first sort all objects of U' based on the counting sort, then we can calculate $U'/IND(D)$ in $O(|D||U'|)$ time [13]. Therefore, in the worst case, the time complexity of step (2) in function Main is $O(|C|^2 \times |U_2|)$. And the overall time complexity of algorithm 1 is also determined by that of the discretization approach in function Main [12].

Moreover, in algorithm 1 we present a new decision tree pre-pruning method, which is based on the concept of knowledge entropy [7]. The basic idea of our pre-pruning method is as follows. Before constructing a sub-tree on the current decision table $T_{current}$, we first calculate the knowledge entropy $I(D)$ of relation $IND(D)$ in $T_{current}$. If $I(D) < \sigma$ (σ is a given threshold value), then we consider that it is not necessary to construct a sub-tree on $T_{current}$, that is, such a sub-tree should be pruned, hence a leaf node instead of a sub-tree will be returned. If $I(D) \geq \sigma$, then we consider that it is necessary to construct a sub-tree on $T_{current}$, that is, such a sub-tree should not be pruned, hence we shall construct a sub-tree on $T_{current}$ and return it.

5   Experimental Results

To evaluate the performance of our algorithm in intrusion detection, we ran it on the KDDCUP99 data set [14]. We compared algorithm DTADE with algorithm ID3. DTADE was implemented in C++, and ID3 is available in Weka [15]. Since the full KDDCUP99 training set is too large for our purposes, a concise subset, known as "10%KDD", will be discussed here. However, the "10%KDD" data set contains 494021 records, which is still very large for our purposes [14]. Hence, in our experiment, we randomly select two subsets: Extracted-Training and Extracted-Test, from the "10%KDD" and KDDCUP99 test set, respectively. Extracted-Training contains 15095 records, which is used as the training set, and Extracted-Test contains 10065 records, which is used as the test set.

There are four steps in our experiments: (1) The first step is to prepare the training set and the test set. (2) Next is the step of discretization. We used four discretization methods: Equal Frequency Binning (EF) [12], Equal Width Binning (EW) [12], Naive Scaler (Naive) [16] and Entropy-based method (EB) [12], to respectively discrete the continuous attributes in Extracted-Training and Extracted-Test, where EW is from Weka and the other three methods are from Rosetta [15-16]. (3) The third step is constructing decision trees on training sets using algorithms DTADE and ID3, respectively. Then different sets of rules can be generated from these trees. (4) The last step is applying the rules generated in (3) to classify the corresponding test sets. For ID3, this process was also implemented in Weka [15]. For the value of threshold σ in algorithm DTADE, we adopted two values, that is, $\sigma = 0$ and $\sigma = 0.5$.

Tables 1 details the experimental results of algorithms DTADE and ID3.

Table 1. Experimental results

Discretization method	ID3			DTADE ($\sigma = 0$)			DTADE ($\sigma = 0.5$)		
	Nodes	Leafs	Precision	Nodes	Leafs	Precision	Nodes	Leafs	Precision
EF15	993	917	92.548%	453	305	92.558%	412	273	92.956%
EF20	1099	1034	92.201%	417	313	92.231%	372	277	92.837%
EF25	2180	2115	89.429%	427	322	92.717%	396	298	92.697%
EF30	1764	1713	87.601%	357	305	90.909%	349	300	90.869%
EW20	1527	1422	93.015%	841	481	93.751%	811	460	93.780%
EW30	1880	1778	92.320%	854	530	92.330%	779	500	92.399%
EW40	2401	2296	92.310%	878	650	92.996%	777	596	93.035%
EW50	2853	2747	91.932%	1031	729	92.588%	897	675	92.817%
Naive	5412	5355	87.611%	462	404	90.939%	443	391	90.949%
EB	3482	3428	88.803%	339	284	92.270%	317	269	92.250%

In Table 1, "EF15" denotes the algorithm EF and the parameter (i.e., the number of bins) of the algorithm is set to 15. "Nodes" and "Leafs" respectively denote the number of nodes and the number of leafs in the decision tree generated by a given algorithm.

From Table 1, we can see that the sizes of the decisions trees generated by our algorithm are much less than those generated by ID3, while the classification accuracies of our algorithm are always higher than those of ID3. Hence, the above experiment results demonstrate the effectiveness of our algorithm for intrusion detection. In addition, from Table 1, we can find that the decision trees generated by algorithm DTADE with $\sigma = 0.5$ are smaller than those generated by DTADE with $\sigma = 0$. However, the former's classification performance is better than the latter in most cases. Therefore, this shows that our method of tree pre-pruning works well, which can effectively reduce the size of the induced decision tree and improve the classification performance of the tree.

6 Conclusion

In this paper, a novel algorithm based on approximation decision entropy was proposed for constructing decision trees. Approximation decision entropy is an extension of Shannon's information entropy in rough sets, which combines the concept of conditional entropy in Shannon's information theory and the concept of approximation accuracy in rough sets. Experimental results on the KDD-CUP99 data set showed that our algorithm generated a better rule learning scheme for intrusion detection than ID3 algorithm.

Acknowledgements. This work is supported by the National Natural Science Foundation of China (grant nos. 60802042, 61103246), the Natural Science Foundation of Shandong Province, China (grant nos. ZR2011FQ005, ZR2011FQ026, ZR2009GQ013), and the Project of Shandong Province Higher Educational Science and Technology Program (grant no. J11LG05).

References

1. Anderson, J.P.: Computer Security Threat Monitoring and Surveillance. James P. Anderson Co., Fort Washington (1980)
2. Li, X.Y., Ye, N.: Decision tree classifiers for computer intrusion detection. Journal of Parallel and Distributed Computing Practices 4(2), 179–190 (2001)
3. Quinlan, R.: Induction of decision trees. Machine Learning 1(1), 81–106 (1986)
4. Quinlan, R.: C4.5: Programs for Machine Learning. Morgan Kaufmann (1993)
5. Shannon, C.E.: The mathematical theory of communication. Bell System Technical Journal 27(3-4), 373–423 (1948)
6. Pawlak, Z.: Rough Sets. Int. J. Comput. Informat. Sci. 11(5), 341–356 (1982)
7. Düntsch, I., Gediga, G.: Uncertainty measures of rough set prediction. Artificial Intelligence 106, 109–137 (1998)
8. Liang, J.Y., Shi, Z.Z.: The information entropy, rough entropy and knowledge granulation in rough set theory. Int. Journal of Uncertainty, Fuzziness and Knowledge-Based Systems 12(1), 37–46 (2004)
9. Miao, D.Q., Hu, G.R.: An Heuristic Algorithm of Knowledge Reduction. Computer Research and Development 36(6), 681–684 (1999)
10. Wang, G.Y., Yu, H., Yang, D.C.: Decision table reduction based on conditional information entropy. Chinese Journal of Computers 25(7), 759–766 (2002)
11. Breslow, L.A., Aha, D.W.: Simplifying decision trees: a survey. Knowledge Engineering Review 12(1), 1–40 (1997)
12. Dougherty, J., Kohavi, R., Sahami, M.: Supervised and Unsupervised Discretization of Continuous Features. In: Proc. of the 12th International Conference on Machine Learning, pp. 194–202. Morgan Kaufmann Publishers (1995)
13. Xu, Z.Y., Liu, Z.P., Yang, B.R., Song, W.: A Quick Attribute Reduction Algorithm with Complexity of $\max(O(|C||U|), O(|C|^2|U/C|))$. Chinese Journal of Computers 29(3), 391–399 (2006)
14. KDD Cup 99 Dataset (1999),
   http://kdd.ics.uci.edu/databases/kddcup99/kddcup99.html
15. Witten, I.H., Frank, E.: Data Mining: Practical Machine Learning Tools and Techniques with Java Implementations. Morgan Kaufmann (2000)
16. Øhrn, A.: Rosetta Technical Reference Manual (1999),
   http://www.idi.ntnu.no/_aleks/rosetta

Application of Rough Set Theory
to Prediction of Antimicrobial Activity
of Bis-quaternary Ammonium Chlorides

Łukasz Pałkowski[1], Jerzy Błaszczyński[2], Jerzy Krysiński[1],
Roman Słowiński[2], Andrzej Skrzypczak[3], Jan Błaszczak[3],
Eugenia Gospodarek[4], and Joanna Wróblewska[4]

[1] Nicolaus Copernicus University, Collegium Medicum,
Department of Pharmaceutical Technology, Jurasza 2, 85-089 Bydgoszcz, Poland
{lukaszpalkowski,jerzy.krysinski}@cm.umk.pl
[2] Poznań University of Technology, Institute of Computing Science,
Piotrowo 3A, 60-965 Poznań, Poland
{jerzy.blaszczynski,roman.slowinski}@cs.put.poznan.pl
[3] Poznań University of Technology, Institute of Chemical Technology,
Skłodowskiej-Curie 2, 60-965 Poznań, Poland
[4] Nicolaus Copernicus University, Collegium Medicum,
Department of Microbiology, Skłodowskiej-Curie 9, 85-094 Bydgoszcz, Poland

Abstract. The paper investigates relationships between chemical structure, surface active properties and antibacterial activity of 70 bis-quaternary ammonium chlorides. Chemical structure and properties of ammonium chlorides were described by 7 condition attributes and antimicrobial properties were mapped by a decision attribute. Dominance-based Rough Set Approach (DRSA) was applied to discover rules exhibiting monotonicity relationships in the data, which are unknown a priori and hold in some parts of the evaluation space. Strong decision rules discovered in this way may enable creating prognostic models of new compounds with the best antimicrobial properties. Moreover, the estimated relevance of the attributes that form the discovered rules allow to distinguish which of the structure and surface active properties describe compounds that have the best and the worst antimicrobial properties.

Keywords: Structure Activity Relationship (SAR), Rough set theory, Dominance-based Rough Set Approach (DRSA), Confirmation measures, Bis-ammonium compounds, Surfactants, Surface active properties.

1 Introduction

In this paper, we propose a new methodology of Structure Activity Relationship (SAR) analysis employing the Dominance-based Rough Set Approach (DRSA) [6]. The type of compounds being analyzed is a group of bis-quaternary ammonium chlorides with a good antimicrobial activity [10]. The quaternary

T. Li et al. (Eds.): RSKT 2012, LNAI 7414, pp. 107–116, 2012.
© Springer-Verlag Berlin Heidelberg 2012

ammonium compounds have good antielectrostatic properties and they are used in cosmetic, textile and pharmaceutical industries [5]. Numerous tests indicated that antimicrobial properties of bis-quaternary ammonium chlorides depend on their structure (e.g. length of alkyl group) [8]. Those tests were based on statistical methods (regression) and non-statistical methods (artificial neural networks) [11,12]. Classical rough set approach was also applied in SAR analysis [9].

In this paper, we apply DRSA to analyse the structure-activity relationship of bis-quaternary ammonium chlorides. Our main goal is identification of relationship between structure and surface active properties of new n-alkyl-bis-N-alkoxy-N-alkyl imidazolium chlorides. In the course of the analysis jRS and jMAF[1] were used. The choice of DRSA is motivated by the aim of discovering synthetic rules that exhibit monotonic relationships between structure and surface active properties of the compounds on the one hand, and their biological activity on the other hand. DRSA is able to deal with possible inconsistencies in the information table prior to discovery of knowledge. It appears to be more suitable for analysis of qualitative data than many of statistical methods that are suited well only for quantitative data. In the analyzed information table surface active properties of compounds were described by quantitative parameters, whereas their chemical structure by qualitative ones. Moreover, using DRSA on a properly (non-invasively) transformed information table, one is able to discover rules showing local monotonic relationships in some parts of the evaluation space, which is not possible using classical rough set approach. While the discovered rules show relationships between condition attributes and the decision attribute, it is also important to inquire what is the relevance of individual condition attributes in these relationships. To measure this relevance, we apply a Bayesian confirmation measure.

The paper is organized as follows. In the next section, material and methods are presented, including description of analyzed compounds, transformation of the information table, rule discovery, and attribute relevance measurement. The last section is devoted to presentation of results and discussion.

2   Material and Methods

2.1   Material

For the sake of the analysis 70 objects that represent bis-ammonium quaternary chlorides were examined. Surface active properties of analyzed chlorides were described by the following parameters:

- CMC - critical micelle concentration (mol/L)
- γCMC - value of surface tension at critical micelle concentration (mN/m)
- $\Gamma \cdot 10^6$ – value of surface excess (mol/m^2)
- $A \cdot 10^{20}$ – molecular area of a single particle (m^2)
- ΔG_{ads} – free energy of adsorption of molecule (kJ/mol)

---

[1] http://www.cs.put.poznan.pl/jblaszczynski/Site/jRS.html

Structure properties of chlorides were described by parameters (see Figure 1):

- n - number of carbon atoms in n-substituent
- R - number of carbon atoms in R-substituent

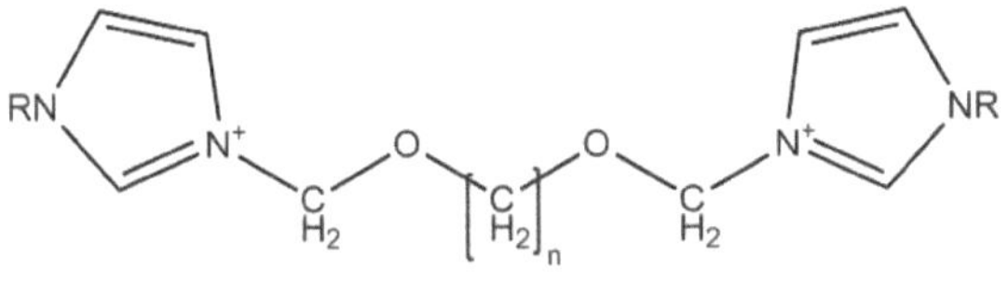

Fig. 1. Structure of analyzed compounds

Staphylococcus aureus ATCC 25213 microorganisms were used to evaluate antibacterial activity of compounds by minimal inhibitory concentration (MIC). In the classical rough-set approach it is necessary to perform discretization procedure for all attributes with continuous values. In the case of this analysis made by DRSA continuous attributes which reflect surface active properties are not discretized. Domains of structure attributes, are presented in Table 1.

Table 1. Numerical coding of the structure condition attributes

Code	n	R	Code	R
1		CH_3	8	C_8H_{17}
2	C_2H_5	C_2H_5	9	C_9H_{19}
3	C_3H_7	C_3H_7	10	$C_{10}H_{21}$
4	C_4H_9	C_4H_9	11	$C_{11}H_{23}$
5	C_5H_{11}	C_5H_{11}	12	$C_{12}H_{25}$
6	C_6H_{13}	C_6H_{13}	14	$C_{14}H_{29}$
7		C_7H_{15}	16	$C_{16}H_{33}$

According to the value of MIC objects were sorted into three decision classes:

- class 1 - good antimicrobial properties: MIC ≤ 0.02 μM/L,
- class 2 - medium antimicrobial properties: $0.02 <$ MIC < 1 μM/L,
- class 3 - weak antimicrobial properties: MIC ≥ 1 μM/L.

2.2 Discovery of Decision Rules Using DRSA

In the Dominance-based Rough Set Approach (DRSA) (for a complete presentation of DRSA see, for example, [6]), information about objects (classification examples) is represented in the form of an *information table*. The rows of the table are labeled by objects, whereas columns are labeled by attributes and entries of the table are attribute-values. The set of attributes is, in general, divided into set C of condition attributes and set D of decision attributes (in most of the cases singleton decision attribute d. Condition attributes whose value sets are ordered are called *ordinal attributes*. Without loss of generality, for ordinal attribute $q \in C$, $\phi : U \to \mathbb{R}$, for all objects $x, y \in U, \phi(x) \geq \phi(y)$ means "x is evaluated at least as high as y on ordinal attribute q", which is denoted $x \succeq_q y$. Ordinal attribute q may have positive or negative monotonic relationship with the decision attribute d (which is also ordinal). Positive relationship means that the greater the value of the condition attribute the higher the class label (i.e. the value of decision attribute), and negative relationship means that the greater the value of condition attribute the lower the class label. Furthermore, values of decision attribute d make a partition of U into a finite number of decision classes, $\mathbf{X} = \{X_t, t = 1, \ldots, n\}$, such that each $x \in U$ belongs to one and only one class $X_t \in \mathbf{X}$. It is supposed that the classes are ordered, i.e. for all $r,s \in \{1, \ldots, n\}$, such that $r > s$, the objects from X_r are in higher class than the ones from X_s.

As it was shown in [1] non-ordinal classification problems can be analyzed by DRSA. Such problems need a proper transformation of information table. This transformation is non-invasive, i.e. it does not bias the matter of discovered relationships. The intuition which stands behind this transformation is the following. Each non-ordinal condition attribute, for which the presence or absence and the possible sign of the monotonicity relationship is not known a priori, is doubled and for the first attribute in the pair it is supposed that the monotonicity relationship is potentially positive, while for the second attribute, that it is potentially negative. Due to this transformation, using DRSA, one will be able to find out if the actual monotonicity is global or local, and if it is positive or negative. The non-ordinal decision attributes are transformed such that:

- in case of a non-ordinal decision attribute, each value of this attribute representing a given feature is replaced by a new decision attribute with two values corresponding to presence and absence of this feature, respectively,
- in case of an ordinal decision attribute, each value of interest t, is replaced by a new decision attribute with two values corresponding to original values under and over t, respectively.

More precisely, given a finite set of objects (universe) U described by condition and decision attributes, we assume that the decision attribute makes a partition of U into a finite set of classes $X_1, X_2, \ldots, X_n$. To discover rules relating values of condition attributes with class assignment, in case of non-ordinal classification problems, we have to consider n ordinal binary classification problems with two sets of objects: class X_t and its complement $\neg X_t$, $t = 1, \ldots, n$, which are number-coded by 1 and 0, respectively. We also assume, without loss of generality, that the value sets of all non-ordinal condition attributes are number-coded. While this is natural for numerical attributes, nominal attributes must be binarized and get 0-1 codes for absence or presence of a given nominal value. In this way, the value sets of all non-ordinal attributes get ordered (as all sets of numbers are ordered). Now, to apply DRSA, we transform the data table such that each number-coded attribute is cloned (doubled). It is assumed that the value set of each original number-coded attribute q' is positively monotonically dependent on the decision, i.e. the greater the value of the condition attribute, the higher the number code (rather 1 than 0) of the class assignment, and the value set of its clone q'' is negatively monotonically dependent on the decision, i.e. the greater the value of the condition attribute, the lower the number code (rather 0 than 1) of the class assignment. Then, using DRSA, we get rough approximations of class X_t and its complement $\neg X_t$, $t = 1, \ldots, n$. These approximations serve to induce decision rules recommending assignment to class X_t (argument pros) or to its complement $\neg X_t$ (argument cons). Due to cloning of attributes with opposite monotonicity relationships, we can have rules that cover a subspace in the condition attribute space, which is bounded from the top and from the bottom. This leads (without discretization) to more synthetic rules than those resulting from induction techniques specific to non-ordinal classification problems. The syntax of decision rules is the following: *if Φ then Ψ*, where Φ and Ψ denote *condition* and *decision* part of the rule, called also *premise* and *conclusion*, respectively. The condition part of the rule is a conjunction of elementary conditions concerning individual attributes, and the decision part of the rule suggests assignment to a decision class or to a union of decision classes. The rules that are considered in DRSA analysis are mainly *certain rules*. This kind of rules cover only consistent objects. Decision rules represent the most important cause-effect dependencies between values of condition attributes and value of decision attribute. These rules are not only confined to important condition attributes, but also they include minimal number of elementary conditions indispensable for presentation of the dependencies. The set of decision rules may be understood as the presentation of cause-effect dependencies in the information table, from which all inessential and redundant information was removed. The rules are characterized by various parameters, such as *strength* (i.e. the proportion of objects covered by premise that are also covered by conclusion), or *confirmation* (i.e. measure that is quantifying the degree to which premise provides evidence for conclusion; see [7]).

2.3   Attribute Relevance

We consider attribute relevance measures that satisfy the property of Bayesian confirmation [3]. These measures take into account interactions between attributes represented by decision rules. In this case, the property of confirmation is related to quantification of the degree to which the presence of an attribute in the premise of a rule provides evidence for or against the conclusion of the rule. The measure increases when more rules involving an attribute suggest a correct decision, or when more rules that do not involve the attribute suggest an incorrect decision, otherwise it decreases.

Let us first give some basic definitions. A *rule* induced from a learning data set L can be denoted as $E \rightarrow H$, which reads as "*if E, then H*". A rule consists of a condition part (called also the premise or the evidence) E, and a conclusion (called also the prediction or the hypothesis) H. Considering a finite set of condition attributes $A = \{a_1, a_2, \ldots a_n\}$, we can define the condition part of the rule as a conjunction of elementary conditions on a particular subset of attributes:

$$E = e_{i_1} \wedge e_{i_2} \wedge \ldots \wedge e_{i_p}, \tag{1}$$

where $\{i_1, i_2, \ldots i_p\} \subseteq \{1, 2, \ldots, n\}$, $p \leq n$, and e_{i_h} is an elementary condition defined on the value set of attribute a_{i_h}, $h \in \{i_1, i_2, \ldots i_p\}$ (e.g., $e_{i_h} \equiv a_{i_h} \geq 0.5$).

The set of rules R induced from data set L can be applied to objects from L or to objects from a testing set T. A rule $r \equiv E \rightarrow H$, $r \in R$, covers object x ($x \in L$ or $x \in T$) if x is satisfying the condition part E. We say that the rule is correctly classifying x if it both covers x and x satisfies the decision part H. If the rule covers x, however, x does not satisfy the decision part H, then we say that the rule classifies x incorrectly. In other words, we say that rule r is true for object x if it classifies this object correctly, and it is not true otherwise.

By $a_i \triangleright E$ we denote the fact that E includes an elementary condition e_i involving attribute a_i, $i \in \{1, 2, \ldots n\}$. An opposite fact will be denoted by $a_i \not\triangleright E$. Let us consider object x ($x \in L$ or $x \in T$), and set of rules R. We use the following notation throughout the paper:

$a = |H \wedge (a_i \triangleright E)|$ - the number of rules that correctly classify x and involve attribute a_i in the condition part,

$b = |H \wedge (a_i \not\triangleright E)|$ - the number of rules that correctly classify x and do not involve attribute a_i in the condition part,

$c = |\neg H \wedge (a_i \triangleright E)|$ - the number of rules that incorrectly classify x and involve attribute a_i in the condition part,

$d = |\neg H \wedge (a_i \not\triangleright E)|$ - the number of rules that incorrectly classify x and do not involve attribute a_i in the condition part.

Given a set of objects T on which the set of rules R is applied, the values of a, b, c, d are defined over set T: a is then interpreted as a number of all rules that correctly classify objects from T and involve attribute a_i. Interpretation of the remaining parameters is analogous. The values of a, b, c, and d can be also treated as frequencies that may be used to estimate probabilities.

Formally, a relevance measure $c(H, (a_i \triangleright E))$ has the property of Bayesian confirmation if and only if it satisfies the following conditions:

$$c(H, (a_i \triangleright E)) = \begin{cases} > 0 & \text{if} \quad \Pr(H|(a_i \triangleright E)) > \Pr(H), \\ = 0 & \text{if} \quad \Pr(H|(a_i \triangleright E)) = \Pr(H), \\ < 0 & \text{if} \quad \Pr(H|(a_i \triangleright E)) < \Pr(H). \end{cases} \tag{2}$$

The specific Bayesian confirmation measure used in this study is called c_1 - its precise definition can be found in [7].

3 Results and Discussion

DRSA allows to present knowledge from the information table as a set of rules without unimportant facts, which could obfuscate cause-effect relationships. The whole experimental procedure, behind the presented DRSA analysis, was conceptually simple but computationally expensive. We constructed multiple sets of decision rules on randomly selected subsets of objects. More precisely, we constructed bagging ensembles [2] with VC-DomLEM rule classifiers [4]. The information table was preprocessed as described in Section 2.2 and in [1]. Thus, the numbers of attributes in presented results may be higher than originally.

From the practical point of view the most interesting are the 1st and the 3rd decision class. The rules that assign to these classes provide guidelines which compounds with preferable antimicrobial properties should be synthesized (class 1) and which compounds should not be taken into consideration (class 3). Results of estimation of attribute relevance in rules are presented on Figure 2. We can observe that attributes: -log CMC (critical micelle concentration) and surface excess have the highest impact on the classification of the most active objects to 1st decision class (see Figure 2). On the other hand, the classification of the most active objects to 3rd decision class is influenced the most by attributes -log CMC and surface tension at CMC. Let us now analyze Figure 3 to observe that the most confirmative attributes are crucial from the classification point of view. In the figure, the percentage of correctly classified objects

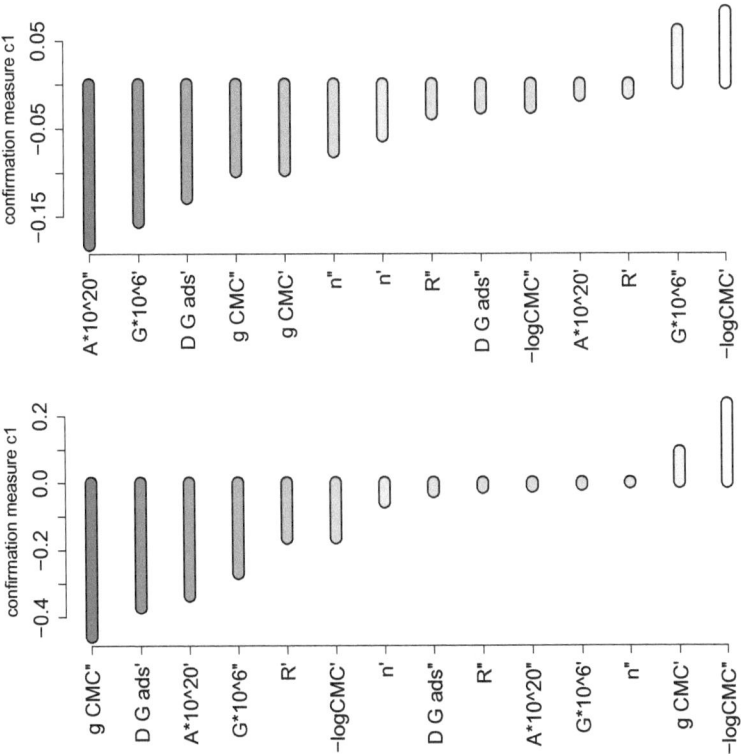

Fig. 2. Predictive attribute confirmation calculated for class 1 (top) and 3 (bottom)

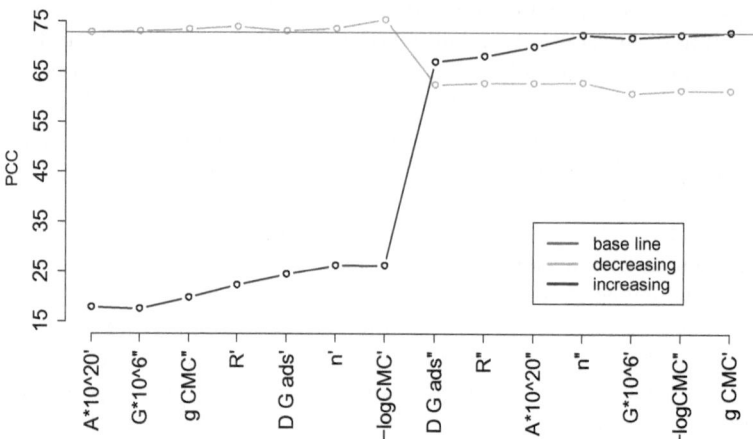

Fig. 3. Classification accuracy of rule classifiers built on consecutive attributes according to the decreasing and increasing evaluation by $c1$ measure

(PCC) by a classifier constructed on all attributes is presented as a base line. All presented values of PCC were estimated in 3-fold stratified cross validation repeated 100 times to get reproducible results. The attributes on the figure are ordered from the least confirmative ($A \cdot 10^{20}$) to the most confirmative (γCMC). Then, by increasing line, we show PCC of the classifiers constructed on set of attributes enlarged by addition of consecutive ones according to their increasing confirmation. We can observe that, for such classifiers, the PCC comparable to the base is achieved after almost all of the attributes are included. Inversely, when we start with one most confirmative attribute (γCMC), and then we add least confirmative ones (decreasing line), we achieve PCC comparable to the base after fewer iterations. Moreover, we achieve PCC better than the base line.

In Table 2, we presented selected strong, most informative, certain decision rules induced from the information table. These are rules that cover at least the half of objects in class, and have high c_1 confirmation. On the basis of these rules, we can recognize the following. The most active compounds are bis-quaternary ammonium chlorides, which posses -log CMC value in the range of 2.71 and 3.13. For these chlorides, value of surface excess is below 2.42 mol/m^2, value of surface tension at CMC is below 46.4 mN/m and the length of R substituent is over C_6H_{13}. The worst class is predominated by bis-quaternary ammonium chlorides, for which value of -log CMC is below 2.45, value of surface tension at CMC is over 44.8 mN/m and value of surface excess is over 2.59 mol/m^2. To this class belong chlorides with short n constituent (below 6).

The results clearly show CMC is the most differentiating attribute which decides whether surface active compound has good antimicrobial properties or not. Ranges of values of condition attributes in decision rules obtained in presented analysis are important from the point of view of the synthesis of new chemical entities with good antimicrobial properties. These ranges present values among which new, effective chemical compounds should be searched for during the synthesis.

Table 2. Decision rules, supporting examples, strength, and value of c_1 measure

ID	Condition attributes							Support	Strength(%)	c_1
	n	R	-logCMC	γCMC	$\Gamma \cdot 10^6$	$A \cdot 10^{20}$	ΔG_{ads}			
			Decision class 1							
1			(2.71, 3.13)	<45.6				12	60	0.7586
2		>6	<3.13	<45.6				12	60	0.7586
3			<3.13	<45.6	<2.41			11	55	0.7330
4			<3.13	<45.6		>71		11	55	0.7330
5		>6	<3.13				>24.8	11	55	0.7330
6			<3.13		<2.41		>24.8	11	55	0.7330
7			(2.84, 3.13)					11	55	0.7330
8			(2.83, 3.13)	<46.4				11	55	0.7330
9			(2.83, 3.13)		<2.42			11	55	0.7330
10			<3.13		<2.40			11	55	0.7330
			Decision class 3							
11				>44.8	>2.59			16	80	0.8703
12					>2.59		<30.2	16	80	0.8703
13				>55.1				15	75	0.8409
14	<6				>2.59			15	75	0.8409
15			<2.45	>44.8				15	75	0.8409
16			<2.45				<30.2	15	75	0.8409
17				>44.8		<62		15	75	0.8409
18						<62	<30.2	15	75	0.8409
19	<6					<62		14	70	0.8125
20	<6		<2.45					14	70	0.8125

Acknowledgment. The second author wish to acknowledge financial support from the Poznań University of Technology, grant no. 91-516/DS-MLODA KADRA. The fourth author wish to acknowledge financial support from the Polish National Science Centre, grant no. N N519 441939.

References

1. Błaszczyński, J., Greco, S., Słowiński, R.: Inductive Discovery of Laws Using Monotonic Rules. Engineering Applications of Artificial Intelligence 25, 284–294 (2012)
2. Błaszczyński, J., Słowiński, R., Stefanowski, J.: Variable Consistency Bagging Ensembles. In: Peters, J.F., Skowron, A. (eds.) Transactions on Rough Sets XI. LNCS, vol. 5946, pp. 40–52. Springer, Heidelberg (2010)
3. Błaszczyński, J., Słowiński, R., Susmaga, R.: Rule-Based Estimation of Attribute Relevance. In: Yao, J., Ramanna, S., Wang, G., Suraj, Z. (eds.) RSKT 2011. LNCS, vol. 6954, pp. 36–44. Springer, Heidelberg (2011)
4. Błaszczyński, J., Słowiński, R., Szeląg, M.: Sequential Covering Rule Induction Algorithm for Variable Consistency Rough Set Approaches. Inf. Sciences 181(5), 987–1002 (2011)
5. Chlebicki, J., Wegrzynska, J.: Surface-active, Micellar, and Antielectrostatic Properties of Bis-ammonium Salts. J. Colloid Interf. Sci. 323, 372–378 (2008)
6. Greco, S., Matarazzo, B., Słowiński, R.: Rough Sets Theory for Multicriteria Decision Analysis. European Journal of Operational Research 129, 1–47 (2001)
7. Greco, S., Słowiński, R., Szczęch, I.: Properties of Rule Interestingness Measures and Alternative Approaches to Normalization of Measures. Inf. Sciences (to appear)

8. Katritzky, A.R., Pacureanu, L.M., Slavov, S.H., Dobchev, D.M., Shah, D.O., Karelson, M.: QSPR Study of the First and Second Critical Micelle Concentrations of Cationic Surfactants. Comput. Chem. Eng. 33, 321–332 (2009)
9. Krysiński, J., Płaczek, J., Skrzypczak, A., Błaszczak, J., Prędki, B.: Analysis of Relationships Between Structure, Surface Properties and Antimicrobial Activity of Quaternary Ammonium Chlorides. QSAR Comb. Sci. 28, 995–1002 (2009)
10. McBain, A.J., Ledder, R.G., Moore, L.E., Catrenich, C.E.: Effects of Quaternary-Ammonium-Based Formulations on Bacterial Community Dynamics and Antimicrobial Susceptibility. Appl. Environ. Microbiol. 70, 3449–3456 (2004)
11. Yoshida, F., Topliss, J.G.: QSAR Model for Drug Human Oral Bioavailability. J. Med. Chem. 43, 2575–2585 (2000)
12. Zupon, J., et al.: Neural Networks for Chemists. Verlag Chemie, Weinheim (1993)

Classification and Decision
Based on Parallel Reducts and F-Rough Sets

Dayong Deng, Lin Chen, and Dianxun Yan

College of Mathematics, Physics and Information Engineering,
Zhejiang Normal University, Jinhua 321004, Zhejiang, China
{dayongd,yandianxun}@163.com, chenlin12345666@126.com

Abstract. F-rough sets are new rough set model, which is consistent
with parallel reducts. In this paper, the methods of classification (decision) with parallel reducts and F-rough sets are discussed. Unlike Pawlak
rough sets or other rough set models, there may be many benchmarks for
classifying(deciding). Three strategies for classifying(deciding) are proposed, including specific decision subsystem, decision subsystem selected
randomly and deciding by a majority vote.

Keywords: F-Rough Sets, Parallel Reducts, Classifying, Deciding.

1 Introduction

Many researchers tried to deal with incremental, dynamical or tremendous data
with rough set theory [1]. Deng and Huang[5] presented a discernibility matrix
and function between two decision tables. Liu[2] introduced an algorithm which
tried to obtain the smallest attribute reducts with incremental data. Wang et
al. [3] proposed a distributed algorithm of attribute reduction based on discernibility matrix and function. In fact, Liu's algorithm [2] and Wang's algorithm
[3] are special cases of Deng's algorithm [5]. Zheng et al. [4] presented an incremental algorithm based on positive region. Kryszkiewicz et al.[6] introduced an
algorithm of attribute reducts in composed decision systems.

Bazan[8] proposed the concept of dynamic reducts to solve the problem of
large amount of data or incremental data. In[7] Deng et al. introduced a new
method to compute stable reducts called parallel reducts in a family of decision
subsystems. In [9] Deng et al. proposed a method of parallel reducts based on
a matrix of attribute significance and obtain both parallel reducts and dynamic
reducts in polynomial time. In[10] Deng et al. introduced a new rough set model
called F-rough sets and attribute significance for parallel reducts. Unlike Pawlak
rough set or other rough set models, F-rough sets are based on a family of
decision subsystems, and consistent with parallel reducts.

Just like Pawlak rough sets and Pawlak reducts, how are parallel reducts and
F-rough sets used to classify(decide) a new data set? Which decision subsystems
should be selected as benchmarks for classifying(deciding)? In this paper, we will
investigate these two issues, and three new strategies for classifying(deciding)
a new data set are proposed. They are specific decision subsystem, decision
subsystem selected randomly and deciding by a majority vote.

T. Li et al. (Eds.): RSKT 2012, LNAI 7414, pp. 117–122, 2012.

2   *F*-Rough Sets and Parallel Reducts

Definition 1. *Let* $DS = (U, A, d)$ *be a decision system, and* $P(DS)$ *the set of all subsystems of* DS, $F \subseteq P(DS)$. *The positive region of* F *(F-positive region in short) is defined as follows:*

$$POS(F, A, d) = \{POS(DT, A, d) : DT \in F\}$$

The *F*-positive region is the set of the positive regions of its elements.

Remark. In this paper, the different forms of $POS()$ always denote the positive region in corresponding situations.

Definition 2. *A subset of condition attributes* $B \subseteq A$ *is a parallel reduct of* F *if and only if it satisfies the following two conditions:*
   (1) $POS(F, B, d) = POS(F, A, d)$,
   (2)For any $S \subset B$, $POS(F, S, d) \neq POS(F, A, d)$.

Example 1. Suppose $F = (DT_1, DT_2)$ corresponding to Table 1 and Table 2 respectively, where a, b, c are condition attributes, d is a decision attribute.

Table 1. Decision Sub-System DS_1

U_1	a	b	c	d
x_1	0	0	1	1
x_2	1	1	0	1
x_3	0	1	0	0
x_4	1	1	0	1

Table 2. Decision Sub-system DS_2

U_2	a	b	c	d
y_1	0	1	0	0
y_2	1	1	0	1
y_3	1	1	0	1
y_4	0	1	0	0
y_5	1	2	0	0
y_6	1	2	0	1

For a concept $X = \{x : d(x) = 1\}$, it is different in the decision subsystems DT_1 and DT_2, and their positive regions are also different. $X|DT_1 = \{x : d(x) = 1\} \cap DT_1 = \{x_1, x_2, x_4\}$, $POS_A(DT_1, X) = \{x_1, x_2, x_4\}$. $X|DT_2 = \{x : d(x) = 1\} \cap DT_2 = \{y_2, y_3, y_6\}$, $POS_A(DT_2, X) = \{y_2, y_3\}$.

It is ease to know $POS_A(DT_1, d) = \{x_1, x_2, x_3, x_4\}$ and $POS_A(DT_2, d) = \{y_1, y_2, y_3, y_4\}$, *F*-positive region $POS(F, A, d) = \{POS_A(DT_1, d), POS_A(DT_2, d)\} = \{\{x_1, x_2, x_3, x_4\}, \{y_1, y_2, y_3, y_4\}\}$. *F*-parallel reducts are $P = \{a, b\}$.

3   Three Types of Classification(Decision) Based on Parallel Reducts and F-Rough Sets

After a parallel reduct P have been obtained, they may be used to classify(decide) a new data set. In the family of decision subsystems there are several decision subsystems. Which decision subsystems are selected as benchmarks for classifying(deciding)? The problem should be solved. The results of classifying(deciding) may be different when decision subsystems for benchmarks are different.

Suppose $F = \{DT_1, DT_2, \cdots, DT_n\}$, where $DT_i = (U_i, A, d)(i = 1, 2, \cdots, n)$. The probability of the decision subsystem DT_i selected is p_i, $\sum_{i=1}^{n} p_i = 1$. X is the new set of data which will be classified(decided). P is a parallel reduct of F. For a specialty decision subsystem DT_i as the benchmark for classifying(deciding), any element $x \in X$, $f(d|x, DT_i)$ denotes the value of decision of $x \in X$ relative to the benchmark DT_i.

$$f(d|x, DT_i) = \begin{cases} \frac{|[x]_P \cap [x]_d|}{|[x]_P|} & if \quad x \in U_i \\ 0 & otherwise. \end{cases}$$

Where $[x]_P$ denotes the equivalence class of x relative to P in DT_i. $|\cdot|$ denotes the cardinality of the set. $f(d|x, DT_i)$ means, if x belongs to U_i, $f(d|x, DT_i) = \frac{|[x]_P \cap [x]_d|}{|[x]_P|}$, otherwise it equals 0.

Example 2. Suppose $F = \{DT_1, DT_2\}$, where DT_1 and DT_2 are corresponding to Table 1 and Table 2 respectively. Parallel reduct $P = \{a, b\}$. Assume that DT_2 is the benchmark for classifying, $x = (1, 2, 0)$ is a new datum, then

$$f(d = 0|x, DT_2) = \frac{|[x]_P \cap [x]_{d=0}|}{|[x]_P|} = \frac{|\{y_5, y_6\} \cap \{y_1, y_4, y_5\}|}{|\{y_5, y_6\}|} = \frac{1}{2}$$

$$f(d = 1|x, DT_2) = \frac{|[x]_P \cap [x]_{d=1}|}{|[x]_P|} = \frac{|\{y_5, y_6\} \cap \{y_2, y_3, y_6\}|}{|\{y_5, y_6\}|} = \frac{1}{2}$$

In the following subsections we will focus on how to select decision subsystems as the benchmarks for classifying(deciding), and three types of strategies will be investigated: specific decision subsystem, decision subsystem selected randomly and deciding by a majority vote.

3.1   Specific Decision Subsystem

Like a single decision system, a specific decision subsystem from the family of decision subsystems is designated as a benchmark when a new set of data should be classified(decided).

How is the specific decision subsystem selected from the family of decision subsystems? There are three ways to select the specific decision subsystem.

(1) The specific decision system is designated by experts. (2) The specific decision subsystem is selected when the data in it are the last. (3) The specific decision subsystem is selected when the data of the specific decision subsystem come from the same or similar situation as the new set of data for classifying(deciding).

After the specific decision subsystem is designated, the new data set can be classified(decided) according to the specific decision subsystem for the benchmark, just like there are only one decision subsystem in the family of decision subsystems.

3.2   Decision Subsystem Selected Randomly

There are two types of randomly selecting when new decisions should be made according to the parallel reduct P. Firstly, a decision subsystem is selected randomly, and it becomes the benchmark for classifying(deciding) all of the new set of data. Secondly, a decision subsystem is selected randomly when a decision should be made in the new set of data. Like specific decision subsystem, the decision subsystem for classifying(deciding)in the first situation is fixed when it has selected randomly. Whenever a decision should be made in the secondly situation, the decision subsystem for classifying(deciding) should be selected randomly.

In the first situation, when a benchmark DT_i from a family of decision subsystems is selected at a probability p_i, the new data set X can be classified(decided) according to the benchmark DT_i.

Example 3. The family of decision subsystem $F = \{DT_1, DT_2\}$ corresponding to Table 1 and Table 2. The new data set for classifying(deciding) is $X = \{z_1, z_2, z_3, z_4\} = \{(0, 1, 0), (1, 1, 0), (0, 0, 0), (0, 0, 1)\}$. DT_1 is selected as a benchmark for classifying(deciding) at the probability 0.6. Then

$$f(d = 0|z_1, DT_1) = \frac{|[z_1]_P \cap [z_1]_{d=0}|}{|[z_1]_P|} = \frac{|\{x_3\} \cap \{x_3\}|}{|\{x_3\}|} = 1$$

similarly,$f(d = 1|z_1, DT_1) = f(d = 0|z_2, DT_1) = 0$, $f(d = 1|z_2, DT_1) = 1$, $f(d = 0|z_3, DT_1) = 0$, $f(d = 1|z_3, DT_1) = 1$, $f(d = 0|z_4, DT_1) = 0$, $f(d = 1|z_4, DT_1) = 1$. From the above computing, $z_1 \to d = 0$,$z_2 \to d = 1$, $z_3 \to d = 1$ and $z_4 \to d = 1$.

In the second situation, when a new datum should be classified(decided), a benchmark from the family of decision subsystems should be selected at a probability. For example, roulette wheel method could be applied when a benchmark is selected.

Example 4. In Example 3, assume that DT_1 is selected at a probability of 50%, and DT_2 50%. Assume z_1, z_2 are classified(decided) by DT_1, and z_3, z_4 by DT_2. The processes of classifying (deciding) are as follows:

$$f(d = 0|z_1, DT_1) = \frac{|[z_1]_P \cap [z_1]_{d=0}|}{|[z_1]_P|} = \frac{|\{x_3\} \cap \{x_3\}|}{|\{x_3\}|} = 1$$

similarly, $f(d = 1|z_1, DT_1) = f(d = 0|z_2, DT_1) = 0$, $f(d = 1|z_2, DT_1) = 1$, $f(d = 0|z_3, DT_2) = f(d = 1|z_3, DT_2) = f(d = 0|z_4, DT_2) = (d = 1|z_4, DT_2) = 0$. From the above computing, $z_1 \to d = 0, z_2 \to d = 1$. However, z_3 and z_4 can not be classified(decided).

3.3   Deciding by a Majority Vote

In this strategy, when a new datum should be classified or a new decision should be made, every decision subsystems are the benchmarks for classifying(deciding). The last decision is made by a majority vote.

$$\max_d \{ \sum_{DT_i \in F} f(d|x, DT_i) \}$$

Any element x for classifying(deciding) is based on the above function.

Example 5. The family of decision subsystem $F = \{DT_1, DT_2\}$ corresponding to Table 1 and Table 2 respectively. The new data set for classifying(deciding) is $X = \{z_1, z_2, z_3, z_4\} = \{(0, 1, 0), (1, 1, 0), (0, 0, 0), (0, 0, 1)\}$.

The values of f for elements of X relative to the benchmark DT_1 are as follows:

$$f(d = 0|z_1, DT_1) = \frac{|[z_1]_P \cap [z_1]_{d=0}|}{|[z_1]_P|} = \frac{|\{x_3\} \cap \{x_3\}|}{|\{x_3\}|} = 1$$

similarly, $f(d = 1|z_1, DT_1) = f(d = 0|z_2, DT_1) = 0$, $f(d = 1|z_2, DT_1) = 1$, $f(d = 0|z_3, DT_1) = 0$, $f(d = 1|z_3, DT_1) = 1$, $f(d = 0|z_4, DT_1) = 0$, $f(d = 1|z_4, DT_1) = 1$

The values of f for elements of X relative to the benchmark DT_2 are as follows: $f(d = 0|z_1, DT_2) = 1$, $f(d = 1|z_1, DT_2) = 0$, $f(d = 0|z_2, DT_2) = 0$, $f(d = 1|z_2, DT_2) = 1$, $f(d = 0|z_3, DT_2) = 0$, $f(d = 1|z_3, DT_2) = 0$, $f(d = 0|z_4, DT_2) = 0$, $f(d = 1|z_4, DT_2) = 0$

So, $\sum_{DT_i \in F} f(d = 0|z_1, DT_i) = 2$, $\sum_{DT_i \in F} f(d = 1|z_1, DT_i) = 0$, $\sum_{DT_i \in F} f(d = 0|z_2, DT_i) = 0$, $\sum_{DT_i \in F} f(d = 1|z_2, DT_i) = 2$, $\sum_{DT_i \in F} f(d = 0|z_3, DT_i) = 0$, $\sum_{DT_i \in F} f(d = 1|z_3, DT_i) = 1$, $\sum_{DT_i \in F} f(d = 0|z_4, DT_i) = 0$, $\sum_{DT_i \in F} f(d = 1|z_4, DT_i) = 1$. According to the above results, the classifications(decisions) for elements of X are, $z_1 \to d = 0$, $z_2 \to d = 1$, $z_3 \to d = 1$, $z_4 \to d = 1$. The element z_3, z_4 can not be classified in DT_2 but can do in DT_1.

4   Conclusion

In this paper, we investigate how to classify(decide) a new set of data with parallel reducts and F-rough sets. Three strategies of classifying(deciding) are proposed, including specific decision subsystem, decision subsystem selected randomly and deciding by a majority vote. Unlike Pawlak rough set or other rough set models, there may be many benchmarks from a family of decision subsystems for classifying(deciding)when a new set of data are classified(decided).

For a new datum, the result of classifying(deciding) may be different when strategies of classifying(deciding)are different. This method is consistent with that of machine learning.

In the future we will investigate more strategies for classifying(deciding).

References

1. Pawlak, Z.: Rough Sets-Theoretical Aspect of Reasoning about Data. Kluwer Academic Publishers, Dordrecht (1991)
2. Liu, Z.: An Incremental Arithmetic for the Smallest Reduction of Attributes. Acta Electronica Sinica 27(11), 96–98 (1999) (in Chinese)
3. Wang, J., Wang, J.: Reduction algorithms based on discernibility matrix: The order attributes method. Journal of Computer Science and Technology 16(6), 489–504 (2001)
4. Zheng, Z., Wang, G., Wu, Y.: A Rough Set and Rule Tree Based Incremental Knowledge Acquisition Algorithm. In: Wang, G., Liu, Q., Yao, Y., Skowron, A. (eds.) RSFDGrC 2003. LNCS (LNAI), vol. 2639, pp. 122–129. Springer, Heidelberg (2003)
5. Deng, D., Huang, H.: A New Discernibility Matrix and Function. In: Wang, G.-Y., Peters, J.F., Skowron, A., Yao, Y. (eds.) RSKT 2006. LNCS (LNAI), vol. 4062, pp. 114–121. Springer, Heidelberg (2006)
6. Kryszkiewicz, M., Rybinski, H.: Finding Reducts in Composed Information Systems. In: Proceedings of International Workshop on Rough Sets and Knowledge Discovery (RSKD 1993), pp. 259–268 (1993)
7. Deng, D., Wang, J., Li, X.: Parallel Reducts in a Series of Decision Subsystems. In: Proceedings of the Second International Joint Conference on Computational Sciences and Optimization (CSO 2009), Sanya, Hainan, China, pp. 377–380 (2009)
8. Bazan, G.J.: Dynamic Reducts and Statistical Inference. In: Proceedings of the Sixth International Conference, Information Processing and Management of Uncertainty in Knowledge Based Systems (IPMU 1996), pp. 1147–1152 (1996)
9. Deng, D., Yan, D., Wang, J.: Parallel Reducts Based on Attribute Significance. In: Yu, J., Greco, S., Lingras, P., Wang, G., Skowron, A. (eds.) RSKT 2010. LNCS (LNAI), vol. 6401, pp. 336–343. Springer, Heidelberg (2010)
10. Deng, D., Yan, D., Chen, L.: Attribute Significance for F-Parallel Reducts. In: Proceedings of 2011 IEEE International Conference on Granular Computing, pp. 156–161 (2011)

Evidential Clustering or Rough Clustering: The Choice Is Yours

Manish Joshi[1] and Pawan Lingras[2]

[1] Department of Computer Science,
North Maharashtra University, Jalgaon, India
joshmanish@gmail.com
[2] Department of Mathematics and Computing Science,
Saint Mary's University, Halifax, Canada
pawan@cs.smu.ca

Abstract. A crisp cluster does not share an object with other clusters. But in real life situations for several applications such rigidity is not acceptable. Hence, Fuzzy and Rough variations of a popular K-means algorithm are proposed to obtain non-crisp clustering solutions.

An Evidential c-means proposed by Masson and Denoeux [6] in the theoretical framework of belief functions uses Fuzzy c-means (FCM) to build upon basic belief assignments to determine cluster membership. On the other hand, Rough clustering uses the concept of lower and upper approximation to synthesize clusters. A variation of popular K-means algorithm namely Rough k-means (RKM) is proposed and experimented with various datasets.

In this paper we analyzed both the algorithms using synthetic, real and standard datasets to determine similarities of these two clustering approaches and focused on the strengths of each approach.

Keywords: Rough Clustering, Fuzzy Clustering, Rough k-means, Fuzzy c-means, belief functions, Evidential c-means.

1 Introduction

The conventional crisp clustering techniques group objects into separate clusters. Each object is assigned to only one cluster. The term crisp clustering refers to the fact that the cluster boundaries are strictly defined and object's cluster membership is unambiguous. Such a requirement is found to be too restrictive in many data mining applications [3]. In practice, an object may display characteristics of different clusters. In such cases, an object should belong to more than one cluster, and as a result, cluster boundaries necessarily overlap.

The conventional clustering algorithm like K-means categorizes an object into precisely one cluster. Whereas, fuzzy clustering [1,7] and rough set clustering [5,8] provide ability to specify the membership of an object to multiple clusters, which can be useful in real world applications.

Fuzzy set representation of clusters, using algorithms such as fuzzy C-means (FCM), make it possible for an object to belongs to multiple clusters with a

T. Li et al. (Eds.): RSKT 2012, LNAI 7414, pp. 123–128, 2012.
© Springer-Verlag Berlin Heidelberg 2012

degree of membership between 0 and 1 [7]. Evidential c-means (ECM) proposed by [6] is an extension of FCM and noise clustering algorithm proposed by Dave et al. [2]. It clearly identifies objects that belong to one or more clusters by the virtue of their position in the problem space. Basic belief assignments values are computed for all possible combinations of k clusters (2^k partitions), which are used to determine cluster membership. Rough k-means algorithm (RKM) [5] groups objects into lower and upper regions of clusters and an object can belongs to an upper region of multiple clusters.

Due to space limitations, readers are referred to the descriptions of RKM and ECM from [6] and [4]. The data sets used for the experimental analysis are Synthetic data set, Library data set, Character Recognition data set, Iris data set and Diamond data set. Except Diamond data set [6] all other data sets' information is available in [4]. Analysis of the two algorithms is completed and the observations regarding similarities, strengths of the ECM algorithm, and the strengths of the RKM algorithm are presented in section 2, 3 and 4 respectively; finally, section 5 concludes the paper.

2   Similarities between ECM and RKM

In almost all experiments with above mentioned data sets we observed similarities between the results obtained using ECM and RKM. We discuss some of the similarities by providing appropriate tabular results.

In case of synthetic data set both the algorithms are good enough to clearly group 60 objects in three distinct clusters correctly. There is a consensus between the results of the two algorithms regarding the remaining five objects too. The RKM algorithm assigns multiple cluster membership to objects 61 to 65 whereas ECM also assigns high 'mass of belief' (m_i) to these objects for a class that consists of more than one clusters as shown in Table 1.

Table 1. Bba values for sample objects of synthetic data set

Objects	ϕ	$\{c_1\}$	$\{c_2\}$	$\{c_1, c_2\}$	$\{c_3\}$	$\{c_1, c_3\}$	$\{c_2, c_3\}$	Ω
61	0.1935	0.019	0.0504	0.0203	0.0545	0.0281	**0.5977**	0.0364
62	0.0461	0.0141	0.0064	0.0077	0.0127	**0.8761**	0.0122	0.0247
63	0.1916	0.0812	0.0729	**0.3798**	0.0226	0.0478	0.0472	0.1569
64	**0.6052**	0.0285	0.0687	0.0237	0.1242	0.0314	0.0915	0.0267
65	0.1657	0.0313	0.0448	0.0368	0.0428	0.0715	**0.3295**	0.2774

For objects 61, 62 and 63 both RKM and ECM assigns membership of cluster ($\{c_2, c_3\}$), $\{c_1, c_3\}$, and $\{c_1, c_2\}$ respectively. RKM clustering result is as shown in Table 2. For objects 64 and 65 however, there is a minute difference in the result. The object 64 has been assigned a membership of two clusters ($\{c_2, c_3\}$) by RKM, whereas ECM reckons the object as an outlier. Similarly, RKM assigns membership of all three clusters to object 65 but ECM evaluates it to be a member of only two clusters ($\{c_2, c_3\}$).

Table 2. RKM result for Synthetic data set

Clusters	Objects in lower approximation	Objects in upper approximation
$\{c_1\}$	1 to 20	62, 63, 65
$\{c_2\}$	21 to 40	61, 63, 64, 65
$\{c_3\}$	41 to 60	61, 62, 64, 65

For a standard Iris data set, the results obtained using the ECM and the RKM are in sync with each other. Both the algorithms correctly classified the first 50 plant instances to 'Iris Setosa' class. The other two classes are not linearly separable and both the ECM and RKM could figure it out. Some of the plant instances in class 'Iris Versicolour' and 'Iris Virginica' are ambiguous and this ambiguity is well recorded by both the algorithms. The information of correctly classified instances (CCI) for Iris data set for both the algorithms is given in Table 3.

Table 3. CCI attained for IRIS Data Set by ECM and RKM

Clusters	ECM	RKM
$\{c_1\}$	50	50
$\{c_2\}$	44	49
$\{c_3\}$	34	44

Masson et al. [6] reported experimental results for Diamond data set. We applied RKM to this data set and demonstrate using Table 4 that both algorithms are generating almost the same results except outliers.

Table 4. ECM and RKM result for Diamond data set

ECM				RKM		
ϕ	$\{c_1\}$	$\{c_2\}$	Ω	Cluster	Lower Approximation	Upper Approximation
$\{12\}$	$\{1,2,3,4,5\}$	$\{7,8,9,10,11\}$	$\{6\}$	$\{c_1\}$	$\{1,2,3,4,5,6\}$	$\{7,12\}$
				$\{c_2\}$	$\{8,9,10,11\}$	$\{7,12\}$

The ECM explicitly identified object 12 as an outlier and hence the cluster centroids are properly positioned to distinctly identify two clusters of five objects each. The RKM on the other hand, has two options with object 12. Either it can be placed in upper approximation of both clusters or in lower approximation of second cluster. In order to accommodate the outlier (12th object) cluster centroids get shifted in RKM and apparently a small difference in cluster members is visible in Table 4.

In subsequent sections we shall discuss experimental results to focus strengths of each algorithm.

3   Strengths of ECM

The obvious strength of the ECM lies in its capability to assign multiple cluster membership to objects that lie on fringes of various clusters using the bba values. Tables 1 shows how objects can be assigned to a class of clusters. For example, object 62 in Table 1 is assigned to two clusters $\{c_1, c_3\}$ as the bba value 0.8761 for this group of clusters is the highest. Even though it seems to be an information overload for data set that has more numbers of clusters (Letter Recognition data set has 8 clusters and hence ECM generated 256 classes), the generated bba values determine membership of objects to multiple clusters.

The other important claim for ECM is its ability to pinpoint outlier objects. For Synthetic data set as shown in Table 1, object 64 is identified as an outlier. And the object 64 corresponds to actually an outlier with respect to other points in the data set. For Diamond data set too, the ECM algorithm outperformed the RKM in determining an outlier object effectively (Table 4).

ECM is good at determining outlier from data set, if any object possesses such characteristics. We tested if ECM wrongly identifies any object as an outlier. We tested for a real world Library data set and a standard Iris data set. We know that all objects in the Library data set are distributed in the range of [0,1] and does not have outlier. Similarly, for Iris data set all 150 plant instances correspond to three given class and none of the instances is an outlier. Table 5 displays the count of how many instances of Library and Iris data set are partitioned by the ECM algorithm in a particular class. We can observe that the ECM identified correctly that none of the objects in the two data sets are outliers.

Table 5. Objects distribution by ECM for Library and Iris data set

Clusters	Library	Iris
ϕ	**0**	**0**
$\{c_1\}$	599	47
$\{c_2\}$	301	21
$\{c_1, c_2\}$	184	15
$\{c_3\}$	281	50
$\{c_1, c_3\}$	184	6
$\{c_2, c_3\}$	194	2
Ω	152	9

4   Strengths of RKM

Similar to the ECM algorithm the RKM algorithm has its primary strength in modeling the ambiguity while assigning the cluster membership to objects.

RKM algorithm is simple and easy to implement. It takes less time for execution as compared to other non-crisp clustering algorithms including the ECM.

Each cluster is represented as a set of lower approximation members and upper approximation members. The objects in the lower approximation belongs to the cluster with certainty while objects in upper approximation of one cluster also additionally belongs to one or more clusters' upper approximation. Instead of assigning descriptive measure of membership (as in FCM or ECM) the cluster boundary is delineated using the approximation concept.

As opposed to FCM and ECM, a cluster can be easily visualized with the help of RKM. Both the FCM and the ECM generates a detailed matrix of size $n \times k$ for FCM and $n \times 2^k$ for ECM. Either the fuzzy membership value generated by FCM or the basic belief assignment value generated by the ECM determines up to what extent an individual object belongs to a particular cluster. Despite of a detailed quantitative measure of individual object's membership, it is difficult to practically formulate a cluster and its members using this descriptive information. In a sense it is an overload of information. A general trend that is followed in FCM and ECM is to pick up the highest value in a row and assign the membership accordingly. But while doing this, we shift our focus from soft clustering to hard clustering.

The strength of RKM lies in its simplicity. We can see in Tables 1 and 2 that cluster formation using RKM is obvious whereas in the ECM for every object the highest bba value decides cluster membership. We have seen that synthetic data set results for both the ECM and RKM are almost similar. Similarly, for all other data sets too the cluster formation using RKM is faster and mostly qualitative. We can see in Table 3 the RKM precision of clustering Iris plant instances is better than that of the ECM. A minimum of 76% of precision is attained using the RKM clustering.

We can see that the RKM can withstand high dimensional data set and gives good results. Both the FCM and ECM could not produce reasonable results for the high dimensional data sets [4]. We tested RKM for the Letter Recognition data set. For the Letter Recognition data set, each cluster is labeled using the character that appears most frequently in the cluster. 'Frequency' indicates the number of events in the cluster that match the cluster label. An Average Precision is defined as the sum of the 'Frequency' for all clusters divided by the total number of events in the data set. Table 6 shows the results obtained when the RKM clustering algorithms is applied to the Letter Recognition data set.

For RKM the uncertainty of an object's cluster membership is discerned and represented with the help of boundary regions of distinct clusters. Hence we can see that more numbers of objects are matching with the label of clusters in RKM. The Average Precision (Average Precision-2) for RKM is more than 75%. If we would have considered only lower approximation objects of the clusters then the average precision (Average Precision-1) would have been only 53%.

It is easy to verify that RKM is good at partitioning large numbers of objects with high dimensionality with reasonable precision.

Table 6. Average precision for the Letter Recognition Data Set

RKM			
Cluster Label	Frequency	Average Precision-1	Average Precision-2
M	496(174)		
H	191(189)		
L	509(338)		
M	526(181)	53.04	75.02
O	710(594)		
A	709(642)		
Z	722(552)		
P	724(573)		

5   Conclusion

In this paper we applied the two non-crisp clustering algorithms namely Evidential c-means (ECM) and Rough k-means (RKM) to various data sets. We briefly introduced the basic concepts of the ECM and the RKM. With the help of several experimental results we analyzed the similarity between the results obtained using the two algorithms. Furthermore, the apparent strengths of each algorithm are stated and verified with the experiments.

The comparison of rough sets and belief functions have helped theoretical enhancements in both the theories. We expect similar benefits by comparing the rough and belief function based clustering and propose to develop a more efficient algorithm that combines strengths of both techniques.

References

1. Bezdek, J.C., Hathaway, R.J.: Optimization of fuzzy clustering criteria using genetic algorithms. In: International Conference on Evolutionary Computation, pp. 589–594 (1994)
2. Dave, R.N.: Clustering relational data containing noise and outliers. Pattern Recogn. Lett. 12, 657–664 (1991)
3. Joshi, A., Krishnapuram, R.: Robust fuzzy clustering methods to support web mining. In: Proc. Workshop in Data Mining and knowledge Discovery, SIGMOD, pp. 15–22 (1998)
4. Joshi, M., Lingras, P., Rao, C.R.: Analysis of Rough and Fuzzy Clustering. In: Yu, J., Greco, S., Lingras, P., Wang, G., Skowron, A. (eds.) RSKT 2010. LNCS, vol. 6401, pp. 679–686. Springer, Heidelberg (2010)
5. Lingras, P., West, C.: Interval set clustering of web users with rough k-means. Journal of Intelligent Information Systems 23, 5–16 (2004)
6. Masson, M., Denoeux, T.: Ecm: An evidential version of the fuzzy c-means algorithm. Pattern Recognition 41, 1384–1397 (2008)
7. Pedrycz, W., Waletzky, J.: Fuzzy clustering with partial supervision. IEEE Transactions on Systems, Man, and Cybernetics, Part B 27(5), 787–795 (1997)
8. Peters, G.: Some refinements of rough k-means clustering. Pattern Recognition 39(8), 1481–1491 (2006)

Heuristic for Attribute Selection Using Belief Discernibility Matrix

Salsabil Trabelsi[1], Zied Elouedi[1], and Pawan Lingras[2]

[1] Larodec, Institut Superieur de Gestion de Tunis, Tunisia
[2] Saint Mary's University Halifax, Canada

Abstract. This paper proposes a new heuristic attribute selection method based on rough sets to remove the superfluous attributes from partially uncertain data. We handle uncertainty only in decision attributes (classes) under the belief function framework. The simplification of the uncertain decision table which is based on belief discernibility matrix generates more significant attributes with fewer computations without making significant sacrifices in classification accuracy.

Keywords: Uncertainty, belief function theory, rough sets, attribute selection, classification.

1 Introduction

A problem of relevant feature selection is one of the important problems in pre-processing stage of the modeling process in machine learning. There are several attempts to solve this problem based on rough set theory [5,14]. Using features from the reduct of the decision system adds to the efficiency of the classification process [8,9]. However, finding optimal reduct is an NP-hard problem. Researchers have proposed credible heuristics that compute acceptable reducts in reasonable time [19,20]. Another issue in real world database is the uncertainty, imprecision or incompleteness. There is some research that adapts exhaustive or heuristic feature selection methods based on rough sets to uncertain environment [16,17].

In this paper, we develop a new heuristic attribute selection method from partially uncertain data based on rough sets. The uncertainty exists only in the decision attribute and is represented by the belief function theory. It is considered as a useful theory for representing and managing total or partial uncertain knowledge because of its relative flexibility. The belief function theory is widely applied in artificial intelligence and to real life problems for decision making and classification. In this paper, we use the Transferable Belief Model (TBM) - one interpretation of belief function theory [13]. To remove the superfluous and the redundant attributes from the uncertain decision table, we adapt the concept of discernibility matrix in the new context to be used in the heuristic algorithm proposed originally in [12] to compute sufficient reduct without costly calculations.

T. Li et al. (Eds.): RSKT 2012, LNAI 7414, pp. 129–138, 2012.
© Springer-Verlag Berlin Heidelberg 2012

This paper is organized as follows: Section 2 provides an overview of the rough set theory. Section 3 introduces the belief function theory as understood in the TBM. Section 4 describes the proposed heuristic for attribute selection based on rough sets under uncertainty. Section 5 details a belief rough set classifier a new classification system able to generate uncertain decision rules from partially uncertain data where the feature selection is one the important steps in the construction procedure. Section 6 reports the experimental results obtained from modified uncertain databases to evaluate the performance of our solution based on two evaluation criteria: the time requirement and the classification accuracy.

2   Rough Set Theory

In this section, we recall some basic notions related to information systems and rough sets [6,7]. An information system is a pair $A = (U, C)$, where U is a non-empty, finite set called the *universe* and C is a non-empty, finite set of attributes. We also consider a special case of information systems called decision tables. A decision table is an information system of the form $A = (U, C \cup \{d\})$, where $d \notin C$ is a distinguished attribute called *decision*. In this paper, the notation $c_i(o_j)$ is used to represent the value of a condition attribute $c_i \in C$ for $o_j \in U$.

For every set of attributes $B \subseteq C$, an equivalence relation denoted by IND_B and called the B-indiscernibility relation, is defined by

$$IND_B = U/B = \{[o_j]_B | o_j \in U\} \tag{1}$$

Where

$$[o_j]_B = \{o_i | \forall c \in B \, c(o_i) = c(o_j)\} \tag{2}$$

Let $B \subseteq C$ and $X \subseteq U$. We can approximate X by constructing the $B - lower$ and $B - upper$ *approximations* of X, denoted $\underline{B}(X)$ and $\bar{B}(X)$, respectively, where

$$\underline{B}(X) = \{o_j | [o_j]_B \subseteq X\} \ and \ \bar{B}(X) = \{o_j | [o_j]_B \cap X \neq \emptyset\} \tag{3}$$

A reduct is a minimal subset of attributes from C that preserves the positive region and the ability to perform classifications as the entire attributes set C. A subset $B \subseteq C$ is a reduct of C with respect to d, iff B is minimal and:

$$Pos_B(\{d\}) = Pos_C(\{d\}) \tag{4}$$

Where $Pos_C(\{d\})$, called a positive region of the partition $U/\{d\}$ with respect to C.

$$Pos_C(\{d\}) = \bigcup_{X \in U/\{d\}} \underline{C}(X) \tag{5}$$

The core is the most important subset of attributes, it is included in every reduct.

$$Core(A, \{d\}) = \bigcap RED(A, \{d\}) \tag{6}$$

Where $RED(A, \{d\})$ is the set of all reducts of A relative to d.

Let's not that finding a minimal reduct (reduct with a minimal number of attributes) among all reducts is NP-hard. This means that computing reducts is not a trivial task. Fortunately, there exist good heuristics that compute sufficiently many reducts in often acceptable time [19,20].

3   Belief Function Theory

The belief function theory is proposed by Shafer [10] as a useful tool to represent uncertain knowledge. Here, we introduce only some basic notations related to the TBM [13], one interpretation of the belief function theory. Let Θ, frame of discernment, be a finite set of exhaustive elements to a given problem. All the subsets of Θ belong to the power set of Θ, denoted by 2^{Θ}. The bba (basic belief assignment) is a function representing the impact of a piece of evidence on the subsets of the frame of discernment Θ and is defined as follows:

$$m : 2^{\Theta} \to [0, 1]$$

$$\sum_{E \subseteq \Theta} m(E) = 1 \tag{7}$$

Where $m(E)$, named a basic belief mass (bbm), shows the part of belief exactly committed to the element E. The bba's induced from distinct pieces of evidence are combined by the cojunctive rule of combination [11].

$$(m_1 \bigcirc m_2)(E) = \sum_{F,G \subseteq \Theta : F \cap G = E} m_1(F) \times m_2(G) \tag{8}$$

To make decisions in the TBM, belief functions can be represented by probability functions called the pignistic probabilities denoted $BetP$ and are defined as [11]:

$$BetP(\{a\}) = \sum_{F \subseteq \Theta} \frac{|\{a\} \cap F|}{|F|} \frac{m(F)}{(1 - m(\emptyset))} \text{ for all } a \in \Theta \tag{9}$$

4   Heuristic for Attribute Selection Method Using Belief Discernibility Matrix

In this section, a new heuristic for simplification of partially uncertain decision system is proposed. We will remove the superfluous and redundant attributes for rules discovery without costly computations. We will keep only the features from the reduct. It is a minimal set of attributes that preserves the ability to classify as much as the entire set of condition attributes. First, we will give an overview of the proposed approach followed by experimental verification.

4.1 Uncertain Decision Table

Our uncertain decision system denoted A contains n objects $\{o_1, o_2, \ldots, o_n\}$, characterized by a set of certain condition attributes $C=\{c_1, c_2,...,c_k\}$ and uncertain decision attribute ud. We propose to represent the uncertainty of each object by a bba m_j expressing belief on decision defined on the frame of discernment $\Theta=\{ud_1, ud_2,...,ud_s\}$ representing the possible values of ud.

Example: Let us use Table 1 to describe our uncertain decision system. It contains eight objects, three certain condition attributes $C=\{$Headache, Muscle-pain, Temperature$\}$ and an uncertain decision attribute ud=Flu with possible value $\{$yes, no$\}$ representing Θ. For example, for the patient o_4, belief of 0.6 is exactly committed to the decision ud_1=yes, whereas belief of 0.4 is assigned to the entire frame of discernment Θ (ignorance).

Table 1. Uncertain decision table

U	Headache	Muscle-pain	Temperature	Flu
o_1	yes	yes	very high	$m_1(yes) = 1$
o_2	yes	yes	high	$m_2(yes) = 1$
o_3	yes	no	high	$m_3(yes) = 0.95 \quad m_3(\Theta) = 0.05$
o_4	no	yes	normal	$m_4(no) = 0.6 \quad m_4(\Theta) = 0.4$
o_5	no	yes	normal	$m_5(no) = 1$
o_6	no	no	high	$m_6(no) = 1$
o_7	yes	no	normal	$m_7(no) = 1$

4.2 Feature Selection with Johnson´s Heuristic Algorithm

This part describes the Johnson´s heuristic algorithm [12] to compute reducts. It sequentially selects features by finding those that are most discernible for a given decision feature. It computes a discernibility matrix M, where each cell $M_{i,j}$ of the matrix corresponding to the set of all condition attributes which discern objects o_i and o_j that do not belong to the same equivalence classes based on decision attribute d. The discernibility matrix is a symmetric n*n matrix with entries $M_{i,j}$ as given below.

$$M_{i,j} = \{c \in C | c(o_i) \neq c(o_j) \ for \ d(o_i) \neq d(o_j)\} \ \forall i, j = 1, ..., n \qquad (10)$$

Given such a matrix M, for each feature, the algorithm counts the number of cells in which it appears. The feature c with the highest number of entries is selected for addition to the reduct R. Then, all the entries $M_{i,j}$ that contain c are removed and the next best feature is selected. This procedure is repeated until M is empty.

Johnson´s Reduct (U,C,{d})
Input: U: objects, C:conditional features, d: decisional features,
Output: R:reduct, R $\subseteq$ C
1. R $\leftarrow \emptyset$
2. M $\leftarrow$ ComputeDiscernibilityMatrix (U,C,{d})
3. **do**
4. c $\leftarrow$ SelectHighestScoringFeature (M)
5. R $\leftarrow$ R $\cup$ {c}
6. **for** (i=0 to $|U|$,j=0 to $|U|$)
7. $M_{i,j}=\emptyset$ **if** c $\in M_{i,j}$
8. C $\leftarrow$ C $\setminus$ {c}
9. **Until** $M_{i,j} =\emptyset$ $\forall$ i,j
10. **Return** R

4.3   Belief Discernibility Matrix

In order to apply the previous heuristic attribute selection method to our uncertain decision table, we should adapt the discernibility matrix under the belief function framework which is originally based on certain decision attribute to be called belief discernibility matrix (M'). The instruction (2.) in the previous algorithm will be changed as follows:

> **2.** M' $\leftarrow$ ComputeBeliefDiscernibilityMatrix (U,C,{ud})

In this case, the belief discernibility matrix will be based on a distance measure to identify the similarity or dissimilarity between decision values of the objects o_i and o_j. The idea is to use the distance measure between two bba's m_i and m_j. The threshold value is used to be more flexible. Hence, belief discernibility matrix M' is defined as follows:

$$M'_{i,j} = \{c \in C | c(o_i) \neq c(o_j) \ for \ dist(m_i, m_j) \geq threshold\} \forall i, j = 1, ..., n \quad (11)$$

Where *dist* is a distance measure between two bba's proposed in [2] which satisfies more properties than many other distance measures proposed in [1,3,4] as defined below.

$$dist(m_1, m_2) = \sqrt{\frac{1}{2}(\| \vec{m_1} \|^2 + \| \vec{m_2} \|^2 -2 < \vec{m_1}, \vec{m_2} >)} \quad (12)$$

$$0 \leq dist(m_1, m_2) \leq 1 \quad (13)$$

Where $< \vec{m_1}, \vec{m_2} >$ is the scalar product defined by:

$$< \vec{m_1}, \vec{m_2} >= \sum_{i=1}^{|2^\Theta|} \sum_{j=1}^{|2^\Theta|} m_1(A_i)m_2(A_j)\frac{|A_i \cap A_j|}{|A_i \cup A_j|} \quad (14)$$

with $A_i, A_j \in 2^\Theta$ for $i, j = 1, 2, \cdots, |2^\Theta|$. $\| \vec{m_1} \|^2$ is then the square norm of $\vec{m_1}$.

Example: To apply our heuristic feature selection method to the uncertain decision system (see Table 1), we start by computing the belief discernibility matrix (see Table 2). We will use the notations H, M and T respectively for Headache, Muscle-pain and Temperature. To obtain Table 2, we have used Equation 11 with a threshold value equal to 0.1. For example, $M'_{1,3} = \emptyset$ because the two objects o_1 and o_3 have $dist(m_1, m_3) = 0.07 \not\geq 0.1$. The decision values of the two objects are considered similar. Next, we compute the reduct according to the Johnson´s heuristic algorithm. First, we find that the attribute Temperature has the highest number of entries which is equal to 11. So, we add Temperature to the reduct. Then, we remove all the cells $M'_{i,j}$ containing T. We still have only H and H, M in the matrix. So, the attribute Headache has now the highest number of entries which is equal to 2. We add the attribute Headache to the reduct. Then, we remove all the cells $M'_{i,j}$ containing H. The matrix is now empty and the process is finished.

Table 2. Belief disecrnibiliy matrix

U	o_1	o_2	o_3	o_4	o_5	o_6	o_7
o_1							
o_2							
o_3							
o_4	H,T	H,T	H,M,T				
o_5	H,T	H,T	H,M,T				
o_6	H,M,T	H,M	H	M,T			
o_7	M,T	M,T	T	H,M			

5   Belief Rough Set Classifier

Belief rough set classifier is a new classification technique proposed originally in [18]. The latter was able to learn belief decision rules for the classification process from partially uncertain data (see Table 1).

The decision rules induced from the uncertain decision table are called belief decision rules where the decision is represented by a bba.

Example: *The belief decision rule relative the object o_3 from the Table 1 is as follows: If Headache = yes and Muscle − pain = no and Temperature = high Then $m_3(yes) = 0.95$ $m_3(\Theta) = 0.05$.*

To create a belief rough set classifier, we need to simplify the uncertain decision system to generate the more significant belief decision rules by means of the following steps:

1. **Step 1. Eliminate the superfluous condition attributes:** We remove the superfluous condition attributes that are not in reduct. We can apply in this step our proposed heuristic for feature selection.

2. **Step 2. Eliminate the redundant objects:** After removing the super-fluous condition attributes, we will find redundant objects. They may not have the same bba on decision attribute. So, we use their combined bbas using a rule of combination.

3. **Step 3. Eliminate the superfluous condition attribute values:** In this step, we compute the reduct value for each belief decision rule R_j of the form: **If** $C(o_j)$ **then** m_j.

4. **Step 4. Generate belief decision rules:** After the simplification of the uncertain decision table, we can generate shorter and significant belief decision rules. With simplification, we can improve the time and the performance of classification of unseen objects.

Once the belief rough set classifier is constructed, the following procedure will be the classification of unseen instances. Our method is able to ensure the standard classification where each attribute value of the new instance to classify is assumed to be exact and certain. We search among all belief decision rules which one corresponds to the unseen object. The new instance's decision will be defined by a bba. In order to make a decision and to get the probability of each singular decision, we apply the pignistic transformation using eqn. (9). We can take only the most probable decision.

6   Experimental Results

In our experiments, several tests were performed on real-world databases obtained from the U.C.I. repository[1] to evaluate the proposed heuristic feature selection method in comparison with exhaustive search proposed originally in [17]. A brief description of the databases is presented in Table 3.

The comparison is based on two evaluation criteria: the time requirement (the number of seconds needed to find the reduct) and the classification accuracy (Percent of Correct Classification (PCC)). To compute PCC, we apply our two methods in the first step of the belief rough set classifier described in Section 5. The belief rough set classifier generated decision rules that were used for classification. These databases were artificially modified in order to include uncertainty in decision attribute. We took different degrees of uncertainty based on increasing values of probabilities P used to transform the actual decision value d_i of each object o_j to a bba $m_j(\{d_i\}) = 1 - P$ and $m_j(\Theta) = P$. A larger P gives a larger degree of uncertainty. Each database is divided into ten parts. Nine parts are used as the training set, the last is used as the testing set. The procedure is repeated ten times, each time another part is chosen as the testing set. This method, called a cross-validation, permits a more reliable estimation of the evaluation criterion. In this paper, we perform ten-fold cross-validation tests with different data splits and we report the average of the evaluation criteria.

---

[1] http://www.ics.uci.edu/~mlearn/MLRepository.html

Table 3. Description of databases

Databases	#instances	#attributes	#decision values
W. Breast Cancer	690	8	2
Balance Scale	625	4	3
C. Voting records	497	16	2
Zoo	101	17	7
Nursery	12960	8	3
Solar Flares	1389	10	2
Lung Cancer	32	56	3
Hyes-Roth	160	5	3
Car Evaluation	1728	6	4
Lymphography	148	18	4
Spect Heart	267	22	2
Tic-Tac-Toe Endgame	958	9	2

Table 4. Experimentation results

Databases	Exhaustive Mean Time (s)	Heuristic Mean Time (s)	Gain in Time (s)	Gain in Time (%)	Exhaustive Mean PCC (%)	Heuristic Mean PCC (%)	Loss of PCC (%)
W. Breast Cancer	154	74	80	52%	86.60	85.46	1.14
Balance Scale	129	51	78	60%	83.43	83.13	0.3
C. Voting records	110	73	37	34%	98.78	98.10	0.68
Zoo	101	46	55	54%	96.41	95.70	0.72
Nursery	380	209	171	45%	96.49	95.65	0.84
Solar Flares	157	113	44	28%	88.69	88.18	0.51
Lung Cancer	48	37	11	23%	75.91	75.69	0.21
Hyes-Roth	91	46	45	49%	97.46	97.23	0.23
Car Evaluation	178	149	29	16%	84.40	83.84	0.56
Lymphography	102	79	23	23%	83.20	82.01	0.95
Spect Heart	109	87	22	20%	85.40	84.70	0.69
Tic-Tac-Toe Endgame	139	109	30	22%	86.37	85.95	0.42
Mean	141	89	52	37%	88.59	85.95	0.6

In Table 4, we report the average time requirement and the classification accuracy for different degrees of uncertainty. From this table, we see that the proposed heuristic feature selection method is faster than the exhaustive search method for attribute selection. It is true for all the databases (see Figure 2). For example, the time requirement for Balance Scale database goes from 129 seconds to 51 seconds. Gain of speed is equal to 78 seconds or 60%. The average gain in speed relative to all databases is 52 seconds or 37%. This gain in computational speed came at very little loss of accuracy (see Figure 1). For example, the classification accuracy for Lung Cancer database goes from 75.91% to 75.69%. The loss of accuracy is equal to 0.21%. The average loss of accuracy relative to all databases is also very small and is equal to (0.6%).

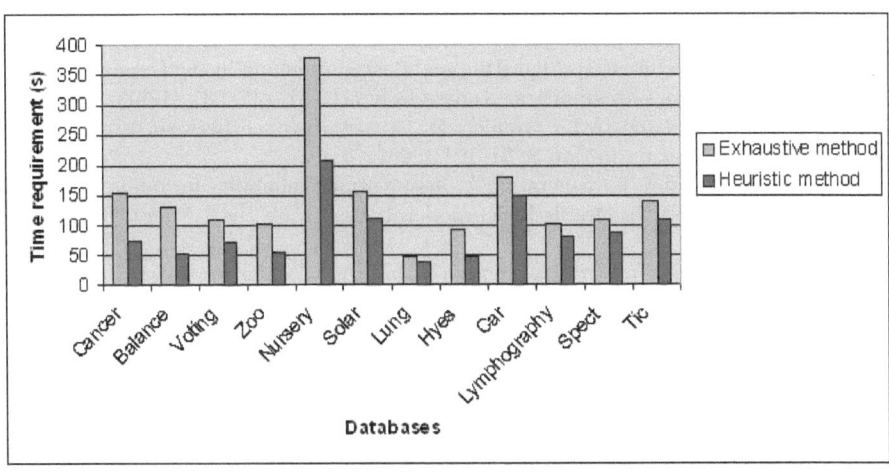

Fig. 1. Classification accuracy for exhaustive and heuristic methods

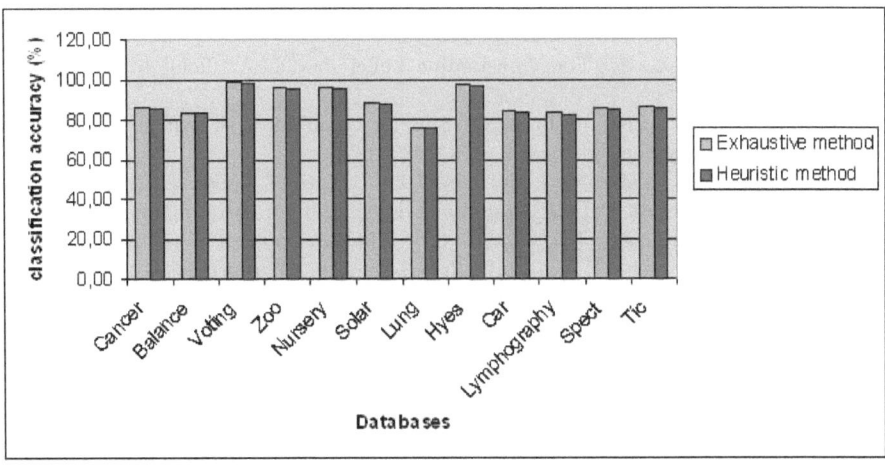

Fig. 2. Time requirement for exhaustive and heuristic methods

7   Conclusion and Future Work

In this paper, we have proposed a heuristic attribute selection method to remove the superfluous attributes from uncertain decision table in less time with minimal loss of classification accuracy. We handle uncertainty in decision attributes using the belief function. The proposed heuristic method was compared with the exhaustive search for optimal set of reduced features. The average gain in computational speed for twelve standard databases was 37%, while the loss of classification accuracy was mere 0.6%.

References

1. Bauer, M.: Approximations algorithm and decision making in the Dempster-Shafer theory of evidence - an empirical study. IJAR 17(2-3), 217–237 (1997)
2. Bosse, E., Jousseleme, A.L., Grenier, D.: A new distance between two bodies of evidence. Information Fusion 2, 91–101 (2001)
3. Elouedi, Z., Mellouli, K., Smets, P.: Assessing sensor reliability for multisensor data fusion within the transferable belief model. IEEE Trans. Syst. Man Cyben. 34(1), 782–787 (2004)
4. Fixen, D., Mahler, R.P.S.: The modified Dempster-Shafer approach to classification. IEEE Trans. Syst. Man Cybern. 27(1), 96–104 (1997)
5. Modrzejewski, M.: Feature selection using rough sets theory. In: Proceedings of the 11th International Conference on Machine Learning, pp. 213–226 (1993)
6. Pawlak, Z.: Rough Sets. International Journal of Computer and Information Sciences 11, 341–356 (1982)
7. Pawlak, Z., Zdzislaw, A.: Rough Sets: Theoretical Aspects of Reasoning About Data. Kluwer Academic Publishing, Dordrecht (1991) ISBN 0-7923-1472-7
8. Pawlak, Z., Rauszer, C.M.: Dependency of attributes in Information systems. Bull. Polish Acad. Sci., Math. 33, 551–559 (1985)
9. Rauszer, C.M.: Reducts in Information systems. Fundamenta Informaticae (1990)
10. Shafer, G.: A mathematical theory of evidence. Princeton University Press, Princeton (1976)
11. Smets, P., Kennes, R.: The transferable belief model. Artificial Intelligence 66, 191–236 (1994)
12. Johnson, D.S.: Approximation algorithms for combinatorial problems. Journal of Computer and System Sciences 9, 256–278 (1974)
13. Smets, P.: The transferable belief model for quantified belief representation. In: Gabbay, D.M., Smets, P. (eds.) Handbook of Defeasible Reasoning and Uncertainty Management Systems, vol. 1, pp. 207–301. Kluwer, Doordrecht (1998)
14. Tanaka, H., Ishibuchi, H., Matuda, N.: Reduction of information system based on rough sets and its application to fuzzy expert system. Advancement of Fuzzy Theory and Systems in china and Japan (1990)
15. Tessem, B.: Approximations for efficient computation in the theory of evidence. Artif. Intell 61(2), 315–329 (1993)
16. Trabelsi, S., Elouedi, Z.: Heuristic method for attribute selection from partially uncertain data using rough sets. International Journal of General Systems 39(3), 271–290 (2010)
17. Trabelsi, S., Elouedi, Z.: Attribute selection from partially uncertain data using rough sets. In: The Third International Conference on Modeling Simulation, and Applied Optimization, UAE, January 20-22 (2009)
18. Trabelsi, S., Elouedi, Z., Lingras, P.: Belief Rough Set Classifier. In: Gao, Y., Japkowicz, N. (eds.) AI 2009. LNCS (LNAI), vol. 5549, pp. 257–261. Springer, Heidelberg (2009)
19. Wroblewski, J.: Finding minimal reducts using genetic algorithms. In: Proceedings of the 2nd Annual Joint Conference on Information Sciences, pp. 186–189 (1995)
20. Zhong, N., Dong, J.Z., Ohsuga, S.: Using Rough Sets with Heuristics for Feature Selection. Journal of Intelligent Information Systems 16(3), 199–214 (2001)

Incremental Rules Induction
Based on Rule Layers

Shusaku Tsumoto and Shoji Hirano

Department of Medical Informatics, School of Medicine,
Faculty of Medicine, Shimane University
89-1 Enya-cho Izumo 693-8501 Japan
{tsumoto,hirano}@med.shimane-u.ac.jp

Abstract. This paper proposes a new framework for incremental learning based on accuracy and coverage. Classification of addition of example into four cases gives two inequalities for accuracy and coverage. The proposed method classifies a set of formulae into three layers: rule layer, subrule layer and non-rule layer by using the inequalities obtained. Then, subrule layer plays a central role in updating rules. The proposed method was evaluated on a dataset on meningitis, whose results show that it outperforms other conventional rule induction methods.

Keywords: incremental rule induction, rough sets, accuracy, coverage, subrule layer.

1 Introduction

There have been proposed several symbolic inductive learning methods, such as induction of decision trees [1–3], and AQ family [4–6]. These methods are applied to discover meaningful knowledge from large databases, and their usefulness is in some aspects ensured. However, most of the approaches induces rules from all the data in databases, and cannot induce incrementally when new samples are derived. Thus, we have to apply rule induction methods again to the databases when such new samples are given, which causes the computational complexity to be expensive even if the complexity is n^2.

Thus, it is important to develop incremental learning systems in order to manage large databases [7, 8]. However, most of the previously introduced learning systems have the following two problems: first, those systems do not outperform ordinary learning systems, such as AQ15 [6], C4.5 [9] and CN2 [4]. Secondly, those incremental learning systems mainly induce deterministic rules. Therefore, it is indispensable to develop incremental learning systems which induce probabilistic rules to solve the above two problems.

Extending concepts of rule induction methods based on rough set theory, we introduce a new approach to knowledge acquisition, which induces probabilistic rules incrementally, called PRIMEROSE-INC2 (Probabilistic Rule Induction Method based on Rough Sets for Incremental Learning Methods).

T. Li et al. (Eds.): RSKT 2012, LNAI 7414, pp. 139–148, 2012.

This method first calculates all the accuracy and coverage values of attributes and induces rules. Then, it classifies attribute into rule layer and subrule layer. Then when an additional example is given, the method is classified into one of the four cases. Then, it updates rule layer and subrule layer and induce rules. The method repeats this process.

The paper is organized as follows: Section 2 makes a brief description about rough set theory and the definition of probabilistic rules based on this theory. Section 3 discusses problems in incremental learning of probabilistic rules. Section 4 gives formal analysis of incremental update of accuracy and coverage, where two important inequalities are obtained. Then, Section 5 presents an induction algorithm for incremental learning based on the above results, which was evaluated in Section 6. Finally, Section 7 concludes this paper.

2   Rough Sets and Probabilistic Rules

2.1   Rough Set Theory

Rough set theory clarifies set-theoretic characteristics of the classes over combinatorial patterns of the attributes, which are precisely discussed by Pawlak [10, 11]. This theory can be used to acquire some sets of attributes for classification and can also evaluate how precisely the attributes of database are able to classify data.

Table 1. An example of dataset

No.	loc	nat	his	nau	class
1	who	per	per	no	m.c.h.
2	who	per	per	no	m.c.h.
3	lat	thr	per	no	migraine
4	who	thr	per	yes	migraine
5	who	per	per	no	psycho

NOTATIONS. loc: location, nat: nature
his: history, nau: nausea, who: whole
lat: lateral, per: persistent, thr: throbbing,
m.c.h.: muscle contraction headache,
migraine: classic migraine,
psycho: psychogenic headache

Let us illustrate the main concepts of rough sets which are needed for our formulation. Table 1 is a small example of database which collects the patients who complained of headache. First, let us consider how an attribute "loc" classify the headache patients' set of the table. The set whose value of the attribute "loc" is equal to "who" is {1,2,4,5}, which shows that the 1st, 2nd, 4th, 5th case (In the following, the numbers in a set are used to represent each record number). This set means that we cannot classify {1,2,4,5} further solely by using the constraint $R = [loc = who]$. This set is defined as the indiscernible set over the relation R and described as follows: $[x]_R = \{1, 2, 4, 5\}$. In this set, $\{1,2\}$ suffer from muscle contraction headache("m.c.h."), $\{4\}$ from classical migraine("migraine"), and $\{5\}$ from psychological headache("psycho"). Hence we need other additional attributes to discriminate between "m.c.h.", "migraine", and "psycho". Using this concept, we can evaluate the classification power of each attribute. For example, "nat=thr" is specific to the case of classic migraine ("migraine"). We can also extend this indiscernible relation to multivariate cases, such as $[x]_{[loc=who]\wedge[nau=no]} = \{1, 2\}$ and $[x]_{[loc=who]\vee[nat=no]} = \{1, 2, 4, 5\}$, where $\wedge$ and $\vee$ denote "and" and "or" respectively. In the framework of rough set theory, the set $\{1,2\}$ is called *strictly definable* by the former conjunction, and also called *roughly definable* by the latter disjunctive formula. Therefore, the classification of training samples D can be viewed as a search for the best set $[x]_R$ which is supported by the relation R. In this way, we can define the characteristics of classification in the set-theoretic framework. For example, accuracy and coverage, or true positive rate can be defined as:

$$\alpha_R(D) = \frac{|[x]_R \cap D|}{|[x]_R|} \quad and \quad (1)$$

$$\kappa_R(D) = \frac{|[x]_R \cap D|}{|D|}, \quad (2)$$

where $|A|$ denotes the cardinality of a set A, $\alpha_R(D)$ denotes an accuracy of R as to classification of D, and $\kappa_R(D)$ denotes a coverage, or a true positive rate of R to D, respectively. For example, when R and D are set to $[nau = yes]$ and $[class = migraine]$, $\alpha_R(D) = 1/1 = 1.0$ and $\kappa_R(D) = 1/2 = 0.50$.

It is notable that $\alpha_R(D)$ measures the degree of the sufficiency of a proposition, $R \rightarrow D$, and that $\kappa_R(D)$ measures the degree of its necessity. For example, if $\alpha_R(D)$ is equal to 1.0, then $R \rightarrow D$ is true. On the other hand, if $\kappa_R(D)$ is equal to 1.0, then $D \rightarrow R$ is true. Thus, if both measures are 1.0, then $R \leftrightarrow D$. For further information on rough set theory, readers could refer to [10, 11].

2.2   Probabilistic Rules

The simplest probabilistic model is that which only uses classification rules which have high accuracy and high coverage.[1] This model is applicable when rules of high accuracy can be derived. Such rules can be defined as:

$$R \overset{\alpha,\kappa}{\rightarrow} d \text{ s.t.} \quad R = \vee_i R_i = \vee \wedge_j [a_j = v_k],$$
$$\alpha_{R_i}(D) > \delta_\alpha \text{ and } \kappa_{R_i}(D) > \delta_\kappa,$$

where δ_α and δ_κ denote given thresholds for accuracy and coverage, respectively. For the above example shown in Table 1, probabilistic rules for m.c.h. are given as follows:

---

[1] In this model, we assume that accuracy is dominant over coverage.

$$[loc = who] \land [nau = no] \to m.c.h.$$
$$\alpha = 2/3 = 0.67, \ \kappa = 1.0,$$
$$[nat = per] \qquad \to m.c.h.$$
$$\alpha = 2/3 = 0.67, \ \kappa = 1.0,$$

where δ_α and δ_κ are set to 0.5 and 0.3, respectively.

It is notable that this rule is a kind of probabilistic proposition with two statistical measures, which is one kind of an extension of Ziarko's variable precision model(VPRS) [11].[2]

3   Problems in Incremental Rule Induction

The most important problem in incremental learning is that it does not always induce the same rules as those induced by ordinary learning systems[3], although an applied domain is deterministic. Furthermore, since induced results are strongly dependent on the former training samples, the tendency of overfitting is larger than the ordinary learning systems.

The most important factor of this tendency is that the revision of rules is based on the formerly induced rules, which is the best way to suppress the exhaustive use of computational resources. However, when induction of the same rules as ordinary learning methods is required, computational resources will be needed, because all the candidates of rules should be considered.

Thus, for each step, computational space for deletion of candidates and addition of candidates should be needed, which causes the computational speed of incremental learning to be slow. Moreover, in case when probabilistic rules should be induced, the situation becomes much severer, since the candidates for probabilistic rules become much larger than those for deterministic rules.

For the above example, no deterministic rule can be derived from Table 1. Then, when an additional example is given as shown in Section 5 (the sixth example), $[loc = lat] \land [nau = yes] \to m.c.h.$ will be calculated. However, in the case of probabilistic rules, two rule will be derived under the condition that $\delta_\alpha = 0.5$ and $\delta_\kappa = 0.3$. If these thresholds are not used, induced probabilistic rules becomes much larger. Thus, there is a trade-off between the performance of incremental learning methods and its computational complexity.

In our approach, we first focus on the performance of incremental learning methods, that is, we introduce a method which induces the same rules as those derived by ordinary learning methods. Then, we estimate the effect of this induction on computational complexity.

---

[2] In VPRS model, the two kinds of precision of accuracy is given, and the probabilistic proposition with accuracy and two precision conserves the characteristics of the ordinary proposition. Thus, our model is to introduce the probabilistic proposition not only with accuracy, but also with coverage.

[3] Here, ordinary learning systems denote methods that induce all rules by using all the samples.

4   Theory for Incremental Learning

[Question] How accuracy and coverage will change when a new sample is added to the dataset ?

Usually, datasets will monotonically increase. Let $[x]_R(t)$ and $D(t)$ denote a supporting set of a formula R in given data an a target concept d at time t.

$$[x]_R(t+1) = \begin{cases} [x]_R(t) + 1 & \textit{an additional example satisfies } R \\ [x]_R(t) & \textit{otherwise} \end{cases}$$

$$D(t+1) = \begin{cases} D(t) + 1 & \textit{an additional example belongs} \\ & \textit{to a target concept } d. \\ D(t) & \textit{otherwise} \end{cases}$$

Thus, from the definition of accuracy (Eqn.(1) and coverage (Eqn. (2)), accuracy and coverage may nonmonotonically change. Since the above classification gives four additional patterns, we will consider accuracy and coverage for each case as shown in Table 2. in which $|[x]_R(t)|$, $|D(t)|$ and $|[x]_R \cap D(t)|$ are denoted by n_R, n_D and n_{RD}.

Table 2. Four patterns for an additional example

t:	$[x]_R(t)$	$D(t)$	$[x]_R \cap D(t)$
original	n_R	n_D	n_{RD}

t+1	$[x]_R(t+1)$	$D(t+1)$	$[x]_R \cap D(t+1)$
Both negative	n_R	n_D	n_{RD}
R: positive	$n_R + 1$	n_D	n_{RD}
d: positive	n_R	$n_D + 1$	n_{RD}
Both positive	$n_R + 1$	$n_D + 1$	$n_{RD} + 1$

4.1   Both: Negative

The first case is when an additional example does not satisfy R and does not belong to d.

In this case,

$$\alpha(t+1) = \frac{n_{RD}}{n_R} \quad \textit{and} \quad \kappa(t+1) = \frac{n_{RD}}{n_D}.$$

4.2   R: Positive

The second case is when an additional example satisfies R, while it does not belong to d.

Table 3. An additional example neither satisfies R nor d

t+1	$[x]_R(t+1)$	$D(t+1)$	$[x]_R \cap D(t+1)$
Both negative	n_R	n_D	n_{RD}

Table 4. An additional example only satisfies R

t+1	$[x]_R(t+1)$	$D(t+1)$	$[x]_R \cap D(t+1)$
R: positive	n_R+1	n_D	n_{RD}

In this case, accuracy and coverage become:

$$\Delta\alpha(t+1) = \alpha(t+1) - \alpha(t) = \frac{n_{RD}}{n_R+1} - \frac{n_{RD}}{n_R} = \frac{-\alpha(t)}{n_R+1}$$

$$\alpha(t+1) = \alpha(t) + \Delta\alpha(t+1) = \frac{\alpha(t)n_R}{n_R+1}.$$

4.3   d: Positive

The third case is when an additional example does not satisfy R, while it belongs to d.

Table 5. An additional example only satisfies d

t+1	$[x]_R(t+1)$	$D(t+1)$	$[x]_R \cap D(t+1)$
d: positive	n_R	n_D+1	n_{RD}

$$\Delta\kappa(t+1) = \kappa(t+1) - \kappa(t) = \frac{n_{RD}}{n_D+1} - \frac{n_{RD}}{n_D} = \frac{-\kappa(t)}{n_D+1}$$

$$\kappa(t+1) = \kappa(t) + \Delta\kappa(t+1) = \frac{\kappa(t)n_D}{n_D+1}.$$

4.4   d: Positive

Finally, the fourth case is when an additional example satisfies R and belongs to d.

$$\alpha(t+1) = \frac{\alpha(t)n_R+1}{n_R+1} \quad and \kappa(t+1) = \frac{\kappa(t)n_D+1}{n_D+1}.$$

Table 6. An additional example satisfies R and d

t+1	$[x]_R(t+1)$	$D(t+1)$	$[x]_R \cap D(t+1)$
Both positive	n_R+1	n_D+1	$n_{RD}+1$

Table 7. Summary of change of accuracy and coverage

Mode				$\alpha(t+1)$	$\kappa(t+1)$
Both negative	n_R	n_D	n_{RD}	$\alpha(t)$	$\kappa(t)$
R: positive	n_R+1	n_D	n_{RD}	$\frac{\alpha(t)n_R}{n_R+1}$	$\kappa(t)$
d: positive	n_R	n_D+1	n_{RD}	$\alpha(t)$	$\frac{\kappa(t)n_D}{n_D+1}$
Both positive	n_R+1	n_D+1	$n_{RD}+1$	$\frac{\alpha(t)n_R+1}{n_R+1}$	$\frac{\kappa(t)n_D+1}{n_D+1}$

Thus, in summary, Table 7 gives the classification of four cases of an additional example.

4.5   Updates of Accuracy and Coverage

From Table 7, updates of Accuracy and Coverage can be calculated from the original datasets for each possible case. Since rules is defined as a probabilistic proposition with two inequalities, supporting sets should satisfy the following constraints:

$$\alpha(t+1) > \delta_\alpha \ \kappa(t+1) > \delta\kappa$$

Then, the conditions for updating can be calculated from the original datasets: when accuracy or coverage does not satisfy the constraint, the corresponding formula should be removed from the candidate of rules. On the other hand, both accuracy and coverage satisfy both constraints, the formula should be included into the candidate. Thus, the following inequalities are important:

$$\alpha(t+1) = \frac{\alpha(t)n_R+1}{n_R+1} > \delta_\alpha, \kappa(t+1) = \frac{\kappa(t)n_D+1}{n_D+1} > \delta\kappa,$$
$$\alpha(t+1) \quad = \frac{\alpha(t)n_R}{n_R+1} < \delta_\alpha, \kappa(t+1) = \frac{\kappa(t)n_D}{n_D+1} < \delta\kappa.$$

Thus, the following inequalities are obtained for accuracy and coverage.

Theorem 1. *If accuracy and coverage of a formula R to d satisfies one of the following inequalities, then R may exclude or include into the candidates of formulae for probabilistic rules.*

$$\frac{\delta_\alpha(n_R+1)-1}{n_R} < \alpha_R(D)(t+1) < \frac{\delta_\alpha(n_R+1)}{n_R}, \tag{3}$$

$$\frac{\delta_\kappa(n_D+1)-1}{n_D} < \kappa_R(D)(t+1) < \frac{\delta_\kappa(n_D+1)}{n_D}. \tag{4}$$

It is notable that the lower and upper bounds can be calculated from the original datasets.

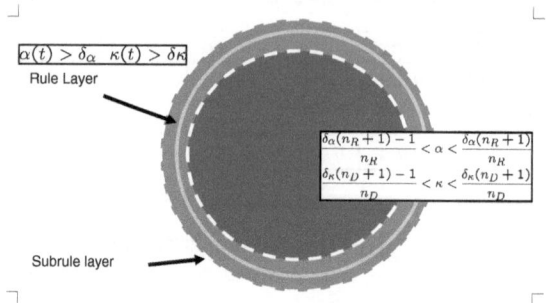

Fig. 1. Intuitive Diagram of Rule and Subrule Layers

Select all the formula whose accuracy and coverage satisfy the above inequalities They will be a candidate for updates. A set of formulae which satisfies the inequalities for probabilistic rules is called *rule layer* and one which satisfies Eqn (3) and (4) is called *subrule layer*. Figure 1 illustrates the relations between rule layer and sublayer.

5   An Algorithm for Incremental Learning

5.1   Algorithm

In order to provide the same classificatory power to incremental learning methods as ordinary learning algorithms, we introduce an incremental learning method PRIMEROSE-INC2 (Probabilistic Rule Induction Method based on Rough Sets for Incremental Learning Methods)[4].

From the results in the above section, a selection algorithm is defined as follows, where the following four lists are used. $List_c$ stores a formula which satisfies the above inequalities shown in Eqn (3) and (4). This $list_c$ is called *subrule layer*. $List_a$ is a list of formulae probabilistic rules which satisfy the condition on the thresholds of accuracy and coverage, which called *rule layer*. Finally, $List_r$ stores a list of formulae which do not satisfy the above condition.

1. Apply rule induction to the initial data.
2. Store formulae in rules into $List_a$ and others in $List_r$.
3. Calculate upper and lower bounds for accuracy and coverage from accuracy, coverage and given thresholds for rules.
4. Check the above inequalities for attributes in $List_a$ and $List_r$. If an attribute satisfies one of the inequalities, then it includes into $List_c$.
5. From an additional example, classify addition from four cases.

---

[4] This is an extended version of PRIMEROSE-INC[12].

(a) An additional example satisfies a target and a formula in $List_c$. Move a formula from $List_r$ to $List_a$.
(b) An additional example only satisfies a formula in $List_c$. Move a formula from $List_a$ to $List_r$.
(c) An additional example only satisfies a target in $List_c$. Move a formula from $List_a$ to $List_r$.
(d) An additional example neither satisfies a target nor a formula in $List_c$. No movement.
6. Generate rules from $List_a$.
7. Return to (3).

6   Experimental Results

PRIMEROSE-INC2[5] was applied to meningitis which consists of 140 examples with two classes and 25 attributes. The proposed method was compared with PRIMEROSE [13], PRIMEROSE-INC[6], C4.5, CN2 and AQ15. The experiments were conducted by the following three procedures. First, these samples were randomly splits into pseudo-training samples and pseudo-test samples. Second, by using the pseudo-training samples, PRIMEROSE-INC2, PRIMEROSE-INC, and PRIMEROSE induced rules and the statistical measures[7]. Third, the induced results were tested by the pseudo-test samples. These procedures were repeated for 100 times and average each accuracy and the estimators for accuracy of diagnosis over 100 trials. Table 8 give the comparison between PRIMEROSE-INC2 and other rule induction methods with respect to the averaged classification accuracy and the number of induced rules. These results show that PRIMEROSE-INC2 outperforms all the other non-incremental learning methods, although they need much larger memory space for running.

Table 8. Experimental results

Method	Accuracy	No. of Rules
PRIMEROSE-INC2	$89.8 \pm 3.4\%$	77.2 ± 2.2
PRIMEROSE-INC	$89.5 \pm 5.4\%$	67.3 ± 3.5
PRIMEROSE	$89.5 \pm 5.4\%$	67.3 ± 3.0
C4.5	$85.8 \pm 2.4\%$	16.3 ± 2.1
CN2	$87.0 \pm 3.9\%$	19.2 ± 1.7
AQ15	$86.2 \pm 2.6\%$	31.2 ± 2.1

---

[5] The program is implemented by using SWI-prolog.
[6] This version is given by setting δ_α to 1.0 and δ_κ to 0.0.
[7] The thresholds δ_α and δ_κ is set to 0.75 and 0.5, respectively in these experiments.

7 Conclusion

Extending concepts of rule induction methods based on rough set theory, we have introduced a new approach to knowledge acquisition, which induces probabilistic rules incrementally, called PRIMEROSE-INC2 (Probabilistic Rule Induction Method based on Rough Sets for Incremental Learning Methods).

The method classifies elementary attribute-value pairs into three categories: rule layer, subrule layer and non-rule layer by using the obtained by inequalities obtained from a proposed framework. This system was evaluated on clinical databases on meningitis. The results show that PRIMEROSE-INC2 outperforms previously proposed methods.

Acknowledgements. This research is supported by Grant-in-Aid for Scientific Research (B) 21300052 from Japan Society for the Promotion of Science (JSPS).

References

1. Breiman, L., Freidman, J., Olshen, R., Stone, C.: Classification And Regression Trees. Wadsworth International Group, Belmont (1984)
2. Cestnik, B., Kononenko, I., Bratko, I.: Assistant 86: A knowledge-elicitation tool for sophisticated users. In: EWSL, pp. 31–45 (1987)
3. Quinlan, J.R.: Induction of decision trees. Machine Learning 1(1), 81–106 (1986)
4. Clark, P., Niblett, T.: The cn2 induction algorithm. Machine Learning 3 (1989)
5. Michalski, R.S.: A theory and methodology of inductive learning. Artif. Intell. 20(2), 111–161 (1983)
6. Michalski, R.S., Mozetic, I., Hong, J., Lavrac, N.: The multi-purpose incremental learning system aq15 and its testing application to three medical domains. In: AAAI, pp. 1041–1047 (1986)
7. Shan, N., Ziarko, W.: Data-based acqusition and incremental modification of classification rules. Computational Intelligence 11, 357–370 (1995)
8. Utgoff, P.E.: Incremental induction of decision trees. Machine Learning 4, 161–186 (1989)
9. Quinlan, J.: C4.5 - Programs for Machine Learning. Morgan Kaufmann, Palo Alto (1993)
10. Pawlak, Z.: Rough Sets. Kluwer Academic Publishers, Dordrecht (1991)
11. Ziarko, W.: Variable precision rough set model. Journal of Computer and System Sciences 46, 39–59 (1993)
12. Tsumoto, S.: Incremental Rule Induction Based on Rough Set Theory. In: Kryszkiewicz, M., Rybinski, H., Skowron, A., Raś, Z.W. (eds.) ISMIS 2011. LNCS, vol. 6804, pp. 70–79. Springer, Heidelberg (2011)
13. Tsumoto, S., Tanaka, H.: Primerose: Probabilistic rule induction method based on rough sets and resampling methods. Computational Intelligence 11, 389–405 (1995)

Optimization of Inhibitory Decision Rules Relative to Length and Coverage

Fawaz Alsolami[1], Igor Chikalov[1], Mikhail Moshkov[1], and Beata Zielosko[1,2]

[1] Mathematical and Computer Sciences & Engineering Division
King Abdullah University of Science and Technology
Thuwal 23955-6900, Saudi Arabia
{fawaz.alsolami,igor.chikalov,
mikhail.moshkov,beata.zielosko}@kaust.edu.sa
[2] Institute of Computer Science, University of Silesia
39, Będzińska St., 41-200 Sosnowiec, Poland

Abstract. The paper is devoted to the study of algorithms for optimization of inhibitory rules relative to the length and coverage. In contrast with usual rules that have on the right-hand side a relation "attribute = value", inhibitory rules have a relation "attribute ≠ value" on the right-hand side. The considered algorithms are based on extensions of dynamic programming.

Keywords: inhibitory decision rules, length, coverage, dynamic programming.

1 Introduction

The paper is devoted to the study of algorithms for inhibitory rule optimization based on extensions of dynamic programming. In contrast with usual rules that have on the right-hand side a relation "attribute = value", inhibitory rules have a relation "attribute ≠ value" on the right hand side.

It was shown in [10] that, for some information systems, usual rules cannot describe the whole information contained in the system. However, inhibitory rules describe the whole information for every information system [7]. Classifiers based on inhibitory rules have often better accuracy than classifiers based on usual rules [4–6]. Greedy algorithms for inhibitory rule construction were studied in [7].

In this paper, we consider algorithms for optimization of inhibitory rules relative to the length and coverage which are based on extensions of dynamic programming. The choice of length is connected with the Minimum Description Length principle. The rule coverage is important to discover major patterns in the data. The considered algorithms allow also sequential optimization of inhibitory rules relative to the length and coverage and vice versa. Similar approach to usual decision rule optimization was studied in [1, 2].

The paper consists of five sections. In Sect. 2, we discuss main notions including the notion of nonredundant inhibitory rule. In Sect. 3, a directed acyclic

T. Li et al. (Eds.): RSKT 2012, LNAI 7414, pp. 149–154, 2012.

graph is considered which allows us to describe the whole set of nonredundant inhibitory rules for each row of a decision table. Section 4 contains descriptions of procedures of optimization relative to the length and coverage. Section 5 contains results of experiments and Sect. 6 – conclusions.

2   Nonredundant Decision Rules

In this section, we consider definitions of notions corresponding to decision tables and inhibitory rules.

A *decision table* T is a rectangular table with n columns labeled with conditional attributes $f_1, \ldots, f_n$. Rows of this table are filled with nonnegative integers which are interpreted as values of conditional attributes. Rows of T are pairwise different and each row is labeled with a nonnegative integer (decision) which is interpreted as a value of the decision attribute d. We denote by $D(T)$ the set of decisions attached to rows of the table T. We denote by $N(T)$ the number of rows in the table T.

A table obtained from T by the removal of some rows is called a *subtable* of the table T. A subtable T' of the table T is called *reduced* if $|D(T')| < |D(T)|$, and *unreduced* otherwise when $|D(T')| = |D(T)|$.

Let T be nonempty, $f_{i_1}, \ldots, f_{i_m} \in \{f_1, \ldots, f_n\}$ and $a_1, \ldots, a_m$ be nonnegative integers. We denote by $T(f_{i_1}, a_1) \ldots (f_{i_m}, a_m)$ the subtable of the table T which contains only rows that have numbers $a_1, \ldots, a_m$ at the intersection with columns $f_{i_1}, \ldots, f_{i_m}$. Such nonempty subtables (including the table T) are called *separable subtables* of T.

We denote by $E(T)$ the set of attributes from $\{f_1, \ldots, f_n\}$ which are not constant on T. For any $f_i \in E(T)$, we denote by $E(T, f_i)$ the set of values of the attribute f_i in T.

The expression

$$f_{i_1} = a_1 \wedge \ldots \wedge f_{i_m} = a_m \to d \neq k \qquad (1)$$

is called an *inhibitory rule over* T if $f_{i_1}, \ldots, f_{i_m} \in \{f_1, \ldots, f_n\}$, $a_1, \ldots a_m$ are nonnegative integers, and $k \in D(T)$. It is possible that $m = 0$. In this case (1) is equal to the rule

$$\to d \neq k. \qquad (2)$$

Let Θ be subtable of T and $r = (b_1, \ldots, b_n)$ be a row of Θ. We will say that the rule (1) is *realizable for* r, if $a_1 = b_{i_1}, \ldots, a_m = b_{i_m}$. The rule (2) is realizable for any row from Θ.

We will say that the rule (1) is *true for* Θ if each row of Θ for which the rule (1) is realizable has the decision attached to it that is different from k. The rule (2) is true for Θ if and only if each row of Θ is labeled with the decision different from k. If the rule (1) is an inhibitory rule over T which is true for Θ and realizable for r, we will say that (1) is an *inhibitory rule for* Θ *and* r *over* T.

We will say that the rule (1) with $m > 0$ is a *nonredundant* inhibitory rule for Θ and r over T if (1) is an inhibitory rule for Θ and r over T and the following conditions hold:

(i) $f_{i_1} \in E(\Theta)$, and if $m > 1$ then $f_{i_j} \in E(T(f_{i_1}, a_1) \dots (f_{i_{j-1}}, a_{j-1}))$ for $j = 2, \dots, m$;

(ii) if $m = 1$ then Θ is unreduced, and if $m > 1$ then the subtable $\Theta' = \Theta(f_{i_1}, a_1) \dots (f_{i_{m-1}}, a_{m-1})$ is unreduced.

The rule (2) is a *nonredundant* inhibitory rule for Θ and r over T if (2) is an inhibitory rule for Θ and r over T, i.e., if each row of Θ is labeled with the decision different from k and $k \in D(T)$.

Let Θ be a subtable of T, τ be a nonredundant rule over T, and τ be equal to (1).

The number m of conditions on the left-hand side of τ is called the *length* of this rule and is denoted by $l(\tau)$. The length of inhibitory rule (2) is equal to 0.

The *coverage* of τ relative to Θ is the number of rows in Θ for which τ is realizable and which are labeled with the decisions other than k. We denote it by $c(\tau)$. The coverage of inhibitory rule (2) relative to Θ is equal to the number of rows in Θ which are labeled with the decisions other than k. If τ is true for Θ then $c(\tau) = N(\Theta(f_{i_1}, a_1) \dots (f_{i_m}, a_m))$.

3 Directed Acyclic Graph $\Lambda(T)$

Now, we consider an algorithm that constructs a directed acyclic graph $\Lambda(T)$ which will be used to describe the set of nonredundant inhibitory rules for T and for each row r of T over T. Nodes of the graph are some separable subtables of the table T. During each step, the algorithm processes one node and marks it with the symbol *. At the first step, the algorithm constructs a graph containing a single node T which is not marked with *.

Let us assume that the algorithm has already performed p steps. We describe now the step $(p+1)$. If all nodes are marked with the symbol * as processed, the algorithm finishes its work and presents the resulting graph as $\Lambda(T)$. Otherwise, choose a node (table) Θ, which has not been processed yet. If Θ is reduced, then mark Θ with the symbol * and go to the step $(p+2)$. Otherwise, for each $f_i \in E(\Theta)$, draw a bundle of edges from the node Θ. Let $E(\Theta, f_i) = \{b_1, \dots, b_t\}$. Then draw t edges from Θ and label these edges with pairs $(f_i, b_1), \dots, (f_i, b_t)$ respectively. These edges enter to nodes $\Theta(f_i, b_1), \dots, \Theta(f_i, b_t)$. If some of nodes $\Theta(f_i, b_1), \dots, \Theta(f_i, b_t)$ are absent in the graph then add these nodes to the graph. We label each row r of Θ with the set of attributes $E_{\Lambda(T)}(\Theta, r) = E(\Theta)$ (this set can be changed during a procedure of optimization). Mark the node Θ with the symbol * and proceed to the step $(p+2)$.

The graph $\Lambda(T)$ is a directed acyclic graph. A node of this graph will be called *terminal* if there are no edges leaving this node. Note that a node Θ of $\Lambda(T)$ is terminal if and only if Θ is reduced.

Later, we will describe procedures of optimization of the graph $\Lambda(T)$ relative to the length and coverage. As a result we will obtain a graph Γ with the same sets of nodes and edges as in $\Lambda(T)$. The only difference is that any row r of each unreduced table Θ from Γ is labeled with a nonempty set of attributes $E_\Gamma(\Theta, r) \subseteq E(\Theta)$.

Let G be the graph $\Lambda(T)$ or a graph Γ obtained from $\Lambda(T)$ by procedures of optimization.

Now for each node Θ of G and for each row r of Θ we describe a set of inhibitory rules $Rul_G(\Theta, r)$ over T. Let Θ be a terminal node of G: Θ is a reduced table. Then $Rul_G(\Theta, r) = \{\to d \neq k : k \in D(T) \setminus D(\Theta)\}$.

Let now Θ be a nonterminal node of G such that for each child Θ' of Θ and for each row r' of Θ' the set of rules $Rul_G(\Theta', r')$ is already defined. Let $r = (b_1, \ldots, b_n)$ be a row of Θ. For any $f_i \in E_G(\Theta, r)$, we define the set of rules $Rul_G(\Theta, r, f_i)$ as follows: $Rul_G(\Theta, r, f_i) = \{f_i = b_i \wedge \alpha \to d \neq k : \alpha \to d \neq k \in Rul_G(\Theta(f_i, b_i), r)\}$. Then $Rul_G(\Theta, r) = \bigcup_{f_i \in E_G(\Theta, r)} Rul_G(\Theta, r, f_i)$.

Theorem 1. *For any node Θ of $\Lambda(T)$ and for any row r of Θ, the set $Rul_{\Lambda(T)}(\Theta, r)$ is equal to the set of all nonredundant inhibitory rules for Θ and r over T.*

An example of the algorithm work can be found in [3].

4   Procedures of Optimization Relative to Length and Coverage

First, we describe the procedure of optimization of the graph G relative to the length l. For each node Θ in the graph G, this procedure assigns to each row r of Θ the set $Rul_G^l(\Theta, r)$ of inhibitory rules with minimum length from $Rul_G(\Theta, r)$ and the number $Opt_G^l(\Theta, r)$ – the minimum length of an inhibitory rule from $Rul_G(\Theta, r)$.

We will move from the terminal nodes of the graph G which are reduced subtables to the node T. We will assign to each row r of each table Θ the number $Opt_G^l(\Theta, r)$ which is the minimum length of an inhibitory rule from $Rul_G(\Theta, r)$ and we will change the set $E_G(\Theta, r)$ attached to the row r in the nonterminal table Θ. We denote the obtained graph by $G(l)$.

Let Θ be a terminal node of G. Then we assign to each row r of Θ the number $Opt_G^l(\Theta, r) = 0$.

Let Θ be a nonterminal node and all children of Θ have already been treated. Let $r = (b_1, \ldots, b_n)$ be a row of Θ. We assign the number $Opt_G^l(\Theta, r) = \min\{Opt_G^l(\Theta(f_i, b_i), r) + 1 : f_i \in E_G(\Theta, r)\}$ to the row r in the table Θ and we set $E_{G(l)}(\Theta, r) = \{f_i : f_i \in E_G(\Theta, r), Opt_G^l(\Theta(f_i, b_i), r) + 1 = Opt_G^l(\Theta, r)\}$.

Now, we describe the procedure of optimization of the graph G relative to the coverage c. For each node Θ in the graph G, this procedure assigns to each row r of Θ the set $Rul_G^c(\Theta, r)$ of inhibitory rules with maximum coverage from $Rul_G(\Theta, r)$ and the number $Opt_G^c(\Theta, r)$ – the maximum coverage of an inhibitory rule from $Rul_G(\Theta, r)$.

We will move from the terminal nodes of the graph G which are reduced subtables to the node T. We will assign to each row r of each table Θ the number $Opt_G^c(\Theta, r)$ which is the maximum coverage of an inhibitory rule from $Rul_G(\Theta, r)$ and we will change the set $E_G(\Theta, r)$ attached to the row r in the nonterminal table Θ. We denote the obtained graph by $G(c)$.

Let Θ be a terminal node of G. Then we assign the number $Opt_G^c(\Theta, r) = N(\Theta)$ to each row r of Θ.

Let Θ be a nonterminal node and all children of Θ have already been treated. Let $r = (b_1, \ldots, b_n)$ be a row of Θ. We assign the number $Opt_G^c(\Theta, r) = \max\{Opt_G^c(\Theta(f_i, b_i), r) : f_i \in E_G(\Theta, r)\}$ to the row r in the table Θ and we set $E_{G(c)}(\Theta, r) = \{f_i : f_i \in E_G(\Theta, r), Opt_G^c(\Theta(f_i, b_i), r) = Opt_G^c(\Theta, r)\}$. Detailed description of the procedure of optimization relative to the coverage can be found in [3].

It is also possible to make sequential optimization relative to the length and coverage. We can find all nonredundant rules with maximum coverage and after that among these rules find all rules with minimum length. We can also change the order of optimization: find all nonredundant rules with minimum length and after that find among such rules all rules with maximum coverage.

Considering complexities of the presented algorithms it is possible to show (see analysis of similar algorithms in [9], page 64) that the time complexities of algorithms which construct the graph $\Lambda(T)$ and make optimization of inhibitory decision rules relative to length or coverage are bounded from above by polynomials on the number of separable subtables of T, and the number of attributes in T.

5 Experimental Results

We considered a number of decision tables from UCI Machine Learning Repository [8]. For each such decision table T we constructed the directed acyclic graph $\Lambda(T)$ and applied to it sequentially the procedure of optimization relative to the length and after that the procedure of optimization relative to the coverage. Average length and coverage of obtained rules (among all rows of T) can be found in Table 1 (column "DP (length+coverage)"). We applied to $\Lambda(T)$ the procedure of optimization relative to the coverage. Average length and coverage of obtained rules (among all rows of T) can be found in Table 1 (column "DP (coverage)"). For some decision tables T, the average values of the length and coverage in the column "DP (coverage)" are different from the average values in the column "DP (length+coverage)" (such values are in bold). It means that

Table 1. Length and coverage of inhibitory rules

Decision table	Rows	Attr	DP (length+coverage)		DP (coverage)	
			Length	Coverage	Length	Coverage
adult-stretch	16	4	1.250	7.000	1.250	7.000
balance-scale	625	4	2.672	11.944	2.672	11.944
breast-cancer	266	9	**2.665**	7.038	**3.429**	**9.534**
cars	1727	6	1.047	543.743	1.047	543.743
hayes-roth-data	69	4	**1.667**	7.580	**1.696**	**7.609**
lymphography	148	18	1.000	141.000	1.000	141.000
monks-2-test	432	6	4.523	12.356	4.523	12.356
monks-2-train	169	6	**3.497**	6.249	**3.568**	**6.379**
nursery	12960	8	1.000	5400.000	1.000	5400.000
shuttle-landing	15	6	**1.400**	1.867	**1.733**	**2.133**
soybean-small	47	35	1.000	37.000	1.000	37.000
teeth	23	8	1.000	16.217	1.000	16.217
zoo	59	16	1.000	50.458	1.000	50.458

for some rows r of the considered decision tables there are no nonredundant inhibitory rules for T and r that have minimum length and maximum coverage among all nonredundant inhibitory rules for T and r.

6 Conclusions

In the paper, we considered algorithms for exact inhibitory rule optimization relative to the length and coverage which are based on extensions of dynamic programming. Presented results show that we can construct optimal rules, i.e., with minimum length and maximum coverage. Further we will study approximate inhibitory rules and construction of classifiers.

References

1. Amin, T., Chikalov, I., Moshkov, M., Zielosko, B.: Dynamic Programming Approach for Exact Decision Rule Optimization. In: Skowron, A., Suraj, Z. (eds.) Rough Sets and Intelligent Systems - Professor Zdzisław Pawlak in Memoriam. Intelligent Systems Reference Library, vol. 42, pp. 209–230. Springer, Heidelberg (2012)
2. Amin, T., Chikalov, I., Moshkov, M., Zielosko, B.: Dynamic programming approach to optimization of approximate decision rules. Information Sciences (submitted)
3. Alsolami, F., Chikalov, I., Moshkov, M., Zielosko, B.: Optimization of inhibitory decision rules relative to coverage. In: Wakulicz-Deja, A. (ed.) Decision Support Systems (to appear, 2012)
4. Delimata, P., Moshkov, M., Skowron, A., Suraj, Z.: Two Families of Classification Algorithms. In: An, A., Stefanowski, J., Ramanna, S., Butz, C.J., Pedrycz, W., Wang, G. (eds.) RSFDGrC 2007. LNCS (LNAI), vol. 4482, pp. 297–304. Springer, Heidelberg (2007)
5. Delimata, P., Moshkov, M., Skowron, A., Suraj, Z.: Comparison of Lazy Classification Algorithms Based on Deterministic and Inhibitory Decision Rules. In: Wang, G., Li, T., Grzymala-Busse, J.W., Miao, D., Skowron, A., Yao, Y. (eds.) RSKT 2008. LNCS (LNAI), vol. 5009, pp. 55–62. Springer, Heidelberg (2008)
6. Delimata, P., Moshkov, M., Skowron, A., Suraj, Z.: Lazy classification algorithms based on deterministic and inhibitory rules. In: Magdalena, L., Ojeda-Aciego, M., Verdegay, J.L. (eds.) IPMU 2008, Torremolinos (Malaga), Spain, June 22-27, pp. 1773–1778 (2008)
7. Delimata, P., Moshkov, M., Skowron, A., Suraj, Z.: Inhibitory Rules in Data Analysis: A Rough Set Approach. SCI, vol. 163. Springer, Heidelberg (2009)
8. Frank, A., Asuncion, A.: UCI ML Repository, http://archive.ics.uci.edu/ml
9. Moshkov, M., Zielosko, B.: Combinatorial Machine Learning. SCI, vol. 360. Springer, Heidelberg (2011)
10. Skowron, A., Suraj, Z.: Rough sets and concurrency. Bulletin of the Polish Academy of Sciences 41(3), 237–254 (1993)

Parallelized Computing of Attribute Core Based on Rough Set Theory and MapReduce⋆

Yong Yang and Zhengrong Chen

Institute of Computer Science & Technology,
Chongqing University of Posts and Telecommunications,
Chongqing, 400065, P.R. China
yangyong@cqupt.edu.cn, 554760686@qq.com

Abstract. In this paper, computing attribute core for massive data based on rough set theory and MapReduce is studied, two novel algorithms for computing attribute core are proposed. A case study proves the correctness of the proposed algorithms, and the proposed algorithms are shown more efficient according to the experiment results on a real massive dataset.

Keywords: rough set, MapReduce, attribute core, positive region, cloud computing.

1 Introduction

Cloud computing [1] is an emerging business computing model. The novelty and advantage of cloud computing is that there is almost unlimited and cheap storage and computing power provided by the cloud computing platform. In the cloud computing environment, the computational tasks are distributed on the resources pool which consists of thousands of computers. Furthermore, the application system can get computing power, storage space and a variety of software services according to its need. Since the massive data would be stored on the cloud computing platform, data mining for the massive data is a valuable research topic among the cloud computing technologies.

Parallel computing is one of the cores of cloud computing technologies. Although there are several different parallel computing models in the applications, MapReduce has become the mainstream since it was proposed in 2004 [2].

Rough set is a valid mathematical theory to deal with imprecise, uncertain, and vague information [3]. It has been applied in many fields such as machine learning, data mining, intelligent data analyzing and control algorithm acquiring successfully since it was proposed by Pawlak in 1982 [4].

In the study of rough set theory, the computation of positive region and attribute core are two basic operations. Since they would affect the efficiencies

---

⋆ Part of this work is supported by National Natural Science Foundation of China (No. 61075019), Scientific Research Foundation of Chongqing Municipal Education Commission (No. KJ110522, KJ110512), Natural Science Foundation of Chongqing University of Posts and Telecommunications(No. A2009-26, No. JK-Y-2010002).

T. Li et al. (Eds.): RSKT 2012, LNAI 7414, pp. 155–160, 2012.

of attribute reduction and value reduction, and accelerate the applications of rough set, it is important to study the highly efficient algorithms of computing positive region and attribute core. Lots of research works have been taken for this research topic. Ye D.Y. defined a new discernibility matrix, and proposed a novel method for computing the core [5]. Xu Z.Y. provided a definition of a simple discernibility matrix, and proposed a quick computing core algorithm [6]. Hu F. combined rough set theory and divide and conquer method, and proposed a method of attribute core computation [7]. However, these methods would be disable or of low efficiency when they faced with massive dataset. Efficient algorithms of computing positive region and attribute core would be researched further.

In this paper, for the purpose of studying high efficient data mining algorithms running on the cloud computing platform, rough set theory combined with MapReduce are studied, and two parallel rough set algorithms based on MapReduce are proposed for computing the positive region and attribute core, the proposed algorithms are proved more efficient according to the experiment results. The rest of this paper is organized as follows. In Sect.2, two novel parallel algorithms for positive region and attribute core is proposed. In Sect. 3, simulation experiments and discussion are introduced. Finally, conclusion is drawn in Sect. 4.

2   Parallel Algorithm for Computing Attribute Core Based on Rough Set and MapReduce

In this section, two parallelized algorithms for positive region and attribute core is proposed based on rough set theory and MapReduce programming mode.

Theorem 1. Given a decision table $S =< U, A = C \cup D, V, f >$, $\forall c_i \in C$, $POS_{\{C-c_i\}}(D) \subseteq POS_C(D)$.

Theorem 2. Given a decision table $S =< U, A = C \cup D, V, f >$, $\forall c_i \in C$, $POS_{\{C-c_i\}}(D) \neq POS_C(D) \Leftrightarrow \exists x \in POS_C(D) \wedge x \notin POS_{\{C-c_i\}}(D)$.

The proof of the theorems are omitted due to the limited number of pages.

The classical algorithm of computing attribute core is based on positive region remained, it is necessary to frequently comparing whether the positive region is remained. Therefore, it must be time consuming. According to theorem 2, a new flag is added for each sample x_i to describe whether $x_i \in Pos_C(D)$ or $x_i \notin Pos_C(D)$. In computing $POS_{\{C-c_i\}}(D)$, if $\exists_{x,y \in U} f(x, \{C-c_i\}) = f(y, \{C-c_i\}) \wedge f(x, D) \neq f(y, D) \wedge (x \in POS_C(D) \| y \in POS_C(D))$, then $POS_{\{C-c_i\}}(D) \neq POS_C(D)$. Therefore, it can avoid comparing $POS_C(D)$ and $POS_{\{C-c_i\}}(D)$, and accelerate computing of *Core*.

According to the definition of positive region, the process of computing positive region $POS_C(D)$ is described as follows. Firstly, computing the partition of U relative to C, that is, $U|IND(C) = \{X_1, X_2, \cdots, X_n\}$. Secondly, in the X_i, if $\forall_{x,y \in X_i} f(x, C) = f(y, C) \wedge f(x, D) = f(y, D)$, then $POS_C(D) = POS_C(D) \cup X_i$. It is found that the processing for each X_i is independent, that is, it can be computed in parallel.

According to the above analysis, a parallel algorithm for positive region based on rough set and MapReduce could be developed. This algorithm consists of three steps mainly. First of all, each sample x_i can be seen as a key/value pair, the key is $f(x_i, C)$, the value is $f(x_i, D)$ and x_i_No, this step is executed in parallel by multiple map operations. Afterwards, the samples which have the same key are merged to generate $U|IND(C) = \{X_1, X_2, \cdots, X_n\}$. Thirdly, it is a step to determine whether each sample x_i belongs to the positive region, that is, $x_i \in Pos_C(D)$ or $x_i \notin Pos_C(D)$, and a new flag $POS_C(D)\_flag$ is added accordingly. This step can be executed parallelized by multiple reduce operations. The output of this algorithm is a new decision table S' in which there is a new flag $POS_C(D)\_flag$ added to the original decision table S.

Algorithm 1. Parallel Algorithm for Computing Positive Region Based on Rough Set and MapReduce
Input: A decision table $S =< U, A = C \cup D, V, f >$
Output: A new decision table S'
Step 1. Map Stage: Computing $U|IND(C) = \{X_1, X_2, \cdots, X_n\}$
    Map_input: $< x\_No, x\_C + x\_D >$
    Map_output: $< x\_C, x\_D + x\_No >$
        Output_key= x_C, Output_value= $x\_D + x\_No$;
        Output: $< Output\_key, Output\_value >$
Step 2. Shuffle Stage: Generate $U|IND(C) = \{X_1, X_2, \cdots, X_n\}$
Step 3. Reduce Stage: In X_j, judge rules whether they are in conflict
    Reduce_input: $< C, X_j >, (X_j \in U|IND(C))$
    Reduce_output: $< x\_No, x\_C + x\_D + POS_C(D)\_flag >$
        $POS_C(D)\_flag = true$;
        if($\underset{x \in X_j, y \in X_j, d \in D}{\exists} d(x) \neq d(y)$) then $POS_C(D)\_flag = false$; end if
        for i=1 to $|X_j|$ do
            Output: $< x_i\_No, x_i\_C + x_i\_D + POS_C(D)\_flag >$;
        end for

According to the definition of attribute core, $\forall c_i \in C$, if $POS_{\{C-c_i\}}(D) \neq POS_C(D)$, then $c_i \in Core$. It is found that the judgment process for each c_i is independent, that is, it can be parallelized computing. Therefore, based on Algorithm 1 and adopted $POS_C(D)\_flag$, whether $POS_{\{C-c_i\}}(D)$ and $POS_C(D)$ are equal can be determined. Therefore, the process of computing attribute core can be parallelized on two levels, one is between the condition attributes and another is for each c_i.

According to the above analysis, a parallel algorithm for attribute core based on rough set theory and MapReduce is proposed. The main steps of the algorithm are described as follows. Firstly, the core of S is initialized to be null. Secondly, Algorithm 1 is called to compute positive region, and a new decision table S' is generated. Thirdly, it is a step to compute attribute core of the S, and can be divided into three stages, that is, a map stage, a shuffle stage and a reduce stage. In the map stage, a key/value pair can be gotten from each sample x in S', in which the key is c_i_No and $f(x, \{C - c_i\})$, the value is $f(x, D) + POS_C(D)\_flag + x\_No$, this stage is executed parallelized by multiple map operations. In the shuffle stage, the samples which has same key are merged to generate $U|IND(\{C - c_i\}) = \{c_i\_X_1, c_i\_X_2, \cdots, c_i\_X_n\}$. Thirdly, whether a condition attribute c_i is core attribute is determined according to the relationship between $POS_{\{C-c_i\}}(D)$ and $POS_C(D)$, that is, if $POS_{\{C-c_i\}}(D) \neq POS_C(D)$, c_i is a core attribute, otherwise c_i is not a core attribute. This stage can be executed parallelized by multiple reduce operations. Finally, the attribute core of S can be gotten.

Algorithm 2. Parallel Algorithm for Computing Attribute Core Based on Rough Set and MapReduce
Input: A decision table $S =< U, A = C \cup D, V, f >$
Output: Attribute Core ($Core$) of S
Step 1. Initialize $Core = \emptyset$;
Step 2. Call Algorithm 1 to computing positive region;
Step 3. Compute attribute core ($Core$);
    Map Stage:
    Map_input: $< x\_No, x\_C + x\_D + POS_C(D)\_flag >$
    Map_output: $< c_i\_No + \{x\_C - c_i\}, x\_D + POS_C(D)\_flag + x\_No >$
        for i=1 to $|C|$ do
            Output: $< c_i\_No + \{x\_C - c_i\}, x\_D + POS_C(D)\_flag + x\_No >$
        end for
    Shuffle Stage: Generate $U|IND(\{C - c_i\}) = \{c_i\_X_1, c_i\_X_2, \cdots, c_i\_X_n\}$
    Reduce Stage:
    Reduce_intput: $< c_i\_No + \{C - ci\}, c_i\_X_j >, (c_i\_X_j \in c_i\_(U|IND(\{C - c_i\})))$
    Reduce_output: $< Core, null >$
        $c_i\_Core\_flag = false$;
        if($\underset{x \in X_i, y \in X_i, d \in D}{\exists} d(x) \neq d(y) \wedge (POS_C(D)\_flag(x) = true||$
        $POS_C(D)\_flag(y) = true)$) then $c_i\_Core\_flag = true$; end if
        if($c_i\_Core\_flag$) then $Core = Core \cup \{c_i\}$; end if
Step 4. Output the attribute core ($Core$) of S.

3   Experiments and Discussion

In this section, firstly, a case study is introduced to verify the correctness of the proposed algorithm, secondly, a series of experiments are taken to test the effectiveness of the proposed algorithms on the massive datasets.

3.1   Case Study

In order to verify the correctness of Algorithm 2, a decision table S [8] is given and shown in Table 1, and a new decision table S' gotten from Algorithm 1 is shown in Table 2. In order to compare the proposed algorithm with existing algorithm, the algorithms in [5–7] and the algebra definition of reduction [3] are chosen. According to the above algorithm, the attribute core of S is listed in Table 3. From Table 3, we can find the result of Algorithm 2 is the same with the referenced algorithms. Therefore, the correctness of Algorithm 2 can be proven.

Table 1. Decision table S

U	c1	c2	c3	D
x1	1	0	1	1
x2	1	0	1	0
x3	1	0	1	2
x4	0	0	1	1
x5	0	0	1	0
x6	1	1	1	1

Table 2. Decision table S'

U	c1	c2	c3	D	$POS_C(D)$_flag
x1	1	0	1	1	FALSE
x2	1	0	1	0	FALSE
x3	1	0	1	2	FALSE
x4	0	0	1	1	FALSE
x5	0	0	1	0	FALSE
x6	1	1	1	1	TRUE

Table 3. The attribute core of S

Algorithm	Algorithm[5]	Algorithm[6]	Algorithm[7]	Algebra definition	Algorithm 2
Core	{c2}	{c2}	{c2}	{c2}	{c2}

3.2   Comparative Experiments for Efficiency of the Proposed Algorithm

In order to test the efficiency of Algorithm 2 on massive datasets, a massive dataset KDDCUP99 [9] is used, which has 4,898,432 records and 41 condition attributes. We randomly selected 10%, 20%, . . . , 100% records from KDDCUP99 to generate 10 datasets. Algorithm 2 is tested on all these datasets, the experiment results are shown in Fig. 1. In the experiments, only one PC is used, which has 2GB memory, Intel Core2 Quad CPU Q8300 2.5GHz, and Ubuntu9.10. Algorithm 2 is runs on the Hadoop-MapReduce platform [10]. The version of Hadoop is Hadoop-0.20.2.

From the experiment results in Fig. 1, we can found that Algorithm 2 can process the KDDCUP99 dataset in short period of time, and the running time of Algorithm 2 is approximately linear in the number of records. The effectiveness of the proposed method can be proved according to this experiment.

Furthermore, a series of comparative experiments are taken. In [7], the processing time of Algorithm[7] for the 10%, 20%, 30% and 40% of KDDCUP99 was approximately 98s, 196s, 294s and 392s. Meanwhile, Algorithm[5] and Algorithm[6]

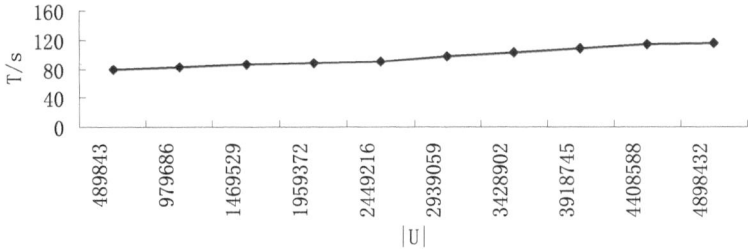

Fig. 1. The experiment results curve

were also tested, unfortunately, both of them were terminated because of the memory overflow. The comparative results are shown in Table 4.

The experimental results and above analysis comparison showed that Algorithm 2 is more efficient for the massive dataset.

Table 4. The comparative experiment results

Dataset	Algorithm[5](s)	Algorithm[6](s)	Algorithm[7](s)	Algorithm 2(s)
KDDCUP10%	-	-	97.969	79.032
KDDCUP20%	-	-	195.937	82.675
KDDCUP30%	-	-	293.906	85.674
KDDCUP40%	-	-	391.874	87.682

4   Conclusion

In this paper, computing attribute core for massive data based on rough set theory and MapReduce is studied. Two algorithms for parallelized computing positive region and attribute core are proposed. A case study verifies the correctness of the proposed algorithm, and comparative experiment results show the effectiveness and high efficiency of the proposed method for data mining on massive dataset. Attribute reduction and knowledge acquisition based on rough set theory and MapReduce will be studied continually in the future.

References

1. Weiss, A.: Computing in Clouds. ACM Networker 11(4), 18–25 (2007)
2. Dean, J., Ghemmawat, S.: MapReduce: a Flexible Data Processing Tool. Communications of the ACM 53(1), 72–77 (2010)
3. Pawlak, Z.: Rough Sets. International Journal of Computer and Information Sciences 11, 341–356 (1982)
4. Guoyin, W.: Rough Set Theory and Knowledge Acquisition. Xi'an Jiaotong University Press, Xi'an (2001) (in Chinese)
5. Dongyi, Y., Zhaojiong, C.: A New Discernibility Matrix and The Computation of a Core. Acta Electronica Sinica 30, 1086–1088 (2002)
6. Zhangyan, X., Bingru, Y., et al.: Quick Computing Core Algorithm Based on Discernibility Matrix. Computer Engineering and Applications 42, 4–6 (2006)
7. Feng, H., Guoyin, W., Ying, X.: Attribute Core Computation Based on Divide and Conquer Method. In: Kryszkiewicz, M., Peters, J.F., Rybiński, H., Skowron, A. (eds.) RSEISP 2007. LNCS (LNAI), vol. 4585, pp. 310–319. Springer, Heidelberg (2007)
8. Guoyin, W.: Calculation Methods for Core Attributes of Decision Table. Chinese Journal of Computers 26, 611–615 (2003)
9. KDDCUP99, http://kdd.ics.uci.edu/databases/kddcup99/
10. Hadoop MapReduce, http://hadoop.apache.org/mapreduce/

Semi-supervised Vehicle Recognition: An Approximate Region Constrained Approach

Rui Zhao[1,2], Zhihua Wei[1,3], Duoqian Miao[1], Yan Wu[1], and Lin Mei[2]

[1] Department of Computer Science and Technology, Tongji University,
201804, Shanghai, China
[2] The Third Research Institute of the Ministry of Public Security,
201204, Shanghai, China
[3] State Key Laboratory for Novel Software Technology, Nanjing University,
210093, Nanjing, China

Abstract. Semi-supervised learning attracts much concern because it can improve classification performance by using unlabeled examples. A novel semi-supervised classification algorithm SsL-ARC is proposed for real-time vehicle recognition. It makes use of the prior information of object vehicle moving trajectory as constraints to bootstrap the classifier in each iteration. Approximate region interval of trajectory are defined as constraints. Experiments on real world traffic surveillance videos are performed and the results verify that the proposed algorithm has the comparable performance to the state-of-the-art algorithms.

Keywords: Semi-supervised learning, object recognition, approximate region interval, constraints.

1 Introduction

Robust object recognition under real-world conditions is still a challenging task and limits the use of state-of-the-art methods in industry (e.g., video surveillance [1]). Recently, object recognition based on semi-supervised learning attracts much attention which exploits both labeled and unlabeled objects in learning classifier [2, 3]. It has been shown that for some kinds of problems, the unlabeled data can dramatically improve the performance of classifier. However, general semi-supervised learning algorithms [4] assume that the objects are independent so that they do not enable to exploit relationship between objects which might contain a large amount of information. For example, in a surveillance video, the certain vehicle location defines a trajectory which represents a kind of relation among the labeling of the video sequence. The objects close to the trajectory are positive examples; objects far away from the trajectory are negative ones.

In semi-supervised learning , more feasible strategy for labeling unlabeled examples is guided by some supervisory information [2]. This information may be in a form of labels associated with some examples or some forms of constraints [5]. The second is more general. Its basic idea is combining detector and

T. Li et al. (Eds.): RSKT 2012, LNAI 7414, pp. 161–166, 2012.
© Springer-Verlag Berlin Heidelberg 2012

tracker [6] where the detector serves as initial model for semi-supervised learning. Object tracking could learn some information from underlying examples, i.e. estimating object location in frame-by-frame fashion. The object to be tracked can be viewed as a single labeled example and the video as unlabeled data. Many authors perform self-learning and co-training for tracking object [7–9]. This kind of approach predicts the position of the objects with a tracker and updates the model with positive examples that are close and negative examples that are far from the current position. The strategy is able to adapt the tracker to new appearances and background, but breaks down as soon as the tracker makes a mistake. In order to avoid above problem, Kalal proposed a bootstrapping method by using structure constraints named P-N learning [10]. This approach integrated tracker and detector and made them correcting each other. The approach demonstrated robust tracking performance in challenging conditions and partially motivated our research.

Basing on the above idea, the paper proposes a new paradigm for learning from dependent unlabeled objects. Relations between objects are used in parallel with classification algorithm to mutually rectify their errors. The relation among the objects is exploited by so called Approximate Region Constraints (ARC). That is, lower approximate region specifies the most frequently acceptable patterns of positive labels, i.e. objects on the moving trajectory. Upper approximate region specifies possibly acceptable pattern of positive labels, i.e. objects near to the trajectory.

The rest of the paper is organized as follows. Section 2 defines the vehicle recognition problem and formulates the semi-supervised algorithm based on ARC. Section 3 validates and analyzes the algorithm on real world traffic videos. The last section concludes and discusses the future works.

2   Semi-supervised Learning Based on Approximate Region Constraints (SsL-ARC)

2.1   SsL-ARC Algorithm Framework

Object detection problem could be regarded as an on-line learning process as follows. Let x be an image patch from a frame set X in videos and y be a label from lable-space $Y = \{1, -1\}$. A set of examples X_l and corresponding set of labels Y_l will be denoted as (X_l, Y_l) and called a labeled set. The task of vehicle recognition is to learn a classifier $f : X \rightarrow Y$ from a prior labeled set (X_l, Y_l) and bootstrap its performance by unlabeled data X_u.

A constrained boosting process is defined to verify the labels by classifier in accordance with the trajectory of object. Its framework is shown in Fig.1. Based on this idea, a SsL-ARC algorithm is proposed.

Classifier. Object detection is performed by classification method that decides about presence of an object in an input frame and determines its location. Based on the "bag of words" ideology, an object could be described by the collection of

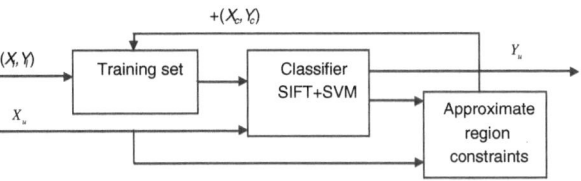

Fig. 1. SsL-ARC algorithm framework

interest points which are perceptually salient [11]. As a result, the Harris-Affine corner measure is used to find the salient feature and SIFT [12] is used for salient point description. In training phase, the Harris-Affine detector detects several salient points and the SIFT descriptor yields a 128 dimensional vector for each of these keypoints. This descriptor along with the scale information of the respective keypoint forms a 131 dimensional pattern. The labeled object represented by the feature vector was assigned a positive label and used for training. Core Vector Machine (CVM) is adopted as classifier which is a recently proposed flavor of SVM [13]. In testing phase, for a patch in a frame, all the salient keypoints were classified. If the number of keypoints belonging to positive class exceeding 50% of the total number of keypoints, the patch is labeled as positive.

Constraints. Selecting a single patch in the first frame as the labeled object, we could draw a trajectory curve in the video volume based on the fact that a single object appears in one location only in a given frame. The curve is obtained by a CamShift tracker which follows the selected object from frame to frame [14]. CamShift essentially climbs the gradient of a back projected probability distribution computed from rescaled color histograms and looks for the nearest peak in an axis-aligned search window. These occur when objects in video sequences are being tracked and the object moves so that the size and location of the probability distribution changes in time. CamShift which is originated from mean shift uses continuously adaptive probability distributions so that it adjusts the size and angle of the target rectangle each time it shifts. It does this by selecting the scale and orientation that are the best fit to the target-probability pixels inside the new rectangle location.

2.2 Approximate Region Constraints

Assume that the location of trajectory in time k is (m_0^k, n_0^k), the centroid of the patch detected by the classifier in time k is (m_1^k, n_1^k), the distance between the centroid of detected patch and the trajectory is as follows.

$$D(k) = \sqrt{(m_1^k - m_0^k)^2 + (n_1^k - n_0^k)^2} \tag{1}$$

Approximate region interval (Ari) based on interval set [15] is defined according to the trajectory obtained by the tracker. The definition of an interval set $Ari_i(a, b)$ of object x_i and its pair of threshold [16] is as follows.

Definition 1. For an vehicle object $x_i \in X_t$, its approximate region interval $Ari_i(a,b) = [\alpha_i^{(a,b)}, \beta_i^{(a,b)}]$, $[a,b]$ is the time span of a video sequence, $k \in [a,b]$.

$D(k) \leq \alpha_i^k$ is defined as the lower approximate region of object x_i which denotes the frequent occurrence region of the vehicle object x_i;

$D(k) \leq \beta_i^k$ is defined as the upper approximate region of object x_i which denotes the possible occurrence region of the vehicle x_i;

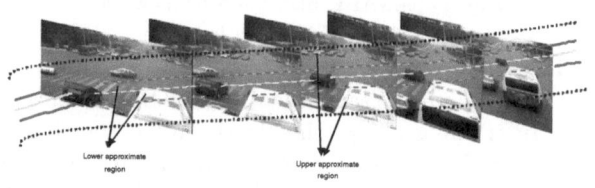

Fig. 2. Rough set approximate region of vehicle trajectory

The lower approximate region and upper approximate region of the trajectory could be described as Fig.3. Approximate region constraint function accepts a group of examples with labels given by the detector (X_u^k, Y_u^k) and output s subset of the group with changed labels (X_c^k, Y_c^k). The lower approximate region constraint which is denoted as L-constraint, is used to identify examples that have been labeled negative by the classifier but it belong to the lower approximate region of trajectory. In iteration k, L-constraint add $n^+(k)$ examples to the training set with labels changed to positive. On the other hand, the upper approximate region constraint which is named U-constraint, is used to identify examples that have been labeled as positive but beyond the upper approximate region of trajectory. In iteration k, U-constraint add $n^-(k)$ examples to the training set with labels changed to negative. These constraints enlarge the pool of positive and negative training set and thus improve the discriminative ability of classifier.

3   Experiments

The proposed algorithm is tested on 6 video sequences originates from surveillance videos of Shanghai Bureau of Urban traffic Management.

The algorithm SsL-ARC is initialized in the first frame by learning the Initial Detector noted as Ini-Detec and sets the initial position of the tracker. For each frame, the detector and the tracker find the location(s) of the object. The tracker finds the trajectory of given object in the continuous frames while the detector recognizes all patches similar to the given object by classifier. The objects close to the trajectory and far away from the trajectory are used as positive examples and negative examples, respectively.

The detector is implemented by integrating the source of Harris-Affine corner measure, SIFT descriptor [17] and Core Vector Machine [13] with the default parameters. Gaussian kernel is selected in CVM. The performance of the SsL-ARC algorithm is evaluated by precision P, recall R and F-measure since the algorithm is a boosted classification algorithm in nature. P is the number of correct detections divided by number of all detections, R is the number of correct detections divided by the number of object occurrences that should have been detected. F combines these two measures.

The performance of Ini-detec, Boost-detec and constrained tracker noted as Cons-Track on 6 sequences is listed in Table1. The performance is measured by P, R and F-measure averaged over time.

From the Table 1, we could observed that the Ini-detec has high precision rate while very low recall rate. the Boost-detec has a significant increase on recall rate. The SsL-ARC algorithm is compared with the famous P-N Learning algorithm [10] on the same video sequences and has better performance on three sequences that occupy half of the dataset. In our experiments, the proposed algorithm shows better performance than P-N Learning.

Table 1. Performance analysis of SsL-ARC

Sequence	Frame	SsL-ARC algorithm(P/R/F)			P-N Learning(P/R/F)
		Ini-detec	Boost-detec	Cons-tracker	
Bus	552	1.00/0.02/0.03	**0.91/0.45/0.60**	0.83/0.41/0.55	0.90/0.38/0.53
Crossing1	705	1.00/0.01/0.02	0.87/0.32/0.47	0.85/0.56/0.67	**0.87/0.45/0.59**
Crossing2	720	1.00/0.06/0.12	**0.90/0.75/0.82**	0.63/0.78/0.70	0.88/0.67/0.76
Evening	487	1.00/0.04/0.08	0.91/0.43/0.58	0.80/0.37/0.51	**0.91/0.48/0.63**
Night1	455	0.76/0.01/0.02	**0.55/0.20/0.29**	0.23/0.22/0.22	0.44/0.18/0.26
Night2	462	1.00/0.06/0.12	0.92/0.65/0.76	0.89/0.8/0.84	**0.90/0.68/0.77**

4 Conclusions

A boosting semi-supervised learning algorithm SsL-ARC constrained by approximate region interval is proposed. The constraints are obtained by defining lower approximate region and upper approximate region of given object trajectory. The initial detector which is based on SIFT features and SVM classifier is bootstrapped by a feedback from these constraints. The algorithm is applied to the problem of real-time vehicle object recognition. Experiments on real world surveillance videos show that the proposed algorithm has better performance compared to the state-of-the-art methods. Further work may focus on defining more refined constraints.

Acknowledgments. This research is supported in part by the National Natural Science Foundation of China under Grants No.60970061, Key Program supported by the Shanghai Committee of Science and Technology under Grant No.10511500700, the opening project of State Key Laboratory for Novel Software Technology in Nanjing University (No.KFKT2011027) and the opening project of Shanghai Key Laboratory of Digital Media Processing and Transmission (No.2011KF03).

References

1. Dee, H., Velastin, S.: How close are we to solving the problem of automated visual surveillance? Mach. Vison. Appl. 19(5-6), 329–343 (2008)
2. Chapelle, O., Schôlkopf, B., Zien, A.: Semi-Supervised Learning. MIT Press, Cambridge (2006)
3. Zhu, X., Goldberg, A.: Introduction to semi-supervised learning. Morgan Claypool Publishers, USA (2009)
4. Nigam, K., McCallum, A., Thrun, S., Mitchell, T.: Text classification from labeled and unlabeled documents using EM. Mach. learn. 39(2), 103–134 (2000)
5. Abu-Mostafa, Y.: Machines that learn from hints. Scientific American 272(4), 64–71 (1995)
6. Li, Y., Ai, H., Yamashita, T., Lao, S., Kawade, M.: Tracking in low frame rate video: A cascade particle filter with discriminative observers of different lifespans. In: 2007 IEEE Computer Society Conference on Computer Vision and Pattern Recognition, pp. 1–8. IEEE Press, Minneapolis (2007)
7. Grabner, H., Bischof, H.: On-line boosting and vision. In: 2006 IEEE Computer Society Conference on Computer Vision and Pattern Recognition, pp. 260–267. IEEE Press, New York (2006)
8. Avidan, S.: Ensemble tracking. IEEE T. Pattern Anal. 29(2), 261–271 (2007)
9. Yu, Q., Dinh, T.B., Medioni, G.: Online Tracking and Reacquisition Using Co-trained Generative and Discriminative Trackers. In: Forsyth, D., Torr, P., Zisserman, A. (eds.) ECCV 2008, Part II. LNCS, vol. 5303, pp. 678–691. Springer, Heidelberg (2008)
10. Kalal, Z., Matas, J., Mikolajczyk, K.: P-N Learning: Bootstrapping Binary Classifiers by Structural Constraints. In: 23rd IEEE Conference on Computer Vision and Pttern Recognition, pp. 13–18. IEEE Press, San Francisco (2010)
11. Sivic, J., Russell, B., Efros, A., Zisserman, A., Freeman, W.: Discovering Objects and their Localization in Images. In: 10th IEEE International Conference on Computer Vision, pp. 370–377 (2005)
12. Lowe, D.: Distinctive image features from scale-invariant keypoints. Int. J. Comput. Vision 60(2), 91–110 (2004)
13. Tsang, I.W., Kwok, J.T., Cheung, P.M.: Core vector machines: Fast SVM training on very large data sets. J. Mach. Learn. Res. 6, 363–392 (2005)
14. Bradski, G.R.: Computer Vision Face Tracking For Use in a Perceptual. User Interface. Intel Technology Journal, 2nd Quarter (1998)
15. Yao, Y.Y.: Interval-set algebra for qualitative knowledge representation. In: 5th International Conference on Computing and Information, pp. 370–374 (1993)
16. Yao, Y.Y.: Three-way decisions with probabilistic rough sets. Information Sciences 180(3), 341–353 (2010)
17. Affine covariant region detectors, http://www.robots.ox.ac.uk/~vgg/research/affine/detectors.html

A Color Image Segmentation Algorithm by Integrating Watershed with Region Merging

Shuangqun Li*, Jiucheng Xu, Jinyu Ren, and Tianhe Xu

College of Computer and Information Technology,
Henan Normal University, Xinxiang 453007, China
`htulsq@163.com`

Abstract. In order to improve the effectiveness of color image segmentation, a color image segmentation algorithm by integrating watershed with region merging is proposed in this paper. First, the image input is divided into many regions by watershed algorithm, and the phenomenon of the image over segmentation emerges for the details and noise of the image information. Second, the integrated regional distance, which is integrated with the factors, such as the image color information, edge strength and adjacency information, is defined. Third, an algorithm of diminutive region merging is designed to remove the diminutive region. As a result, the color image segmentation is more efficiently realized. Finally, a simulation experiment is implemented with the algorithm proposed, then the analysis is also given in this paper, and it is proved that the algorithm proposed is more effective in the color image segmentation and solves the problem of the over segmentation generated by the watershed algorithm segmentation.

Keywords: Neighbourhood, Image Segmentation, Watershed Algorithm, Region Merging.

1 Introduction

Image segmentation is a technique to extract the interesting part from the digital image and plays a key role in many image processing fields, such as pattern recognition, image understanding and computer vision [1, 2]. It is also one of the difficult problems in the image processing technology. A reasonable and effective image segmentation can provide very available information for image understanding and object analysis, and is also helpful to understand the high-level semantic information of the image [3].

Watershed segmentation algorithm based on mathematical morphology is time-consuming. An image segmentation algorithm [4] based on immersion simulations was proposed by Vincent et al. Although this method is good for segmentation in a short time, yet the segmentation results may generate a large number of diminutive regions (over-segmentation), which is not conducive to the image processing and analysis. ZHANG Jian-ming et al. presented a new algorithm of

---

* Corresponding author.

T. Li et al. (Eds.): RSKT 2012, LNAI 7414, pp. 167–173, 2012.
© Springer-Verlag Berlin Heidelberg 2012

watershed segmentation method based on gradient modification and hierarchical region merging to deal with over-segmentation [5], which merges diminutive regions and similar regions based on criterion of the average gray value of the region. YU Wang-sheng et al. proposed a color image segmentation algorithm combined marker-based watershed and region merger [6], which merges diminutive regions based on real sense of region-similarity of human-vision. However, the segmentation results were not good enough without considering the color distribution information of the image region edge. Recently the related research indicates that integrating numerous kinds of the information benefits legitimately for image segmentation, that is to say, the image segmentation needs to consider the image color information as well as the information of image region edge and image region adjacency [5–7].

Based on the above analysis, this paper presents an improved image segmentation algorithm integrating watershed algorithm and region merge. The regional distance integrating the color information, edge strength and adjacency information of the region is defined in the paper. Based on it, the color image segmentation is designed. Experiment results show that the image segmentation algorithm proposed is more coincident to the human visual perception.

2   Watershed Algorithm

To perform image initial segmentation, the image segmentation algorithm based on immersion simulations is proposed, which can quickly calculate the every region of the watershed segmentation [4] and includes two steps: sorting and immersion [8].

Step 1. First, calculate the pixel gradients of the image, and then scan the image to get the probability density of each gradient. Calculate the sort position of all pixels and store them in the sorted array by ascending gradients.

Step 2. Process these pixels according to ascending gradients, and pixels of the same gradient are seemed as a gradient level.

Step 3. Put pixels of the entire neighborhood identified into a FIFO queue when a gradient level h (current level) is processed.

Step 4. If the FIFO queue is not empty, then delete the first element of the queue and take the first element as current processing pixel. The adjacent pixel of current pixel is processed sequentially. If the adjacent pixel has been identified, then refresh the identity of the current pixel by the identity of the adjacent pixel. If the adjacent pixel has not been identified, then put the adjacent pixel into the FIFO queue. Algorithm performs Step 4 repeatedly until the FIFO queue is empty.

Step 5. Check whether there is unidentified pixel by scanning pixels of the current gradient level again. If unidentified pixels are detected, the identity values of the current regional will be increased by 1, finally algorithm performs Step 4 from the unidentified pixel.

Step 6. Return to Step 3 to process the next gradient level until all the gradient levels have been processed.

Due to the details and noise of the image information, the image segmentation algorithm based on immersion simulations can lead to the image over-segmentation problem.

3   Merger of the Image Regions

3.1   Distance Metric between Graph Regions

Lots of diminutive regions will be produced when the image has been divided using watershed algorithm, and for the sake of the better segmentation results, we need to merge these diminutive regions. However, distance metric between image regions is an important standard for region merging, the method of distance metric determines the results of image region merging and final segmentation results. The necessary conditions for the image region merging are: the two regions are similar in color, their space must be adjacent, and their edge strength is weak. So, we integrate color information, edge strength and adjacency relationship to define the method of distance metric.

G is set as a digital image indicated with RGB color space, its resolution is MN. Union of image regions produced along with the pre-segmenting of G image is set as $R = \{R_1, R_2, \cdots, R_k\}$, R_i is the i-th image region $(i = 1, 2, \cdots, k)$, the color mean of the three color components is defined as follows:

Definition 1. The color mean of the image region R_i in each color component can be defined as:

$$F_c(R_i) = \frac{1}{K} \sum_{F(m,n)\in R_i} F(m, n, c), \quad c \in \{r, g, b\}. \tag{1}$$

$F(m,n,c)$ is the value of the pixel $F(m,n)$ on the color component c, K is the number of pixels in the image region R_i.

Definition 2. Supposing R_i and R_j are two image regions, the color distance of R_i and R_j in RGB color space can be defined as:

$$D_c(R_i, R_j) = \frac{|R_i| \cdot |R_j|}{|R_i| + |R_j|} \sqrt{\sum_{c=r,g,b} (F_c(R_i) - F_c(R_j))^2}. \tag{2}$$

$|R_i|$ and $|R_j|$ are separately the numbers of pixels contained in the image regions of R_i and R_j.

In (2), the product of $|R_i|$ and $|R_j|$ can reduce the color distance between the diminutive regions and other regions. Once color means is equal, small regions will have priority to merge.

Definition 3. Supposing pixel $F(m,n)$ is the edge pixel of image regions of R_i and R_j, whose circular neighborhood is $\delta(m, n) = \{(x, y) | \sqrt{(x-m)^2 + (y-n)^2} \le d\}$, d is the radius of the circular neighborhood, then the distance between pixel $F(m,n)$ and pixel in circular neighborhood can be defined as:

$$D_\delta(m,n) = \frac{1}{|\delta(m,n)|} \sum_{x,y \in \delta(m,n)} \sqrt{\sum_{c=r,g,b} (F(x,y,c) - F(m,n,c))^2}. \quad (3)$$

$|\delta(m,n)|$ is the cardinality of the circular neighborhood of pixel $F(m,n)$.

Definition 4. Supposing E_{ij} is the edge of image regions of R_i and R_j, then the edge strength can be defined as:

$$D_e(R_i, R_j) = \frac{1}{|E_{ij}|} \sum_{F(m,n) \in E_{ij}} D_\delta(m,n). \quad (4)$$

$|E_{ij}|$ is the pixel number of image region edge, $F(m,n)$ is an edge pixels.

Definition 5. [9] Adjacency relation of image regions of R_i and R_j in space can be defined as:

$$\Delta_{ij} = \begin{cases} 1, & if\ R_i\ is\ adjacent\ to\ R_j \\ +\infty, & if\ R_i\ is\ not\ adjacent\ to\ R_j \end{cases} \quad (5)$$

Definition 6. Integrated regional distance between image regions of R_i and R_j can be defined as:

$$D(R_i, R_j) = (D_c(R_i, R_j))^p \cdot (D_e(R_i, R_j))^q \cdot \Delta_{ij}. \quad (6)$$

The diminutive areas can be merged through the integrated regional distance, and the values of p and q can be changed to adjust the roles of the color distance and the edge strength. In this simulation experiment, we let $p=q=1$.

3.2   Merger Algorithm of the Image Regions

Region merging is to merge diminutive regions and similar regions, through which the number of image regions can be reduced to achieve better segmentation results. When integrated regional distance of two regions is less than threshold T, they should be merged. Integrated regional distance used in this paper considers several factors such as similarity of the color characteristics, the edge strength and spatial information of the edge of the region. In conclusion, a new method of region merging is proposed.

   Algorithm 1. Merger Algorithm of the Image Regions.
   Input. Regional sequence R_k and regional adjacency relationship (RAG) generated by initial segmentation.
   Output. Regional sequence R_m generated by region merging.
   Step 1. Calculate the color distance and edge strength among these image regions;
   Step 2. Calculate the integrated regional distance on the basis of color distance, edge strength and adjacency relations among the segmented regions;

Step 3. Seeking such two regions in regional sequence R_k, that their distance is the shortest;

Step 4. If the minimum of the integrated regional distances is less than a threshold T, the region containing less pixels will be merged into another region, and this region will be deleted from the regional sequence R_k, following with the modification of the adjacency relationship and integrated regional distances of the merged regions, then goto Step 3; Otherwise, goto Step 5;

Step 5. Output the merged regional sequence R_m.

4   Experimental Results and Analysis

In this section, a simulation experiment is implemented, to evaluate the efficiency of the algorithm proposed in this paper, with the algorithm proposed by Matlab 2009. And the experimental hardware environment is as follows: Pentium (R) D CPU 2.80GHz1GB memoryWindows XP. In this experiment, we let $d=4$, i.e., the circle region radius of the edge pixel is 4, and the threshold value of the integrated regional distance is 24.

To analyze the experimental results, five kinds of color images, each of which extracts randomly twenty color images, are applied to the experiment. And the segmentation results are more corresponding with the visual character of people. It is shown that some images used for the simulation experiment and the segmentation images which are obtained by the algorithm proposed and the watershed algorithm respectively are in Fig. 1.

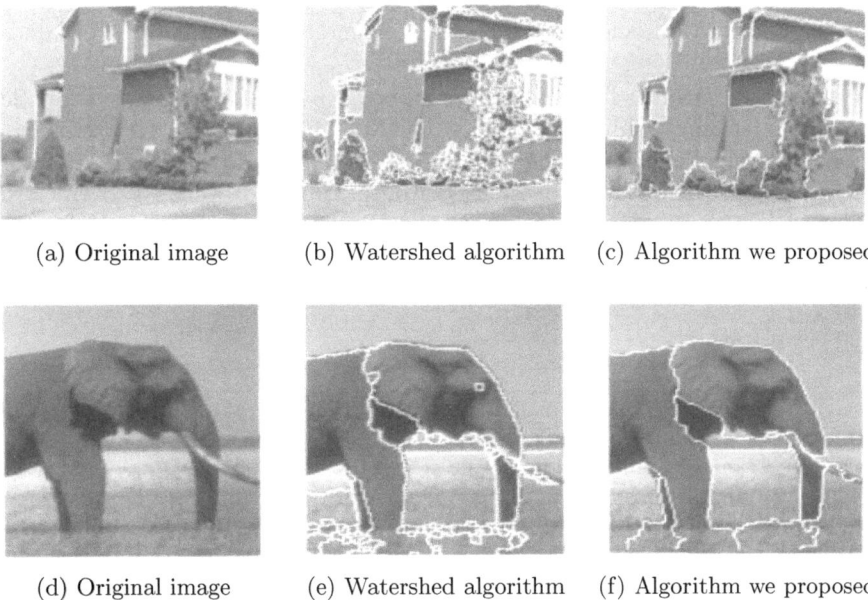

(a) Original image   (b) Watershed algorithm   (c) Algorithm we proposed

(d) Original image   (e) Watershed algorithm   (f) Algorithm we proposed

Fig. 1. Comparison chart of image segmentation results

The effectiveness of the segmentation images obtained by the algorithm proposed and the watershed algorithm is compared. The algorithm proposed considers the factors, such as the color feature of the image region, the edge information, and adjacency relationship between the regions, hence, its segmentation results are more coincident to the human visual perception. From Fig. 1, obviously, many diminutive image regions appear in (b) and (e), and the perception vision of people is not satisfied by the over-segmentation. However, in (c) and (f) the change of the leaves' color is slowly and the edge information is not evident. Therefore, those leaves are combined into a large region by using the algorithm proposed, the image divided is more coincident to the human visual perception. The algorithm proposed is more effective in the color image segmentation, and solves the problem of the over-segmentation generated by the watershed algorithm segmentation.

5   Conclusion

In this paper, a novel image segmentation algorithm based on the watershed and the region merger is presented. The phenomenon that the details and noise of the image information cause the image over-segmentation, can be availably avoided by considering the factors of the color feature, the edge information, and adjacency relationship of the diminutive image regions when merging the diminutive image regions. And the simulation experiment illustrates that the algorithm proposed can get a better segmentation results. The further work is to investigate the automatic-stop region integration algorithm based on the watershed algorithm, and a much better segmentation results can be obtained.

Acknowledgement. This paper is supported by the National Natural Science Foundation of China (No. 60873104, No. 61040037), the Key Project of Science and Technology of Henan Province of China (No. 112102210194), the Natural Science Foundation of Education Department of Henan Province (No. 2011A520022), and the Youth Science Fund of Henan Normal University (No. 2010qk20).

References

1. Gonzalez, R.C., Woods, R.E.: Digital Image Processing, 2nd edn. Publishing House of Electronics Industry, Beijing (2006)
2. YuJin, Z.: Image Segmentation. Science Press, Beijing (2001)
3. YuJin, Z.: Image Engineering, 2nd edn. Tsinghua University Press, Beijing (2007)
4. Vincent, L., Soille, P.: Watersheds in Digital Spaces: An Efficient Algorithm Based on Immersion Simulations. IEEE Transactions on Pattern Analysis and Machine Intelligence 13(6), 583–598 (1991)
5. Jianming, Z., Ju, Z., et al.: Watershed segmentation algorithm based on gradient modification and region merging. Journal of Computer Applications 31(2), 369–371 (2011)

6. Wangsheng, Y., Zhiqiang, H., et al.: Color Image Segmentation Based on Marked-Watershed and Region-Merger. Acta Electronica Sinica 39(5), 1007–1012 (2011)
7. Qu, G., Yumin, Y., et al.: Image segmentation based on watershed and improved fuzzy C-means clustering. Application Research of Computers 28(12), 4773–4775 (2011)
8. Xiaoqi, L., Yunzhou, F., et al.: A Study of Medical Image Segmentation Based on Human Interactive Region Merges of Watershed. Chinese Journal of Medical Imaging 18(6), 516–520 (2010)
9. Qixiang, Y., Wen, G., et al.: A Color Image Segmentation Algorithm by Using Color and Spatial Information. Journal of Software 15(4), 522–530 (2004)

A Mixed Strategy Multi-Objective Coevolutionary Algorithm Based on Single-Point Mutation and Particle Swarm Optimization

Xin Zhang[1,2], Hongbin Dong[1,*], Xue Yang[1], and Jun He[3]

[1] Department of Computer Science and Technology,
Harbin Engineering University, Harbin, China
[2] The 54th Research Institute of CETC, Shijiazhuang, China
[3] Department of Computer Science, Aberystwyth University
Aberystwyth, SY23 3DB, UK
donghongbinbjtu@gmail.com

Abstract. The particle swarm optimization algorithm has been used for solving multi-objective optimization problems in last decade. This algorithm has a capacity of fast convergence; however its exploratory capability needs to be enriched. An alternative method of overcoming this disadvantage is to add mutation operator(s) into particle swarm optimization algorithms. Since the single-point mutation is good at global exploration, in this paper a new coevolutionary algorithm is proposed, which combines single-point mutation and particle swarm optimization together. The two operators are cooperated under the framework of mixed strategy evolutionary algorithms. The proposed algorithm is validated on a benchmark test set, and is compared with classical multi-objective optimization evolutionary algorithms such as NSGA2, SPEA2 and CMOPSO. Simulation results show that the new algorithm does not only guarantee its performance in terms of fast convergence and uniform distribution, but also have the advantages of stability and robustness.

Keywords: multi-objective optimization, single-point mutation, particle swarm optimization, mixed strategy, coevolutionary algorithms.

1 Introduction

In many real world optimization problems, it is important to optimize multiple competing objectives simultaneously; therefore multi-objective optimization algorithms have been developed to solve such kind of optimization problems. Evolutionary algorithms (EAs), which simulate the process of natural evolution, are one of most efficient and popular algorithms for solving multi-objective optimization problems[1]. Coevolutionary algorithms are a special form of EAs, which involve two or more populations concurrently. Besides evolution at the individual level, evolution will also happen at the population level through competition and collaboration among different populations[4].

---

* Corresponding author.

T. Li et al. (Eds.): RSKT 2012, LNAI 7414, pp. 174–184, 2012.
© Springer-Verlag Berlin Heidelberg 2012

Compared with normal EAs, coevolutionary algorithms are able to adapt to more complex dynamic environment. This potentially will lead to better stability and robustness.

At present, a lot of search operators have been proposed for improving the efficiency of evolutionary algorithms, such as Gaussian mutation, Cauchy mutation, single-point mutation[6], and particle swarm optimization. Each search operator has its own advantages and disadvantages, however none of them will be efficient for all problems according to no free lunch theorem. The superiority of one operator is dependent on the type of problems. For example, particle swarm optimization (PSO) is well known for its fast convergence; but this capability strongly relies on unimodality of objective functions. For the multi-modal objective functions, the fast convergence capability will become a disadvantage which leads to converge quickly to local optimum[7]. Single-point mutation (SP) has very good global exploration ability which is efficient in multi-modal objective functions[6]. But for unimodal objective functions, its convergence speed is usually slow.

This paper presents a multi-objective mixed strategy coevolutionary algorithm which combines both single point mutation and particle swarm search together (denoted by (MO)MS-SP&PSO). The single-point mutation is taken for making effective global exploration, and particle swarm optimization for improving the convergence speed. Single-point mutation and particle swarm optimization work together in the framework of mixed strategy coevolutionary algorithms[5]. This paper will verify the effectiveness of the algorithm on a standard test set such as ZDT, DTLZ, and compare it with other classic multi-objective optimization algorithms including NSGA2[3], SPEA2[11], CMOPSO[2].

2 Multi-Objective Optimization Problems

In this paper, we consider a problem composed of n decision variables, M objective functions, and k constraint conditions. It is described as follows:

$$\min y = F(\boldsymbol{x}) = [f_1(\boldsymbol{x}), f_2(\boldsymbol{x}), f_M(\boldsymbol{x})], \tag{1}$$
$$s.t.\ g_i(\boldsymbol{x}) \leq 0, i = 1, 2, \ldots, p,$$
$$h_i(\boldsymbol{x}) = 0, i = 1, 2, \ldots, q,$$

where $\boldsymbol{x} = (x(1), x(2), \ldots, x(n)) \in D$ is called decision vector, D the decision space, $y = (f_1, f_2, \ldots, f_M) \in Y$ objective vector, and Y objective space. In the problem (1), there are $p + q$ constraints, where q constraints are inequality and p constrains are equality.

For most multi-objective optimization problems, it is impossible to make all targets optimal simultaneously because sub-objectives may conflict with each other. Therefore, in multi-objective optimization, the aim is to find a solution set rather than a single solution.

3 Multi-Objective Coordination Operator Evolutionary Programming Based on Single Point and PSO Operator

In the particle swarm optimization algorithm (CMOPSO) for multi-objective optimization, two sets are used, i.e., one internal set P(particle swarm), and another external archive A. The contribution of PSO search in the CMOPSO algorithm is not always positive. It can be seen that the number of particles in A is increasing in the course of evolution. In the early evolutionary stage, there are only a few particles in A, so the process of P randomly selecting a g_{best} from the archive A will probably generate a lot of non-dominated solutions near the selected g_{best}. However, in the later evolutionary stage, there are more particles in the archive A, it is hard to find anymore non-dominated solutions through the PSO procedure. At this time, the behavior of PSO is like a blind search on a local Pareto front and contributes little.

(MO)MS-SP&PSO aims at improving the above disadvantage existing in CMOPSO through introducing a new mixed strategy based on single-point mutation and particle swarm optimization. The details of the algorithm are described in the following.

3.1 Mixed Strategy

The idea is to combine two different algorithms together. Particle swarmoptimization is taken for acceleratingconvergence, while the single-point strategy forglobal exploration. The two search operators do notact on the same population. PSO search acts on the population P while the single-point mutation acts on the external archive A. A is used to store non-dominated solutions. Single-point mutation acts on it and this may probably help to find better solutions even if PSO search get into trapped. The two operators are applied in the framework of mixed strategy coevolutionary algorithms: each time one strategy (PSO search or SP mutation) is chosen according to a predefined probability distribution; and then the strategy is applied to the related population. Denote particle swarmoptimization searchby PSO strategy, single-point mutation by SP strategy. The basic flowchart of (MO)MS-SP&PSOis shown in Fig.1 and Fig.2.

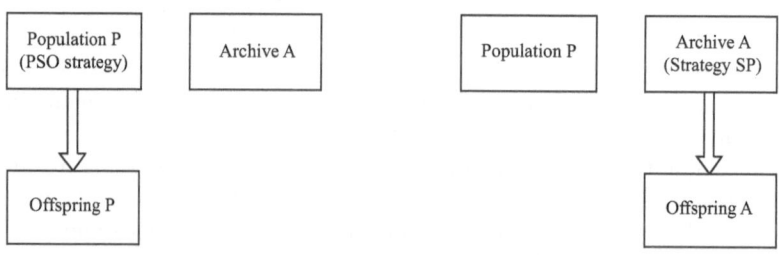

Fig. 1. PSO strategy   **Fig. 2.** SP strategy

3.2   Maintenance of External Archive

There are two different ways to update the archive.

(1) During the application of the PSO strategy, once a new offspring population P' is generated, then all non-dominated individuals will be added into the archive A, and form the next generation $A(t+1)$. This is the same as that in SPEA2.
(2) During the application of the SP strategy, once a new archive A' is produced through mutation and selection, then all non-dominated individuals in the union of archives A and A' will be used to form the next generation archive $A(t+1)$. In this procedure, there is no change on the population P.

After maintenance of the archive, $P(t+1)$ randomly select a non-dominated individual from $A(t+1)$ as the global best particle in next iteration.

Individuals in the external archive compose the final Pareto frontier. In the process of evolution, the non-dominated solutions from the population and the previous archive need to be added into a new external archive continuously. If all individuals are in non-inferior relationship, according to the density of information, high density individuals will be removed except border ones.

3.3   Algorithm Description

The (MO)MS-SP&PSO algorithm for solving the multi-objective optimization problem (1) is described as following: Denote μ the size of population P, and $|A|$ the archive size, which is fixed in the algorithm.

Each particle (i.e. individual) in the population are defined by a triad $(x_i^{(0)}, \eta_i^{(0)}, v_i^{(0)})$, $\forall i \in \{1, \ldots, \mu\}$, where x_i the decision vector, η_i the step-size of single point mutation, and v_i the speed of particle i. Each vector in the triad is n-dimensional.

$$\begin{cases} x_i^{(0)} = (x_i^{(0)}(1), x_i^{(0)}(2), \ldots, x_i^{(0)}(n)) \\ \eta_i^{(0)} = (\eta_i^{(0)}(1), \eta_i^{(0)}(2), \ldots, \eta_i^{(0)}(n)) \\ v_i^{(0)} = (v_i^{(0)}(1), v_i^{(0)}(2), \ldots, v_i^{(0)}(n)) \\ \quad i = 1, \ldots, \mu \end{cases} \tag{2}$$

The mixed strategy is defined by a vector: $\rho = (\rho_1, \rho_2)$, where ρ_1 is the probability of applying single-point mutation strategy and ρ_2 the probability of applying particle swarm optimization strategy.

Initialization. Initialize population $P(0)$: set the generation counter $t = 0$, for each particle i, the value of $x(d)$ is initialized by a uniform distribution.

$$x(d) \in [a_d, b_d];$$
$$\eta_i^{(0)}(d) = 0.5(b_d - a_d);$$
$$v_i^{(0)}(d) = 0, d = 1, \ldots, n.$$

Initialize the external archive $A(0)$, P_{bi}, P_g: evaluate the fitness of each particle in $P(0)$, then chose those non-dominated solutions and add them into the non-dominated solution set called *Leaders*:

$$Leaders = \{x_i | x_i \in P^{(0)} \; non\text{-}dominated\}$$

Randomly select a particle from the set *Leaders* as P_g. Set $P_{bi} = x_i$, $A(0) = P(0)$.

Initialization mixed strategy vector ρ: set $\rho^{(0)} = (0, 1)$.

The Variation of the Mixed Strategy. Given the population $P(t)$ or the external archive $A(t)$, randomly select a strategy to be implemented according to the mixed strategy $\rho = (\rho_1, \rho_2)$.

Single-point Mutation Strategy. Single point mutation strategy acts on the external archive $A(t)$, while the population $P(t)$ is unchanged. In this mutation, only one of the n components is altered in each iteration. For example, choose one component d randomly $\{1, \ldots, n\}$, then parent individual $(x_i^{(t)}, \eta_i^{(t)}, v_i^{(t)})$ will generate the offspring individual $(x_{i'}^{(t)}, \eta_{i'}^{(t)}, v_{i'}^{(t)})$ by the following formulas:

$$\begin{cases} \eta_{i'}^{(t)}(d) = \eta_i^t(d) \exp(-\alpha) \\ x_{i'}^{(t)}(d) = x_i^t(d) + \eta_{i'}^{(t)}(d) N_d(0, 1) \\ v_{t'}^{(t)} = v_i^{(t)} \end{cases} \tag{3}$$

The experimental result in paper [5] shows that the value of α will affect the quality of the final solution, but the performance of the mixed strategy algorithm will not be greatly affected. Thus, we set $\alpha = 2.1$ here. The initial value of η is $\eta_i^{(0)}(d) = 0.5(b_d - a_d)$. If for one generation $\eta_i^{(t)}(d) < 10^{-4}$, then we reset $\eta_i^{(t)}(d) = 0.5(b_d - a_d)$. $N_d(0, 1)$ means the standard deviation of the Gaussian mutation. After the single-point mutation strategy, the external archive $A(t)$ will generate an offspring set $A'(t)$ and $A(t+1)$ will be generated from $A(t) \cup A'(t)$ by the maintenance process.

PSO Mutation Strategy. PSO mutation strategy act on the population $P^{(t)}$. Each individual in $P^{(t)}$ $(x_i^{(t)}, \eta_i^{(t)}, v_i^{(t)})$ generates an offspring as follows:

$$\begin{cases} v_{i'}^{(t)} = \omega v_i^{(t)} + c_1 R_1 (Pb_i^t - x_i^t) + c_2 R_2 (Pg^t - x_i^t) \\ x_{i'}^{(t)} = x_i^{(t)} + v_{i'}^{(t)} \\ \eta_{i'}^{(t)} = \eta_i^{(t)} \end{cases} \tag{4}$$

where the meaning of parameters in Equation (4) are:
c_1, c_2: learning factor.
R_1, R_2: uniformly distributed random number in $[0, 1]$.
ω: PSO operator inertia weight.

After PSO mutation, the memory Pb_i of each individual in new population will be updated as follows: first we compare the new location with the original Pb_i; if the new location dominated Pb_i, then update Pb_i to be the new location of individual i, otherwise, Pb_i remain unchanged.

Interval Constraints Handling. If the applied strategy is single point mutation, then we restrict each individual i' in the offspring $I'(t)$ population, with the following formula:

$$x_{i'}(d) = \begin{cases} [x_{i'}(d) - a_d] \bmod (b_d - a_d) + a_d, & x_{i'}(d) > b_d \\ [x_{i'}(d) - b_d] \bmod (a_d - b_d) + b_d, & x_{i'}(d) < a_d \\ x_{i'}(d), & x_{i'}(d) \in [a_d, b_d] \end{cases} \quad (5)$$

If the applied strategy is particle swarm optimization, then we restrict the particles which are flying out of the boundary, change its speed direction.

Generate the Next Generation Population $P^{(t+1)}$. If a single point mutation strategy is selected, then a new offspring archive set A' will be generated; but the particle swarm will not be changed;

If the strategy is PSO strategy, then new particle swarm P' will be generated.

The Fitness Assignment. Given the union of $A^{(t)}$ and offspring A' (denoted by $UA^{(t)}$), and measure the fitness of individuals in the union $UA^{(t)}$.

Archive Update. Let $A^{(t+1)} = \{x_i | x_i \in UA^{(t)} \ non\text{-}dominated\}$, if the size of $A^{(t+1)}$ is equal to $|A|$, accepted; if the size of $A^{(t+1)}$ is less than $|A|$, the best individuals $|A| - A^{(t+1)}$ in UA^t will be added to $A^{(t+1)}$; if the size of $A^{(t+1)}$ is larger than $|A|$, an archive truncation procedure is invoked which iteratively removes individuals from $A^{(t+1)}$ until $|A^{(t+1)}| = |A|$. This truncate procedure is the same as that in SPEA2 and the details are described below:

Individual i is chosen for removal should satisfy the condition: $\forall j \in \overline{P_{t+1}}$, $i \leq_d j$.

$$i \leq_d j :\Leftrightarrow \forall 0 < k < |\overline{P_{t+1}}| : \sigma_i^k = \sigma_j^k \ \vee$$
$$\exists 0 < k < |\overline{P_{t+1}}| : [(\forall 0 < l < k : \sigma_i^l = \sigma_j^l) \wedge \sigma_i^k < \sigma_j^k]$$

Update the Memory of the Whole Particle Swarm Pg. Clear the *Leaders* set, and select non-dominated solutions from $A^{(t+1)}$ then add them into the *Leaders* set. Finally, randomly select an individual from the *Leaders* set as Pg.

Adjustment of Mixed Strategy. In order to fine-tune optimization, it is better to apply a large size of *Leaders* set, which means a big probability of choosing the single-point mutation strategy. In order to accelerate the convergence, it is better to apply a small size of *Leaders* set, which means a big probability of choosing the PSO strategy.

In each iteration, *Leaders* will be cleared first and then rebuild by:

$$Leaders = \{x_i | x_i \in A^{(t+1)} \ non\text{-}dominated\}.$$

Mixed strategy vector is adjusted as follows:

$$\begin{cases} \rho_1 = \lambda \times |Leaders|/|A| \\ \rho_2 = 1 - \rho_1 \end{cases} \quad (6)$$

where λ is a range parameter to prevent the mixed strategy from degenerating into pure strategy. In the experiment, $\lambda = 0.8$. It follows an observation: the probability of a single point mutation strategy ρ_1 is determined by the size of *Leaders*.

In the early evolutionary stage, there are only a few particles in the external archive A, the size of *Leaders* is small. So the PSO strategy should be played as a dominant strategy in order to accelerate convergence;

In the later evolutionary stage, there are more particles in the archive A, the size of *Leaders* becomes larger, it is hard to find anymore non-dominated solutions through the PSO procedure, and the behavior of PSO is like a blind search on a local Pareto front and contributes little. So the SP strategy should play as a dominant strategy to help find better solutions. In addition, when the algorithm converges, the single-point operator can also fine-tune the parato-frontier and make the distribution of solutions more uniformly.

End of the Judge. Set the current generation $t = t + 1$. If $t \geq t_{max}$ or another stopping criterion is satisfied, then stop and output the result; otherwise return to step 2.

4   The Experimental Results

In the experiment, we compare the new algorithm with several classical multi-objective evolutionary algorithms such as NSGA2, SPEA2, and CMOPSO. We test our algorithm on several benchmark functions such as ZDT and DTLZ.

4.1   Experiment Setting

Test Function. In this paper, we select 5 instances of ZDT problems and two instances of DTLZ problem to test the performance of the new algorithm. The characteristics of these instances are shown in Table 1:

Table 1. Characteristics of test functions

Test problem	Target number	Characteristic
ZDT1	2	Protruding PFture
ZDT2	2	Protruding PFture
ZDT3	2	Protruding, incontinuity PFture
ZDT4	2	209 local PFture
ZDT6	2	ununitary PFture
DTLZ2	3	Continuity PFture
DTLZ7	3	nonincontinuity PFture

As the number of goals and the dimension of decision variable vector can be extended, the parameters of DTLZ used in the experiment are shown in Table 2:

Table 2. DTLZ problem parameter selection

Test problem	parameter k	parameter $|xk|$
DTLZ2	3	10
DTLZ7	3	20

Algorithm Parameters. The population size P in the (MO)MS-SP&PSO algorithm is $\mu = 100$, and the archive size $|A|$ is 100.

For comparison, this paper selects three classical multi-objective evolutionary algorithms: CMOPSO, NSGA2 and SPEA2. The population size in NSGA2 is set to be 100 and the archive size 100. In SPEA2, the internal and external archives sizes are 100. The parameters of PSO operator in CMOPSO and (MO)MS-SP&PSO are random selected following the uniform distribution. The limit range is

$$\omega \in (0.1, 0.5), \quad c_1, c_2 \in (1.5, 2.0), \quad R_1, R_2 \in (0, 1).$$

The termination condition of each algorithm is a maximum number of iterations $t_{max} = 500$ (i.e. 50,000 times the function evaluation). Other parameters used in the algorithm are given in Table 3.

Table 3. Parameter setting of NSGA2, SPEA2

parameter	NSGA2	SPEA2
Crossover probability	0.8	0.8
Distribution index for crossover	15	15
Mutation probability	1/n	1/n
Distribution index for mutation	20	20

Performance Measures. We use HV[9], IGD[8], EPSILON[10] as the performance indicators in this paper where HV, IGD are taken as integrated indicators, and EPSILON as a convergence indicator.

4.2 Results and Analysis

The Simulation Results. The mean and variance of performance indicators for each algorithm are shown in Table 4, Table 5 and Table 6. The dark gray area is the optimal value in the table, while the light gray area is the second optimal value.

It can be seen from Table 4, Table 5 that: based on the performance indicators HV and IGD, CMOPSO is the best for ZDT1 and ZDT2, however, for ZDT4 and all the multi-objective problem DTLZ2, DTLZ7, CMOPSO's performance is the worst; (MO)MS-SP&PSO algorithm's performance is either the best, or the second and the it is little bad from the best for all test problems.

It can be seen from Table 6 that: based on the performance indicator EP-SILON, Except ZDT3, (MO)MS-SP&PSO's performance is also either the best,

Table 4. The Mean Best and Standard Deviation of HV

	NSGA2	SPEA2	CMOPSO	(MO)MS-SP&PSO
ZDT1	6.59e-01,$_{3.2e-04}$	6.60e-01,$_{3.0e-04}$	6.61e-01,$_{4.7e-04}$	6.61e-01,$_{2.2e-04}$
ZDT2	3.26e-01,$_{3.0e-04}$	3.26e-01,$_{3.9e-03}$	3.28e-01,$_{2.7e-04}$	3.28e-01,$_{2.4e-04}$
ZDT3	5.15e-01,$_{1.3e-04}$	5.14e-01,$_{6.8e-04}$	5.14e-01,$_{1.1e-03}$	5.15e-01,$_{2.4e-04}$
ZDT4	6.55e-01,$_{2.5e-03}$	6.43e-01,$_{1.3e-02}$	0.00e+00,$_{0.0e+00}$	6.59e-01,$_{1.9e-03}$
ZDT6	3.88e-01,$_{1.5e-03}$	3.79e-01,$_{2.2e-03}$	4.01e-01,$_{8.9e-05}$	4.00e-01,$_{3.9e-04}$
DTLZ2	3.75e-01,$_{5.3e-03}$	4.04e-01,$_{2.0e-03}$	1.64e-02,$_{6.3e-02}$	4.05e-01,$_{3.0e-03}$
DTLZ7	2.80e-01,$_{4.5e-03}$	2.90e-01,$_{2.4e-03}$	8.86e-03,$_{4.8e-02}$	2.98e-01,$_{2.6e-03}$

Table 5. The Mean Best and Standard Deviation of IGD

	NSGA2	SPEA2	CMOPSO	(MO)MS-SP&PSO
ZDT1	1.86e-04,$_{7.9e-06}$	1.52e-04,$_{3.0e-06}$	1.38e-04,$_{3.7e-06}$	1.47e-04,$_{2.0e-06}$
ZDT2	1.93e-04,$_{1.0e-05}$	2.36e-04,$_{4.3e-04}$	1.43e-04,$_{3.0e-06}$	1.46e-04,$_{2.2e-06}$
ZDT3	2.51e-04,$_{8.3e-06}$	3.99e-04,$_{8.7e-04}$	2.18e-04,$_{1.8e-05}$	2.23e-04,$_{7.3e-06}$
ZDT4	2.31e-04,$_{3.4e-05}$	1.48e-03,$_{1.4e-03}$	1.86e-01,$_{8.7e-02}$	1.63e-04,$_{2.9e-06}$
ZDT6	3.71e-04,$_{4.0e-05}$	6.45e-04,$_{7.1e-05}$	1.20e-04,$_{7.2e-06}$	1.16e-04,$_{5.2e-06}$
DTLZ2	7.76e-04,$_{3.5e-05}$	5.92e-04,$_{1.4e-05}$	4.93e-03,$_{8.4e-04}$	5.96e-04,$_{1.4e-05}$
DTLZ7	2.25e-03,$_{1.5e-04}$	1.86e-03,$_{7.9e-05}$	1.02e-01,$_{2.0e-02}$	1.79e-03,$_{4.8e-05}$

Table 6. The Mean Best and Standard Deviation of EPSILON

	NSGA2	SPEA2	CMOPSO	(MO)MS-SP&PSO
ZDT1	1.31e-02,$_{1.7e-03}$	9.00e-03,$_{6.9e-04}$	6.31e-03,$_{5.1e-04}$	8.80e-03,$_{9.0e-04}$
ZDT2	1.33e-02,$_{1.9e-03}$	1.92e-02,$_{5.3e-02}$	6.11e-03,$_{3.4e-04}$	8.16e-03,$_{7.1e-04}$
ZDT3	8.12e-03,$_{1.5e-03}$	2.04e-02,$_{5.5e-02}$	8.16e-03,$_{3.6e-03}$	8.85e-03,$_{1.2e-03}$
ZDT4	1.45e-02,$_{2.3e-03}$	8.08e-02,$_{6.6e-02}$	6.18e+00,$_{2.8e+00}$	9.17e-03,$_{1.1e-03}$
ZDT6	1.50e-02,$_{1.4e-03}$	2.45e-02,$_{2.9e-03}$	4.95e-03,$_{5.5e-04}$	6.38e-03,$_{5.9e-04}$
DTLZ2	1.29e-01,$_{1.7e-02}$	8.12e-04,$_{6.1e-03}$	5.82e-01,$_{9.6e-02}$	8.36e-02,$_{5.5e-03}$
DTLZ7	1.45e-01,$_{3.5e-02}$	9.94e-03,$_{9.6e-03}$	1.05e+01,$_{2.0e+00}$	9.11e-02,$_{7.3e-03}$

or the second and the results are little different from the best. NSGA2 is the best for ZDT3, but its performance is just a little better than that of (MO)MS-SP&PSO.

The Simulation Results. If the target functions are unimodal (like ZDT1 and ZDT2), the CMOPSO algorithm is the best during the comparison. It confirms that PSO strategy has an advantage in searching unimodal objective functions. The performance of the (MO)MS-SP&PSO algorithm is very close to CMOPSO. This means that adding single-point mutation doesn't bring too many benefits, but it doesn't bring too much negative effect too.

If the target functions are multimodal or discrete, the CMOPSO algorithm is not efficient since its global exploration ability is relatively weak, and the search is easy to fall into local Pareto frontiers. The (MO)MS-SP&PSO algorithm is the best for solving these problems. So adding single-point mutation may bring a great benefit since the single-point mutation can enhance the global search capacity.

It is hard for the CMOPSO algorithm to converge to the true Pareto front (PFture). But the (MO)MS-SP&PSO algorithm always converges to the front very quickly. This indicates that the mixed strategy has generated a successfully improvement on the CMOPSO algorithm.

If we compare the (MO)MS-SP&PSO algorithm with NSGA2 and SPEA2, the performance of (MO)MS-SP&PSO is either the best, or a runner close to the best. This demonstrates that the (MO)MS-SP&PSO algorithm is very competitive.

To sum up, the (MO)MS-SP&PSO algorithm take the advantages of both particle swarm optimization and single-point mutation as complementary strategies, we get a new algorithm with fast convergence, dispersion, stability and robustness.

5 Summary

In this paper, we propose a mixed strategy coevolutionary algorithm for solving multi-objective optimization problems. The algorithm can be regarded as an improved version of the multi-objective particle swarm algorithm CMOPSO. We add single-point mutation into the CMOPSO algorithm and propose a new mixed strategy which chooses either particle swam optimization strategy or single-point mutation strategy based a probability distribution. This algorithm combines the advantages of both particle swarm optimization and single-point mutation. The mixed strategy can adaptively adjust the percentage of applying particle swam optimization and single-point mutation operators during running the algorithm.

The experiment results have shown that this algorithm does not only keep the fast convergence properties and dispersion properties very well, but also has the advantage of stability and robustness.

Acknowlegement. This work is partially supported by the National Natural Science Foundation of China under Grant (60973075), the Natural Science Foundation of Heilongjiang Province of China under Grant (F200937), and Ministry of Industry and Information Technology (B0720110002).

References

1. Coello Coello, C.A., Van Veldhuizen, D.A., Lamont, G.B.: Evolutionary Algorithms for Solving Multi-Objective Problems (2002)
2. Coello Coello, C.A., Pulido, G.T., Lechuga, M.S.: Handling multiple objectives with particle swarm optimization. IEEE Transactions on Evolutionary Computation 3(8), 256–279 (2004)

3. Deb, K., Pratap, A., Agarwal, S., Meyarivan, T.: A fast and elitist multiobjective genetic algorithm: Nsga-ii. IEEE Transactions on Evolutionary Computation (6), 182–197 (2002)
4. Hongbin, D., Houkuan, H., Guisheng, Y., Jun, H.: An overview of the research on coevolutionary algorithms. Journal of Computer Research and Development 45(3), 454–463 (2008)
5. Hongbin, D., Jun, H., Houkuan, H., Wei, H.: Evolutionary programming using a mixed mutation strategy. Information Sciences 1(177), 312–327 (2007)
6. Mingjun, J., Hualong, Y., Yongzhi, Y., Zhihong, J.: A single component mutation evolutionary programming. Applied Mathematics and Computation 215(10), 3759–3768 (2010)
7. Shi, Y.H., Eberhart, R.C.: A modified particle swarm optimization. In: IEEE International Conference on Evolutionary Computation, Anchorage, Alaska, pp. 69–73 (1998)
8. Van Veldhuizen, D.A., Lamont, G.B.: Multiobjective Evolutionary Algorithm Research: A History and Analysis (1998)
9. Van Veldhuizen, D.A., Lamont, G.B.: Multiobjective evolutionary algorithm test suites. In: Proc. Symp. Appl. Comput., San Antonio, TX, pp. 351–357 (1999)
10. Zitzler, E., Laumanns, M., Bleuler, S.: A Tutorial on Evolutionary Multiobjective Optimization (2004)
11. Zitzler, E., Laumanns, M., Thiele, L.: Spea2: Improving the strength pareto evolutionary algorithm, Gloriastrasse 35, CH-8092 Zurich, Switzerland (2001)

A Novel Distributed Machine Learning Method for Classification: Parallel Covering Algorithm

Yanping Zhang, Yuehua Wang, and Shu Zhao[*]

Anhui University, Computer Science and Technology Institute,
Key Laboratory of Intelligent Computing and
Signal Processing of Ministry of Education,
Hefei, Anhui Province, 230039, P.R. China
zhaoshuzs2002@hotmail.com

Abstract. In this paper, we propose a novel distributed machine learning method: Parallel Covering Algorithm, which is inspired by the module feature of CA (Covering Algorithm). Classic method of CA is presented, and we analyze its independent part. Then we develop the Parallel CA by utilizing its modularity as well as data-set decomposition. Detailed implementation of the parallel computing process is described. In the experiment, three data sets are used to evaluate the Parallel CA, and the comparison with classic CA is also shown in the paper. Speedup and efficiency are two criterions to evaluate the performance of the algorithm. Both the analysis and the comparison indicate that the Parallel CA is more effective than CA. We also empirically compare the results obtained by Parallel SVM on a large data set, and it shows that our proposed algorithm is effective.

Keywords: Machine Learning, Covering Algorithm, Parallel CA, Classification, Distribution.

1 Introduction

Covering algorithm was firstly proposed by Ling Zhang in 1998[1][2], as an constructive algorithm for MLP, similar to other popular machine learning algorithms, it can be used to solve classification problems, and it was extensively applied in text categorization [3], image recognition [4], stock forecast [5], etc. Based on Zhang's theory, many improved algorithms have been proposed to solve many pattern recognition problems effectively [6,7].

Classification is very important in machine learning field. Classic algorithm can only solve small-scale problems efficiently, but it is hard to solve large-scale problems because the training time is usually unbearable to us. Distributed parallel computer's high performance makes it effective to solve massive amount of data, so it is essential for us to design parallel machine learning algorithm to solve large scale problems when the data-set becomes increasingly larger.

---

[*] Corresponding author.

T. Li et al. (Eds.): RSKT 2012, LNAI 7414, pp. 185–193, 2012.
© Springer-Verlag Berlin Heidelberg 2012

So far, many classic machine learning algorithms such as SVM and BP have been designed to paralleled algorithms correspondingly. Thanks to parallel computing, it greatly reduced the train time elapsed in these algorithms. Parallel SVM is essentially based on the method of decomposing sample space [8,9], and cutting large problem into smaller independent problems, then solving it simultaneously. Parallel neural network and parallel genetic algorithms are also based on the decomposition of the problem [10,11,12]. Speeding up the training process through parallelization is often difficult due to dependencies between the computation steps, and the concurrent process requires no interference between the various sub-problems.

CA is an efficient machine learning algorithms. It is known for high training speed and high classification accuracy. But when encountered large data-set, like other classic machine learning algorithms, CA also requires a very long time to train a model. In the training process of CA, data-set is trained in the order of category, and all the categories are mutually independent. Therefore, we can make the data-set divided by category into series of subsets, and then train these subsets respectively. Due to the independency among these subsets in the training process, we can train these subsets simultaneously. In the paper, we implemented the parallel CA in memory shared multi-core parallel computing environment. Our method is also applicable in cluster system to accelerate the training.

This paper is organized as follows. Section 2 gives a brief overview of classic CA. Section 3 describes our proposed Parallel CA in detail. Then in Section 4, experiment and analysis are presented. Finally, Section 5 is the conclusion of the paper.

2   Covering Algorithm Overview

For briefness of description, the symbols used in this paper are defined as follows:

In a classification problem, $d=(\boldsymbol{x}, y)$ denotes a sample in a data-set, $\boldsymbol{x} \in R^d$, y represents m classes. Assume the size of the data-set is n, the dimension is h.

Let $D=\{d_i|d_i=(\boldsymbol{x}_i, y_i), y_i=1,2,\ldots\ldots m \ i=1,2,\ldots,n\}$

D_i denotes a subset of D, which contains all the samples of class i $=1,2,\ldots,m$.

Classification Model:

The model created by CA is a cover-set: $C=\{c_i|c_i=(\boldsymbol{x},r),i=1,2\ldots\ldots\}$, $\boldsymbol{x}$ denotes the center of the cover, r denotes its radius.

2.1   Learn Procedure

Based on the theory, learning procedure of CA is described as follows.

Input: Train set D,

Output: Cover set C

Define a transform:

$$T(\boldsymbol{x}) = (\boldsymbol{x}, \sqrt{R^2 - |\boldsymbol{x}|^2}), \quad R \geq \max\{|\boldsymbol{x}| \ | \ \boldsymbol{x} \in D\}$$

Step1. Use $T(\boldsymbol{x})$ to do a projection: D->S^h, S^h is an h dimensional sphere in $h+1$ dimensional space.

Step2. Modeling

$C=\Phi$

for each D_i in D

    while D_i is NOT Empty

        Select an $\boldsymbol{x}$ in D_i randomly;

        $d^1 = \max\{<\boldsymbol{x}\ \boldsymbol{x}'> \mid \boldsymbol{x}\in D_i,\ \boldsymbol{x}\notin D_i\}$

        $d^2 = \min\{<\boldsymbol{x}\ \boldsymbol{x}'> \mid \boldsymbol{x},\boldsymbol{x}'\in D_i,\ <\boldsymbol{x},\ \boldsymbol{x}'>\ > d^1\}$

        $r = (d^1+d^2)/2$

        Create a cover c with $\boldsymbol{x}$ as its center and r as its radius, do coverage operation, then remove $\boldsymbol{x}$ and the samples that are covered by c in D_i ;

        Add c to C ;

    loop

  end for

Note: d_1 maximizes the product of $\boldsymbol{x}$ and $\boldsymbol{x}'$, i.e. the minimum distance among the samples that do not belong to D_i. d_2 minimizes the product of $\boldsymbol{x}$ and $\boldsymbol{x}'$, i.e. the maximum distance of $\boldsymbol{x}$ with samples that in the same class.

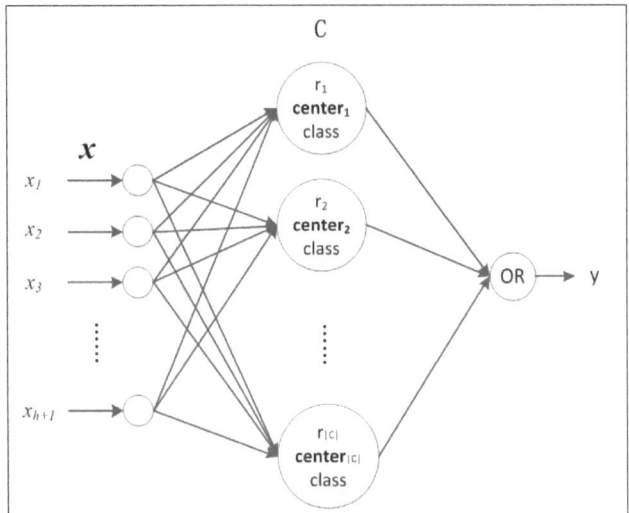

Fig. 1. A three layer neural network constructed by Covering Algorithm

2.2 Classification

By using CA to train a data-set, we can obtain a cover set C, i.e. the model. Then we use C to construct a three-layer feed forward neural network for classification, the network is shown in Fig. 1.

    Input Layer: $h+1$ nodes in all(after projection, the dimension increased by 1), this layer doesn't do anything with the input vector;

Hidden Layer: $|C|$ nodes in total, every element in C produces a hidden node, and every hidden nodes consists of three parameters, i.e. cover center, radius and its class. For a node $i(i=1,2,\ldots\ldots |C|)$, the transfer function is:

$$y_i = f(<\boldsymbol{x}, \textbf{center}_i > - r_i)^* class , f \text{ is a sign function.}$$

When a sample falls into a cover, the output of the corresponding node is *class*, otherwise zero.

Output layer: the output of hidden nodes converge to the output layer that makes up with only one node, and this node do 'OR' operation and output the classification result.

Note: It is very easy for us to use the constructed neural network mentioned above to classify, but there is a small problem. A new sample might fall in several overlapped covers, or falls outside of all the covers. In this case, the network cannot classify a sample directly. Then we can use the nearest cover to make a decision.

3   Parallel Covering Algorithm

In order to parallelize the classic CA, we have to find out the independent part in the training process. Let C_i denote cover-set of D_i, every cycle of for-loop produce a cover set. C_i only covers samples in D_i. Therefore, each class of training set is trained independently by the algorithm. This character offers us the basis to parallelize the classic CA. Here, our parallel computing environment is a memory shared multi-core computer.

Assume we have p processing cores available, p<m, and we need another extra core to serve as a scheduler.

Step1: Split D into subsets $D_1, D_2, \ldots\ldots D_m$, each subset contains a class of samples, set $C = $ NULL;

Step2: According to the training algorithm mentioned in section 2, launch p sub training sets to processing cores simultaneously, and each core deal with a subset independently. If all the subsets complete training, turn to step4 ;

Step3: After completing these p training tasks, p sub cover-sets return and merge them into C, then turn to step2 to train the untrained subsets;

Step4: return cover-set C, finish the training procedure.

Fig. 2 visually depict our method, it is based on memory shared multi-core parallel computing environment. We can see that parallel CA can make full use of the computation resources, all available processing cores are not in idle status. If p>m, under ideal conditions, the total train time could be reduced to $1/p$ of that by non-parallel implementation theoretically. Due to the schedule and the communication cost, usually, the ideal speedup cannot be attained.

Although our implementation is based on multi-core computer, the algorithm itself is also applicable to the cluster system. Due to the page limitations, the cluster system is discussed in this paper.

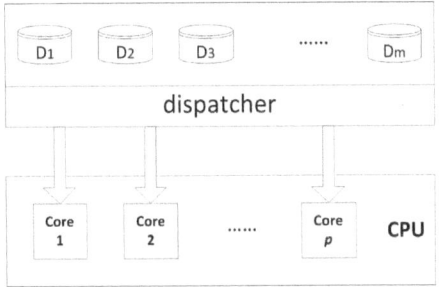

Fig. 2. Training data-set in memory shared multi-core parallel computing environment

4 Experiments and Analysis

Experiments are carried on three public data-sets: Optical Recognition of Handwritten Digits [13], Pen-Based Recognition of Handwritten Digits [14], and MNIST handwritten digit [15]. These three data-sets are all about Handwritten digit recognition, in this paper, we name them OCR1, OCR2 and OCR3 respectively. Detailed information of these data-sets are described in Table 1 .

Table 1. data-set used in the experiments

Data Name	Instances	Number of attribute	Number of Classes	Train size	Test size
OCR1	5620	64	10	3823	1797
OCR2	10992	16	10	7494	3498
OCR3	70000	784	10	60000	10000

The hardware setting is a HP Z800 workstation, it equipped with two CPU of Xeon5660 2.8GHz, 12 cores in all, 24GB memory. The operating system is Windows 7, we use C++ language to implement the algorithm.

The aim of the experiment is to show the high performance of our proposed parallel CA. Firstly, we carried a group of experiment in non-parallel conditions, i.e. we use only one thread to implement the experiment, which is equivalent to the batch mode. Then we carried another experiment, and run the algorithm on parallel conditions, we set the number of threads increased from 2 to 10 with step of 2, and 10 is the maximum degree of parallelism.

To evaluate its performance, the following two criterions are used: speedup and efficiency. They are defined respectively by :

$$speedup = \frac{\text{the elapsed train time of CA}}{\text{the elapsed train time of parallel CA}} \tag{1}$$

$$efficiency = \frac{speedup}{\text{number of processors}} \tag{2}$$

4.1   Uniprocessor Environment

The first experiment is carried in batch mode, the data-set is trained in sequences. Actually, it is a special situation of the parallel algorithm, as we assign only one processing core to train the data-set. The result is shown in Table 2.

Table 2. Experiment result of CA in batch mode

Data Name	Train Time(s)	Number of Covers	Accuracy
OCR1	0.195	364	96.43%
OCR2	0.177	362	96.15%
OCR3	556.794	7116	95.31%

We can see that CA gets a very high classification accuracy. From the result in the table, we can see that when the data-set size is less than ten thousands, CA trained very fast, but when the data-set scales to tens of thousands, the training speed declined.

Table 3. Experiment result of Parallel CA

Data Name	Processors	Train Time(s)	Speedup	Efficiency	Accuracy
OCR1	2	0.122	1.59	0.80	96.43%
OCR1	4	0.079	2.46	0.62	96.45%
OCR1	6	0.062	3.14	0.52	96.47%
OCR1	8	0.055	3.54	0.44	96.42%
OCR1	10	0.052	3.75	0.38	96.45%
OCR2	2	0.114	1.55	0.78	96.12%
OCR2	4	0.071	2.49	0.62	96.15%
OCR2	6	0.054	3.28	0.55	96.10%
OCR2	8	0.048	3.69	0.46	96.14%
OCR2	10	0.045	3.93	0.39	96.15%
OCR3	2	319.811	1.74	0.87	95.31%
OCR3	4	212.383	2.62	0.65	95.36%
OCR3	6	159.982	3.48	0.58	95.39%
OCR3	8	140.817	3.95	0.49	95.33%
OCR3	10	130.313	4.27	0.42	95.36%

4.2   Multi-core Parallel Computing Environment

The second experiment is carried in parallel conditions. In order to analyze the speedup and efficiency, we did 5 groups of experiments, each increases the number of processing cores by 2. The experimental result is shown in Table 3. The value in column 'speedup' is relative to that of in Table 2.

The speedup and efficiency of the parallel CA with respect to different number of processing cores are illustrated in Fig. 1 and Fig. 2. They show that the speedup of the parallel CA improves with the increase of the number of processing cores, demonstrating the efficiency of the parallel CA. The efficiency curves gradually decline as the number of processing core increases, the reason may lie in that more processing cores lead to more communicate cost.

It is noteworthy that the result of OCR3 reflects the truth better, because the data-set itself is very large, and the training time is relatively long, so compare to the very short time consuming on OCR1 and OCR2, the error on OCR3 is smaller.

Due to the limitations of the experimental conditions, without high-performance supercomputers and cluster systems, we only complete our experiment in multi-core parallel computing environment. In [16], MNIST data-set is trained on high-performance supercomputer by Parallel SVM, it spends 2623.58 seconds, greatly longer than our method. But our proposed parallel CA only uses 130.3s to train the data-set while the classification accuracy is maintained.

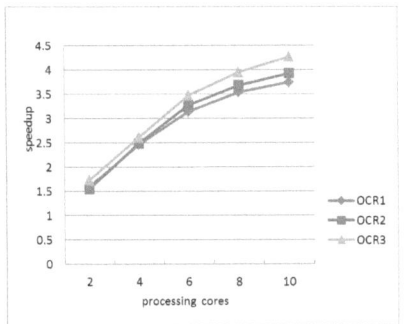

Fig. 3. Speedup with respect to the number of processing cores

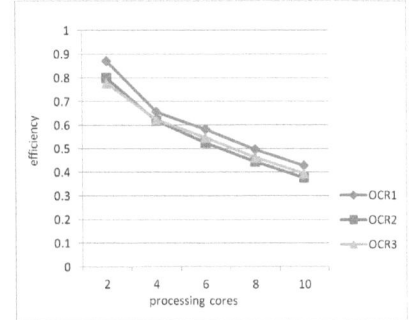

Fig. 4. Efficient with respect to the number of processing cores

5   Conclusion

This paper proposes the Parallel CA implemented in parallel computing environment. The Parallel CA uses multiple CPU processing cores to train large data-set by partitioning the whole data-set into smaller subsets, and launch them into parallel computing environment to speedup the training process.

Experiments on three data-sets demonstrate the efficiency of the Parallel CA. Our proposed method greatly improves the training speed of Covering Algorithm, and at the same time ensures the classification accuracy. Limited to the experimental conditions, we completed the experiment in memory shared multi-core computing environment. However, our experiment showed high performance on the data-set. The proposed parallel CA is equally applicable in the distributed cluster parallel computing environment.

The current parallel covering algorithm only achieves class-level subset parti-tion, the parallel speedup is limited to some extent. In the future work, we will research on how to divide the data-set in other ways to find a better way to speed up the parallel cover algorithm.

Acknowledgements. This work is partially supported by National Grand Fundamental Research 973 Program of China under Grant #2007CB311003, and supported by National Natural Science Foundation of China under Grant #61073117 and Grant #61175046, and supported by Natural Science Foundation of Anhui Province under Grant #11040606M145, and supported by Outstanding Young Talents in Higher Education Institutions of Anhui Province under Grant #2010SQRL021, and supported by Academic Innovative Research Projects of Anhui University graduate student of 2010 under Grant #yqh100147.

References

1. Zhang, L., Zhang, B.: A forward propagation learning algorithm of multilayered neural networks with feed-back connections. Journal of Software 8(4), 252–258 (1997)
2. Zhang, L., Zhang, B.: A geometrical representation of McCulloch Pitts neural model and its applications. IEEE Transactions on Neural Networks 10(4), 925–929 (1999)
3. Jie, M., Zheng, G.: Application of Cover Algorithm in Text Categorization. Computer Technology and Development 17(7), 183–189 (2007)
4. Zhang, Y., Zhang, L., Duan, Z.: A Constructive Kernel Covering Algorithm and Applying It to Image Recognition. Journal of Image and Graphics 9(11), 1304–1308 (2004)
5. Zhang, C., Zhang, Y.: Stock Prediction Based on Alternative Covering Design Algorithm. Microcomputer Development 15(12), 35–37 (2005)
6. Zhang, L., Zhang, B., Yin, H.: An Alternative Covering Design Algorithm of Multilayer Neural Networks. Journal of Software 10(7), 737–742 (1999)
7. Wu, T., Zhang, L., Zhang, Y.: Kernel Covering Algorithm for Machine Learning. Chinese Journal of Computers 28(8), 1295–1301 (2005)
8. Lei, Y.-M., Yan, Y., Chen, S.-J.: Parallel Training Strategy Based on Support Vector Regression Machine. In: 15th IEEE Pacific Rim International Symposium on Dependable Computing, pp. 159–164 (2009)
9. Zanni, L., Serafni, T., Zanghirati, G.: Parallel Software for Training Large Scale Support Vector Machines on Multiprocessor Systems. Journal of Machine Learning Research 7, 1467–1492 (2006)
10. de Llano, R.M., Bosque, J.L.: Study of neural net training methods in parallel and distributed architectures. Future Generation Computer Systems 26, 267–275 (2010)
11. Scanzio, S., Cumani, S.: Parallel implementation of Articial Neural Network training for speech recognition. Pattern Recognition Letters 31, 1302–1309 (2010)
12. Eklund, S.E.: A massively parallel architecture for distributed genetic algorithms. Parallel Computing 30, 647–676 (2004)

13. Optical Recognition of Handwritten Digits data-set,
    `http://archive.ics.uci.edu/ml/machine-learning-databases/optdigits/`
14. Pen-Based Recognition of Handwritten Digits data-set,
    `http://archive.ics.uci.edu/ml/machine-learning-databases/pendigits/`
15. MNIST data-set, `http://yann.lecun.com/exdb/mnist/`
16. Cao, L.J., Keerthi, S.S., Ong, C.J., Zhang, J.Q.: Parallel Sequential Minimal Optimization for the Training of Support Vector Machines. IEEE Transactions on Neural Networks 17 (2006) 1045–9227

A Spatial Clustering Method
for Points-with-Directions

Jing Wang and Xin Wang

Department of Geomatics Engineering, University of Calgary,
Calgary, Alberta, Canada
{wangjing,xcwang}@springer.com

Abstract. In this paper, we first formalize a points-with-directions clustering problem. Then we propose a points-with-directions clustering (PDC) and an improved PDC+ methods. The proposed methods can handle point data with direction and trajectory data. The trajectory data is divided into a set of points with directions. The proposed methods are evaluated with a hurricane dataset.

Keywords: data mining, spatial clustering, trajectory mining.

1   Introduction

Spatial cluster analysis is an important research topic for knowledge discovery[1]. Spatial clustering algorithms exploit spatial relationships among the data objects in determining inherent groupings of the input data.

Most of traditional spatial clustering algorithms only work with point dataset. How-ever, clustering based on the sequences of the locations and timestamps of moving objects in term of trajectory is an emerging area of clustering research. For example, find interesting patterns from the track of the vessels or the track of animals. The trajectories are represented as a series of points with directions. For example, a hurri-cane track can be represented as a set of observation locations with moving directions. To solve the problem of clustering points-with-directions, we propose two generic clustering methods PDC and PDC+. The PDC method is a density-based spatial clustering method considering the direction at each point. It starts to group data points from an arbitrary point and labels each point as cluster or noise based on point density and vector similarity of neighbourhood. PDC+ is an improved version of PDC. It considers the effect of neighbourhood when initial a cluster as well as the length of direction vector.

The rest of the paper is organized as follows. Section 2 gives an overview of clus-tering algorithm. Section 3 proposes our points-with-directions clustering methods. Section 4 presents the results of experimental evaluation. Section 5 discusses the future work and concludes the paper.

2   Related Work

The traditional spatial clustering methods can be classified into five categories in term of clustering techniques[2], Partitioning method, Hierarchical method,

T. Li et al. (Eds.): RSKT 2012, LNAI 7414, pp. 194–199, 2012.

Density based method, Grid-based method, and Model-based method. The proposed clustering method is extended from DBSCAN (Density Based Spatial Clustering of Applications with Noise)[3]. DBSCAN is the first and the most classic density-based spatial clustering method.

Trajectories clustering methods can be classified into two classes, whole trajectory clustering and partial trajectory clustering. Whole trajectory clustering takes the whole trajectory as an inseparable part. Most of them [4] [5] [6] derived from traditional point clustering algorithm. These methods may not be able to find similar portions of trajectories. Such as the hurricane tracks. Partial Trajectory clustering proposed to discover common sub-trajectories by divided the trajectory into small segments. The most famous one is the partition-and-group Framework [7]. This algorithm has two phases: portioning and grouping. However, this method is too complicated and may not find clusters in small scale. Other research [8][9] divided the whole trajectory into points, but they mainly focus on individual object.

The distance function is the key component of clustering methods in that it measures the similarity among spatial objects. It is a numerical description of how similar two objects are in space. According to [10], we usually use geometric distance as the scale of measurement in the ideal model. The geometric distances are defined by exact mathematical formulas to reflect the physical length between two objects in defined coordinate systems, such as Euclidean Distance and Manhattan Distance. Wang and Wang[11] gives a list of geometric distance functions. Spatial objects may have significantly different non-spatial attributes that distinguish them from each other and influence the clustering result.

3   Points-with-Directions Clustering

3.1   Problem Definition

The input data for this clustering framework includes two types of spatial data: a set of points or a set of trajectories. For the point data, a given point p can be represented as $\{x, y, \boldsymbol{D}\}$, where x and y are the spatial coordinates. $\boldsymbol{D}$ is a none-spatial attribute that can be described as a vector placed in a metric space. For the trajectory data, a given trajectory T can be represented as a series of points with direction if it can be divided at each vertex v_i. The direction at each vertex v_i is defined as the direction from v_i to the next vertex v_{i+1}. Thus a point with direction of the trajectory is denoted as $p_i\{x_i, y_i, \boldsymbol{v_i v_{i+1}}\}$. The given trajectory T can be defined as a sequence of points $\{ p_1, p_2, \ldots p_n \}$, where n = $Count$(vertexes)-1.

The output data is a set of clusters $O = \{C_1, ..C_m\}$, where m is the total number of the generated clusters. A cluster is a set of vectors with similar directions and they are close to each other according to the distance measure.

In the paper, the similarity of the vectors can be defined by the cosine value of the angle between two vectors. Thus, $Sim(\boldsymbol{D_1}, \boldsymbol{D_2}) = \cos\theta = \frac{\boldsymbol{D_1}\boldsymbol{D_2}}{|\boldsymbol{D_1}||\boldsymbol{D_2}|}$. For the spatial distance measure, we use the Minkowski distance. For the two dimensional data, we use the Euclidean distance.

3.2   Spatial Clustering Method for Points-with-Directions

Given a dataset D, the distance function *dist,* vector similarity measure *cos,* and parameters *Eps, Minpts,* and *MinCosAgl,* a neighbourhood around a point of a given radius (*Eps*) must contain at least a minimum number of points (*MinPts*), i.e., the density in the neighborhood is determined by the distance function for two points p and q, denoted by $dist(p,q)$ and the similarity measure of the vector at two points, denoted by $cos(p,q)$. The following definitions are used to formalize the problem.

Definition 1: The *Eps-cos-neighborhood* of a point p, denoted by $N_{Eps-cos}(p)$, is defined by $N_{Eps-cos}(p) = \{q \in D |\ dist(p,q) \le Eps \text{ and } cos(p,q) \ge MinCosAgl\}$.

Definition 2: A point p is *directly density-reachable* from a point q wrt. *Eps, MinCosAgl, MinPts* if 1) $p \in N_{Eps-cos}(q)$ 2) $|N_{Eps-cos}(q)| \ge MinPts$.

  q is called a core point if the number of its neighborhood is greater than *Minpts.* q is a border point if q is not a core point, but a neighbor of a core point. Otherwise, q is a noise. Directly density-reachable is symmetric for pairs of core points. In general, however, it is not symmetric if one is core point and the other one is on the border of the searching circle.

Definition 3: A point p is *density-reachable* from a point q wrt. *Eps, MinCosAgl,* and *MinPts* if there is a chain of points $p_1, \ldots, p_n$, $p_1 = q$, $p_n = p$ such that p_{i+1} is directly density-reachable from p_i.

Definition 4: A point p is *density-connected* to a point q wrt. *Eps, MinCosAgl,* and *MinPts* if there is a point o such that both, p and q are density-reachable from o wrt. *Eps, MinCosAgl,* and *MinPts.*

Definition 5: Let D be a set of data points. A cluster C wrt. *Eps, MinCosAgl,* and *MinPts* is a non-empty subset of D satisfying the following conditions: 1) $\forall p$, q: if $p \in C$ and q is density-reachable from p wrt. *Eps, MinCosAgl,* and *MinPts,* then $q \in C$. 2) $\forall p$, $q \in C$: p is density-connected to q wrt. *Eps, MinCosAgl,* and *MinPts.*

  Once the parameters *Eps, MinCosAgl,* and *MinPts* are defined, PDC algorithm starts to group data points from an arbitrary point q. It begins by finding the *neighbourhood* of point q. If the neighbourhood contains points fewer than *MinPts,* then the point q is labelled as a noise. Otherwise, a cluster is created and all points in q's neighbourhood are placed in this cluster. Then the neighborhoods of all q's neighbors are examined to see if they can be added to the cluster. If a cluster cannot be expanded from this point with respect to the density reachable and density-connected definitions, PDC chooses another arbitrary ungrouped point and repeats the process. This procedure is iterated until all data points in the dataset have been placed in clusters or labelled as noise.

  To better fit for the trajectory data, we add a new rule for the cluster merge. If two core points p and q are adjacent from the same trajectory and almost in the same direction, they should be merged into the same cluster.

Definition 6: A point p is *density-connected* to a point q wrt. *Eps, MinCosAgl,* and *MinPts,* if 1) p' is *directly density-reachable* from a point p and q' is *directly density-reachable* from a point q; 2) $q, p \in T_i$ and q at v_i, p at v_{i+1}; 3) $cos(p,q) \geq MinCosAgl$.

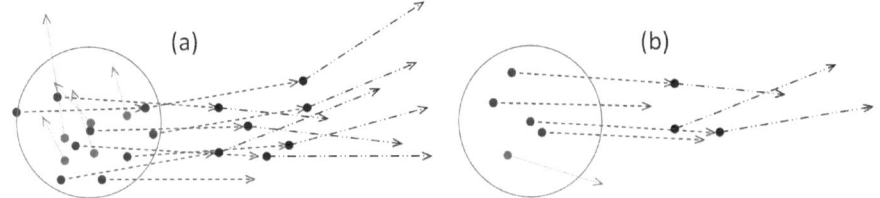

Fig. 1. Limitations of PDC

3.3   The Limitations of PDC

Although PDC gives good results and is efficient in many cases, it may not be suitable for the cases in Figure 1(a). We can identify two directions in the search circle. One is going up in vertical; the other one is more like going horizontal. If the *MinPts*=5, the circle is the *Eps* search area, and the *MinCosAgl* are satisfied in both two directions, then there will be two core clusters initialized. In Figure 1(b), there is one main direction going horizontal. There are 4 vectors almost parallel except one may be not satisfy the *MinCosAgl*. So the algorithm may fail to find any clusters here. So, one of the drawbacks of the PDC is that it does not consider the effect of neighbourhood when initial a core cluster. The percentage of the vector direction may affect the main direction of the cluster initiation. In addition, PDC does not consider the length of vectors. If we consider the length of the vector as an indicator of the direction obvious or not in the two cases mentioned above, it will be much easier to find out the directions for the initial clusters.

3.4   PDC+

Based on the analysis in Section 3.3, we propose an improved version of Points-with-Directions Clustering method, called PDC+.

We introduce a new concept main direction when initial clusters, which indicate the most popular direction in one initial cluster and we change the definition of neighbour corresponding.

Definition 7: The main direction of a *Eps-neighborhood* of a point p, denoted by a vector $\boldsymbol{G}$, is defined as $\boldsymbol{G} = \sum\limits_{i=1}^{n} \boldsymbol{D_i}$, where $\boldsymbol{G_i}$ is the vector value of q and $n = Count(N_{Eps}(p))$.

So, we redefine the Definition 1 as new Definition 8 in the PDC+.

Definition 8: The *Eps-cos-neighborhood* of a point p, denoted by $N'_{Eps-cos}(p)$, is defined by $N'_{Eps-cos}(p) = \{q \in D| \ dist(p,q) \leq Eps \ \& \ avg(\boldsymbol{G},q) \geq MinCosAgl$ $\}$. $avg(\boldsymbol{G},q) = \frac{1}{n} \sum\limits_{i=1}^{n} cos(q, \boldsymbol{G_i})$, $n = Count(N_{Eps}(p))$.

As the second version considers the length of the vectors, it can represent attributes' weight.

After we introduce the main direction, when we merge the clusters belong to one trajectory, we will consider the cosine angle between the two main directions. So the redefine the Definition 6 is Definition 9 in PDC+.

Definition 9: A point p is *density-connected* to a point q wrt. *Eps, MinCosAgl,* and *MinPts,* if 1) p' is *directly density-reachable* from a point p and q' is *directly density-reachable* from a point q;2)$q, p \in T_i$ and q at v_i ,p at v_{i+1}; 3) $cos(\boldsymbol{G_p}, \boldsymbol{G_q}) \geq MinCosAgl$.

4 Experimental Evaluation

We implement the PDC and PDC+ using C# and ArcObject into an extension in ArcMap. We use a Hurricane dataset to test our methods. This hurricane dataset include all the Atlantic hurricanes from the year 1971 through 2008. This dataset has 429 trajectories, which have divided into 12871 points with direction based on the same time interval, 6 hours.

Figures 2(a) and (b) show the PDC and PDC+ result when the parameters *Eps* is set to 0.4, *Minpts* to 6 and *MinCosAgl* to 0.91. In the result of PDC, 35 clusters are detected. In the result of PDC+, totally 64 clusters are identified.

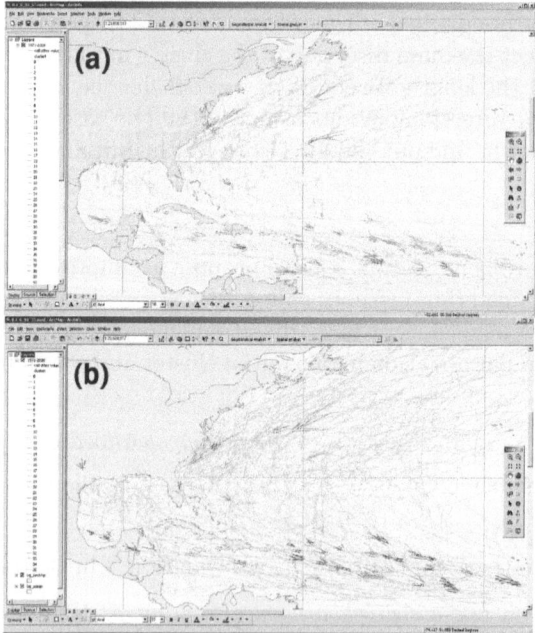

Fig. 2. PDC and PDC+ clustering result

The result shows the most active hurricane movement area with the similar movement directions. The result should be interpreted by the specialist in weather forecast area. The clusters which indicate the sudden direction change from previous track may imply some useful information.

5   Conclusions and Future Work

In this paper, we first formalize the points-with-directions clustering problem. Then we propose two clustering methods PDC and PDC+ to solve the problem. A hurricane dataset is used to evaluate the two methods.

This research could be extended in the future in the following ways: First, new merge rules should be added into our methods in the future. After separating a trajectory into a set of points-with-directions, the current methods only consider limited connection among the points from the same trajectory during clustering. Second, the current method can not give the extracted representative vector for each final cluster. For the general user, a representative vector should be more helpful for them to discover hidden knowledge behind the data.

References

1. Han, J., Kamber, M., Tung, A.K.H.: Spatial Clustering Methods in Data Mining: A Survey. In: Miller, H.J., Han, J. (eds.) Geographic Data Mining and Knowledge Discovery (2001)
2. Han, J., Kamber, M., Pei, J.: Data Mining: Concepts and Techniques. Elsevier (2011)
3. Ester, M., Kriegel, H., Sander, J., Xu, X.: A Density-Based Algorithm for Discovering Clusters in Large Spatial Databases with Noise. In: KDD 1996, pp. 226–231 (1996)
4. Gaffney, S., Smyth, P.: Trajectory Clustering with Mixtures of Regression Models. In: KDD 1999, pp. 63–72 (1999)
5. Camargo, S.J., Robertson, A.W., Barnston, A.G., Ghil, M.: Clustering of eastern North Pa-cific tropical cyclone tracks: ENSO and MJO effects. Geochem. Geophys. Geosyst. 9, 23 (2008)
6. Nanni, M., Pedreschi, D.: Time-focused clustering of trajectories of moving objects. J. Intell. Inf. Syst. 27(3), 267–289 (2006)
7. Lee, J.G., Han, J., Whang, K.Y.: Trajectory Clustering: A Partition-and-Group Framework. In: SIGMOD 2007, pp. 593–604 (2007)
8. Kalnis, P., Mamoulis, N., Bakiras, S.: On Discovering Moving Clusters in Spatio-temporal Data. In: Medeiros, C.B., Egenhofer, M., Bertino, E. (eds.) SSTD 2005. LNCS, vol. 3633, pp. 364–381. Springer, Heidelberg (2005)
9. Li, Y., Han, J., Yang, J.: Clustering moving objects. In: KDD 2004, pp. 617–622 (2004)
10. Tobler, W.: A Computer Movie Simulating Urban Growth in the Detroit Region. Economic Geography 46(2), 234–240 (1970)
11. Wang, X., Wang, J.: Using Clustering Methods in Geospatial Information Systems. Geomatica 64(3), 347–361 (2010)

An Argumentation Framework for Non-monotonic Reasoning in Description Logic

Geng Wang and Zuoquan Lin

School of Mathematic Science, Peking University, Beijing, 100871, China
{wanggeng,lz}@pku.edu.cn

Abstract. With the development of semantic web technology, ontology level reasoning among various agents has become an important topic. To handle the inconsistency between different knowledge bases, a non-monotonic reasoning method is always in need. However, for most approaches to non-monotonic ontology reasoning, new mechanism must be added into the reasoning procedure. In this paper, we will try to perform a type of ontology level non-monotonic reasoning by using argumentation theory, and show that such argumentation framework can be completely carried out under description logic. Also, the framework we present can be used for other types of reasoning, for example, inductive reasoning.

Keywords: Argumentation, Description Logic, Non-monotonic reasoning.

1   Introduction

Since the development of semantic web technology and description logic, ontology based reasoning has been widely used in all kinds of professional fields. However, the complexity of such problems is always too high, and such reasoning tasks always take place where there is no adequate information and/or computability. Sometimes we wish to get a partial, non-strict result, and revise it when there is more information or more time.

We will try to solve this problem by using argumentation theory, which introduced into non-monotonic reasoning by Dung with his abstract framework [6] in 1993 but have developed greatly since then. An argumentation procedure can be considered as a reasoning task between agents, which is exactly the problem we need to solve for reasoning upon semantic web.

In this paper, we will give out a definition of DL argumentation framework, under which all reasoning procedure can be completely carried out by a DL inference engine with little modification. We will first give out an abstract framework, and show that when applying a specific language into this abstract framework, there will be no additional parts the reasoning task must be taken into consideration other than the language itself. In other words, argumentation is a "separated" non-monotonic method which takes no interaction with the monotonic logic itself under our framework.

T. Li et al. (Eds.): RSKT 2012, LNAI 7414, pp. 200–206, 2012.

We mainly focus on our discussion on $\mathcal{ALC}$, but it will not be hard for some results to expand into other DL languages.

First, we shall briefly talk about the description logic $\mathcal{ALC}$. Then, we will define defeasible arguments and the relationship among them within $\mathcal{ALC}$. Next, we will show that the defeasibility comes from the exception of defeasible terminologies, and with the definition, we will give out two new types of attack, and how the ordering among arguments can be defined using exceptions. Finally, we will talk about how to reason inductively under our framework.

2   The Description Logic $\mathcal{ALC}$

Ontology level reasoning used on the semantic web [4], is usually based on description logics, for example, the web ontology language OWL [3]. Description Logics (DLs) [2] are a group of sub-languages of predicate logic, which trade expressibility for efficiency. A DL language contains mainly concepts (unary predicates); individuals (constants); roles (binary predicates), and the construction of formulas is usually restricted to reduce complexity.

We will describe a simple form of description logic called $\mathcal{ALC}$ [12], attributive language with complements, which only accept atomic roles and simple constructions of concepts are allowed. Like other description logics, $\mathcal{ALC}$ has an efficient tableau algorithm for the decision problems [10].

Definition 1. *An $\mathcal{ALC}$ concept is constructed from atomic concept by the rules below:*

$$C, D \to A|\top|\bot|\neg C|C \sqcap D|C \sqcup D|\forall R.C|\exists R.C$$

where A is an atomic concept and R is an atomic role.

An $\mathcal{ALC}$ knowledge base consists of terminologies (or TBox) and assertions (or ABox). Terminologies are knowledge of concepts and roles with the form of $C \sqsubseteq D$ or $C \equiv D$, and assertions are knowledge of individuals with the form of $C(a)$ or $R(a, b)$.

A $\mathcal{ALC}$ knowledge base has an interpretation in set theory:

Definition 2. *For an interpretation $\mathcal{I}$, individuals, concepts and atomic roles are interpreted by:*

$$a^{\mathcal{I}} \in \Delta^{\mathcal{I}} \qquad\qquad A^{\mathcal{I}} \subseteq \Delta^{\mathcal{I}}$$

$$R^{\mathcal{I}} \subseteq \Delta^{\mathcal{I}} \times \Delta^{\mathcal{I}} \qquad\qquad \top^{\mathcal{I}} = \Delta^{\mathcal{I}}$$

$$\bot^{\mathcal{I}} = \emptyset \qquad\qquad (\neg C)^{\mathcal{I}} = \Delta^{\mathcal{I}} \backslash C^{\mathcal{I}}$$

$$(C \sqcap D)^{\mathcal{I}} = C^{\mathcal{I}} \cap D^{\mathcal{I}} \qquad\qquad (C \sqcup D)^{\mathcal{I}} = C^{\mathcal{I}} \cup D^{\mathcal{I}}$$

$$(\forall R.C)^{\mathcal{I}} = \{a \in \Delta^{\mathcal{I}}|\forall b.(a, b) \in R^{\mathcal{I}} \to b \in C^{\mathcal{I}}\}$$

$$(\exists R.C)^{\mathcal{I}} = \{a \in \Delta^{\mathcal{I}}|\exists b.(a, b) \in R^{\mathcal{I}} \wedge b \in C^{\mathcal{I}}\}$$

Terminologies and assertions are interpreted by:

$$(C \sqsubseteq D)^{\mathcal{I}} \Leftrightarrow C^{\mathcal{I}} \subseteq D^{\mathcal{I}} \qquad (C \equiv D)^{\mathcal{I}} \Leftrightarrow C^{\mathcal{I}} = D^{\mathcal{I}}$$

$$C(a)^{\mathcal{I}} \Leftrightarrow a^{\mathcal{I}} \in C^{\mathcal{I}} \qquad R(a, b)^{\mathcal{I}} \Leftrightarrow (a^{\mathcal{I}}, b^{\mathcal{I}}) \in R^{\mathcal{I}}$$

3   Defeasible $\mathcal{ALC}$

The concept of abstract argumentation framework is first introduced by P.Dung [6] in 1993, and there are lots of varieties. Our argumentation framework for description logic we discussed below will base on the abstract argumentation framework defined by Martínez et al [11], and the reasoning procedure under it has a form similar to the definition of defeasible logic programming by García and Simari [7].

Definition 3. *A defeasible knowledge base is a pair $\langle \mathcal{D}, \mathcal{K} \rangle$, where $\mathcal{K}$ is a knowledge base of $\mathcal{ALC}$ (contains TBox $\mathcal{T}$ and ABox $\mathcal{A}$), and $\mathcal{D}$ is a set of terminologies, each called a* defeasible terminology. *For $\Sigma \subseteq \mathcal{D}$, we use $\Sigma \models_{\mathcal{K}} p$ for $\Sigma \cup \mathcal{T} \cup \mathcal{A} \models p$.*

A defeasible terminology can be considered as a rule holds at most times, but not always. For example, water boils at 100 degree centigrade, but only when under standard atmospheric pressure. However, if we consider a certain pot of water when such condition is not mentioned, we suppose that it is on normal ground, and it does boil at 100 degree centigrade.

This introduces the concept for exception. With given exceptions upon defeasible terminologies, we can express that under which cases the terminologies cannot be used.

Definition 4. *An exception function $\mathcal{E}$ among a set of defeasible terminologies $\mathcal{D}$ is a mapping from $\mathcal{D}$ to all concepts, and $\{C \sqcap \mathcal{E}(C \sqsubseteq D) \sqsubseteq D | C \sqsubseteq D \in \mathcal{D}\} \not\models_{\mathcal{K}} \bot$. $\mathcal{E}$ is called the exception function among defeasible terminologies.*

For a defeasible terminology $C \sqsubseteq D$, $\mathcal{E}(C \sqsubseteq D)$ is called the exception concept of the terminology.

We can use defeasible terminologies absolutely the same as those strict terminologies inside a reasoning task, the difference only occurs when the validity of such reasoning task is called in question (or *attacked*). Each task is called an argument.

Definition 5. *A defeasible argument structure (defeasible argument or argument for short) is a pair $\langle \Sigma, p \rangle$ where $\Sigma \subseteq \mathcal{D}$ and p is $C(\alpha)$ and α is an individual, such that:*

(1) $\Sigma \not\models_{\mathcal{K}} \bot$;
(2) $\Sigma \models_{\mathcal{K}} p$;
(3) There is no $\Sigma' \subset \Sigma$ which $\Sigma' \models_{\mathcal{K}} p$.
If $\Sigma = \emptyset$, it is called a strict argument.
An argument $\langle \Sigma_1, p_1 \rangle$ is a sub-argument of $\langle \Sigma, p \rangle$, if $\Sigma_1 \subseteq \Sigma$.

An argument itself is a reasoning procedure inside the knowledge base, although may not completely sound as defeasible terminologies are involved. For a subsumption task, which determines whether $C \sqsubseteq D$, it can be written by $\neg C \sqcup D(x)$ where x is an individual which does not occur in the ABox. Note that below we shall only use one single individual (named x) for individual occur in the argument but not in ABox.

Not all defeasible terminologies are equal, one might be (strictly) more applicable then another. For example, the theory of relativity by Einstein can be used on objects moving near the speed of light, while Newton's laws can only be used on objects with a much lower speed. As an argument can contain multiple defeasible terminologies, it is also necessary to give out a partial order among all arguments.

Definition 6. *Suppose that $(\mathcal{D}, \preceq)$ is a poset, and let $(\mathcal{D}^*, \vee, 0)$ be the minimal bounded join-semilattice where $\mathcal{D} \subseteq \mathcal{D}^*$ and $0 \notin \mathcal{D}$. For an argument $A = \langle \Sigma, p \rangle$, let $U(A) = \bigvee \Sigma$ be the* unreliability *of argument A. If $U(A_1) \preceq U(A_2)$, we call that A_1 is stronger than A_2.*

It is easy to see that if A is a strict argument, then $U(A) = 0$, which means a strict argument is strictly stronger than any other argument. We now introduce the first type of attack: using a stronger argument to attack another.

Definition 7. *Two arguments $\langle \Sigma_1, p_1 \rangle$ and $\langle \Sigma_2, p_2 \rangle$ disagree, if p_2 is logically equivalent to $\neg p_1$.*

$\langle \Sigma_1, p_1 \rangle$ attacks $\langle \Sigma, p \rangle$ if there is a sub-argument $\langle \Sigma_2, p_2 \rangle$ of $\langle \Sigma, p \rangle$ which disagrees with $\langle \Sigma_1, p_1 \rangle$ (also called a normal attack). If $U(\Sigma_2) \nprec U(\Sigma_1)$, $\langle \Sigma_1, p_1 \rangle$ is called a defeater to $\langle \Sigma, p \rangle$.

Furthermore, if $U(\Sigma_1) \prec U(\Sigma_2)$, it is called a proper defeater, otherwise, it is called a block defeater.

If p_2 is an assertion $C(a)$ where a is an individual a occurs in ABox, and $\langle \Sigma_2, C(x) \rangle$ is also an argument, it is called a case attack.

Two arguments might be not comparable, but still, they can attack each other. However, there will be no winner, as they only talk about a same problem from different kinds of view.

It should be noticed that there are problems for case attacks: for example, we cannot use an object moving near the speed of light to attack Newton's laws, which is in fact widely used in low-speed such as architectural cases. When can case attack be used? The answer is related to our definition of exception function.

Definition 8. *With a given exception function $\mathcal{E}$, $\mathcal{E}^*$ is an exception function among arguments, if for an argument $A = \langle \Sigma, p \rangle$, $\mathcal{E}(\Sigma) \cup \{\neg \mathcal{E}^*(A)\} \models_{\mathcal{K}} \Sigma$.*

One can simply take the exception function to be the union of all exceptions of defeasible terminologies which are used in the argument, while it can be more complex, with more precision. In fact, if the exception function is perfectly given (for example, sometimes when working on scientific terminologies), we can use deduction as a partial order to define the relationship $\preceq$ as below:

Definition 9. *Let A and A' be two defeasible arguments, and $\mathcal{E}^*$ be an exception function. Then, $U(A) \preceq U(A')$ iff $\mathcal{E}^*(A) \models_{\mathcal{K}} \mathcal{E}^*(A')$.*

Of course, the exception of an argument should not be provable, otherwise the argument itself will become invalid. For example, if on a mountain of 5km height, where the pressure is lower than standard can be proved, it can be used to attack the argument which claimed that a certain pot of water here boils at 100 degree centigrade. This is a type of attack other than the normal attack, we call it *exception attack* defined below:

Definition 10. $A' = \langle \Sigma', p' \rangle$ *is an exception attack to A iff $p' \models_\mathcal{K} \mathcal{E}^*(A)$.*

Back to the problem above: a case attack is illegal when the special case falls into the exception concept. This makes a new type of attack, which we called *attack to case attack*:

Definition 11. *If $A' = \langle \Sigma', p' \rangle$ is a case attack where p is $C(a)$ to an argument A, $A'' = \langle \Sigma'', p'' \rangle$ is an attack to $\langle \Sigma, p \rangle$ if $p \models_\mathcal{K} E_A(x/a)$, where $E_A(x/a)$ is the assertion obtained by substitute all occurrence of x by a in E_A.*

For the two type of attacks defined above and a given partial order, we can define defeaters and proper defeaters as normal attacks. We shall pass the details.

4   Inductive Terminologies

When we observe that some humans are mortal without knowing that there is a human who is immortal, it should be natural for us to reason that all humans are mortal. However, such inductive reasoning is not supported by classical deductive logic, as soundness is required in classical logic, but the result of inductive reasoning might be questionable. However, we can introduce inductive knowledge as the same as defeasible knowledge, it can be attacked, thus the unsoundness of inductive reasoning will no longer be fatal.

Definition 12. *A defeasible knowledge base with inductive terminologies $\langle \mathcal{K}, \mathcal{D}, \mathcal{I}, \mathcal{E} \rangle$ where $\mathcal{D}$ and $\mathcal{I}$ are both sets of terminologies, $\mathcal{E}$ is a function from $\mathcal{D} \cup \mathcal{I}$ to all concepts, and for each $C \sqsubseteq D \in \mathcal{I}$, $\neg \mathcal{E}(C \sqsubseteq D)$ is a nominal set $\{a_1, ..., a_n\}$ which for each a_i, $\models_\mathcal{K} \neg C \sqcup D(a_i)$.*

Other than defeasible terminologies which hold on most cases, inductive terminologies hold only on few cases. For a inductive terminology $C \sqsubseteq D$, $\neg \mathcal{E}(C \sqsubseteq D)$ is where it holds, and other is not. Note that we allow nominal sets in the definition, so the reasoning task will be under $\mathcal{ALCO}$, which does not really affect the reasoning procedure as a terminology with nominal set can be turned into assertions.

   An argument with inductive terminology is called an inductive argument. By the definition of exception functions, we give inductive terminologies a similar form as defeasible terminologies, and so do inductive arguments. However, it is not equal between defeasible and inductive arguments: defeasible arguments are not as reliable as strict arguments, and inductive arguments are even less reliable than defeasible arguments. Specifically, because a single counter example can be used to overthrow an inductive terminology, all case defeaters against an inductive argument will be valid.

We give an example to show the difference between the two types of terminologies.

Example 1. If $Cat \sqsubseteq Mammal$ and $Mouse \sqsubseteq Mammal$ are in TBox, and $Cat(Tom)$, $Mouse(Jerry)$, $\neg Aquatic(Tom)$ and $\neg Aquatic(Jerry)$ are in ABox, then we can construct an inductive terminology $Mammal \sqsubseteq \neg Aquatic$, which says all mammals are not aquatic, and the exception of this terminology is $\{Tom, Jerry\}$. If there is another assertion $Mammal(Peter)$ in ABox, We can say that, $\langle \{Mammal \sqsubseteq \neg Aquatic\}, \neg Aquatic(Peter) \rangle$ is an inductive argument.

However, if another agent on the web can show that there is a whale Moby-Dick which is aquatic and is a mammal, (which means that, $Aquatic(Moby)$ and $Whale(Moby)$ in ABox and $Whale \sqsubseteq Mammal$ in TBox), it can form an attack $\langle \emptyset, Mammal(Moby) \sqcup Aquatic(Moby) \rangle$ which is a case attack to the former inductive argument. This case attack is legal, for by given a counterexample, the inductive argument is thus overthrown.

Now, let $Mammal \sqsubseteq Viviparity$ be a defeasible terminology with $Monotrem$ ata as the exception concept, and thus $\langle \{Mammal \sqsubseteq Viviparity\}, Viviparity(P$ $eter) \rangle$ is a defeasible argument. Under this case, we cannot use a non-viviparity platypus (which belongs to Monotremata) to disprove the argument, because it is only a special case and is not common.

5   Conclusion and Future Work

The research on ontology level argumentation began from the field of multi-agent system, in which argumentation is a negotiation method rather than a reasoning procedure, but has soon combined with argumentation theory within the field of knowledge representation to improve the ontology level reasoning.

However, most researchers have not developed the potentiality of ontology based argumentation. In DR-Prolog and some other appraoches [1, 8, 13], the argumentation procedure is translated into the form of other non-monotonic reasoning such as logic programming. In the works of some others [5,14], though with a similar form of our framework, the argumentation procedure cannot handle non-monotonicity very well. In our framework, we give full scale of non-monotonicity while neither losing the expressibility of the DL language nor using mechanism other than description logic itself. Heymans and Vermeir [9] extended a specific description language with partial order among the knowledge base which provides a defeasible reasoning method, but it cannot be used for argumentations take place between different knowledge bases.

The framework is not only expressive, but also good for practical usage since the argumentation procedure can be carried out with little modification to the DL inference engine. As we have mentioned above, by using exception function for the definition of order among arguments, whether an attack is a defeater (or a proper defeater) can also be calculated by a DL inference engine.

In fact, when under a certain DL language, the abstract argumentation framework may seem insufficient on reasoning ability. We have introduced "exception attack" and "attack to case attack" as complements for arguing under $\mathcal{ALC}$ other than normal attacks defined under the abstract framework. How to improve the framework so it can use better on OWL, or maybe on the more expressive first-order language, is still a challenge for us.

References

1. Antoniou, G., Bikakis, A.: Dr-prolog: A system for defeasible reasoning with rules and ontologies on the semantic web. IEEE Transactions on Knowledge and Data Engineering 19(2), 233–245 (2007)
2. Baader, F., McGuiness, D.L., Nardi, D., Patel-Schneider, P. (eds.): Description Logic Handbook: Theory, implementation and applications. Cambridge University Press, Cambridge (2002)
3. Bechhofer, S., van Harmelen, F., Hendler, J., Horrocks, I., McGuinness, D.L., Patel-Schneider, P.F., Stein, L.A.: Owl web ontology language reference (2004), http://www.w3.org/TR/owl-ref/
4. Berners-Lee, T., Hendler, J., Lassila, O.: The semantic web. Scientific American 284(5), 34–43 (2001)
5. Black, E., Hunter, A., Pan, J.Z.: An Argument-Based Approach to Using Multiple Ontologies. In: Godo, L., Pugliese, A. (eds.) SUM 2009. LNCS (LNAI), vol. 5785, pp. 68–79. Springer, Heidelberg (2009)
6. Dung, P.M.: On the acceptability of arguments and its fundamental role in non-monotonic reasoning and logic programming. In: Proc. of the 13th IJCAI, Chamber, France (1993)
7. García, A.J., Simari, G.R.: Defeasible logic programming: An argumentative approach. Theory and Practice of Logic Programming 4(1), 95–138 (2004)
8. Gómez, S., Chesnevar, C., Simari, G.: An argumentative approach to reasoning with inconsistent ontologies. In: Knowledge Representation and Ontologies Workshop, pp. 11–20 (2008)
9. Heymans, S., Vermeir, D.: Integrating ontology languages and answer set programming. In: 14th International Workshop on Database and Expert Systems Applications, pp. 584–588. IEEE Computer Society (2003)
10. Horrocks, I.R.: Using an expressive description logic: Fact or Fiction? In: Proc. of KR 1998, pp. 636–647 (1998)
11. Martínez, D., García, A., Simari, G.: Modelling well-structured argumentation lines. In: Proc. of IJCAI 2007, pp. 465–470 (2007)
12. Schmidt-Schauß, M., Smolka, G.: Attributive concept descriptions with complements. Artificial Intelligence 48, 1–26 (1991)
13. Wang, K., Billington, D., Blee, J., Antoniou, G.: Combining Description Logic and Defeasible Logic for the Semantic Web. In: Antoniou, G., Boley, H. (eds.) RuleML 2004. LNCS, vol. 3323, pp. 170–181. Springer, Heidelberg (2004)
14. Zhang, X., Lin, Z.: An Argumentation-Based Approach to Handling Inconsistencies in DL-Lite. In: Mertsching, B., Hund, M., Aziz, Z. (eds.) KI 2009. LNCS (LNAI), vol. 5803, pp. 615–622. Springer, Heidelberg (2009)

Analysis of Symmetry Properties
for Bayesian Confirmation Measures

Salvatore Greco[1], Roman Słowiński[2,3], and Izabela Szczęch[2]

[1] Department of Economics and Business, University of Catania, Italy
salgreco@unict.it
[2] Institute of Computing Science, Poznan University of Technology, Poland
{Izabela.Szczech,Roman.Slowinski}@cs.put.poznan.pl
[3] Systems Research Institute, Polish Academy of Sciences, Warsaw, Poland

Abstract. The paper considers symmetry properties of Bayesian confirmation measures, which constitute an important group of interestingness measures for evaluation of rules induced from data. We demonstrate that the symmetry properties proposed in the literature do not fully reflect the concept of confirmation. We conduct a thorough analysis of the symmetries regarding that the confirmation should express how much more probable the rule's hypothesis is when the premise is present rather than when the premise is absent. As a result we point out which symmetries are desired for Bayesian confirmation measures and which are truly unattractive. Such knowledge is a valuable tools for assessing the quality and usefulness of measures.

Keywords: Bayesian confirmation measures, symmetry properties, rule evaluation.

1 Introduction

Discovering knowledge from data is the domain of inductive reasoning. Knowledge patterns induced from data are often expressed in a form of "*if, then*" rules. They are consequence relations representing correlation, association, causation between independent and dependent attributes. To measure the relevance and utility of the discovered rules many measures of interestingness have been proposed and studied [7], [9], [11]. Among these measures, an important role is played by Bayesian confirmation measures, which express in what degree a premise confirms a conclusion [1], [10], [12], [13]. Such measures are extremely valuable due to the fact that they indicate disconfirmatory rules, i.e. rules which are completely misleading and should be discarded from further use. In this context, the group of confirmation measures should be regarded as a useful tool for evaluation of rules. Analysis of confirmation measures with respect to their properties is an active research area. Properties express the user's expectations towards the behavior of measures in particular situations. They group the measures according to similarities in their characteristics. Using the measures which satisfy the desirable properties one can avoid considering unimportant rules.

T. Li et al. (Eds.): RSKT 2012, LNAI 7414, pp. 207–214, 2012.
© Springer-Verlag Berlin Heidelberg 2012

Therefore, knowledge of which commonly used interestingness measures satisfy certain valuable properties, is of high practical and theoretical importance [14]. Among widely studied properties for Bayesian confirmation measures, there is a group of symmetry properties [2-6], monotonicity properties [1], [8], [14], weak Ex_1 and weak logicality L property [15]. In this article we consider symmetry properties for Bayesian confirmation measures. Though this issue has been taken up by many authors before (e.g., [2], [3], [4]), we propose a new set of desirable symmetry properties that exploits the deep meaning of the confirmation concept. The confirmation measures should express how much more probable the hypothesis is when the premise is present rather than when the premise is absent. Regarding such an interpretation, we justify that evidence symmetry, hypothesis symmetry and the combination of them both, are the only truly desirable symmetry properties. The paper is organized as follows. In the next section there are preliminaries on rules and their quantitative description. In section 3, we discuss the concept of confirmation and its interpretation. Section 4 describes the approaches to symmetry properties in the literature. The proposition of a new set of desirable symmetries is introduced in section 5. Finally, the last section provides conclusions.

2   Preliminaries

A rule induced from a dataset U shall be denoted by $E \rightarrow H$ (read as "if E, then H"). It consists of a premise (evidence) E and a conclusion (hypothesis) H. Throughout the paper we shall use the following notation corresponding to a 2x2 contingency table of the premise and the conclusion (Table (1)):

- a is the number of positive examples to the rule, i.e., the number of objects in U satisfying both the premise and the conclusion of the rule,
- b is the number of objects in U not satisfying the rule's premise, but satisfying its conclusion,
- c is the number of counterexamples i.e. objects in U satisfying the premise but not the conclusion of the rule,
- d is the number of objects in U that do not satisfy neither the premise nor the conclusion of the rule.

The cardinality of the dataset U, denoted by $|U|$, is the sum of a, b, c and d.

Table 1. Contingency table of E and H

	H	$\neg H$	Σ
E	a	c	$a+c$
$\neg E$	b	d	$b+d$
Σ	$a+b$	$c+d$	$|U|$

Reasoning in terms of a, b, c and d is natural and intuitive for data mining techniques since all observations are gathered in some kind of an information table describing each object by a set of attributes. However, a, b, c and d can also be regarded as frequencies that can be used to estimate probabilities: e.g., the probability of the premise is expressed as $Pr(E) = (a + c)/|U|$, and the probability of the conclusion as $Pr(H) = (a + b)/|U|$. Moreover, conditional probability of the conclusion given the premise is $Pr(H|E) = a/(a + c)$.

3   Property of Bayesian Confirmation

Formally, an interestingness measure $c(H, E)$ has the property of Bayesian confirmation if and only if it satisfies the following BC (1) conditions:

$$c(H, E) \begin{cases} > 0 & if \quad Pr(H|E) > Pr(H), \\ = 0 & if \quad Pr(H|E) = Pr(H), \\ < 0 & if \quad Pr(H|E) < Pr(H). \end{cases} \tag{1}$$

The BC definition identifies confirmation with an increase in the probability of the conclusion H provided by the premise E, neutrality with the lack of influence of the premise E on the probability of conclusion H, and finally disconfirmation with a decrease of probability of the conclusion H imposed by the premise E [6], [10].

A logically equivalent way to express the BC conditions is [6], [10]:

$$c(H, E) \begin{cases} > 0 & if \quad Pr(H|E) > Pr(H|\neg E), \\ = 0 & if \quad Pr(H|E) = Pr(H|\neg E), \\ < 0 & if \quad Pr(H|E) < Pr(H|\neg E). \end{cases} \tag{2}$$

To avoid ambiguity, we shall denote the above formulation (2) as BC'. Notice that according to BC, E confirms H when E raises the probability of H, while, according to BC', E raises the probability of H if the probability of H given E is higher than the probability of H given non E.

Measures that possess the property of Bayesian confirmation are referred to as *confirmation measures* or *measures of confirmation*. For a given rule $E \rightarrow H$, interestingness measures with the property of confirmation express the credibility of the following proposition: H is satisfied more frequently when E is satisfied, rather than when E is not satisfied.

Let us stress that the BC conditions (or BC' equivalently) do not impose any constraints on the confirmation measures except for requiring when the measures should obtain positive or negative values. As a result many alternative, non-equivalent measures of confirmation have been proposed [3], [5], [8], [14].

To help to handle the plurality of Bayesian confirmation measures, many authors have considered desirable properties of such measures. Analysis of measures with respect to their properties is a way to distinguish measures that behave according to user's expectations. An important group of properties constitute symmetry properties considered by many authors e.g., Carnap [2], Eells and Fitelson [4], Crupi et al. [3]. Though the literature on symmetries is rich, we claim that there is a need to propose a new approach to the analysis of the symmetry properties that exploits the deep meaning of the confirmation concept. In fact, a confirmation measure should give an account of the credibility that it is more probable to have the conclusion when the premise is present, rather than when the premise is absent. Following that interpretation we propose a new set of desirable symmetry properties for Bayesian confirmation measures.

4   Properties of Symmetry

The work of Carnap [2] had inspired many authors that took up the symmetry topic. In particular, Eells and Fitelson have analysed in [4] a set of well-known confirmation measures from the viewpoint of the following four properties of symmetry:

- evidence symmetry ES: $c(H, E) = -c(H, \neg E)$
- hypothesis symmetry HS: $c(H, E) = -c(\neg H, E)$
- commutativity (inversion) symmetry IS: $c(H, E) = c(E, H)$
- total (evidence-hypothesis)symmetry EHS: $c(H, E) = c(\neg H, \neg E)$

Let us observe that the above symmetries are formed by applying the negation operator to the evidence (ES), to the hypothesis (HS), or both (EHS), as well as switching the position of the evidence and the hypothesis (IS). The research of Eells and Fitelson [4] implies that only hypothesis symmetry HS is a desirable property, while evidence symmetry ES, commutativity symmetry IS and total symmetry EHS are not. As an illustration of their reasoning, they used rules concerning drawing cards from a standard deck. They claim e.g. that the evidence symmetry should be discarded due to the following counterexample: *if the drawn card is the seven of spades then the card is black*. Obviously, the *seven of spades* confirms that *the card is black* to a greater extent than *the not-seven of spades* disconfirms the same hypothesis. As a result the equality in evidence symmetry is found unattractive by Eells and Fitelson. Thus, in their opinion, an acceptable measure of Bayesian confirmation should not satisfy the evidence symmetry (i.e. for some situation $c(H, E) \neq -c(H, \neg E)$). Analogous reasoning can be conducted with respect to the other symmetries analyzed in [4].

Recently, Crupi et at. [3] have argued for an extended and systematic treatment of the issue of symmetry properties. They propose to analyse a confirmation measure $c(H, E)$ with respect to seven symmetries being all combinations obtained by applying the negation operator to the premise, hypothesis or both, and/or by inverting E and H:

- $ES(H, E) : c(H, E) = -c(H, \neg E)$
- $HS(H, E) : c(H, E) = -c(\neg H, E)$
- $EIS(H, E) : c(H, E) = -c(\neg E, H)$
- $HIS(H, E) : c(H, E) = -c(E, \neg H)$
- $IS(H, E) : c(H, E) = c(E, H)$
- $EHS(H, E) : c(H, E) = c(\neg H, \neg E)$
- $EHIS(H, E) : c(H, E) = c(\neg E, \neg H)$

Moreover, Crupi et al. [3] go even further as the analysis is conducted separately for the case of confirmation (i.e. when $Pr(H|E) > Pr(H)$) and for the case of disconfirmation (i.e. when $Pr(H|E) < Pr(H)$). Such approach results in 14 symmetry properties. Using examples (analogical to Eells and Fitelson) of drawing cards from a standard deck, Crupi et al. point out which of the symmetries are desired and which are definitely unwanted. For instance, they concur with the results of Eells and Fitelson regarding the inversion symmetry only in case of confirmation. Crupi et al. claim that IS is desirable in case of disconfirmation, as for an exemplary rule: *if the drawn card is an Ace, then it is a face*, the strength with which an *Ace* disconfirms *face* is the same as the strength with which the *face* disconfirms an *Ace*, i.e. $c(H, E) = c(E, H)$. The results obtained by Crupi et al. point that in case of confirmation only the HS, HIS and $EHIS$ are the desirable properties. In case of disconfirmation, they favour HS, EIS and IS properties, finding all other symmetries as unattractive.

5   A New Set of Symmetry Properties

Let us observe that the approaches of Eells and Fitelson [4] as well as Crupi et al. [3] mainly concentrate on entailment and refutation of the hypothesis by the premise. This, however, boils the concept of confirmation down only to situations where there are no counterexamples (entailment) and where there are no positive examples to a rule (refutation).

In our opinion, the concept of confirmation is much broader than a simple analysis whether there are counterexamples to a rule or not. In fact, according to the BC' interpretation of the confirmation concept, a confirmation measure should give an account of the credibility that it is more probable to have the conclusion when the premise is present, rather than when the premise is absent. This means that we should look at confirmation from the perspective of passing from a situation where the premise is absent to the situation where the premise is present. Then, the increase of confirmation (i.e. the difference in conditional probabilities $Pr(H|E)$ and $Pr(H|\neg E)$) becomes important, not just the absence or presence of counterexamples.

Analogically, for disconfirmation a confirmation measure $c(H, E)$ should express how much it is less probable to have the hypothesis when the premise is present rather than when the premise is absent. Again, we should, thus, pass from the situation where the premise is absent to the situation where the premise is present, just like we did in case of confirmation. Therefore, we postulate to consider the symmetry properties together for cases of confirmation and disconfirmation. There is no need to treat them differently as they both consider passing from $Pr(H|\neg E)$ to $Pr(H|E)$.

Let us now conduct the analysis aiming at determining which symmetry properties are desirable and which are unattractive regarding the deep meaning of confirmation concept. First, let us consider the evidence symmetry (ES). Analyzing ES we need to verify whether the equation $c(H, E) = -c(H, \neg E)$ is desirable or not. Let us examine both sides of this equation using an exemplary scenario α where the values of contingency table of E and H are: $a = 100, b = 99$, $c = 0, d = 1$. Let us observe, that for $c(H, E)$ we have that $Pr(H|\neg E) = 0.99$ and $Pr(H|E) = 1$, which gives us a 1% increase of confirmation. On the other hand, for $c(H, \neg E)$ we get exactly the same components but the other way around: $Pr(H|E) = 1$ and $Pr(H|\neg E) = 0.99$, which results in 1% decrease of confirmation. Thus, clearly the confirmation of a rule $E \to H$ should be of the same value but of the opposite sign as the confirmation of a $\neg E \to H$ rule. Therefore, we can conclude that the evidence symmetry is desirable.

This result is in opposition to what Eells and Fitelson [4], and Crupi et al. [3] advocated for. It is due to the fact that they treat the entailment of the conclusion by the premise (i.e. situation where there are no counterexamples to the rule) as the maximal confirmation, whereas we consider the increase of confirmation when passing from the absence of the premise to its presence. For the exemplary rule of Eells and Fitelson: *if the drawn card is the seven of spades then the card is black*, the conditional probabilities are the following $Pr(H|E) = 1$ and $Pr(H|\neg E) = 51/103$. They claim that just because E definitely confirms H, while E does not definitely disconfirms H, we should regard the ES as unattractive. However, for the rules $E \to H$ and $\neg E \to H$, if we interpret the concept of confirmation as expressing how much more probable is the hypothesis (to have *the black card* both for $E \to H$ and $\neg E \to H$), when the evidence (we have drawn *the seven of spades* for $E \to H$ and we have drawn *not-the seven*

Table 2. New symmetry properties

ES	YES *for any* (H, E) $c(H, E) = -c(H, \neg E)$
HS	YES *for any* (H, E) $c(H, E) = -c(\neg H, E)$
EIS	NO *for some* (H, E) $c(H, E) \neq -c(\neg E, H)$
HIS	NO *for some* (H, E) $c(H, E) \neq -c(E, \neg H)$
IS	NO *for some* (H, E) $c(H, E) \neq c(E, H)$
EHS	YES *for any* (H, E) $c(H, E) = c(\neg H, \neg E)$
$EHIS$	NO *for some* (H, E) $c(H, E) \neq c(\neg E, \neg H)$

of spades for ¬*E* → *H*) is realized rather than when it is not realized (we have drawn *not-the seven of spades* for *E* → *H* and we have drawn *the seven of spades* for ¬*E* → *H*), the confirmation has the same absolute value but opposite sign. Thus, we claim that using the deep meaning and interpretation of the confirmation concept, evidence symmetry is a desirable property for Bayesian confirmation measures.

We have conducted analogous analysis for all other symmetries. The results are gathered in Table 2. The set of desirable properties contains only the evidence symmetry, the hypothesis symmetry and their composition i.e. the evidence-hypothesis symmetry. This implies that a valuable Bayesian confirmation measure should satisfy only those symmetry properties. By defining the new set of symmetry properties we gain a tool for assessing the quality of confirmation measures.

6   Conclusions

Bayesian confirmation measures constitute an important group of measures for evaluation of rules induced from data. A valid research area concerns the properties of confirmation measures. Analysis of measures with respect to their properties allows to determine measures that behave according to the user's expectations. It is also a way to handle the plurality of measures and point the most appropriate ones for particular applications.

This article concentrated on the group of symmetry properties. Our analysis was conducted regarding that confirmation measures should reflect how much more it is probable to have the conclusion *H* when the premise *E* is present rather than when it is absent. Such interpretation of the confirmation concept led to our proposition of a new set of desirable symmetry properties. We claim that only symmetries formed by applying the negation operator to the rule's premise, conclusion or both (i.e. *ES*, *HS* and *EHS*) are desirable. Properties *IS*, *EIS*, *HIS*, *EHIS* are unattractive. Thus, valuable confirmation measures should only satisfy *ES*, *HS* and *EHS*.

Consequently, our future research will concentrate on verification which of the commonly used confirmation measures satisfy *ES*, *HS* and *EHS*, not enjoying the other symmetries at the same time. Moreover, experiments on real datasets shall be performed to show the advantages of using such measures.

References

1. Brzezinska, I., Greco, S., Slowinski, R.: Mining Pareto-optimal rules with respect to support and anti-support. Engineering Applications of Artificial Intelligence 20(5), 587–600 (2007)
2. Carnap, R.: Logical Foundations of Probability, 2nd edn. University of Chicago Press, Chicago (1962)
3. Crupi, V., Tentori, K., Gonzalez, M.: On Bayesian measures of evidential support: Theoretical and empirical issues. Philosophy of Science 74, 229–252 (2007)

4. Eells, E., Fitelson, B.: Symmetries and asymmetries in evidential support. Philosophical Studies 107(2), 129–142 (2002)
5. Fitelson, B.: The Plurality of Bayesian Measures of Confirmation and the Problem of Measure Sensitivity. Philosophy of Science 66, 362–378 (1999)
6. Fitelson, B.: Studies in Bayesian Confirmation Theory. Ph.D. Thesis, University of Wisconsin, Madison (2001)
7. Geng, L., Hamilton, H.J.: Interestingness Measures for Data Mining: A Survey. ACM Computing Surveys 38(3), article 9 (2006)
8. Greco, S., Pawlak, Z., Slowinski, R.: Can Bayesian confirmation measures be useful for rough set decision rules? Engineering Applications of Artificial Intelligence 17, 345–361 (2004)
9. Hilderman, R., Hamilton, H.: Knowledge Discovery and Measures of Interest. Kluwer Academic Publishers (2001)
10. Maher, P.: Confirmation Theory. The Encyclopedia of Philosophy, 2nd edn. Macmillan Reference, USA (2005)
11. McGarry, K.: A survey of interestingness measures for knowledge discovery. The Knowledge Engineering Review 20(1), 39–61 (2005)
12. Słowiński, R., Brzezinska, I., Greco, S.: Application of Bayesian Confirmation Measures for Mining Rules from Support-Confidence Pareto-Optimal Set (Invited Paper in). In: Rutkowski, L., Tadeusiewicz, R., Zadeh, L.A., Żurada, J.M. (eds.) ICAISC 2006. LNCS (LNAI), vol. 4029, pp. 1018–1026. Springer, Heidelberg (2006)
13. Słowiński, R., Szczęch, I., Urbanowicz, M., Greco, S.: Mining Association Rules with Respect to Support and Anti-support-Experimental Results. In: Kryszkiewicz, M., Peters, J.F., Rybiński, H., Skowron, A. (eds.) RSEISP 2007. LNCS (LNAI), vol. 4585, pp. 534–542. Springer, Heidelberg (2007)
14. Szczęch, I.: Multicriteria Attractiveness Evaluation of Decision and Association Rules. In: Peters, J.F., Skowron, A., Wolski, M., Chakraborty, M.K., Wu, W.-Z. (eds.) Transactions on Rough Sets X. LNCS, vol. 5656, pp. 197–274. Springer, Heidelberg (2009)
15. Szczęch, I., Greco, S., Slowinski, R.: New property for rule interestingness measures. In: Ganzha, M., Maciaszek, L., Paprzycki, M. (eds.) Proceedings of the Federated Conference on Computer Science and Information Systems FedCSIS, pp. 103–108 (2011)

Applying Verbal Decision Analysis in Selecting Specific Practices of CMMI

Thais Cristina Sampaio Machado, Plácido Rogério Pinheiro,
Adriano Bessa Albuquerque, and Marcelo Marcony Leal de Lima

University of Fortaleza (UNIFOR), Graduate Program in Applied Computer Sciences
Av. Washington Soares, 1321, Bl J Sl 30, 60.811-905, Fortaleza, Brazil
thais.sampaio@edu.unifor.br, {placido,adrianoba}@unifor.br,
marcelomarcony@hotmail.com

Abstract. Although many activities are available in order to satisfy the Project Management area, Software Development Organizations face difficulties to implant functional and effective practices of Project Management. The main objective of this work is to provide an analysis of Capability Maturity Model Integration (CMMI) Specific Practices applied in the Software Development. This work intends to examine them toward to identify which are the most preferable ones, according to elicitation of preferences of a decision maker, in order to improve a software development process for a stated type of project and Company. The process of decision making is supported by the application of a classification methodology named ZAPROS-LM of Verbal Decision Analysis (VDA), classifying the mentioned Specific Practices (SP), as alternatives.

Keywords: VDA, ZAPROS-LM, CMMI, SP, Elicitation of Preferences.

1 Introduction

Project Management started being practiced informally and disorganized. Project Management Institute (PMI) [12] concerns about the concepts standardization of Project Management and its application in practice.

Although all the standardization and consolidated concepts determined by PMI, Software Development Organizations face large difficulties to implant functional and effective practices of Project Management. Aiming to attend the processes defined by PMI, Software Engineering Institute (SEI) developed the model CMMI [15], ongoing over 20 years, providing several practices to be applied in order to help Software Development Organizations to define and improve their processes producing high quality projects. The problem is that, usually, the organizations are not capable of implementing every Specific Practice from CMMI. Hence, which would be the best practices of it to be implemented by the organization?

Specific Practices from CMMI can be described qualitatively, based on a set of multiple criteria. Therefore, the paper is an approach to support the process of decision making [6]. The characteristics were evaluated qualitatively, applying

T. Li et al. (Eds.): RSKT 2012, LNAI 7414, pp. 215–221, 2012.

verbal decision analysis. The System [6] ZAPROS-LM, which belongs to the VDA framework, is used for solving problems that has qualitative nature and difficult to be formalized, called unstructured [11]. The application intends to separate the SP of CMMI level 2 in groups. The division into groups will be responsible to identify which SP should be considered by the organization to implement part of this project management process.

According to [15], Capability Maturity Model Integration (CMMi) is a process improvement maturity model for the development of products and services. Moreover, it consists of covering the Software Product lifecycle with best practices to address the development activities. Different from the guide PMBOK, CMMI is composed by definitions to address to Software Development project. Created by SEI, CMMis main purpose is to guide organizations to improve their process of product development or maintenance. 5 levels of maturity are stated to Companies in accordance to processes attendance to Generic and Specific goals, therefore, attendance to Specific and Generic Practices.

2   Verbal Decision Analysis: Overview and Application

Decision making is a special kind of human activity aimed at the conclusion of an objective for people and for organizations. In the human world, emotions and reason become hard to separate. According to [2] in the majority of multi-criteria problems, exists a set of alternatives, which can be evaluated against the same set of characteristics (called criteria or attributes). These multi-criteria (or multi-attribute) descriptions of alternatives will be used to define the necessary solution.

The Verbal Decision Analysis (VDA) framework is structured on the assurance that most decision making problems can be qualitatively described. The Verbal Decision Analysis supports the decision making process by the verbal representation of problems [3] [5] [7] [8] [13] [14] [16]. Figure 1 introduces an easy visualization of Verbal Decision Analysis methodologies from the VDA framework according to their objectives. As long as the other methods that belong to the Verbal Decision Analysis framework, methodology ZAPROS-LM is also applied to solve problems described qualitatively and supports the decision making process [6]. According to [8], figure 2 presents the structure to apply the VDA method ZAPROS-LM. As the first step to apply ZAPROS-LM, there were defined the criteria, which the alternatives are going to be evaluated against. For each criterion, there is a scale of values associated [8] [9] [10]. For applying the methodology, a pair of criteria is selected from the list to be compared. A simple rule is formulated to generate questions and conclude the comparison between the selected criteria, according to definition in [6].

Thus, according to the decision makers answers, the scale of criteria is created [6]. To construct the joint ordinal scale, it will be compared all possible pairs of values upon all criteria.

Below are listed the criteria and criteria values which will be base to apply the methodology. The criteria values are described from the naturally most preferable to the less preferable one.

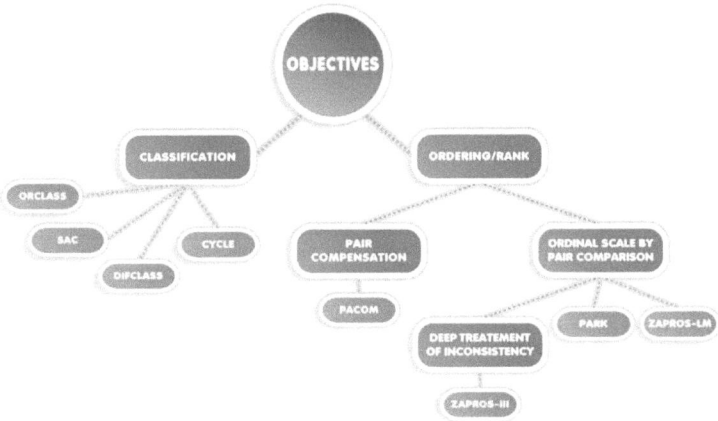

Fig. 1. Methodologies from VDA framework visualization

- Criterion A: Deadline accomplishment:
* A1: Interferes positively totally in deadline accomplishment
* A2: Interferes positively partially in deadline accomplishment
* A3: Does not interfere in deadline accomplishment
- Criterion B: Product Quality in Product Quality
* B1: Interferes positively totally in Product Quality
* B2: Interferes positively partially in Product Quality
* B3: Does not interfere
- Criterion C: Team Motivation
* C1: Interferes positively totally in team motivation
* C2: Interferes positively partially in team motivation
* C3: Does not interfere in team motivation

Analyzing each alternative and having the right support by the decision maker, an experienced professional in Software Engineering, it was possible to classify the alternatives in criterion values. The alternatives for the application are Specific Practices of CMMI from level 2 directed to Project Management, the Process Area Project Planning.

Table 1 presents the list of alternative and the classification made by the decision maker. Then, there is created a vector for each alternative to be evaluated, in accordance to its characterization, with the criteria value related to each criterion.

Afterwards, the application of System ZAPROS-LM ought to consider projects and organizations with characteristics described as follows.

- Project scope: Iteration 2 weeks iteration, Active and involved client: Product Owner presence in the end of each iteration, Team population: 3 - 5 participants, plus ScrumMaster/project manager, Team stability: Changes in Team may occur in the end of iteration, Project complete time estimative: An estimated product of less than 5 months.

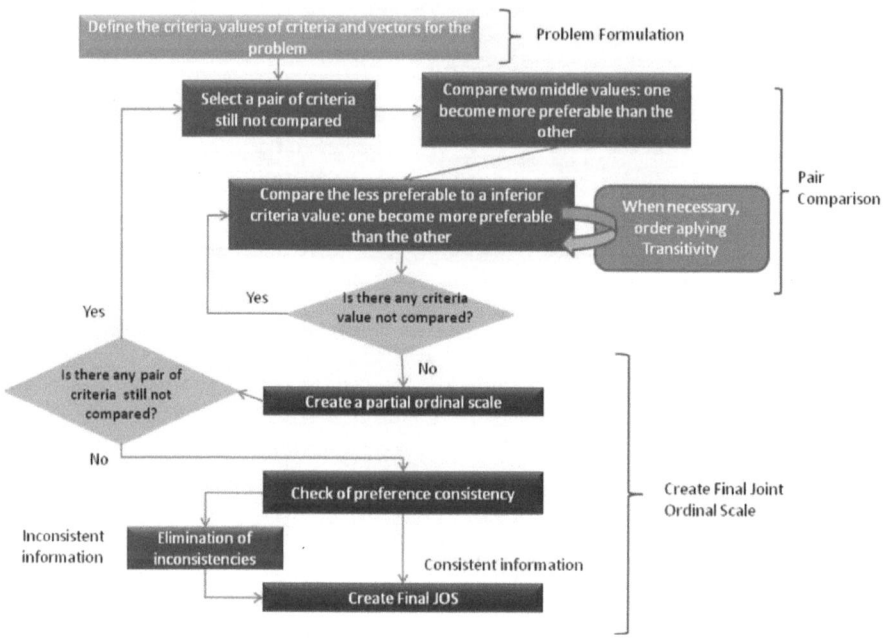

Fig. 2. Procedure to apply the ZAPROS-LM method

- Organization scope: Small Software Development Companies, composed by less than 50 functionaries.

As the second phase, a pair comparison is started. The comparison is made for hypothetical alternatives, next to the best possible alternative (A1B1C1). The vector composed by the hypothetical alternatives is: V = (211, 311, 121, 131, 112, 113).

For the first comparison, in order to construct the Joint Ordinal Scale, the criteria A and B were chosen. Then, the pairs of criteria B/C and A/C were compared. After all paired comparison, according to the decision maker responses, partial scales of preferences are formulated:

Pair compared AxB : A2≺B2≺A3≺B3
Pair compared BxC : C2≺C3≺B2≺B3
Pair compared AxC : C2≺C3≺A2≺A3

Concluded the comparisons and created the partial rank of preferences, the next step for the application is the definition of a Joint Ordinal Scale (JOS). Follows the generated JOS:
A1B1C1≺C2≺C3≺A2≺B2≺A3≺B3

The JOS is composed by all of hypothetical alternatives. For all the vectors of hypothetical alternatives, one rank value is assigned. Using this ranked scale, below is assigned a relation between values from the scale and a rank of preference.

Vector 111: Rank Value 1, Vector 112: Rank Value 2, Vector 113: Rank Value 3, Vector 211: Rank Value 4, Vector 121: Rank Value 5, Vector 311: Rank Value 6, Vector 131: Rank Value 7

Each real alternative is composed by the vectors of hypothetical alternatives as defined above. For example, real alternative SP1 (vector 123) is composed by:

- The hypothetical alternative 111 for criterion A; the hypothetical alternative 121 for criterion B; the hypothetical alternative 113 for criterion C;

Analogous, Table 1 shows the real alternatives and its respective rank values, followed by the ordered rank value (from the lower value to the bigger one).

Table 1. Board of alternatives and characterization

ID	Specific Practices	A	B	C	Vectors	Rank value	Final Rank
SP1	SP 1.1 Estimate the Scope of the Project	A1	B2	C3	111 / 121 / 113	1 / 5 / 3	135
SP2	SP 1.3 Define Project Lifecycle	A2	B2	C3	211 / 121 / 113	4 / 5 / 3	345
SP3	SP 1.4 Determine Estimates of Effort and Cost	A1	B1	C2	111 / 111 / 112	1 / 1 / 2	112
SP4	SP 2.1 Establish the Budget and Schedule	A1	B1	C2	111 / 111 / 112	1 / 1 / 2	112
SP5	SP 2.2 Identify Project Risks	A1	B1	C3	211 / 111 / 113	4 / 1 / 3	134
SP6	SP 2.4 Plan for Project Resources	A2	B1	C2	211 / 111 / 112	4 / 1 / 2	124
SP7	SP 2.5 Plan for Needed knowledge and Skills	A2	B2	C2	211 / 121 / 112	4 / 5 / 2	245
SP8	SP 2.6 Plan Stakeholder Involvement	A1	B2	C2	111 / 121 / 112	1 / 5 / 2	125
SP9	SP 2.7 Establish the Project Plan	A1	B2	C3	111 / 121 / 113	1 / 5 / 3	135
SP10	SP 3.1 Review Plans That Affect the Project	A1	B1	C3	111 / 111 / 113	1 / 1 / 3	113
SP11	SP 3.2 Reconcile Work and Resource Levels	A1	B1	C3	111 / 111 / 113	1 / 1 / 3	113
SP12	SP 3.3 Obtain Plan Commitment	A2	B2	C1	211 / 121 / 111	4 / 5 / 1	145

The next step for applying the methodology is a pair comparison between the rank values of real alternatives. A new rank needs to be created from the minor final rank to the major one. A Final Scale of Preferences can be done.

As result, the final scale of preferences composed by real alternatives is:

SP3/SP4≺SP10/SP11≺SP6≺SP8≺SP5≺SP1/SP9≺SP12≺SP7≺SP2

3 Conclusions, Future Works and Acknowledgment

Project Management is an area of Software Development and Information Technology applied in every Organization to coordinate and monitory projects. Development Software Organizations face large difficulties to implant functional and effective practices of Project Management.

Specific Practices from CMMi were evaluated qualitatively, applying verbal decision analysis. The ZAPROS-LM method, from VDA framework [6], was applied to rank alternatives existent from the less preferable to the most preferable one to identify which SP should be considered more preferable by the organization to implement part of this project management process.

The paper proves that verbal decision analysis methodologies can be applied in real problems of elicitation of preferences and decision making, helping Software Development Companies that would like to implant Specific Practices from CMMI.

As future works, more research can be done applying another criteria and criteria values to evaluate the alternatives. A new research will apply hybrid methodologies for solving the problem [1] and rank approaches from the Specific Practices. More research will be done when applied the selected Specific Practices in a real software development organization, to study the gain obtained as results in projects. The second author is thankful to the National Counsel of Technological and Scientific Development (CNPq) for the support received on this project.

References

1. Machado, T.C.S., Menezes, A.C., Tamanini, I., Pinheiro, P.R.: A Hybrid Model in the Selection of Prototypes for educational Tools: An Applicability In Verbal Decision Analysis. In: IEEE Symposium Series on Computational Intelligence SSCI (2011)
2. Gomes, L.F.A., Moshkovich, H., Torres, A.: Marketing decisions in small businesses: how verbal decision analysis can help. Int. J. Management and Decision Making 11(1), 19–36 (2010)
3. Larichev, O.I., Brown, R.: Numerical and verbal decision analysis: comparison on pratical cases. Journal of Multicriteria Decision Analysis 9(6), 263–273 (2000)
4. Marçal, A.S.C.: SCRUMMI: Um processo de gestão ágil baseado no SCRUM e aderente ao CMMI. Master Thesis I Graduate Program in Applied Computer Sciences, University of Fortaleza (2009)
5. Larichev, O.I.: Method ZAPROS for Multicriteria Alternatives Ranking and the Problem of Incomparability. Informatica 12, 89–100 (2001)
6. Larichev, O.I., Moshkovich, H.M.: Verbal decision analysis for unstructured problems. Kluwer Academic Publishers, The Netherlands (1997)
7. Tamanini, I., Pinheiro, P.R.: Challenging the Incomparability Problem: An Approach Methodology Based on ZAPROS. In: Le Thi, H.A., Bouvry, P., Pham Dinh, T. (eds.) MCO 2008. CCIS, vol. 14, pp. 338–347. Springer, Heidelberg (2008)
8. Tamanini, I.: Improving the ZAPROS Method Considering the Incomparability Cases. Master Thesis – Graduate Program in Applied Computer Sciences, University of Fortaleza (2010)
9. Machado, T.C.S., Meneze, A.C., Pinheiro, L.F.R., Tamanini, I., Pinheiro, P.R.: Toward The Selection of Prototypes For Educational Tools: An Applicability In Verbal Decision Analysis. In: 2010 IEEE International Joint Conferences on Computer, Information, and Systems Sciences, and Engineering, CISSE (2010)
10. Machado, T.C.S., Menezes, A.C., Pinheiro, L.F.R., Tamanini, I., Pinheiro, P.R.: Applying Verbal Decision Analysis in Selecting Prototypes for Educational Tools. In: 2010 IEEE International Conference on Intelligent Computing and Intelligent Systems (ICIS), Shanghai, China (2010)
11. Simon, H., Newell, A.: Heuristic Problem Solving: The Next Advance in Operations Research. Oper. Res. 6, 4–10 (1958)
12. PMI - Project Management Institute. A Guide to the Project Management Body of Knowledge, 3rd edn., USA (2004)
13. Moshkovich, H., Larichev, O.I.: ZAPROS-LM- A method and system for ordering multiattribute alternatives. European Journal of Operational Research (1995)
14. Larichev, O.I.: Ranking Multicriteria Alternatives: The Method ZAPROS III. European Journal of Operational Research 131 (2001)

15. Chrissis, M.B., Konrad, M., Shrum, S.: Guidelines for Process Integration and Product Improvement. CMMI for Development, version 1.2., 2nd edn. Addisson-Wesley, EUA (2007)
16. Tamanini, I., Machado, T.C.S., Mendes, M.S., Carvalho, A.L., Furtado, M.E.S., Pinheiro, P.R.: A Model for Mobile Television Applications Based on Verbal Decision Analysis. In: Sobh, T. (org.) Advances in Computer Innovations in Informations Sciences and Engineering, Berlin, Heidelberg (2008)

Belief Networks in Classification
of Laryngopathies
Based on Speech Spectrum Analysis

Teresa Mroczek[1], Krzysztof Pancerz[1], and Jan Warchoł[2]

[1] University of Information Technology and Management in Rzeszów, Poland
{tmroczek,kpancerz}@wsiz.rzeszow.pl
[2] Medical University of Lublin, Poland
jan.warchol@am.lublin.pl

Abstract. The paper is devoted to classification of laryngopathies on the basis of a family of coefficients reflecting spectrum disturbances around basic tones and their multiples in patients' speech signals. In experiments, a special computer tool called BeliefSEEKER is tested. BeliefSEEKER is capable to generate belief networks and also to generate sets of belief rules. The paper presents feature selection and classification mechanisms as well as the experiments carried out on real-life data.

Keywords: belief networks, belief rules, laryngopathy, decision support system.

1 Introduction

Our research concerns designing methods for classification of patients with selected larynx diseases using a non-invasive diagnosis. Two diseases are taken into consideration: Reinke's edema (RE) and laryngeal polyp (LP). In general, the classification is based on selected parameters of a patient's speech signal (phonation). There exist various approaches to analysis of bio-medical signals (cf. [17]). In general, we can distinguish three groups of methods according to a domain of the signal analysis: analysis in a time domain, analysis in a frequency domain (spectrum analysis), analysis in a time-frequency domain (e.g., wavelet analysis). Therefore, in our research, we are going to build a specialized computer tool for supporting diagnosis of laryngopathies with a hierarchical structure based on multiple classifiers working on signals in time and frequency domains.

In the previous papers, we have tested several approaches for the computer tool being developed both based on the speech spectrum analysis (e.g. [13], [14]) and based on speech signal analysis in a time domain (e.g. [18], [19]). Paper [14] was devoted to the rule-based classification of patients on the basis of a family of coefficients reflecting spectrum disturbances around basic tones and their multiples. Nine coefficients have been extracted. A classification ability of these coefficients has been tested using classification algorithms available in the popular data mining and machine learning software tools: WEKA [21], Rough Set Exploration System (RSES) [3], and NGTS [10].

T. Li et al. (Eds.): RSKT 2012, LNAI 7414, pp. 222–231, 2012.
© Springer-Verlag Berlin Heidelberg 2012

One of the important problems in a classification process is extraction of proper attributes describing classified cases. In this paper, we try to supplement a classification process described in [14] with the approach based on belief networks. A special computer tool called BeliefSEEKER [8] (a belief network and belief rule induction system) has been used in our investigations. Belief-SEEKER has been developed at the University of Information Technology and Management in Rzeszów, Poland, in cooperation with the University of Kansas. BeliefSEEKER was applied to different classification and prediction problems, especially, to decision support in diagnosis of melanocytic skin lesions (cf. [11]).

2 Laryngopathy Data

Data for experiments were extracted from sound samples of the subjects. Two groups were taken into consideration [20]. The first group included persons without disturbances of phonation - the control group (CG). They were confirmed by a phoniatrist opinion. The second group included patients of Otolaryngology Clinic of the Medical University of Lublin in Poland. They had clinically confirmed dysphonia as a result of Reinke's edema (RE) or laryngeal polyp (LP). Experiments were carried out by a course of breathing exercises with instruction about the way of articulation. The task of all examined patients was to utter separately different Polish vowels with extended articulation as long as possible, without intonation, and each on separate expiration.

In [14], the approach based on extracting features (parameters) reflecting patient's speech spectrum disturbances around a basic tone and its multiples (harmonics) has been presented. That approach has been extended in [13] by adding some new features. Clinical experience shows that harmonics in the speech spectrum of a healthy patient are distributed approximately steadily. However, larynx diseases may disturb this distribution [20] (see Figures 1 and 2 to compare some exemplary spectrums). Therefore, the analysis of a degree of disturbances can support the diagnosis of larynx diseases.

Disturbances are expressed by a family of coefficients computed for neighborhoods of a basic tone f_0 and its four multiples (f_1, f_2, f_3, f_4). In a real situation frequencies f_1, f_2, f_3, etc., are not distributed steadily (cf. [20]). It means, that we need to find a real distribution of harmonics. In the presented approach, it is done on the basis of the resultant spectrum (a sum of spectrums calculated for selected N-point time windows into which an original speech signal is divided). For each original frequency f, a maximum magnitude is searched in the interval $[f - d_1, f + d_1]$. This maximum value is assumed as a real harmonic. The original basic tone f_0 has been obtained for each patient from histogram created in the Multi-Dimensional Voice Program (MDVP). It is a software tool for quantitative acoustic assessment of voice quality, calculating various parameters on a single vocalization (see [1]). On the basis of f_0, its harmonics (for ideal case) have been calculated. Each coefficient expresses the distribution of a spectrum around a given frequency f. We can distinguish two types of coefficients: (1) the regularity coefficient R determining a degree of slenderness of this distribution, (2) the

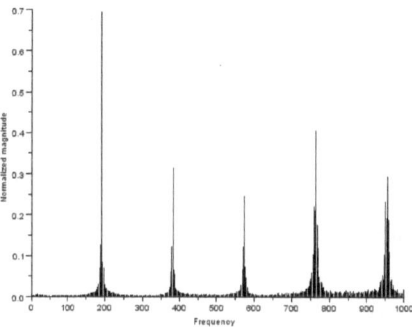

Fig. 1. An exemplary spectrum obtained using DTFT for a patient from the control group

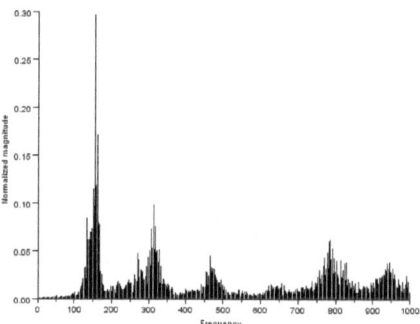

Fig. 2. An exemplary spectrum obtained using DTFT for a patient with a laryngeal polyp

deviation coefficient D determining a relative difference between a real multiple derived from the spectrum and a multiple calculated on the basis of the basic tone f_0.

After execution of the procedure described in detail in [13], we obtain a Pawlak's decision table [15] which is an input for classification algorithms. In this table, each case is described by nine attributes. Each attribute corresponds to one coefficient. A sample of the decision table is presented in Table 1.

In the decision table being a training set of cases for classifiers, we assign, to each patient, one of the two classes: *norm* - a norm - for the patient from the control group, i.e., without disturbances of phonation confirmed by a phoniatrist opinion, *path* - pathology - for the patient either with laryngeal polyp or with Reinke's edema (both clinically confirmed).

Table 1. A sample of the decision table

Patient ID	R_0	R_1	R_2	R_3	R_4	D_1	D_2	D_3	D_4	$CLASS$
#1	0.90	0.89	0.89	0.82	0.75	0.01	0.00	0.00	0.00	*norm*
#2	0.85	0.82	0.75	0.59	0.58	0.04	0.03	0.08	0.08	*path*
...		...	...	...	...	...	...	...	...	...

3   Theoretical Foundations of BeliefSEEKER

In this section, we briefly describe theoretical foundations of a belief network development module implemented in our own computer tool called BeliefSEEKER [8]. The BeliefSEEKER system generates a learning model from data by means of the heuristic algorithm using a Bayesian function, called a matching function. This function matches a network structure to distribution of probability. The algorithm monitors a process of creating the learning model by a specific parameter called marginal likelihood (ML) [5].

A belief network has two components: a directed acyclic graph and a probability distribution. Nodes in the directed acyclic graph represent stochastic variables and arcs represent directed dependencies among variables that are quantified by conditional probability distributions. Let M_h belong to the set $M = \{M_1, \ldots, M_g\}$ of belief networks generated for variables $X_1, \ldots, X_m$, where each network from M represents a hypothesis on the dependency structure relating the variables. In our approach, variables correspond to attributes describing cases (see Section 2). A marginal likelihood function $p(D|M_h)$, where D denotes a sample of data, is calculated by means of averaging out h from a likelihood function $p(D|h)$, where h is a vector parameterizing distribution of variables $X_1, \ldots, X_m$ conditioned by M_h. Hence $p(D|M_h) = \int p(D|h)p(h)dh$, where $p(h)$ is a prior density of h conditioned by M_h. The computation of the marginal likelihood requires the specification of a parameterization of each model M_h and the elicitation of a prior density for h. The application of the prior Hyper-Dirichlet distribution for h [6] with hyper-parameters α_{ijk} leads to the following solution for $p(D|M_h)$, called further ML:

$$ML = \prod_{i=1}^{v} \prod_{j=1}^{q_i} \frac{\Gamma(\alpha_{ij})}{\Gamma(\alpha_{ij} - n_{ij})} \prod_{k=1}^{c_i} \frac{\Gamma(\alpha_{ijk} + n_{ijk})}{\Gamma(\alpha_{ijk})},$$

where:

- v is the number of nodes in the network,
- q_i is the number of possible combinations of parents of the node X_i (if a given attribute does not contain nodes of the type "parent", then q_i is equal to 1),
- c_i is the number of classes within the variable (attribute) X_i,
- n_{ijk} is the number of rows in the decision table, for which parents of the variable (attribute) X_i have value j, and this variable has value k,

- α_{ijk}, α_{ij} are parameters of the prior Dirichlet distribution [9],
- Γ is a function which is calculated for natural numbers as $\Gamma(n) = (n-1)!$.

Data visualization in BeliefSEEKER allows us to present a set of attributes in the form of the set of nodes $V = \{v_1, \ldots, v_n\}$, for which we can define a set of parents of individual nodes $P = \{P_1, \ldots, P_n\}$. Additionally, we assume that a maximal number of parents (*parent_limit*) of an individual node, restricting the cardinality of the vector P_i, is known.

Algorithm 1. Algorithm for creating the learning model.

$V \leftarrow \{v_1, \ldots, v_n\}$;
$E \leftarrow \emptyset$;
$P \leftarrow \{P_1, \ldots, P_n\}$;
$P_1 \leftarrow \emptyset$, ..., $P_n \leftarrow \emptyset$;
create the network $G = (V, E)$;
$max\_ml \leftarrow ML(G)$;
$i \leftarrow n$;
while $i > 0$ **do**
  if $card(P_i) \leq parent\_limit$ **then**
    $opt\_parent \leftarrow v_i$;
    $j \leftarrow i - 1$;
    while $j > 0$ **do**
      $k \leftarrow i - 1$;
      while $k > 0$ **do**
        if $v_k \notin P_i$ **then**
          $P_i \leftarrow P_i \cup \{v_k\}$;
          $E \leftarrow E \cup \{(v_k, v_i)\}$;
          $tmp\_ml \leftarrow ML(G)$;
          if $tmp\_ml > max\_ml$ **then**
            $max\_ml \leftarrow tmp\_ml$;
            $opt\_parent \leftarrow v_k$;
          end
          $P_i \leftarrow P_i - \{v_k\}$;
          $E \leftarrow E - \{(v_k, v_i)\}$;
        end
        $k \leftarrow k - 1$;
      end
      if $opt\_parent = v_i$ or $opt\_parent \in P_i$ **then**
        break a current loop;
      end
      $P_i \leftarrow P_i \cup \{opt\_parent\}$;
      $E \leftarrow E \cup \{(opt\_parent, v_i)\}$;
      $j \leftarrow j - 1$;
    end
  end
  $i \leftarrow i - 1$;
end

The process of creating the learning model is controlled by a matching function, i.e., the marginal likelihood ML, and its goal is to find such a structure of the network that maximizes a value of the matching function. Formally, the process can be depicted as in Algorithm 1. Searching for the network structure begins with the estimation of an initial value of the marginal likelihood ML for a network without arcs. The main goal of this operation is to determine an initial value of the matching function. The algorithm analyzes consecutively the whole set V of nodes to determine relationships between nodes. In each step of the algorithm, such a parent node is searched that its addition to the network maximizes ML. We can distinguish two stages of searching for the optimal parent. Firstly, a node v_k is added temporarily to the network and an arc between v_k and v_i is created (v_k becomes a parent node of v_i). For such a temporary structure, a value of ML (tmp_ml) is calculated and if it is greater than the current value of ML, then it becomes a new value of ML taken into consideration in the next step of the algorithm and the node v_k becomes an optimal parent (opt_parent). Otherwise, a next parent is searched within nodes from the set V. Secondly, the node v_k and the arc (v_k, v_i) are permanently added to the network if v_k does not belong to the vector P_i of parents and $v_k \neq v_i$. Addition of parent nodes is finished if the cardinality of P_i achieves the value $parent\_limit$.

It is worth noting that a space of networks searched by the algorithm is dependent on the order of attributes a priori imposed. It is a basic disadvantage of this algorithm which can cause generation of a model corresponding to the local maximum. Therefore, a capability of BeliefSEEKER has been extended by modification of the K2 algorithm. This extension consists in searching for a network which maximizes the matching function among networks obtained for different orders of descriptive attributes.

4 Experiments

In experiments, we have used the BeliefSEEKER system outlined in Section 3. This system works on text data recorded in the form of a Pawlak's decision table. A sample of the decision table for considered data has been shown in Section 2. In BeliefSEEKER, input data are presented in the form of a decision table as well as in the graphical form as nodes.

The examined data table consists of 77 cases with balanced distribution of decision classes (38 - norm and 39 - pathology). Cases are described by continuous attributes. Therefore, experiments have been carried out for different numbers of intervals determined in a discretization process. Models obtained in this way have been compared in terms of classification accuracy evaluated using a standard 10 cross-validation test (CV-10). Comparison has been made for two discretization methods, i.e., intervals with a constant width and intervals with an equal number of cases. Intervals obtained in a discretization process have been used to create learning models in the form of belief networks. A set of belief networks using the K2 algorithm supervised by parameters controlling a model generation process (automatic incrementation of the Dirichlet parameter α [8],

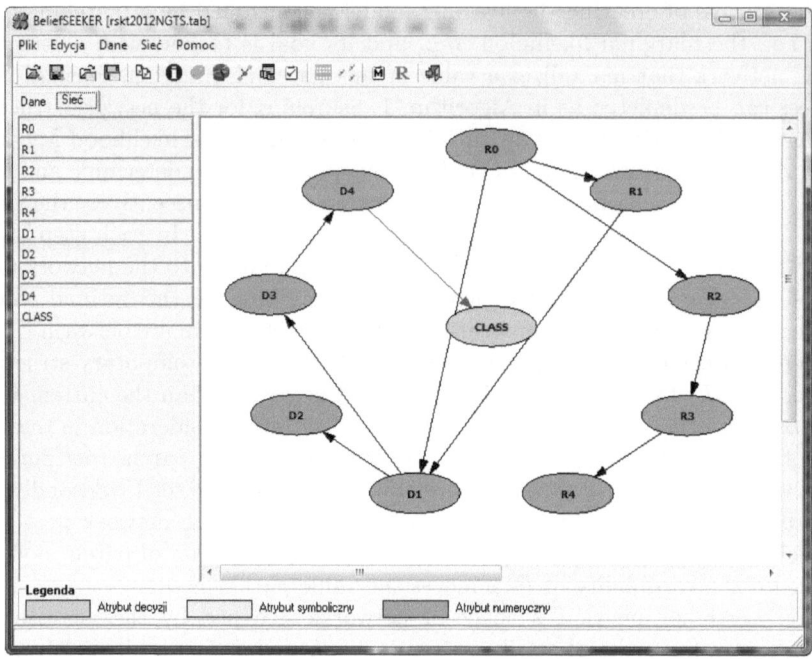

Fig. 3. An example of the belief network generated in BeliefSEEKER for laryngopathy data

changes of the order of descriptive attributes) has been generated. An exemplary belief network is shown in Figure 3.

Evaluation of the obtained networks in terms of classification accuracy of unseen cases showed that the discretization method does not influence the classification error. Table 2 includes results obtained for different numbers of discretization intervals and different values of the Dirichlet parameter α. The number of discretization intervals plays an important role in the classification process. The best results have been obtained for five discretization intervals.

The qualitative analysis of belief networks generated by BeliefSEEKER enabled us to define the set of the most important attributes (in terms of classification accuracy) among nine attributes describing cases (the subjects), see Tables 3 and 4.

The approach presented in this paper has been tested using classification algorithms available in the popular data mining and machine learning software tools: WEKA [21], Rough Set Exploration System (RSES) [3]. Three of them are rule-based algorithms: exhaustive (RSES) [2], LEM2 (RSES) [7], genetic (RSES) [22]. Two of them are decision-tree based algorithms: J48 (WEKA) - an implementation of C4.5 [16], CART (WEKA) [4]. On the basis of Tables 3 and 4, the first

Table 2. Classification errors

No. of intervals	$\alpha = 1$	$\alpha = 10$	$\alpha = 20$	$\alpha = 30$	$\alpha = 40$	$\alpha = 70$	$\alpha = 80$	$\alpha = 100$
10	25.00%						41.43%	
9	25.00%		36.43%					
8	25.00%		31.43%			35.00%		
7	23.57%		31.43%					
6	30.00%					33.57%		
5	23.57%	21.43%			32.14%			
4	26.43%		29.29%			40.00%		
3	31.43%	30.71%		33.57%				37.14%
2	30.00%				38.57%			

Table 3. The most important attributes

No. of intervals	$\alpha = 1$	$\alpha = 10$	$\alpha = 20$	$\alpha = 30$	$\alpha = 40$	$\alpha = 70$	$\alpha = 80$	$\alpha = 100$
10	D4						R1,D4	
9	D4		D4,D1					
8	D4		D4,D1			D4,D1,R4		
7	D4		D4,D1					
6	D4,R4					D4,D1,R4		
5	D4	D4,R4			D4,R1,R4			
4	D4		D4,R2			D4,R2,R1		
3	D4	D4,R1		D4,D3,R1				D4,D3,R1,D1
2	D4,R3				D4,R3,R4			

Table 4. Frequency of attribute occurrence

Attribute	Occurrence frequency [%]
D4	100
D3	26.08
R4	26.08
R1	26.08
D5	7.69
R2	7.69
R3	7.69

seven attributes as well as the first four attributes (with the highest occurrence frequency) have been selected for case description. Classification accuracy has been compared with results obtained for the whole set of descriptive attributes (nine attributes described in Section 2) [13].

Experiments showed that classification accuracy has been improved after attribute selection on the basis of belief networks obtained using BeliefSEEKER. Moreover, it is easy to see that classification accuracy obtained in the popular data mining and machine learning software tools is significantly better than that one obtained directly in BeliefSEEKER using belief rules. BeliefSEEKER includes a unique algorithm for conversion belief networks into belief rules [8].

Table 5. Results of experiments: classification accuracy

Algorithm	Classification accuracy		
	All attributes (cf. [13])	7 attributes	4 attributes
Exhaustive (RSES)	0.8430	**0.9140**	**0.8710**
LEM2 (RSES)	0.8700	**0.8930**	**0.9420**
Genetic (RSES)	0.8430	**0.9000**	0.8430
J48 (WEKA)	0.8441	0.8311	**0.8571**
CART (WEKA)	0.7922	0.7922	0.7922

5   Conclusions

Research presented in this paper has been based on a probabilistic approach to searching information hidden in data. A data set consisting cases classified for determining laryngopathy has been used in experiments. The main goal of experiments was to test usefulness of the approach based on belief networks in the classification process of laryngopathies. Experiments carried out using a special computer tool called BeliefSEEKER showed that, in the classification process of laryngopathies, the tested approach is suitable for selection of attributes describing the subjects. However, a proper classification should be performed using external data mining approaches. In the further research, we will try to use constructive induction mechanism (CIM). CIM usually uses operators to combine two or more features (descriptive attributes) to generate new features. Some results of improvement of learning models using CIM, based on belief networks, has been shown in [12].

Acknowledgments. This research has been supported by the grant No. N N516 423938 from the National Science Centre in Poland.

References

1. Multi-Dimensional Voice Program (MDVP) (2011),
   http://www.kayelemetrics.com
2. Bazan, J.G., Nguyen, H.S., Nguyen, S.H., Synak, P., Wroblewski, J.: Rough set algorithms in classification problem. In: Polkowski, L., Tsumoto, S., Lin, T.Y. (eds.) Rough Set Methods and Applications, pp. 49–88. Physica-Verlag, Heidelberg (2000)
3. Bazan, J.G., Szczuka, M.S.: The Rough Set Exploration System. In: Peters, J.F., Skowron, A. (eds.) Transactions on Rough Sets III. LNCS, vol. 3400, pp. 37–56. Springer, Heidelberg (2005)
4. Breiman, L., Friedman, J., Olshen, R., Stone, C.: Classification and Regression Trees. Chapman & Hall, Boca Raton (1993)
5. Cooper, F., Herskovits, E.: A Bayesian method for the induction of probabilistic networks from data. Machine Learning 9(4), 309–347 (1992)

6. Cowell, R., Dawid, A., Lauritzen, S., Spiegelhalter, D.: Probabilistic Networks and Expert Systems. Springer, New York (1999)

7. Grzymala-Busse, J.: A new version of the rule induction system LERS. Fundamenta Informaticae 31, 27–39 (1997)

8. Grzymała-Busse, J.W., Hippe, Z.S., Mroczek, T.: Deriving Belief Networks and Belief Rules from Data: A Progress Report. In: Peters, J.F., Skowron, A., Marek, V.W., Orłowska, E., Słowiński, R., Ziarko, W. (eds.) Transactions on Rough Sets VII. LNCS, vol. 4400, pp. 53–69. Springer, Heidelberg (2007)

9. Heckerman, D.: A tutorial on learning with bayesian networks. Tech. rep., Learning in Graphical Models (1996)

10. Hippe, Z.: Machine learning - a promising strategy for business information processing? In: Abramowicz, W. (ed.) Business Information Systems, pp. 603–622. Academy of Economics Editorial Office, Poznan (1997)

11. Hippe, Z., Mroczek, T.: Melanoma classification and prediction using belief networks. In: Kurzynski, M., Puchała, E., Wozniak, M. (eds.) Computer Recognition Systems, pp. 337–342. University of Technology Publishing Office, Wrocław (2003)

12. Paja, W., Pancerz, K., Wrzesień, M.: A New Hybrid Method of Generation of Decision Rules Using the Constructive Induction Mechanism. In: Yu, J., Greco, S., Lingras, P., Wang, G., Skowron, A. (eds.) RSKT 2010. LNCS, vol. 6401, pp. 322–327. Springer, Heidelberg (2010)

13. Pancerz, K., Paja, W., Szkoła, J., Warchoł, J., Olchowik, G.: A rule-based classification of laryngopathies based on spectrum disturbance analysis - an exemplary study. In: Van Huffel, S., et al. (eds.) Proc. of the BIOSIGNALS 2012, Vilamoura, Algarve, Portugal, pp. 458–461 (2012)

14. Pancerz, K., Szkoła, J., Warchoł, J., Olchowik, G.: Spectrum disturbance analysis for computer-aided diagnosis of laryngopathies: An exemplary study. In: Proc. of the International Workshop on Biomedical Informatics and Biometric Technologies (BT 2011), Zilina, Slovak Republic (2011)

15. Pawlak, Z.: Rough Sets. Theoretical Aspects of Reasoning about Data. Kluwer Academic Publishers, Dordrecht (1991)

16. Quinlan, J.: C4.5. Programs for machine learning. Morgan Kaufmann Publishers (1993)

17. Semmlow, J.: Biosignal and Medical Image Processing. CRC Press (2009)

18. Szkoła, J., Pancerz, K., Warchoł, J.: Computer diagnosis of laryngopathies based on temporal pattern recognition in speech signal. Bio-Algorithms and Med-Systems 6(12), 75–80 (2010)

19. Szkoła, J., Pancerz, K., Warchoł, J.: Recurrent neural networks in computer-based clinical decision support for laryngopathies: An experimental study. Computational Intelligence and Neuroscience 2011, article ID 289398 (2011)

20. Warchoł, J.: Speech Examination with Correct and Pathological Phonation Using the SVAN 912AE Analyser. Ph.D. thesis, Medical University of Lublin (2006) (in Polish)

21. Witten, I.H., Frank, E.: Data Mining: Practical Machine Learning Tools and Techniques. Morgan Kaufmann (2005)

22. Wróblewski, J.: Genetic algorithms in decomposition and classification problem. In: Polkowski, L., Skowron, A. (eds.) Rough Sets in Knowledge Discovery 2, pp. 471–487. Physica-Verlag, Heidelberg (1998)

Comparing Similarity of Concepts Identified by Temporal Patterns of Terms in Biomedical Research Documents

Shusaku Tsumoto and Hidenao Abe

Department of Medical Informatics, School of Medicine, Faculty of Medicine
Shimane University
89-1 Enya-cho Izumo 693-8501 Japan
{tsumoto,abe}@med.shimane-u.ac.jp

Abstract. In this paper, we present an analysis of a relationship between temporal trends of automatically extracted terms in medical research document and their similarities on a structured vocabulary. In order to obtain the temporal trends, we used our temporal pattern extraction method that combines an automatic term extraction, an importance index of the terms, and clustering for the values in each period. By using a set of medical research documents that were published every year, we extracted temporal patterns of the automatically extracted terms. Then, we calculated their similarities on the medical taxonomy by defining a distance on the tree structure. For analyzing the relationship between the terms included in the patterns and the similarity of the terms on the taxonomy, the differences of the averaged similarities of the terms in each pattern are compared between the two trends of the temporal patterns.

Keywords: Temporal Text Mining, Knowledge Base, Temporal Clustering.

1 Introduction

In recent years, information systems in medical field have developed rapidly, and the amount of stored data has increased year after year. Documents are also accumulated not only in clinical situations, but also in worldwide repositories by various medical studies. Such data now provide valuable information to medical researchers, doctors, engineers, and related workers. Hence, the detection of new, important, and remarkable phrases and words has become very important to aware valuable evidences in the documents.

With respect to biomedical research documents, the MeSH vocabulary provides overall concepts and terms for describing them in a simple and accurate way. New concepts, which appear as new concepts every year, are usually added to the vocabulary if the concepts are useful. One criterion for adding new concepts is related to how attention paid to them by researchers appears as an emergent pattern in published documents. Around few hundred of new concepts

T. Li et al. (Eds.): RSKT 2012, LNAI 7414, pp. 232–241, 2012.
© Springer-Verlag Berlin Heidelberg 2012

are added every year, and the maintenance of the concepts and their related structure has been done by manually. Thus, MeSH has another aspect as an important knowledge base for the biomedical research field. However, the relationships between some data-driven trends and the newly added concepts did not be clarified.

In order to clarify the relationship, we developed a method for analyzing the similarity of terms on the structured taxonomy and the trend of a data-driven index of the terms.

In this paper, we describe a result of the analysis by using the method for identifying similar terms based on the temporal behavior of usages of each term [1]. The temporal pattern extraction method on the basis of term usage index consists of automatic term extraction methods, term importance indices, and temporal clustering in the next section. Then, in Section 3, we performed a case study showing the differences between similar terms detected by the temporal patterns of medical terms related to migraine drug therapy in MEDLINE documents. Finally, we conclude the analysis result in Section 4.

2   A Method for Analyzing Distances on Taxonomy and Temporal Patterns of Term Usage

In order to analyze the relationships between usages of words and phrases in temporally published documents and the difference on a taxonomy, we used the temporal pattern extraction method based on data-driven indices [1]. By using the similar terms identified on the basis of temporal patterns of the indices, we measure their similarities between each term on the taxonomy that can be assumed as the sets of tree structures of concepts on a particular domain.

2.1   Obtaining Temporal Patterns of Data-Driven Indexes Related to Term Usages

In order to discover various trends related to usages of the terms in temporally published corpus, we developed a method for obtaining temporal patterns of an importance index. This framework obtains some temporal patterns based on the importance index from the given temporally published sets of documents. It consists of the following processes.

- Automatic term extraction in overall documents
- Calculation of importance indices
- Obtaining temporal clusters for each importance index
- Assignment of some meanings for the obtained temporal patterns

Automatic Term Extraction in a Given Corpus. Firstly, a system determines terms in a given corpus. We consider the difficulties of term extraction without any dictionary and apply a term extraction method [2] that is based on the adjacent frequency of compound nouns. This method involves the detection of technical terms by using the following values for a candidate compound noun CN:

$$FLR(CN) = f(CN) \times (\prod_{i=1}^{L}(FL(N_i) + 1)(FR(N_i) + 1))^{\frac{1}{2L}}$$

where $f(CN)$ means frequency of a candidate noun CN separately, and $FL(N_i)$ and $FR(N_i)$ indicate the frequencies of different words on the right and the left of each noun N_i in bi-grams included in each CN.

Calculation of Data-Driven Indices for Each Term. After determining terms in the given corpus, the system calculates importance indices of these terms in the documents in each time period for representing the usages of the terms as the values.

Some importance indices for words and phrases in a corpus are well known. Term frequency divided by inverse document frequency (tf-idf) is one of the popular indices used for measuring the importance of terms [3]. The tf-idf value for each term $term_i$ can be defined for the documents in each year, D_{year}, as follows:

$$TFIDF(term_i, D_{year}) =$$

$$tf(term_i, D_{year}) \times log\frac{|D_{year}|}{df(term_i, D_{year})}$$

where $tf(term_i, D_{year})$ is the frequency of each term $term_i$ in a corpus with $|D_{year}|$ documents. Here, $|D_{year}|$ is the number of documents included in each year, and $df(term_i, D_{year})$ is the frequency of documents containing term.

In the proposed framework, we suggest treating these indices explicitly as a temporal dataset. This dataset consists of the values of the terms for each time point by using each index $Index(;D_{year})$ as the features. Fig. 1 shows an example of such a dataset consisting of an importance index for each year. The value of the term $term_i$ is described as $Index(term_i, D_{year})$ in Fig. 1.

Generating Temporal Patterns for Detecting Trends of Terms. After obtaining the dataset, the framework provides the choice of an adequate trend extraction method to the dataset. A survey of the literature shows that many conventional methods for extracting useful time-series patterns have been developed [4,5]. Users can apply an adequate time-series analysis method and identify important patterns by processing the values in the rows of Fig. 1. By considering these patterns with temporal information, we can understand the trends related to the terms such as transition of technological development with technical terms. The temporal patterns as the clusters also provide information about similarities between the terms at the same time. We denote the similar terms based on the temporal cluster assignments as $term_i \in c_k$.

Selecting m time periods as the features for the dataset

		$Index(\bullet, D_{2000})$		$Index(\bullet, D_{year})$	
$term_1$	$\cdots$	$Index(term_1, D_{2000})$	$\cdots$	$Index(term_1, D_{year})$	$\cdots$
$\vdots$		$\vdots$		$\vdots$	
$term_i$	$\cdots$	$Index(term_i, D_{2000})$	$\cdots$	$Index(term_i, D_{year})$	$\cdots$
$\vdots$		$\vdots$		$\vdots$	
$term_n$	$\cdots$	$Index(term_n, D_{2000})$	$\cdots$	$Index(term_n, D_{year})$	$\cdots$

Fig. 1. Example of dataset consisting of an importance index

Assigning Meanings of the Trends of the Obtained Temporal Patterns.
After obtaining the temporal patterns c_k, in order to identify the meanings of
each pattern by using trends of the extracted terms for each importance index,
we apply linear regression analysis. The degree of the centroid of a temporal
pattern c is calculated as follows:

$$Deg(c) = \frac{\sum_{j=1}^{M}(c_j - \bar{c})(x_j - \bar{x})}{\sum_{j=1}^{M}(x_j - \bar{x})^2}$$

where $\bar{x}$ is the average of $t_j - t_1$ for M time points and $\bar{y}$ is the average of the values
c_j. Each value of the centroid, c_j, is a representative value of the importance
index values of assigned terms in the pattern as $Index(term_i \in c_k, D_{year})$. Each
time point t_j corresponds to each year, and the first year assigns to the first time
point as t_1.

Simultaneously, we calculate the intercept $Int(c)$ of each pattern c_k as follows:

$$Int(c) = \bar{y} - Deg(c)\bar{x}$$

Then, by using the two linear trend criteria, we assigned some meanings of the
temporal patterns related to the usages of the terms.

2.2   Defining Similarity of Terms on a Structured Taxonomy

In this paper, we assume a tree structure of concepts that are defined with a
relation such as is-a as 'structured taxonomy'. In the biomedical domain, MeSH
(Medical Subjects Headings) [6] is one of the important structured taxonomy for
representing key concepts of biomedical research articles. MeSH consists of 16

categories including not only proper categories for biomedicine but also general categories such as information science. It contains 25,588 concepts as 'Descriptor', and 464,282 terms as 'Entry Terms'. Each concept has one or more entry terms and the tree numbers as the identifier in the hierarchy structure.

For this structure, we defined similarity of each pair of terms represented by using distance in the tree structure of MeSH as shown in Figure 2.

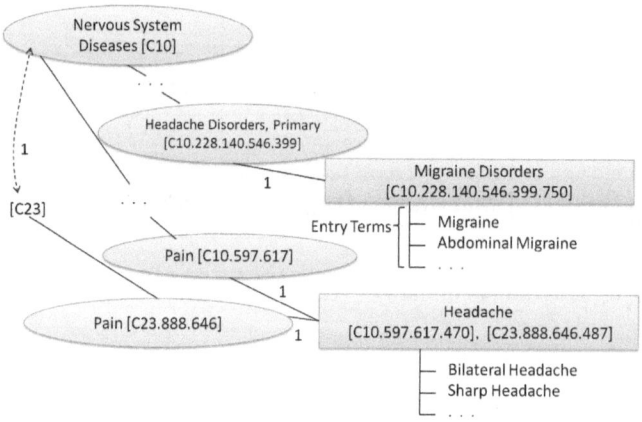

Fig. 2. Example of MeSH hierarchy structure for migraine disorders and headache

For example, when the distance between the two terms, $term_{i1}$ and $term_{i2}$, denotes as $Dist(term_{i1}, term_{i2})$, the distance between 'migraine' and 'sharp headache', $Dist(migraine, sharpheadache)$, is calculated as 8 or 9. By using this distance, the similarity between each pair of terms is defined as the following:

$$Sim(term_{i1}, term_{i2}) = \frac{1}{1 + Dist(term_{i1}, term_{i2})}$$

where the similarity can be calculated when the both terms have tree numbers in MeSH.

For overall terms belonging to some group g, we also defined their averaged similarity in the group as the following:

$$Avg.Sim(g) = \frac{1}{numPair} \sum_{term_i \in g} Sim(term_{i1}, term_{i2})$$

where $numPair$ is the number of matched pairs of the terms included in the group g, defined as $numPair =_{|term_i \in g \cap hasTreeNumber(term_i)|} C_2$.

3   Analyzing Similarity of Terms in Temporal Patterns of Medical Research Documents

In this section, we describe a case study for analyzing similarity of terms detected some temporal patterns in medical research documents. For obtaining the temporal patterns, we used an importance index of the terms in each set of documents that ware published year by year. The medical research documents are retrieved from MEDLINE by using a search scenario over time. The scenario is related to migraine drug therapy similar to the first one in a previous paper on MeSHmap [7].

In this case study, we consider the search scenario and three meanings of trends as temporal clusters by using the degrees and intercepts of the trend lines for each term as follows. As for the meanings, we assigned "emergent" to ascending trend lines with negative intercepts and "popular" to ascending trend lines with positive intercepts. Subsequently, we calculated the similarity of the medical terms included in the temporal patterns on the MeSH structure.

3.1   Analysis of Disease over Time

In this scenario, a user may be interested in exploring the progression of ideas in a particular domain, say, corresponding to a particular disease. By performing the search such that the disease is represented according to the year, one may obtain a temporal assessment of the changes in the field.

Let us assume that the user wants to explore the evolution of ideas about drugs used to treat migraine. The user performs a search for abstracts of articles "migraine/drug therapy [MH:NOEXP] AND YYYY [DP] AND clinical trial [PT] AND english [LA]" through PubMed. The string "YYYY" is replaced with the four digits necessary for retrieving articles published each year.

With this search query, we obtain articles published between 1980 and 2009 with the abstract mode of PubMed. Fig. 3 shows the numbers of article titles and abstracts retrieved by the query. In this study, we assume each title and abstract text to be one document.

From all of the retrieved abstracts, the automatic term extraction method identifies 61,936 terms. Similarly, from all of the titles, the system extracts 6,470 terms.

3.2   Obtaining Temporal Patterns of Medical Terms about Migraine Drug Therapy Studies

By calculating the document frequency and the tf-idf values as the importance indices for each year on titles and abstracts respectively. By using the two document sets and the two indices, we obtained four datasets to obtain temporal clusters that consist of temporal behavior of each index year by year for each term.

As for the clustering algorithm, we used k-means clustering algorithm implemented in Weka[8](weka-3-6-2). Since the implementation search better cluster

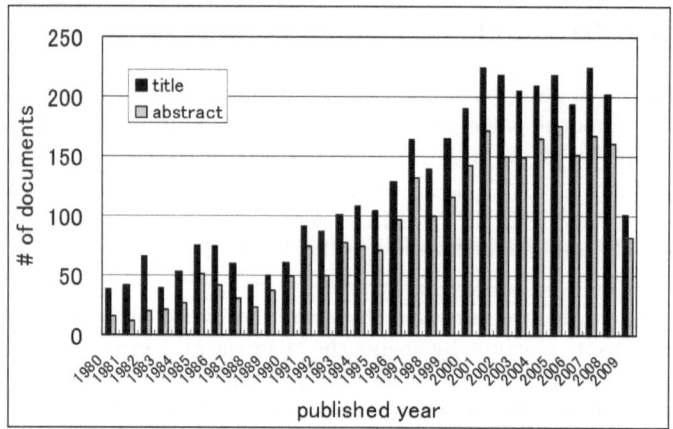

Fig. 3. Numbers of documents with titles and abstracts related to migraine drug therapy published from 1980 to 2009

assignments by minimizing the sum of squared errors (SSE), we set up 1% of the number s of terms as the upper limits of the number of clusters. And we also set up the maximum iteration to search better assignment, is 500 times.

Table 1 shows the result of k-means clustering on the four dataset.

Table 1. Overall result of temporal clustering on titles and abstracts about migraine drug therapy by using the three importance indices

Dataset		# of Clusters	SSE	Total trend		# of patterns		
				Avg.Deg	Avg.Int	Emergent	Popular	Subsiding
Abstrtacts	tf−idf	129	216.81	0.17	−0.44	81	37	0
	df	129	36.67	0.04	−0.06	103	25	1
Titles	tf−idf	14	125.69	0.10	−0.03	5	9	0
	df	14	10.14	0.03	0.02	4	9	1

Fig. 4 shows the emergent cluster centroid and the top ten emergent terms on the abstracts on the basis of tf-idf. The cluster is selected with the following conditions: including phrases, highest linear degree with minimum intercepts to y-axis by sorting the average degrees and the average intercepts of the 14 clusters.

As shown in Fig. 4, we can detect the emergent terms included in the emergent pattern that related to triptans drug therapy. The cluster also includes some terms related to the time for the therapy. The drugs including triptans, which are appeared in this pattern, are approved later 1990s in US and European countries, and early 2000s in Japan. Based on the result, the method obtained the temporal patterns related to the topics that attract interests of researchers in

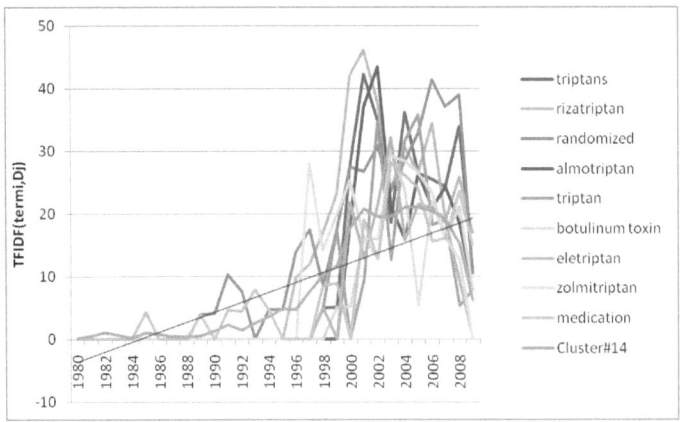

Fig. 4. The detailed result of Cluster #14 of tf-idf on the abstracts

this field. In addition, the degree of the increasing and the shapes of the temporal patterns of each index show some aspects the movements of the research issue.

3.3 Similarity of the Terms in Obtained Temporal Patterns on MeSH

By using the similarity measure as described in Section 2, we calculated the averaged similarity of the medical terms included in each temporal pattern. In order to analyze the relationship between the trends and the similarities, we compare the representative values of the averaged similarities of the term.

As shown in Table 2, the similarities for each temporal pattern are calculated. Smaller similarity value means that the terms included in the temporal pattern are defined separately on the MeSH structure. Besides, greater similarity value means that the similar terms on the temporal pattern are also defined similarly on the MeSH structure.

Then, for clarifying the relationships between the temporal patterns and the similarity on the taxonomy, we compare the difference of the similarity values by separating the meanings of the linear trends; emergent or not emergent. As for the first representative values, we compare the two groups of the average values by using t-test. Then, we compare the two groups of the similarity values by using Wilcoxon rank sum test. Table 3 shows the averages and the medians of the similarity values.

The similarity values around 0.13 means that the pair of terms defined in each place with from 6 to 7 paths. By testing the difference between the two groups based on the linear trends, for the abstracts, the similarities of the terms included in the emergent temporal patterns are significantly smaller than the terms included in the popular patterns based on the tf-idf values. This result indicates that the tf-idf index detects new combinations of the concepts as its emergent trend. Besides, based on the temporal patterns by using the document frequency,

Table 2. Temporal patterns obtained for the tf-idf dataset on the titles and the similarities of the terms in each temporal pattern

k	term_ck	Deg(ck)	Int(ck)	Meaning	Avg.Sim(ck)
2	clinical efficacy	0.063	−0.142	Emergent	0.133
3	placebo−controlled study	0.069	−0.299	Emergent	0.135
5	5−ht1 b/1 dagonists	0.044	−0.139	Emergent	0.131
8	migraine therapy	0.339	−1.870	Emergent	0.141
14	acute treatment of migraine	0.789	−3.566	Emergent	0.135
1	migraine patients	0.059	3.874	Popular	0.120
4	cluster headache	0.050	0.203	Popular	0.144
6	migraine	1.816	3.589	Popular	0.174
7	double−blind study	0.058	0.206	Popular	0.121
9	patients	0.748	1.363	Popular	0.138
10	management of migraine	0.025	0.816	Popular	0.133
11	tension−type headache	0.058	0.196	Popular	0.133
12	oral sumatriptan	0.255	0.955	Popular	0.131
13	migraine prophylaxis	0.025	0.523	Popular	0.160

Table 3. Comparison of the representative values. (a)Averages, (b)Medians. ∗ means significant difference on $\alpha = 0.05$.

(a) Averages

	Abstracts		Titles	
	Emergent	Not Emergent	Emergent	Not Emergent
tf-idf	0.126∗	0.130∗	0.134	0.139
df	0.129∗	0.125∗	0.134	0.141

(b) Medians

	Abstracts		Titles	
	Emergent	Not Emergent	Emergent	Not Emergent
tf-idf	0.126∗	0.130∗	0.132	0.138
df	0.131∗	0.125∗	0.133	0.139

the terms included in the emergent patterns are defined more similarly. More frequently used terms in the recently published documents are defined nearer than the other popular terms. This can be understandable by considering the process for maintaining the structure of the concepts manually.

4   Conclusion

In this paper, we show a comparison of the similarity of the terms that are grouped up by using their temporal behavior of the two importance indices;

document frequency and tf-idf index. Then, we obtained the temporal patterns of the biomedical terms by using the two importance indices related to the usages of the terms on the biomedical research documents as the temporal corpus. By using MeSH as the structured taxonomic definition of the medical terms, we compared the averaged similarity based on the distances between the terms included in each temporal pattern.

By separating the trends of the temporal patterns based on the linear regression technique, the averaged similarities of the terms in each pattern show significant differences on the larger corpus. Based on the temporal patterns with the emergent trend of the tf-idf, the terms included in such patterns are not similar compared to the terms included in the popular patterns. This indicates that the index detects new combination of the concepts with its trend. Besides, the similarity of the different index shows the opposite relationship between its trend and the similarity on the taxonomic definition.

In the future, we will introduce more importance indices for detecting various aspects of term usages. Then, we will also obtain some predictive models such as numerical prediction models for analyzing the places of new concepts.

Acknowledgment. This research is supported by Grant-in-Aid for Scientific Research (B) 24300058 from Japan Society for the Promotion of Science(JSPS).

References

1. Abe, H., Tsumoto, S.: Trend detection from large text data. In: Proceedings of the 2010 IEEE International Conference on Systems, Man and Cybernetics, pp. 310–315. IEEE (2010)
2. Nakagawa, H.: Automatic term recognition based on statistics of compound nouns. Terminology 6(2), 195–210 (2000)
3. Sparck Jones, K.: A statistical interpretation of term specificity and its application in retrieval. Document Retrieval Systems, 132–142 (1988)
4. Keogh, E., Chu, S., Hart, D., Pazzani, M.: Segmenting time series: A survey and novel approach. In: Data Mining in Time Series Databases, pp. 1–22. World Scientific (2003)
5. Liao, T.W.: Clustering of time series data: a survey. Pattern Recognition 38, 1857–1874 (2005)
6. Medical subject headings, http://www.nlm.nih.gov/mesh/
7. Srinivasan, P.: Meshmap: a text mining tool for medline. In: Proc. of AMAI Symposium 2001, pp. 642–646 (2001)
8. Witten, I.H., Frank, E.: Data Mining: Practical Machine Learning Tools and Techniques with Java Implementations. Morgan Kaufmann (2000)

Extracting Incidental and Global Knowledge through Compact Pattern Trees in Distributed Environment

K. Swarupa Rani, V. Kamakshi Prasad, and C. Raghavendra Rao

Department of Computer and Information Sciences,
University of Hyderabad, Hyderabad
School of Information Technology, JNTU Hyderabad, Hyderabad, India
{swarupacs,crrcs}@uohyd.ernet.in, kamakshiprasad@yahoo.com

Abstract. This paper proposes to extract incidental and global knowledge through Compact Pattern Trees in a hierarchical structure through distributed and parallel computing paradigm. This method also facilitates privacy preserving with a minimal communication load. We present the experiments on different kinds of benchmark datasets for proposed mechanism.

Keywords: Transactional Tree, Restructure Tree, Organize Tree, Privacy Preserving, Frequent Patterns.

1 Introduction

Extracting rules for hierarchal layers in order to obtain multi-level information for decision making in various situations. Many business organizations need collaborative knowledge to make decisions. Several distributed and parallel algorithms were proposed with excellent scalability. An algorithm have been developed for privacy preserving in distributed environment [3], in addition to that some algorithms developed for discovering unknown patterns through distributed [1,2] and parallel [6] environment.

Privacy preserving data mining also plays key role in data mining. In Year 2000 by Agrawal et al. [3] proposed a reconstruction procedure to estimate the distribution of original data. In this paper we modified and extended the works of Tanbeer et al. [4,6] and Vadivel et al. [5] in order to derive incidental and global knowledge through Compact Pattern Trees.

Tanbeer et al. proposed CP_Tree (Compact Pattern Tree) [4] by repeating the procedure (insertion and restructure) several times in order to mine the frequent patterns by insertion phase and restructuring phase. Insertion phase, that inserts transactions into CP-tree according to current order of I-list (Item list) and updates frequency count of respective items in I-list. In restructuring phase, rearranges the I-list according to frequency-descending order of items and restructures the tree nodes according to the new I-list. Where as Vadivel et al. proposed [5] modified Compact Pattern Tree. In this approach instead

T. Li et al. (Eds.): RSKT 2012, LNAI 7414, pp. 242–247, 2012.

of repeating the procedure several times the entire database is scanned once in order to construct the tree and restructure the tree based on sorted items list.

The nature of business is distributed across the continents geographically and the transactions at different locations (zone) themselves are sufficient. There is a need to extract location-wise patterns and consolidating those patterns at global level for minimizing data transmission load. This implicitly results in privacy preserving. This demands distributed computing strategies.

In this paper we focused to mine local and global rules from abstraction of the distributed data by preserving the privacies of distributed data sources and addressed the constraints of communication cost. This paper is organized as follows. Section 2 introduces Compact Pattern Tree. In Section 3 we describe our proposed method, while Section 4 presents experimental results and its interpretations. Finally in Section 5 we draw conclusions.

2   An Overview of Compact Pattern Tree

In this Section we discuss the construction and restructure mechanism of Compact Pattern Tree. Table 1 shows the sample transactions. The approach starts with constructing tree by maintaining ilist (item appearance order list (unsorted list)). Scanning the entire transactions once and inserting into the tree by maintaining ilist as shown in Fig 1 (a) (TT-Transactional Tree). Further, ilist sorted in descending order and maintained as isort (ilist sorted in frequency descending order). Based on isort the restructuring of the tree takes place and shown in Fig 1 (b) (RT-Restructured Tree)

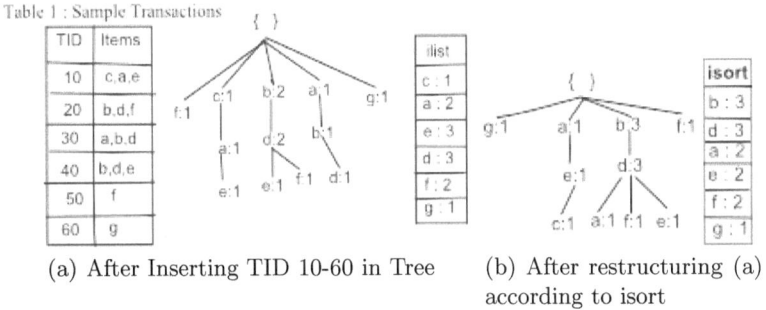

Table 1 : Sample Transactions

TID	Items
10	c,a,e
20	b,d,f
30	a,b,d
40	b,d,e
50	f
60	g

(a) After Inserting TID 10-60 in Tree   (b) After restructuring (a) according to isort

Fig. 1. Compact Pattern Tree

3   Global Rules from Distributed Compact Pattern Tree

We proposed the model to obtain Multilevel information by deriving local and global knowledge for mining frequent pattern in distributed environment.

3.1   Extraction of Local and Global Rules

Let 'n' be the number of distributed sites. $LDB_1, LDB_2, \cdots LDB_n$ are the Local DataBases. $LAR_1, LAR_2, \cdots LAR_n$ are the Local Association Rules and GAR are the Global Association Rules. GCPT is the Global Compact Pattern Tree, $LCPT_i$ is the ith node Local Compact Pattern Tree, TL_i i^{th} node local database item appearance order list. TS_i is TL_i sorted in frequency descending order, Table 2 are the sample transactions to illustrate section 3.2

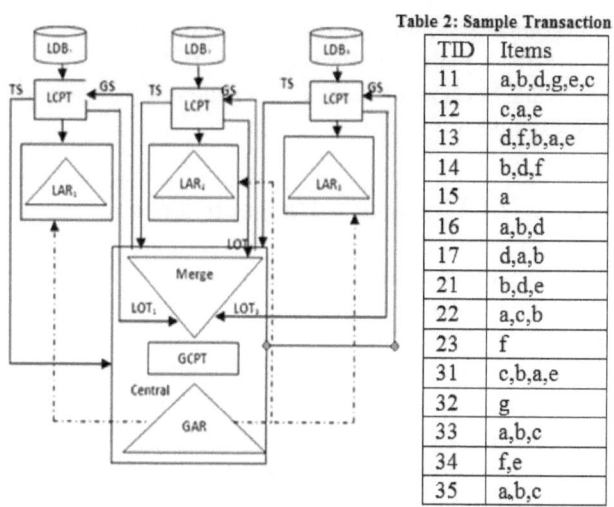

Table 2: Sample Transaction

TID	Items
11	a,b,d,g,e,c
12	c,a,e
13	d,f,b,a,e
14	b,d,f
15	a
16	a,b,d
17	d,a,b
21	b,d,e
22	a,c,b
23	f
31	c,b,a,e
32	g
33	a,b,c
34	f,e
35	a,b,c

Fig. 2. The Local and Global Association Rules Architecture

The Figure 2 depicts architecture for 'n' is 3, local node uses local database such as $LDB_{1,2,3}$ and builds $LCPT_{1,2,3}$ in respective nodes and communicates TS to Central Server. On receipt of the GS from central server local node reorganized LCPT as LOT and communicates to Central Server. Local node derives the LAR and plugs GAR from Central Server for business strategy building. The Central Server compiles TS received from different local nodes, generates GS and distributes to all the local nodes. On the receipt of LOTs from the local nodes it merges to obtain GCPT. By applying FPTree mining technique [7] algorithm to GCPT generates all frequent patterns which satisfy minsup and generates (GAR) Global Association Rules with minconf. The Central node facilitates GAR for each local node as well as global policy making system. By sending the abstraction of the distributed data to central location, we achieved privacy preserving without any information loss while generating rules.

3.2   Algorithms and Illustrations of Proposed Methodology

The proposed method contains two phases: (1) Organizing (2) Merging of Compact Pattern Trees. Assuming 'n' is three (three different local nodes). LDB_1

contains TID's from 11 to 17 and LDB_2 contains TID's from 21 to 23 and LDB_3 contains TID's from 31 to 35.

In order to compute GCPT, it involves two phases: The first phase is the distributed activity(reordering of LCPT's) at each location based on trigger raised by the central server and sending back LOT's to the central location. The second phase is to Merge LOT's.

The following are the illustrations of constructing, organizing and merging the Compact Pattern Trees. As mentioned in the Section 2 about the construction and restructuring of the tree, the same procedure is followed for construction and restructuring of TID's from 11 to 17 (LDB_2) , TID's from 21 to 23(LDB_2) and TID's from 31 to 35(LDB_3).

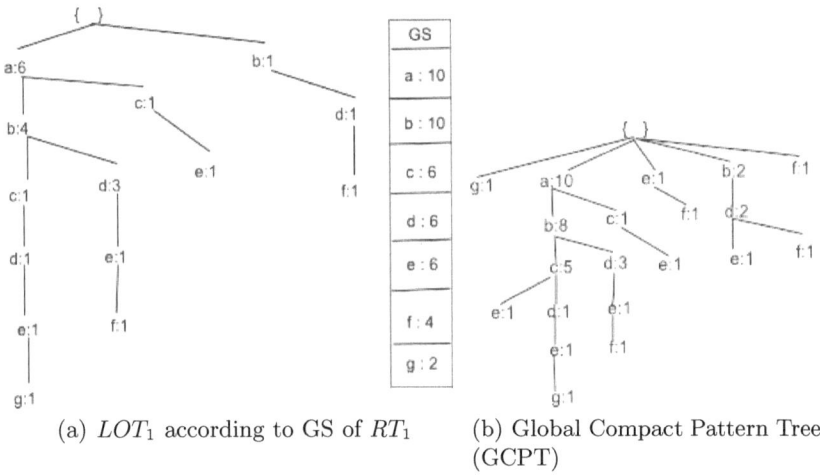

(a) LOT_1 according to GS of RT_1 (b) Global Compact Pattern Tree (GCPT)

Fig. 3. Deriving Global Knowledge from GCPT

In the following, we describe how to organized the LCPT's in order to achieve GCPT. From each LCPT we can derive incidental knowledge by applying FPTree mining technique [7]. In order to obtain GCPT the approach begins by combining $TS_{1,2,3}$ and sorted in frequency descending order we obtain GS as shown in fig 3(a). The Central Server distributes GS to three local nodes.

The first local node reorganizes the RT_1 by using GS, reorganized tree is nothing but the LOT_1 as shown in fig 3(a). The same procedure is followed for the remaining two local nodes for obtaining the $LOT_{2,3}$ and merging $LOT_{1,2,3}$ of $LDB_{1,2,3}$ the resultant GCPT obtained and shown in fig 3(b). From Fig. 3(b) we can derive global knowledge by applying FPTree mining technique [7].

TT is constructed by applying the algorithm [5]. We proposed algorithms for restructuring, organizing and merging, but only GCPT and merging were shown in following sections.

Algorithm for Global Compact Pattern Tree

```
Input:  TS_i ,  LCPT_i  where i=1 to n
Output:GCPT
Method:
```
(1)a)Combine $TS_1, TS_2 \cdots TS_n$ as GS
    b)Sort the GS in frequency descending order
(2) for i=1 to n
       LOT_i = OrganizeLCPT($LCPT_i, GS$)
(3) GCPT=MergeLOT (LOT_1, LOT_2, $LOT_3 \cdots LOT_n$)

An algorithm MergeLOT(Merging Compact Pattern Trees)

```
Input:  LOT_i  where i=1 to n
Output:GCPT
Method:
```
(1) if n=2
(2) GCPT=Merge(LOT_1, LOT_2)
(3) Else
(4) $GCPT_1 = MergeLOT(LOT_1, \ LOT_2, \ LOT_3 \ \cdots \ LOT_{[n/2]}))$
(5) $GCPT_2 = MergeLOT(LOT_{[n/2]+1}, \cdots LOT_n)$
(6) GCPT=Merge($GCPT_1$, $GCPT_2$)
Function Merge
Input: LOT_1, LOT_2
Output:GCPT
(1) GCPT = LOT_1
(2) for each branch b_i in LOT_2
(2) s = $maxmatch(GCPT, \ b_i)$
(3) if s.length == 0
(4) GCPT.add-at-root(b_i)
(5) else $b_i sub = b_i.subpath(1, s.length)$
(6) $s.frequency = s.frequency + b_i sub.frequency$
(7)s.add-at-end($b_i.subpath(s.length + 1, b_i.length)$)

4   Experimental Results

The implementation of the existing and proposed algorithms is done using Java Programming Language. The Mushroom dataset and Pima Indian Diabetes Dataset is taken from UCI Machine Learning Repositories (with 1000 transactions) is used in the experiment. The data is converted into transactional database suitable to our experiments. In merging phase we used OpenMP(multiprocessing tool) mechanism in order to achieve GCPT in less time. In Fig 4(b). Dotted lines indicates number of transactions are executed in single processor environment. Straight line indicates number of transactions are executed in number of processors in parallel environment.

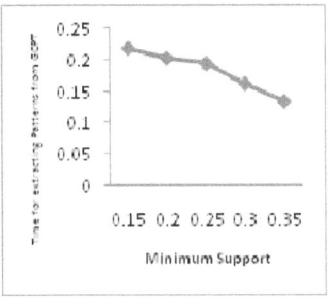

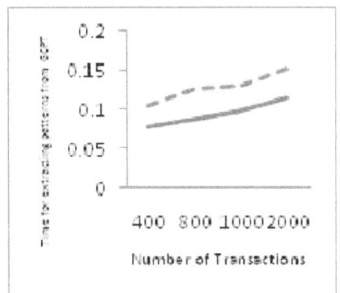

(a) Extracting patterns for various thresholds

(b) Execution Time for extracting patterns from GCPT

Fig. 4. Extracting Patterns from various thresholds in Sequential and Parallel Environment

5   Conclusion

This approach a local site posses local rules as well as global rules for making decision. Similarly, a central site has global as well as all local rules for making business policies. This paper discovered global rules from abstraction of the distributed data including the privacy preservation of local sites i.e., additionally no other local site or global site receives original data. From this paper, we attained global knowledge through multithreaded and distributed environment.

References

1. Guo, Y., Rueger, S., Sutiwaraphun, J., Forbes-Millot, J.: Meta-learning for parallel data mining. In: Proceedings of the Seventh Parallel Computing Workshop (1997)
2. Subramonian, R., Parthasarathy, S.: A framework for distributed data mining. In: Proceedings of Workshop on Distributed Data Mining, alongwith KDD 1989 (August 1998)
3. Agrawal, R., Srikant, R.: Privacy-Preserving Data Mining. In: ACM SIGMOD Int. Conf. on Management of Data, Dallas (2000)
4. Tanbeer, S.K., Ahmed, C.F., Jeong, B.S., Lee, Y.K.: Efficient single-pass frequent pattern mining using a prefix tree. Information Sciences 179, 559–583 (2008)
5. Vishnu Priya, R., Vadivel, A., Thakur, R.S.: Frequent Pattern Mining Using Modified CP-Tree for Knowledge Discovery. In: Cao, L., Feng, Y., Zhong, J. (eds.) ADMA 2010, Part I. LNCS, vol. 6440, pp. 254–261. Springer, Heidelberg (2010)
6. Syed, K.T., Chowdhury, F.A., Jeong, B.S.: Parallel and Distributed Frequent Pattern Mining in Large Databases. In: Proc. HPCC 2009, pp. 407–414. IEEE, Seoul (2009)
7. Han, J., Pei, J., Yin, Y.: Mining Frequent Patterns without Candidate Generation. In: International Conference on Management of Data (2000)

Hierarchical Path-Finding
Based on Decision Tree

Yan Li, Lan-Ming Su, and Wen-Liang Li

Machine Learning Center, Faculty of Mathematics and Computer Science,
Hebei University, Baoding 071002, China
`ly@hbu.cn, hbueducn@126.com, slmslsx@163.com`

Abstract. Path-finding is a fundamental problem in computer games, and its efficiency is mainly determined by the number of nodes it will expand. A* algorithm is unsuitable for path-finding on large map under limited computer sources and real-time demand, because the number of nodes it will expand grows fast with the size of the search space. HPA* can greatly improve the efficiency by generating abstract graph of the given map to memorize the map information before doing pathfinding. Through evenly partitioning the map as preprocessing, it can also reduce the influence of terrain factor on the output. As a result, it finds near optimal paths instead of optimal ones. And the evenly partition on the map doesnt consider the terrain distribution, which may still cause resource waste to some extent. In this paper, we present DT-HPA* (Hierarchical Path-Finding A* based on Decision Tree), a hierarchical path-finding approach on the map which has been divided by decision tree. This approach views each point on the map as an instance, and divides the map according to cut-points of continuous valued decision tree. The result of division is that the map is cut into some rectangular regions in different size, and retains the regions contain a kind of terrain. The experimental results show that, compared to HPA*, DT-HPA* can find more optimal paths with fewer detected nodes.

Keywords: Game intelligence, Hierarchical path-finding, Information entropy, Decision tree, DT-HPA*.

1 Introduction

With the development of computer games, high quality animation and sound are required, and game intelligence is also become important. Path-finding is an important component of computer game and the foundation of other game components. However, the resource allocated to path-finding is very limited, and real-time response is a very strict requirement in a computer game. Therefore, a high efficiency path-finding algorithm is necessary under those restraints.

Path-finding is searching a path form a given start state to a goal state. The searching process involves two aspects: search strategy and search space abstraction. A number of typical algorithms are A*[1, 2], IDA*, LPA*[3], etc. There are also numerous enhancements to these algorithms. One of them is considering an

T. Li et al. (Eds.): RSKT 2012, LNAI 7414, pp. 248–256, 2012.

abstraction of map as the search space, such as gird-based abstraction[4] which is the most widely used, framed quad-trees based[5, 6], waypoint based[7] and triangles based[8]. Graph abstraction has become an important way of improving path-finding efficiency [9, 10]. For large state spaces, the effectiveness of such algorithms is unable to meet games requirement. Hierarchical path-finding [11-14], a technique reduces problem complexity by abstracting a large map into linked local clusters, can find a nearly optimal path effectively.

Terrain is one of the most important factors should be considered. However, there is few such algorithms take terrain into account. In this paper, we presents DT-HPA*(Hierarchical Path-Finding A* based on Decision Tree), a hierarchical approach which divides maps according to the distribution of terrain. Our method treats nodes in search space as samples of a classification problem. Map is divided into small rectangular clusters with uniform sizes by cut-points of continuous valued attributes decision tree. Each cluster corresponds to a leaf node of the tree. The process of path-finding is similar to that of HPA*[11].

The rest of the paper is organized as follows. We begin by introducing several related path-finding methods in section 2. Section 3 shows the relationship between terrain and A* and formulates the problem of map division. Our method is presented in section 4. Empirical comparisons are given in section 5.

2   Related Work

A* search [1,2] has become a standard approach for path-finding problems. If time and space permit, it can always find the shortest path between two states. A* repeatedly examines the node with lowest cost, and the expansion is guided by the cost heuristic function $f(n)=g(n)+h(n)$, where $g(n)$ is the cost form the starting node to the current node and $h(n)$ is the cost form the current node to the goal node. Algorithms search efficiency is mainly determined by the number of nodes it detects.For large search spaces, the efficiency of A* can not satisfy games demands because too many nodes to be detected.

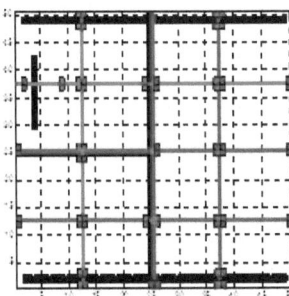

Fig. 1. Map division

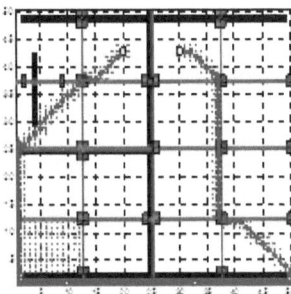

Fig. 2. HPA* path-finding

Hierarchical approach can find a nearly optimal path on large map effectively. This technique transforms a complex problem into several simple sub problems, reduces the complexity of path-finding by abstracting a large map into linked local clusters. HPA*[11] is a typical hierarchical method including two processes: abstract graph construction and online path-finding. Firstly, it divides the map into small rectangular clusters with uniform size, and then chooses entrances [11] between two adjacent clusters. Finally, it computes the distance between the nodes in the same cluster, and the abstract graph is generated. This completes the offline process (see Fig.1., the red small cubes are entrances). The online process includes inserting start and goal points and searching a path on the abstract graph, and finally refining this path to a local path as the blue line in Fig.2., where the red points are expanded nodes.

3   Problem Formulation

3.1   Influence of Terrain to A*

In this part, we will research the impact of terrain on the efficiency of A* algorithm. We make a comparison of A* path-findings on different maps. Fig.3.(a) is the search result on the map without obstacle. 3 (b) and 3(c) are the search results on one map with different start and goal points.

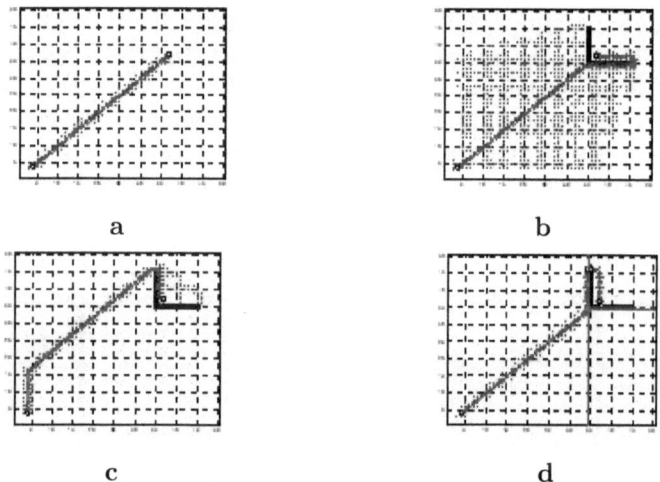

Fig. 3. Path-finding on different maps: (a) search without obstacle (b) search from left to right (c) search from right to left (d) search on divided map

Comparing nodes A* expands in Fig.3.(a) and 3(c), we found that, on the map without obstacle, the number of expanded nodes is linear to the distance between start and goal points. In this case, A* is efficiency even if the map is large. Fig.3.(b) and 3(c) shows that A* needs expand much more nodes if the terrain around the goal point is complex. Observe the lower left of Fig.2., A* need expand all node of the cluster. One approach that can reduce this influence is separating the obstacle from the traversable area. Fig.3.(d) shows the result on the map where obstacle has been divided into a cluster, where the search process is divided into two parts: searching on left map without obstacle and searching on the upper right corner map containing obstacle.

3.2   Map Division Is a Classification Problem

HPA* divides the map into smaller rectangular clusters with uniform size. It does not consider the information of terrain distribution which has a great influence on A* algorithm (see Fig.3.). In order to reduce the influence from terrain, we propose a method which can automatically separate the obstacle and the traversable area into different clusters.

Here, the separation can be viewed as a classification problem. We should use a linear classifier which can divide a map into rectangular areas. For a given grid map, the only message we can get is terrain information. Each tile can be viewed as a sample, Horizontal and vertical ordinates are condition attributes and traversable or not is the category attribute. Since the condition attributes are continuous valued, we should choose a classifier which can handle continuous value. Therefore, decision tree is used as the classifier and cut-points are used as the boundaries of clusters.

4   Hierarchical Path-Finding Based on Decision Tree

In this section, we will present our hierarchical approach on the divided map by decision tree (DT-HPA*). It differs from HPA* mainly on the map division in

```
1  data preparation
2  division(map)
3  get map size S
4  if S < threshold then
5  return
6  end if
7  get maps entropy E
8  if E is zero then
9  return
10 end if
11 measure all cut-points, get[axis, value]
12 if axis is x then        %optimal cut-point is on x axis
13 division(mapleft)
14 division(mapright)
15 else                     %optimal cut-point is on y axis
16 division(mapup)
17 division(mapbelow)
18 end if
```

Fig.4. Division algorithm

abstract graph construction, i.e., DT-HPA* divides the map based on the terrain information.

Just as we can see from Fig.4. above, we introduce map division in detail, which is a recursive process of building decision tree using map data.

Practical issues in map division include handing continuous attributes, stopping criterion and choosing an appropriate attribute selection measure. We discuss each of these issues as follows.

4.1   Handing Continuous-Valued Attributes

For a given continuous-valued attribute, examples are sorted by the values on this attribute and divided by a cut-point. Cut-points selection is a key step of constructing continuous-valued attributes decision tree. In map data set, the attributes of each node are all integer, the midpoint between two adjacent integers is evaluated as a potential cut-point. Information gain [15] is used to measure those cut-points and the one with highest information gain will be selected.

$$\text{Gain}(T, \text{map})$$

$$= ent(map) - |\tfrac{map_1}{map}|ent(mp_1) - |\tfrac{map_2}{map}|ent(mp_2) \tag{1}$$

$$ent(map) = -p_+logp_+ - p_-logp_- \tag{2}$$

Where $|\cdot|$ denotes the size of a data set, ent(map) is the entropy of map, cut-point T divides map into map_1 and map_2 , p_+ and p_- are the proportions of two kinds of terrains.

One special feature of map data is that there are always several samples with each attribute value. So the cut-point selection method proposed by Fayyad [18] is unfit for this situation. We need evaluate each candidate cut-point and select the optimal one. If the size of the map is m×n, (m-1)×(n-1) evaluations will take place.

4.2   Stopping Criterion

Each cluster corresponds to a leaf node of the tree. For hierarchical path-finding, the search space of abstract graph is not large, and so does the number of clusters. Therefore the size of the tree should be controlled. Two stopping criteria are given as follows: (1) The entropy of the map is zero; (2) The size of the map is less than a given threshold.

When maps entropy is zero, one cluster only contains a kind of terrain. If the terrain is traversable, the efficiency of path-finding is very high. If the terrain is obstacle, there is no way in this area and further division is unnecessary. When the map only contains a few nodes even if the maps terrain is complex, the efficiency of path-finding will not be very poor. A threshold can be used to control the number of clusters, which determines the number of clusters.

4.3   Attribute Selection Measure

The heuristic function plays an essential role in cut-point selection. Many heuristic functions have been proposed in the literature. Information gain is a measure based on information entropy which is used to measure the impurity of a data set. Beside information entropy, one alternative impurity measure that has been used successfully is Gini index proposed by Breiman in CART [16]:

$$Gini(S) = 1 - \sum_{i=1}^{k} p^2(C_i|S) \tag{3}$$

There is a sample set S and several classes C_i (i = 1,...,k). p(C_i|S) is the probability of samples in S falling in Ci. If all data in S belong to the same class, Gini index is zero. If p(C_i|S) to each class is equal, Gini index reaches its maximum 1-1/k. Both (2) and (3) are used to build decision tree, the same result is obtained.

4.4   Online Path-Finding

After using decision tree to partition a given map, the on-line path-finding will be done on the clustered map. This includes three steps:

1. Inserting start and goal points into abstract graph.
2. Finding a path on abstract graph.
3. Finding the real path according abstract path in each cluster.

Thus completes the whole process of path-finding using DT-HPA*. Fig.5. is the path-finding result on the map divided by decision tree. The start and goal point are same to those in Fig.2., and the red points are entrances; blue line is the found path. It is showed that the number of expanded nodes is reduced obviously.

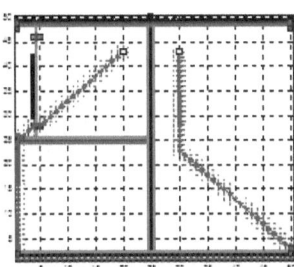

Fig.5. DT-HPA* path-finding

5   Empirical Study

We compare our method with HPA* on random maps and use Matlab7.0 as the simulation tool. In the experiments, we define the cost of going through diagonal squares is 1.4, and the cost of going to non-diagonal square is 1.

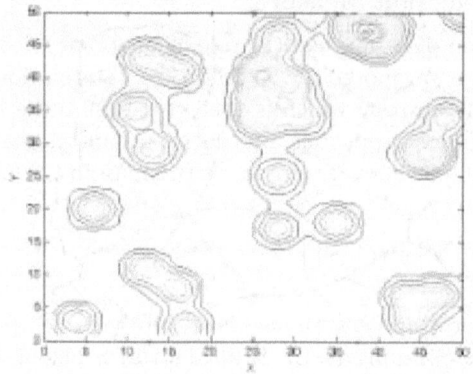

Fig.6. A random map

Fig.6. is a random 50×50 map. The start point is at left lower and the goal point is at upper right. They have a comparatively long distance, which may be better to show the effectiveness of DT-HPA*. Fig.7. shows the results.

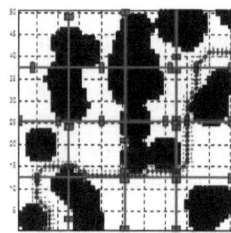 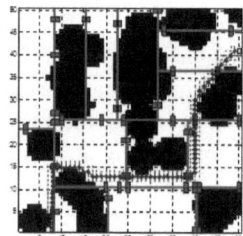

Fig.7. Results of HPA*(left) and DT-HPA*(right)

There are two performance criteria: the length of the path which affects the path quality, and the number of expanded nodes which determines the efficiency. Detailed comparison results are shown in Table 1, where clusters is the number of clusters obtained by the two algorithms; abstract graph is the size of abstract graph; num expanded is the number of expanded nodes.

Table 1. A detailied comparison of the two mehtods

algorithm	cluster	abstract graph	num. expanded	path length
HPA	16	66	223	84
DT-HPA	14	70	212	80.4

The results on random map indicate our method can find shorter path with fewer expanded nodes compared with HPA*. When the path is longer, the advantage of DT-HPA* will be more obvious.

6 Conclusions and Future Work

We have presented a hierarchical path-finding method based on decision tree division DT-HPA*, which can be regarded as an improvement of HPA*. It differs form other hierarchical methods in that it takes account of terrain distribution. The clustering method uses cut-points of decision tree as the border lines. The number of abstract clusters can be controlled by a threshold also. The results on random game map are obviously superior to those of HPA*.

There are still some possible extensions to DT-HPA*. Entrances selection can be improved by identifying more representative nodes. Another issue is to improve the speed of map division through checking fewer cut-points.

Acknowledgment. This paper is supported by the national natural science foundation of China (60903088, 61170040), 100-Talent Programme of Hebei Province(CPRC002).

References

1. Hart, P.E., Nilsson, N.J., Raphael, B.: A Formal Basis for the Heuristic Determination of Minimum Cost Paths. IEEE Transactions on Systems Science and Cybernetics, 100–107 (1968)
2. Patrick, L.: A* Path-finding for Beginners,
http://www.policyalmanac.org/games/aStarTutorial.html
(updated July 18, 2005)
3. Koening, S., Likhachev, M., Furcy, D.: Lifelong planning A*. Artificial Intelligence Journal 155(1-2), 93–146 (2004)
4. Kenny, D., Nash, A., Koenig, S.: Theta*: Any-Angle Path Planning on Grids. Journal of Artificial Intelligence Research 39, 533–579 (2010)
5. Samet, H.: An overview of Quad trees, Octrees, and Related hierarchical data structures. NATO ASI Series, vol. 40, pp. 51–68 (1988)
6. Yahja, A., Stentz, A., Singh, S., et al.: Framed-Quad tree path planning for mobile robots operating in sparse environments. In: Proceedings of IEEE Conference on Robotics and Automation (ICRA), Leuven, Belgium (May 1998)
7. Choset, H., Lynch, K.M., Hutchinson, S., et al.: Principles of Robot Motion. MIT Press (2004)
8. Demyen, D., Buro, M.: Efficient triangulation-based path-finding. In: Proceedings of AAAI (2006)
9. Sturtevant, N., Buro, M.: Partial pathfinding using map abstraction and refinement. In: Proceedings of AAAI, pp. 1392–1397 (2005)
10. Bulitko, V., Sturtevant, N.: Graph abstraction in real-time heuristic search. Journal of Artificial Intelligence Research 30, 51–100 (2007)
11. Botea, A., Muller, M., Schaeffer, J.: Near optimal hierarchical path-finding. Journal of Game Development 1, 7–28 (2004)

12. Rabin, S.: A* Aesthetic Optimizations. Game Programming Gems, 264–271 (2000)
13. Jansen, M.R., Buro, M.: HPA* Enhancements. In: Proceedings of the Third Artificial Intelligence and Interactive Digital Entertainment Conference, Stanford, California, USA, pp. 84–87 (2007)
14. Harabor, D., Botea, A.: Hierarchical path planning for multi-size agents in heterogeneous environments. In: IEEE Symposium on Computational Intelligence and Games, pp. 258–265 (2008)
15. Quinlan, J.R.: Improved use of continuous attributes in C4.5. Journal of Artificial Intelligence Research 4, 77–90 (1996)
16. Breiman, L., Friedman, J.H., Olshen, R.A., et al.: Classification and Regression Tree. Wadsworth International Group, Monterey (1984)

Human Activity Recognition with Trajectory Data in Multi-floor Indoor Environment

Xu Zhang[1], Goung-Bae Kim[2], Ying Xia[3], and Hae-Young Bae[1]

[1] Department of Computer Science, Inha University, South Korea
zhangxu.jn@gmail.com, hybae@inha.ac.kr
[2] Department of Computer Education, Seowon University, South Korea
gbkim@seowon.ac.kr
[3] College of Computer Science and Technology,
Chongqing University of Posts and Telecommunications, China
xiaying@cqupt.edu.cn

Abstract. In pervasive and context-awareness computing, transferring user movement to activity knowledge in indoor is an important yet challenging task, especially in multi-floor environments. In this paper, we propose a new semantic model describing trajectories in multi-floor environment, and then N-gram model is implemented for transferring trajectory to human activity knowledge. Our method successfully alleviates the common problem of indoor movement representation and activity recognition accuracy affected by wireless signal calibration. Experimental implementation and analysis on both real and synthetic dataset exhibit that our proposed method can effectively process with indoor movement, and it renders good performance in accuracy and robustness for activity recognition with less calibration effort.

Keywords: Semantic Trajectory Model, Indoor Moving Object, Activity Recognition, Knowledge Discovery.

1 Introduction

With the rapid progress of wireless communication, sensor technology and wide spread use of smart phone, it is much easier to acquire human location of daily activity. However, recorded sample locations are difficult to be understood and analyzed as a result of ignoring semantic information. Furthermore, it is more difficult to extract implicit patterns from these ambiguous raw data.

Trajectory-based activity recognition builds upon some fundamental functions of location estimation and machine learning, and can provide new insights on how to infer high-level goals and objectives from low-level sensor readings [1]. Various types of probabilistic models have been proposed on human activity discovery based on the inference of hidden information in movement in literature [2-4]. However, there are two important problems in the existing work. First, it is impossible to receive GPS signal in buildings. Second, estimation from raw sensor data suffers from noise and need great effort for calibration.

T. Li et al. (Eds.): RSKT 2012, LNAI 7414, pp. 257–266, 2012.

Our research aims to understand human activity based on trajectory without intervention. To alleviate the existing problems, we share a common methodology with general trajectory analysis and knowledge discovery process [1], and propose our method: (I) Raw information collection and management, which management information from variety sensor device and process data clean on them. (II) Inference of hidden information, which enrich trajectory with semantic information according to indoor environment and other context information. (III) Activity recognition, which is a process to extract high-level activity and goal related information from low-level sensor readings through machine learning and data mining techniques.

The remainder of the paper is organized as follows. The next section offers an insight into related work. Section 3 states the problem definition and challenges. Semantic indoor trajectory model is first introduced and we describe the activity recognition method performed on semantic indoor trajectories in Section 4. Finally we show our experiments and analysis in Section 5 and draw a conclusion in Section 6.

2   Related Work

We review previous work related to our trajectory-based activity recognition in two sections: (I) Semantic trajectory model for movement representation [5-8,13-15]. (II) Probabilistic models used for activity recognition [2,3,9-11,16].

A formal conceptual view on trajectories has been given by [5], which decomposes trajectory into a sequence of stops and moves. The proposed model is defined for enriching trajectory semantics based on the observation of geographic knowledge(e.g. point of interests, POI) and velocity change of moving objects. A hybrid trajectory model and computing platform is presented in [6] for developing a semantic overlay-analyzing and transforming raw mobility data (GPS) to meaningful semantic abstractions, starting from raw feeds to semantic trajectories. Recently, indoor data [7,8] management and knowledge discovery has emerged as a hot topic. However, semantic indoor trajectory is new still less of clear definition and common understanding.

In recent years, many attempts have been made to integrate high-level behavior inference with low-level sensor modeling. Statistical models like Hidden Markov Models(HMMs) [9], Dynamic Bayesian Networks(DBNs) [2,3,10] and Conditional Random Fields(CRF) [11] are mainly used to bridge the gap between low-level sensor data and high-level activities.

3   Problem and Challenges

There is a strong emphasis on developing techniques for higher level, semantic events inferred at varying semantic abstractions from raw location feeds. Solutions used today mostly require human intervention for such semantic (and contextual) abstractions of trajectory data [6]. Indoor is a new frontier area involved certain characteristics, which is different from outdoor trajectory analysis method. We give the obvious challenges in modeling indoor trajectory and performing activity recognition.

(1) General trajectory conceptual model considers raw position as a serial of location with timestamp, e.g. (x, y, t). However, indoor environment generally involved with multi-floors, which need a new model to describe. As it is shown in Fig. 1, people in different floors can be located at place A, B, C, which are recognized with same latitude/longitude ignoring floor information according to previous research. This will definitely provide an ambiguous understanding of human location and activity.

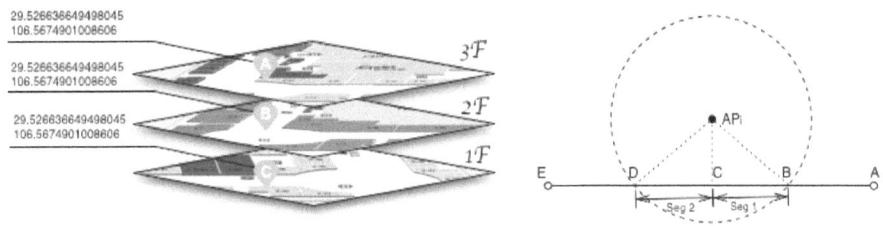

Fig. 1. Location problem and received signal strength problem

(2) Despite the power and flexibility of the probabilistic model in providing a coherent modeling framework in activity recognition, the complexity of the model is exponential in the number of hidden states [2].

(3) Probabilistic models are sensitive to raw sensor data calibration, which needs great effort. SAR method [3] proposed to partition user trace into several signal segments with the consideration that each segment consists of signals that exhibit consistent characteristics in the signal space. This is easy to be understood from Fig. 1, the signal-strength values within a longer period of time will be accumulated into one observation corresponding to one time slice. When moving from location A to E, signal strength at B and D should be same and strength achieves the maximum at location C under ideal condition. According to SAR, B-C and C-D should be two trace segments. However, people walking from A to E maybe in a constant status, besides, it is not necessary to look into the detail. This problem becomes serious when signal collection interval increases, which is shown in experiments.

With the proposed challenges above, we first give an extended definition of trajectory conceptual model for multi-floor indoor environment. Then, N-gram method is used for activity recognition towards a series of actions acquired from semantic indoor trajectory model.

4   Human Activity Recognition from Trajectory Data

Recording continuous movement is the foundation of managing and understanding movement. A number of data preprocessing steps need to be considered before activity recognition, which can render data easy to handle and ready to reveal profound movement patterns.

In this section, we describe our method in two phases: (I) Low-level sensor data are processed with semantic indoor trajectory model (II) Episode trajectory data are analyzed with N-gram to build activity models and perform activity recognition.

4.1   Semantic Indoor Trajectory Model

People usually tend to visit a place to do certain activity which is closely linked to the place [8]. However, discrete location calculated from received signal strength is not accurate due to the signal variant problem. The following definitions explain our method from sensor signal data to semantic indoor trajectory data. We are motivated from [6,14] and segment indoor trajectory into *episodes* with semantic annotations, and finally transfer raw sensor data to action episodes in trajectory.

Definition 1. (Raw Sensor Data) A sequence of AP information and corresponding received signal strength(RSS) collected from mobile devices at each location.

With consideration of the characteristic of multi-floor indoor environment, we propose to locate human with RSS-based method and match RSS-location to indoor map database with floor information.

Definition 2. (Raw Location and Trajectory) A sequence of points recording the trace of a moving object with spatial and temporal information, i.e. $T = L_1, , L_m$, where $L_i = (x, y, f, t)$ is a tuple representing current location at timestamp t.

(x, y) is used for location and f indicate the height(floor) of current location. In multi-floor indoor environment, we are more interested with floor and shop/area information that can identify our location in the building. Indoor-location[8] is used to describe which floor and where exactly the person is, while indoor-area is a clue to exhibit human activity. Lack of indoor-location and indoor-area information prevents the accurate inference of hidden information while most people usually perform a certain activity in the building. According definition 2, human movement in different floors with same longitude and latitude could be considered as two different trajectories in a building. We have solved the first challenge proposed in Section 3.

Generally, people in indoor do not continuously move during a trajectory lifespan. Raw trajectory consists of all discrete location points gathered from mobile devices, which covers entire spatial, temporal features of a moving object. However, semantic and other hidden information is not obvious or even not covered. We proposed to enrich trajectory with semantic places in definition 3.

Definition 3. (Semantic Places) A set of meaningful indoor areas/objects used for annotating trajectory data. Each semantic place SP_i has extent attributes describing the place, which can be partitioned into three categories according to the geometric shape:

SP_{point}, where the semantic place is point location, e.g. elevator, escalator, stairs

SP_{region}, where the semantic place is region location, e.g. shop, rest area

SP_{line}, where the semantic place is line location, e.g. walking-path

All human movements are considered to be in semantic places, and the status will change between each pair of semantic places. We consider three more statuses in indoor environment beside *move* and *stop* in existing conceptual trajectory model [5]: *enter*, *exit* and *inactive*. These status are important for structure semantic trajectory episodes.

Existing research [2] proposed to use a sensor model to estimate the locations and then perform an activity recognition method. In activity recognition, continuous movement or action is more important however localization techniques focus more on discrete position of an object. We propose to segment trajectory to identify specific semantically meaningful part for domain knowledge. An *episode* [6] is a maximal subsequence of a trajectory such that all its spatio-temporal positions comply with a given predicate that bears on the spatio-temporal positions and/or their annotations. We consider that human perform a stable action during a trajectory episode, which corresponding to *move, stop, enter, exit, inactive* with a timestamp or duration.

Definition 4. (Semantic Trajectory Episodes) A representation of a semantic trajectory as a sequence of episodes ep_1, ep_2, ... , ep_m, such that each episode corresponds to a subsequence of the original trajectory and is represented as a tuple $ep_i = (sp, TS\_enter, TS\_exit, status)$ where sp is a semantic place ($sp \in SP$), TS_enter, TS_exit are the time the moving object enters and exits sp, and status is *move, stop, enter, exit, inactive*.

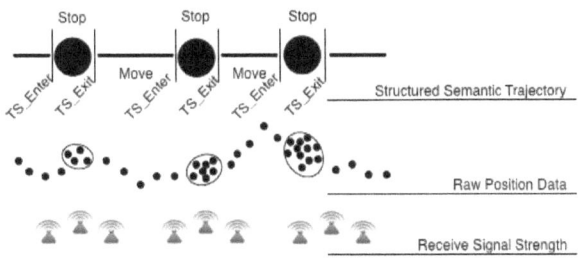

Fig. 2. Structured semantic trajectory

A structured semantic trajectory in Fig. 2 is a sequence record of position evolution of an object moving in space during a given time interval in order to achieve a designed goal. Semantic trajectory episode is composed of one or several segments and the object status during each *episode* do not change. Human activities in a building are more or less rough-grained in that the precise location information is not needed. With the definition 3 and 4, we can segment trajectory into three categories according to the semantic places: Point of Interests(POI), Region of Interests(ROI) and Line of Interests(LOI).

Then a structured semantic trajectory is segmented into a list of *episodes* of POIs, ROIs and LOIs to infer indoor-area of human movement. In some situation, we are suddenly run out of RSS area, which could be marked as *inactive*. This is ignored in previous work, and will definitely affect recognition accuracy. According to our definition, we can model this situation in one *episode* according previous and future location. Then we propose to perform activity recognition on semantic trajectory episodes.

4.2   Activity Recognition Model

In this section, we focus on activity recognition based on semantic indoor trajectory model. Our work is motivated from previous work LAR [2] and SAR [3]. As it is shown in Fig. 3, our proposed recognition method has two levels: a low-level semantic indoor trajectory model and a high-level N-gram model. Action episodes E_1, E_2, ... , E_{n-1} from semantic indoor trajectory model are considered as input and E_n is the predicted output of activities. E_n is only dependent on E_{n-1}, ... , E_1.

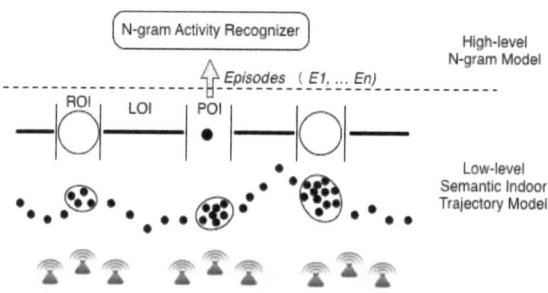

Fig. 3. Activity recognition model

According to our semantic indoor trajectory model, each *episode* can serve as atomic units of a users movement to represent a human movement in ROI, POI and LOI as it is shown in Fig. 3 . Then we can infer the most likely goal from episode actions as follows according to N-gram model definition [12], and perform calculation with Bayes's Rule:

$$G^* = argmax P\left(G_m | E_1, ..., E_n\right) \qquad (1)$$

$$= argmaxP\left(G_m|E_{1:n}\right) \tag{2}$$

$$= argmax\frac{P\left(E_{1:n}\right)P\left(G_m\right)}{P\left(E_{1:n}\right)} \tag{3}$$

According to our definitions, transitions and time durations between *episode* actions are independent. The computational complexity is linear in the number of goals and in the length of an *episode* action sequence. Then we can further use Bigram model which provides the conditional probability of an *episode* given the preceding *episode*:

$$G^* = argmaxP\left(G_m\right)P\left(E_1|G_m\right)\cdot\prod_{i=2}^{k}P\left(E_i|E_{i-1},G_m\right)\prod_{i=1}^{k}P\left(D_i|E_i\right) \tag{4}$$

With the consideration of independent *episode* action durations, $P(Di|Ei)$ can be computed with nonparametric or parametric representations of time distribution. The duration time D_i is easy to be understood according to our semantic indoor trajectory in which it preserve a constant status *move/stop*, and D_i is easy to be calculted by $(TS\_exit\text{-}TS\_enter)$. The activity goal can be recognized with a sequence of *episode* actions according to equation (4). The advantage of our proposed two-layer recognition model is that it can alleviate the performance decrease caused by inaccuracy indoor locations and RSS signal noise. It is obvious that all the inputs to N-gram are continuous movement sections, in which location is not needed to be precise. Indoor-area like POI, ROI and LOI is important for accuracy understanding of human activities.

5   Experiments and Evaluation

We set up our experimental test-bed in Wanda Plaza of Chongqing. We have 10 users equipped with Android phone to collect Wifi signals and trace human trajectories. As it is shown in Fig. 4, we have our test floor divided into 1.2m*1.2m grid and AP locations represented as triangle. Then the collected RSS is used to estimate current user location and further model them to our semantic indoor trajectory. To evaluate our method, we follow the criteria proposed by [3]: Accuracy and Robustness. We compared our method with SAR and LAR with respect to the observed data in different test environment.

Real Data: 50 trajectories collected with different RSS intervals varying from 1s to 7s, in which 40 for training and 10 for testing.

As it is shown in Fig. 5, SAR and LAR accuracy sharply decrease when time interval increase. This is easy to understand that location estimated in LAR describes human movement discontinuous, which weakens the discriminative power of signals towards different locations, and in turn reduces the recognition accuracy. SAR also suffers from AP signal strength problem shown in Fig. 1. Our proposed method exhibit a stable performance because semantic indoor trajectory guarantees that human movement is continuous. So, episode actions do

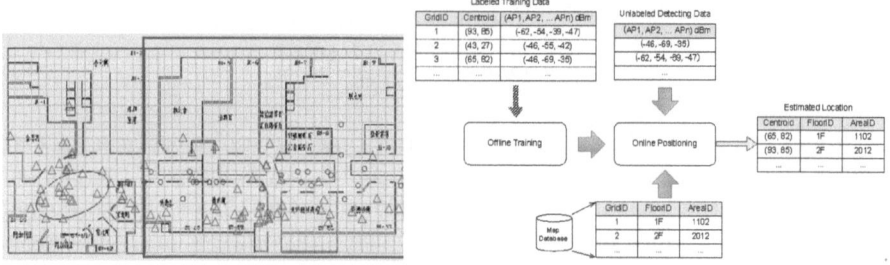

Fig. 4. Test-bed environment : floor/AP information and location estimation

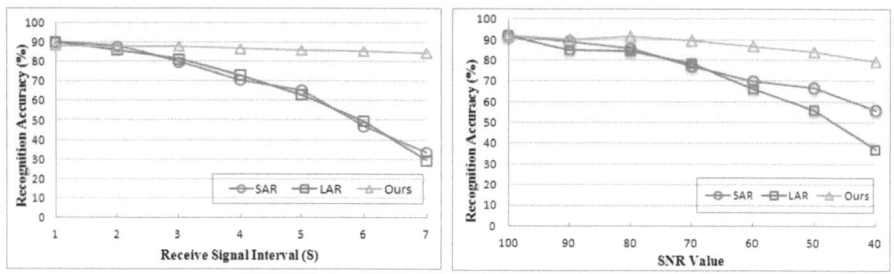

Fig. 5. Recognition evaluation I : affections from RSS time interval and noise

not change too much with different test time interval, which output a stable accuracy.

Synthetic Data: simulated signal strength variations by adding different levels of Gaussian noise to received signal sequences, totally 50 trajectories collected with time interval 2 second, in which 40 for training and others for testing.

In this test, we follow the definition of SNR [3], which is the ratio of variance of signal to variance of noise. We can see from Fig. 5 that when the noise level increases(SNR value decrease), all of the three method show a decrease in performance, however, our proposed method suffers little what proves its robustness. Estimated location from RSS data is greatly sensitive to noise. However, we are interested with continuous movement and sequence actions from semantic view. Noise in RSS data is not seriously affecting our episode actions as input, and sure it cannot harm our recognition accuracy.

Real Data: We have 60 trajectories collected by separated hour period per day, while another 70 trajectories separated by day in a week.

It is obvious that all these three method decrease in accuracy at 18:00-20:00, 20:00-22:00 and weekends in Fig. 6. Specially, SAR and LAR sharply decreased. That is reasonable with the consideration that during these times, there are much more people in the plaza. Crowded people will significantly affect received signal strength indoor, which will further influence recognition accuracy. However, RSS problem is alleviated in our semantic indoor trajectory model as a benefit of our input parameters-semantic indoor trajectory episode actions.

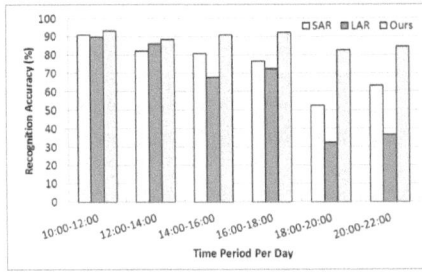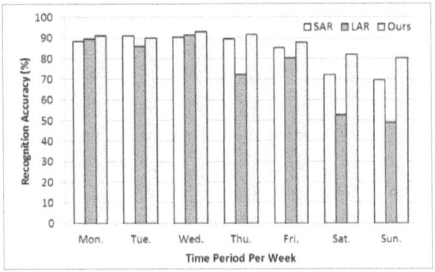

Fig. 6. Recognition evaluation II : affections from complex environment

6 Conclusions

In this paper, we research on trajectory-based activity recognition that processes the sequences of sensor readings and context, and output the prediction of actions and goals of human movement in multi-floor indoor environment. We proposed a new semantic indoor trajectory model to describe indoor movement. With the involved environment information, the proposed model alleviates the accuracy problem caused by signal strength. Finally, N-gram method is applied to trajectory episodes for human activity recognition. Our experimental results on real and synthetic dataset demonstrate that our proposed method is both accurate and robust for activity recognition with less calibration effort. The future research should pay more attention to consider the user-user relationship in multi-user activity recognition.

Acknowledgments. This research is supported by a grant(07KLSGC05) from Cuttingedge Urban Development - Korean Land Spatialization Research Project funded by Ministry of Construction & Transportation of Korean Government.

References

1. Zheng, Y., Zhou, X.F. (eds.): Computing with Spatial Trajectory. Springer (2011)
2. Yin, J., Chai, X.Y., Yang, Q.: High-level Goal Recognition in a Wireless Lan. In: Proceedings of the 19th National Conference on Artificial Intelligence (2004)
3. Yin, J., Yang, Q., Shen, D., Li, Z.N.: Activity Recognition via User-Trace Segmentation. ACM Transactions on Sensor Networks (TOSN) 4(4) (2008)
4. Zeng, Z., Ji, Q.: Knowledge Based Activity Recognition with Dynamic Bayesian Network. In: Daniilidis, K., Maragos, P., Paragios, N. (eds.) ECCV 2010, Part VI. LNCS, vol. 6316, pp. 532–546. Springer, Heidelberg (2010)
5. Spaccapietra, S., Parent, C., Damiani, M.L., Macedo, J.A., Porto, F., Vangenot, C.: A Conceptual View on Trajectories. Journal of Data & Knowledge Engineering (65), 126–146 (2008)
6. Yan, Z.X., Parent, C., Spaccapietra, S., Chakraborty, D.: A Hybrid Model and Computing Platform for Spatio-semantic Trajectories, pp. 65–70. Springer (2010)

7. Jensen, C.S., Lu, H., Yang, B.: Indoor-A New Data Management Frontier. In: Mokbel, M. (ed.) Special Issue on New Frontiers in Spatial and Spatio-temporal Database Systems, IEEE Data Engineering Bulletin, vol. 33(2), pp. 12–17 (2010)
8. Noh, H.Y., Lee, J.H., Oh, S.W., Hwang, K.S., Cho, S.B.: Exploiting Indoor Location and Mobile Information for Context-Awareness Service. Information Processing and Management (2011)
9. Bui, H., Venkatesh, S., West, G.: Policy recognition in the abstract hidden Markov model. J. Art. Intel. Res. 17, 451–499 (2002)
10. Liao, L., Fox, D., Kautz, H.: Learning and inferring transportation routines. In: Proceedings of the 19th National Conference in Artificial Intelligence (AAAI), San Jose, CA, pp. 348–353 (2004)
11. Liao, L., Fox, D., Kautz, H.: Extracting Places and Activities from GPS Traces Using Hierarchical Conditional Random Fields. International Journal of Robotics Research 26(1) (2007)
12. Chen, S.F., Goodman, J.: An Empirical Study of Smoothing Techniques for Language Modeling. In: Proceedings of the Thirty-Fourth Annual Meeting of the Association for Computational Linguistics, pp. 310–318 (1996)
13. Alvares, L.O., Bogorny, V., Kuijpers, B., Macedo, J., Meolans, B., Vaisman, A.: A Model for Enriching Trajectories with Semantic Geographical Information. In: GIS (2007)
14. Yan, Z., Chakraborty, D., Parent, C., Spaccapietra, S., Karl, A.: SeMiTri: A Framework for Semantic Annotation of Heterogeneous Trajectories, In: EDBT (2011)
15. Yan, Z., Spremic, L., Chakraborty, D., Parent, C., Spaccapietra, S., Karl, A.: Automatic Construction and Multi-level Visualization of Semantic Trajectories. In: GIS (2010)
16. van Kasteren, T.L.M., Englebienne, G., Kröse, B.J.A.: Transferring Knowledge of Activity Recognition across Sensor Networks. In: Floréen, P., Krüger, A., Spasojevic, M. (eds.) Pervasive Computing. LNCS, vol. 6030, pp. 283–300. Springer, Heidelberg (2010)

Remote Sensing Image Data Storage and Search Method Based on Pyramid Model in Cloud

Ying Xia and Xuanlun Yang

School of Computer Science and Technology,
Chongqing University of Posts and Telecommunications, Chongqing, 400065, China
`xiaying@cqupt.edu.cn`, `yangxuanlun@163.com`

Abstract. With the development of satellite remote sensing technology, the volume of remote sensing image data grows exponentially, while the processing capability of common computer system is hard to satisfy the requirements of remote sensing image data accessing. In this paper, we propose a storage and search method Based on MapReduce mechanism in cloud computing environment, called BMR. It is a distributed and parallel storage method which combined with Pyramid Model and MapReduce Thinking. It recodes the tiles of each remote sensing image and defines the storage rule to ensure the tiles can be stored and searched in parallel. Experiments show that BMR method achieves good I/O performance.

Keywords: Cloud Computing, MapReduce, Pyramid Model, Image Data.

1 Introduction

With the development of satellite remote sensing technology, remote sensing image data are easily collected and play an irreplaceable role in many fields such as navigation, land planning, environment management, etc. However, the size of remote sensing image data is growing exponentially, for example, in the 15th level of global satellite map, data size has reached 3 TB, even more than 63 PB in the 22nd level. Due to the limitation of the processing capability of a single node, traditional centralized image data management system is hard to support the efficient storage and accessing of remote sensing image data through Internet [1]. In the real applications, massive remote sensing images are stored but each query is only focus on a small part [2]. The research of efficient remote sensing image data storage and search is necessary.

In recent years, the development of cloud computing technology provide a new way for big data processing. Cloud computing provides a scalable platform by dynamic scheduling the computing and storage resources [3]. Thus it is suitable for efficient storage and query processing of remote sensing images.

The rest of the paper is organized as follows. In Section 2, basis concepts and related works are introduced. In Section 3, the method of image data storage and search method based on MapReduce is discussed. Section 4 presents the experimental results. Finally, in section 5, we conclude this paper.

T. Li et al. (Eds.): RSKT 2012, LNAI 7414, pp. 267–275, 2012.

2   Background

2.1   Multiresolution Pyramid and Image Block Technology

General speaking, the size of a satellite image is more than 300MB, and even may exceed 10GB or 100GB after mosaic. For efficient image accessing such as browsing and enhancement, data storage and search methods are crucial. The multi-resolution pyramid model based on the image block technology is an effective method for remote sensing image data organization [4], see in Fig. 2. The original image is tiered to some layers according to different resolutions first, and then the each layer is stratified to blocks, called tiles. During image browsing, corresponding tiles are accessed according to user interest [5].

The image pyramid model puts the original image data as the bottom of pyramid, and establishes a series of image layer with the same scope but different resolutions by resampling. Among them, the resolution of the bottom layer is highest, and it is reduced gradually for the upper layers [6,7], the image pyramid is shown as Fig. 1.

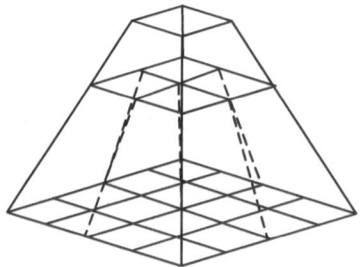

Fig. 1. Pyramid build

At present, most of the storage methods for tiles are layer-based. For example, SuperMap creates corresponding folder for each layer firstly, each folder consists of some sub-folders, each sub-folder used to store the tiles in each column of the sane row. During the access process, given the layer, row and column number, the target tile could be searched.

2.2   File Organization and Programming Model in Cloud

Hadoop is a distributed system widely used in cloud computing environments. HDFS (Hadoop Distributed File System) is one of the file systems realized by Hadoop, it is generally used for massive data storage. HDFS stores data in various DataNodes with the unit of 64MB (default) and record corresponding block information in the NameNode. As a result, if there are too many small files (smaller than 64MB) need to be stored, the write performance of Hadoop will be reduced greatly and large storage space of NameNode will be took up [8].

MapReduce, as a computing and programming model emerging in recent years, is suitable for handling large data sets [8,9]. Large-scale data analysis task of a MapReduce application can be completed in a distributed and parallel computing environment.

Aiming to improve the accessing performance of massive image data, a distributed and parallel storage solution based on MapReduce mechanism in cloud computing environment, called BMR Stroage, is proposed.

2.3   Related Work of Massive Image Data

Aiming to solve the storage and accessing problems of image data, many experts have conducted the research from different viewpoints. Lindstrom presents the out-of-core technology to simplify terrain [10]. Xincai Wu studies the physical storage methods of raster data and organized massive raster data by directory structure [11]. Zhongmin Li uses 2-D tile pair of base vectors to improve the performance of terrain roaming system [12]. The above methods run in common computer system and mainly use index or caching technology to improve the performance of image data processing.

Xuhui Liu merges small tile data into big ones and thus improves the I/O performance in Hadoop by reducing the disk's seeking time [14]. Shuming Huo improves the speed of pyramid combination based on MapReduce thinking, the image data are wrote into HDFS frist, then the MapReduce task is executed and data are stored into HBase (Hadoop Database) [5]. The above two methods use the cloud computing technology, which is similar as the idea in this paper.

3   The Image Data Storage and Search Method Based on MapReduce

3.1   The MapReduce Programming Paradigm

MapReduce is a programming model which consists of Map and Reduce functions. Map function is used to produces a set of intermediate key/value pairs, and Reduce function is performed to merge all the values with the same key.

In this paper, Key refers to the number of tiles of read or write, value refers to the read or write status of tiles. Value is assigned to 1 if read or write is success, otherwise value is 0. In the Reduce function, the sum of all values is calculated first, if the result is same as the request number, the read or write is success, otherwise is failed.

3.2   Tiles Encoding Method

In general, a tile of the Pyramid Model can be uniquely identified as $(level, tx, ty)$ [13]. Here $level$ is the layer number, tx is the column number, and ty is the row number. In this paper, we studied the corresponding relation between the layers, and recode the tiles. It is well known that, the definition of tile size would make a great impact to the I/O performance. In this paper, the size of tile is set to 256×256 by experience. Assuming there are row rows and col columns after blocking, and then there are $(row * col)$ tiles. So tiles are named as 0, 1, 2, $\cdots$ $(row * col)$ from top to bottom and from left to right. Then the number Num of tile $(level, tx, ty)$ will be calculated as $Num = (ty * col + tx)$. Fig. 2 is one image data after stratifying and blocking, tile numbers are shown as Fig. 3 after coding.

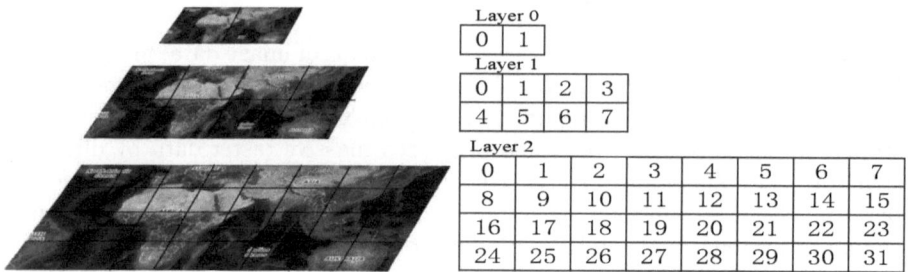

Fig. 2. Stratifying and blocking based on the Pyramid Model

Fig. 3. Tiles encoding

3.3 Mapping Method between Tile Layers

According to the stratifying and blocking rules of Pyramid Model, one tile in layer N corresponding to four tiles in the layer $N + 1$. Therefore, we will try to make the neighbor tiles store in different nodes to improve the performance and increase the throughput. When tiles are accessed, they could be sended in parallel. In every layer, the tiles number is a even number, for this reason, it easily and fast for us to store the tiles in even nodes. Assuming there are $2k$ nodes, and they are named as 0, 1, 2, $\cdots$ $2k - 1$. Accroding to the coding rules we can get the tile existing in row Num/col, and the tiles of even row stored on node $Num\%K$ (Num is the tile number), and the tiles of odd row stored on node $Num\%K + K$. For example, there are 4 DataNodes, the tiles Pyramid are shown as the Fig. 3. There are 2 rows and 4 columns in the first layer, the tile 4 is in row $4/4 = 1$ (the number 4 of numerator is tile number, the number 4 of denominator is the column number in the first layers). So, the tile should be stored on node $4\%2 + 2 = 2$ (4 is the tile number, 2 is half of the number of nodes), other tiles are shown as Fig. 4.

As what shown in Fig. 4, the neighbor tiles 0, 1, 4, 5 or 1, 2, 5, 6 are stored in different DataNode. If we need to access the tiles 0, 1, 4, 5, the search tasks could be executed on the entire DataNodes, and these tiles are sent to the target location in parallel. In this case, read performance is improved effectively.

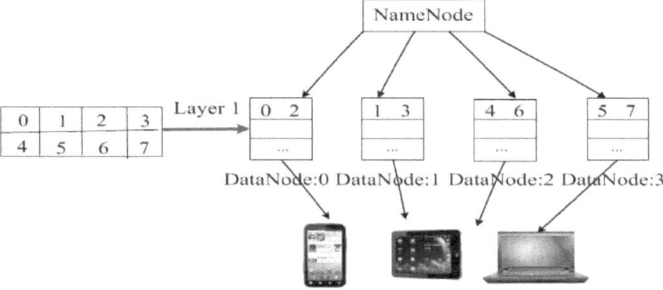

Fig. 4. Mapping between tiles and DataNodes

3.4   Process of Image Data Storage and Search

In this paper, the local file system of Hadoop is used to store tiles. It mainly used to interact with the HDFS, and supported by MapReduce framework. In this paper, all of the design is built based on MapReduce, and big tasks are divided into many small tasks which could be executed parallel and independent.

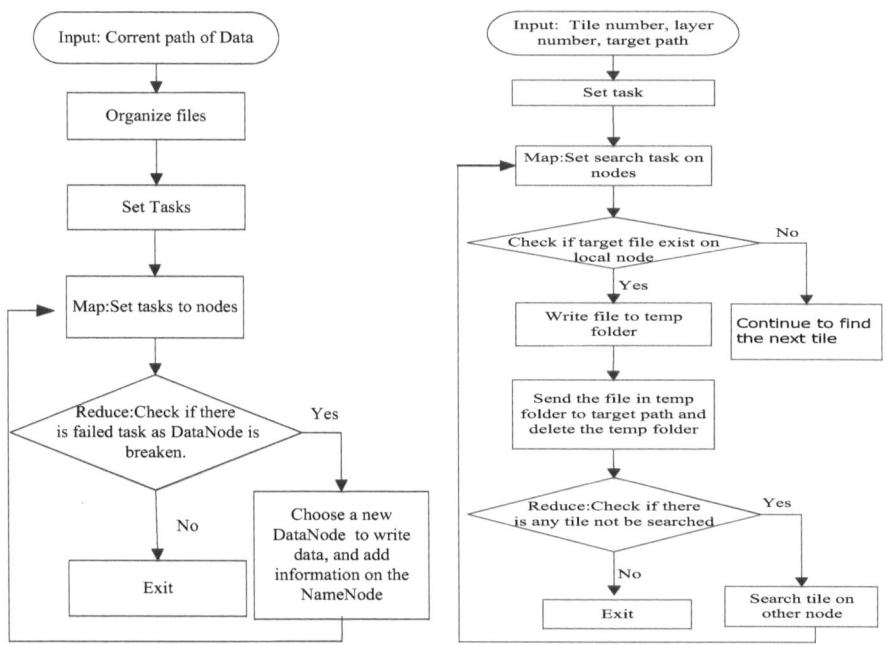

Fig. 5. Flow chart of storage   **Fig. 6.** Flow chart of search

Fig. 5 is the flow chart of storage method. At first, the current path of the file which will be saved should be input. In the second step, all tiles should be reorganized and recoded. In the third step, the upload task is divided into many small tasks and executed in parallel. Then we upload the data to Hadoop by MapReduce programs. At last, the integrity of tasks would be checked.

Fig. 6 is the flow chart of search. First of all, we should input three parameters: search tile's number, layer number, and the target path which intend to store tiles. The second step is for assign task, it similar with the third step of storage process. In the process of task running, the check program of the target tiles existence will be executed on each DataNode. If the target tiles exist in DataNode, they are written into a temp folder. After the search task finished, all of the tiles in the temp folder are written into the target path which is input at first step, then delete the temp folder. At last, we check the integrity of the task, if the search task failed, we search it on other node, otherwise program exit.

3.5 Algorithm Complexity Analysis

Assuming that the count of files that need to be stored is n. In paper [14], when m tiles are made up a big data, the writing operation of both index and data should be executed n times, and then n/m big data will be formed. At last, n/m data and one index file should be written to complete the final file writing, so the time complexity is $T(n) = O(n + n + n/m + 1) = O(3n)$. In this paper, assuming that there are k nodes in our experiment, each tile needs to be recoded and there are n times writing operation are used for encoding. In addition, n/k data should be put into each of the K DataNodes, then the time complexity is $T(n) = O(n + n/k) = O(2n)$. In paper [14],during the search process, n times retrieve need to be executed for indexing the exact record from n files, therefore, the time complexity is $T(n) = O(n)$. However,we obtain a time complexity $T(n) = O(1)$ with our proposed storage rules that guarantee a direct access to tiles.

4   Experiment Environment and Datasets Analysis

Usually, the data security, space accountability and storage efficiency are used for performance evaluation. In this paper, the BMR Storage method is proposed to improve image data storage efficiency, therefore the time is used as the evaluation criterion. We evaluate the BMR Storage method against two existing methods, one is the method in paper [14], the other is that we write data to HDFS by the interface provided by Hadoop directly.

4.1   Experiment Environment and Datasets

Our test platform is built on a cluster with four nodes which are named as ubuntu, ubtuntu01, ubuntu02, ubuntu03. In these four nodes, ubuntu acts as both NameNode and a DataNode, the others are DataNode. The node ubtuntu01 and ubuntu03 have 2GB memory, and others have 1GB memory. The operating system is ubuntu10.10. Hadoop version is 0.20.2 and eclipse version is eclipse 3.3.1.1. The number of replications is set to 1 during the test.

In this paper, test data are the open satellite image data including 22 layers, we only choose from third layer to ninth layer which are comprised of 87311 tiles, the biggest tile is 164KB and the smallest file is 3.66KB. Fig. 7 shows the distribution of file sizes.

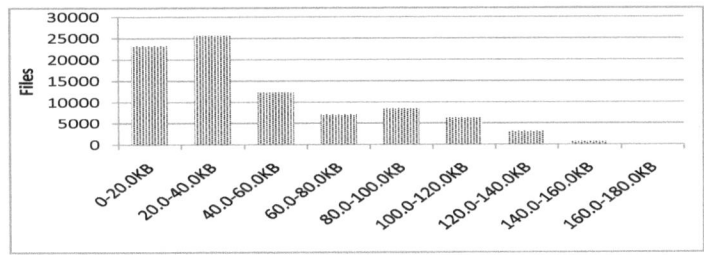

Fig. 7. Distribution of test file sizes

4.2   The I/O Performance Test

Experiment 1, we use the eighth layer that includes 16384 tiles to test write performance. Fig. 8 shows the experiment results.

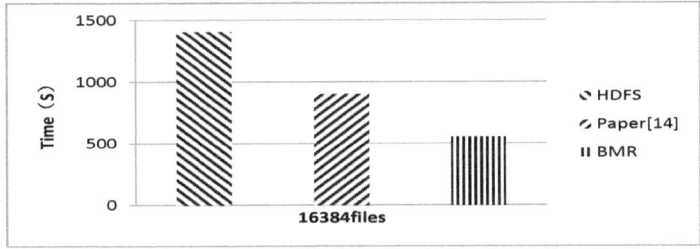

Fig. 8. Comparation of write performance

In the process of writing data, if using interface provided by Hadoop to write the tiles to HDFS directly, the block information have to be written to the NameNode after one tile is written to DataNode. So, a lot of block information have to be written to the NameNode, and the efficiency will be greatly affected, and when the volume of tiles up to 2 billion, the size of the block information would exceeded the ordinary Computer storage capacity. In paper [14], many small tiles are merged into a big one. In this case, the size of block information which write to NamNode is reduced greatly, and write performance is improved. In this paper, we design a storage rule. Using this rule, the tiles data could be written to DataNodes in parallel, and there is no need to write the block information to the NameNode. So, writing performance is improved to some extent.

We read the 5000 and 1500 tiles file in Experiment 2 respectively, the results are shown as Fig. 9.

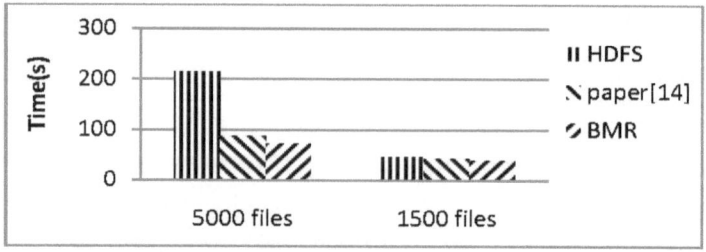

Fig. 9. Read performance comparation

In the process of reading data, compared with paper [14] in which search and read tiles through index file, the proposed method search and read tiles through MapReduce framework in parallel. In our method, the Map task and the Reduce task take up a certain time. So, when the amount of data is small, the read efficiency performance is not very clear, but with the increase in the number of files, performance has improved greatly. The result is shown in Fig. 9.

5 Conclusion

Aiming to improve the access efficiency of massive remote sensing image, we analyze the characters of image data, reorganize tiles and define a storage rule. In this way, accessing efficiency can enhanced since tiles could be searched and stored in parallel. The experimental results have proved that the proposed BMR Storage method can improve the tile data's I/O performance to some certain extents in Hadoop. However, due to lack of consideration the data safe problem, the data copy number is set to 1. More research works will be done about data copies in the future.

Acknowledgements. This paper is supported by National Natural Science Foundation of China (No.61 073146), National Natural Science Foundation of China (No.41101432) and Natural Science Foundation of Chongqing (CSTC 2010BB2416).

References

1. Li, Z.M., Wu, Y.Z.: Terrain Data Object Storage Mode and Its Declustering Method. Acta Geodaetica e 37(4), 490–494 (2008)
2. Wang, K.M., Tang, X.A., Chen, M., Xie, Y.H., Yang, C.D.: Terrain Grid Dataset Research Based on Pyramid Structure. Modern Electronic Technology 31(21), 39–42 (2008)

3. Armbrust, M., Fox, A.: Above the Clouds: A Berkeley View of cloud Computing. Technical Report. University of California at Berkley (2009)
4. Zhang, W.: Research on the Technology and Application of Mobile GIS Based on Distributed Stroage. Unpublished Master dissertation, Information Engineering University of the People's Liberation Army (2010)
5. Huo, S.M.: Research on the Key Techniques of Massive Image Data Management Based on Hadoop. Unpublished Master dissertation, National University of Defense Technology (2010)
6. Wan, Y.W., Cheng, C.Q., Song, S.H.: Research on Rapid Showing Mass RS Images Based on Global Subdivision Grid. Geography and Geo-Information Science 25(3), 33–56 (2009)
7. Yu, F.X., Wang, G.X., Wan, G.: Rapid Allocation and Display of Massive Remote Sensing Image. Hydrographic Surveying and Charting 26(2), 26–30 (2006)
8. Tom, W.: Hadoop: The Definitive Guide. Tsinghua University Press, Beijing (2010)
9. Jeffrey, D., Sanjay, G.: MapReduce: Simplied Data Processing on Large Clusters. In: Proceedings of the 6th Conference on Symposium on Opearting Systems Design & Implementation, San Francisco, CA, pp. 1–13 (2004)
10. Zhu, L., Pan, M., Li, L.Q., Wu, H.P.: Research on Massive Grid Data Treatment Techniques in GIS. Application Research of Computers 23(1), 66–68 (2006)
11. Zhang, J.B., Liu, D., Wu, X.C.: Research and relization of raster data storage management in GIS. Journal of Guilin Uiversity of Technology 26(1), 54–58 (2006)
12. Li, Z.M., Gao, L.: 2-D tiles pair of base vectors twice mapping declustering method. Comput. Eng. Appl. 46(10), 20–22 (2010)
13. Zhou, P.: The Research and Application of Tile Map Technology in Intelligent Transportation System. Unpublished Master dissertation. Tongji University (2008)
14. Liu, X.H., Han, J.Z., Zhong, Y.Q., Han, C.D.: Implementing WebGIS on Hadoop: A Case Study of Improving Small File I/O Performance on HDFS. In: Proceedings of 2009 IEEE International Conference on Cluster Computing and Workshops, pp. 1–8. IEEE Press, New Orleans (2009)

Learning to Classify Service Data
with Latent Semantics

Le Luo, Li Li*, and Ying Wang

Faculty of Computer and Information Science,
Southwest University, 400715 Chongqing, China
{luole,lily,wangying}@swu.edu.cn

Abstract. Service discovery plays an important role in service composition. In order to achieve better performance of service discovery, often,
service classification should be in place to group available services into
different classes. While having a powerful classifier at hand is essential
for the task of classification, existing methods usually assume that the
class-labels of services are available in prior, which is not true. Traditional clustering methods consume a great deal of time and resources
in processing Web service data and result in poor performance, because
of the high-dimensional and sparse characteristics of WSDL documents.
In this paper, Latent Semantic Analysis (LSA) is combined with the
Expectation-Maximization (EM) algorithm to compensate for the poor
performance of a single learning model. The obtained class-labels are
then used by the Support Vector Machine (SVM) classifier for further
classification. We evaluate our approach based on real world WSDL files.
The experimental results reveal the effectiveness of the proposed method
in terms of accuracy and quality of service clustering and classification.

Keywords: Web Service Clustering, Web Service Classification, Latent
Semantic Analysis, Support Vector Machine.

1 Introduction

Web service discovery plays an important role in improvement of QoS of service
composition, which now becomes increasingly important in applications ranging
from healthcare to industry. The first step in service discovery requires service
classification results to be ready. Different machine learning algorithms can be
thought as candidates. However, most of them are based on the same assumption
that class-labels of services are given in prior, which is generally not true.

Web Service Description Language (WSDL)[1] is the de facto language to describe Web services. WSDL documents are normally available from UDDI. Solving service classification hence lies in performing WSDL document classification,
and of course clustering is needed in the first place. Two problems can arise when
dealing with WSDL documents. First, since WSDL document clustering relates

---

* Corresponding author.

[1] Web Services Description Language (WSDL) 1.1, http://www.w3.org/TR/wsdl

T. Li et al. (Eds.): RSKT 2012, LNAI 7414, pp. 276–281, 2012.
© Springer-Verlag Berlin Heidelberg 2012

to processing very sparse and high-dimensional data, traditional clustering methods are barely adequate to cope with it. Many existing methods may result in
poor performance with extremely skewed outcome without conforming to the
correct data distribution. Second, when performing multiclass classification of
WSDL data, only a few studies investigate the impact of the number of classes
of training data on classification accuracy and how many classes are appropriate
for a particular issue. The correlation between the number of classes and classification accuracy will be useful to evaluate reliability of classification results.

To address the first problem, we combine Latent Semantic Analysis (LSA) [1]
with the Expectation-Maximization (EM) Algorithm [2] by first carrying out
LSA analysis and then applying the EM clustering method to the obtained
class-labels (LSA-EM for short in the rest of the paper). We evaluate the LSA-
EM method and the results show that our method is far superior in achieving
accuracy than the traditional ones. For the second one, we utilize Support Vector
Machines (SVM) [3] for document classification, and by trial and error, we evaluate the results and obtain an optimum solution, which is the concrete number
of class-labels regarding a particular training set. This can serve as an evaluation
criterion of classification accuracy in multiclass classification.

2   Our Approach

Initially, a document-term matrix is constructed using the Vector Space Model [4]
whose rows represent documents, columns correspond to all the words from the
vocabulary, and cells hold values of the weight of each word in a document where
we adopt the tf-idf [4] metric for the wight because of its good performance.

Subsequently, we employ LSA to find out the latent structure of the matrix
by Singular Value Decomposition (SVD). Initially, the term-document matrix
denoted by A is decomposed through SVD as follows: $A = USV^T$ where U
describes the row vectors representing terms, V^T describes the column vectors
representing documents. On performing SVD, for filtering out the noise of the
matrix, part of row vectors of U, diagonal matrix S and column vectors of V^T
are eliminated and thus U, S and V^T reduce to $\hat{U}$, $\hat{S}$ and $\hat{V}^T$ respectively, then
we obtain $\hat{A} = \hat{U}\hat{S}\hat{V}^T$. Note that $\hat{U}$ and $\hat{V}^T$ represent terms and documents
without noise. Since what we want to label are documents but not terms, we
should focus on matrix $\hat{V}^T$. Having finished SVD of the term-document matrix
$\hat{A}^T$ and obtained $\hat{V}$, we then do an EM clustering analysis of row vectors of $\hat{V}$
so as to label the documents, and finally can retrieve the label of any documents
according the corresponding labeled row vectors of $\hat{V}$.

Finally, since the document data is labeled, they can be trained with the
supervised learning algorithm SVM. In practice, the multiclass SVM is always
involved. Clearly, Multiclass SVMs aim to process data with two more labels in
a dataset. Among the approaches handling multiclass problems, the one-versus-
one [6] is applied widely in the case of practical applications because of their
better accuracy of classification and higher efficiency.

3   Evaluation

3.1   Dataset Description

For analysis of clustering and classification, we adopt a set of unclassified WSDL documents available in some UDDI registries[2]. Such a set is composed of 3609 documents. Then, the WSDL documents are preprocessed. Words are extracted from the WSDL documentation tag and operation name tag, which are more representative of the service. The extracted composite words are split in general case. After that words are filtered by a stop-list. Thus a WSDL document is condensed into a document in which only representative words are retained. Finally, for comparing the clustering method we proposed with others, we construct a sparse document-term matrix model which holds 3609 rows representing the documents and 2470 columns representing terms in the vocabulary.

3.2   Comparison of Clustering Methods

To compare some clustering methods (KMeans and EM) with the LSA-clustering (firstly using LSA, then using other clustering methods) algorithm on WSDL data, we apply the KMeans and EM algorithm to the data processed by LSA and the data not processed by LSA, and then validate the difference between the general clustering algorithms and the LSA-clustering algorithms. Results are presented in Table 1, in which the data is clustered for seven times through four different methods, and each time the data is clustered into k clusters ($k = 10$ for the first time, $k = 15$ for the second, etc.). Then the proportions of top three clusters are respectively listed for each k and each method in the table.

Table 1. Proportion of top 3 clusters for different ks and methods

Clustering methods	Number of clusters (k)						
	10	15	20	25	30	35	40
KMeans	90%	79%	86%	79%	79%	75%	74%
	5%	5%	3%	5%	5%	5%	5%
	2%	4%	3%	5%	2%	5%	3%
LSA-KMeans	82%	52%	55%	50%	75%	54%	69%
	5%	25%	12%	16%	5%	11%	5%
	5%	5%	10%	6%	5%	5%	4%
EM	25%	18%	20%	14%	22%	22%	18%
	23%	15%	17%	14%	17%	14%	14%
	19%	12%	13%	13%	14%	13%	14%
LSA-EM	18%	22%	13%	14%	12%	13%	13%
	17%	14%	11%	8%	10%	8%	11%
	14%	13%	10%	7%	8%	7%	11%

---

[2] http://www.uoguelph.ca/~qmahmoud/qws/index.html

From the clustering results in Table 1, we can clearly see that utilizing the KMeans and LSA-KMeans algorithm on WSDL data yields much skewer clustering results than using the EM and LSA-EM do; on the contrary, the results yielded by the latter two methods distribute more evenly. It is evident that the LSA-clustering method outperforms the general method—the LSA-KMeans is better than the Kmeans and the LSA-EM is better than the EM. Moreover, the likelihood function of EM converges slower for high-dimensional data and thus the EM is much more time-consuming with WSDL data. Although the criterion function of the KMeans converges more quickly than that of the EM, the EM performs much better than the KMeans in clustering accuracy. It is evident that the LSA-EM is the best at processing WSDL data.

The LSA-EM system is in line with the empirical study thus proving its effectiveness. We randomly select 3 services manually from all the documents— namely, *version* containing 94 documents, *bind* containing 36 documents and *autocomplete* containing 57 documents—for observation of the four methods. When the 3609 documents are clustered into 40 groups, we count and record documents clustered into wrong groups using four methods respectively. Results are presented in Table 2, in which minus numbers indicate that documents belonging to the 3 groups (*version*, *bind* and *autocomplete*) are misclustered into other groups and positive numbers indicate documents belonging to other groups are misclustered into the above 3 groups. Apparently, the LSA-EM is the best.

Table 2. Number of documents clustered into wrong groups

Clustering methods	services		
	version	bind	autocomplete
KMeans	-3	2628	170
LSA-KMeans	47	2490	139
EM	526	170	84
LSA-EM	-3	0	83

To sum up, the EM algorithm performs well for general data rather than WSDL data characterized by high-dimensionality and sparsity. The LSA-EM reduces the data to low-dimension, which is equivalent to transforming the WSDL data to "general data" which is suitable for the EM method to process, and furthermore mines latent semantics by LSA. Thus when the data is then clustered by the EM, the results are more precise than by the EM.

3.3   Analysis of Multiclass Classification Performance

In our experiment, we utilize the classification tool *libsvm*[3] which adopt one-versus-one strategy for multiclass classification and choose the radial basis function kernel when training data. Virtually, although one-versus-one multiclass

---
[3] http://www.csie.ntu.edu.tw/~cjlin/libsvm/

algorithm is better than others, we can prove in experiments that if data with too many labels is trained and predicted, the accuracy of classification is not as high as expected. The more labels the data has, the lower accuracy of classification results we get. We performed clustering analysis using LSA-EM and obtained 50 clusters, and then selected randomly, from the 50 clusters, 11 blocks of data, each respectively containing 3 clusters, 5 clusters, 10 clusters, etc. After that, we divided every block into thirds, two thirds for training and one third for predicting. Finally, the comparison results are obtained shown in Fig. 1.

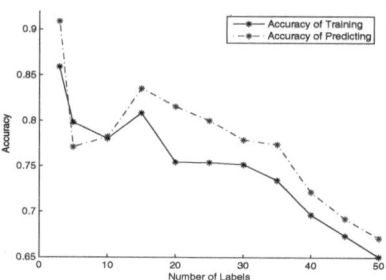

Fig. 1. The relation between the number of labels and classification accuracy

Obviously, the accuracy decreases as the number of labels increases. Although the accuracy does not strictly decrease monotonically with the number of labels, it does as a whole. As illustrated in Fig. 1, when the number of labels is 5 or 10, the accuracy does not conforms to the strictly decreasing monotonically rule. Basically, the reason is that the clusters in every data blocks are selected randomly, and accidentally the instances in the 5 or 10 different clusters mingle with each other so as to make the boundary of different clusters not clear. However, all the possible reasons cannot change the fact that when the number of labels exceeds to some extent, the accuracy of classification decreases sharply. As for WSDL data, the accuracy decreases dramatically when the number of classes of training data exceeds 35. Therefore, it's not a good idea to train data with more labels in classification with SVMs.

3.4   Discussion

Compared with other supervised learning algorithms, the SVM converges slow in large dataset, but it solves the over-fitting and feature redundancy problem, gives a great performance at generalization, and naturally yields better classification results than others [7,8,9,10]. In [7] Joachims develops a theoretical learning model for SVMs and explains why SVMs perform well for document classification. Debole [10] conducts dozens of experiments in different situation as to compare the results from Rocchio, KNN, and SVM respectively, and the experiments showe SVMs outperform other two methods clearly. Apparently, SVMs appear to be most promising in classifying documents. Hence, although there exist many methods for document classification, the SVM is adopted in this study.

4   Conclusions

Web service discovery relies on service classification. The quality of classification is determined by the provided class-labels, which are usually needed to learn in practice. Unlike the traditional classification algorithms, in which class-labels are given in prior, in this paper, those class-labels are instead learnt from the training set step by step. Benefits of this approach are two-fold: (1) the latent semantics between services and their belonging categories are exposed, which can be used to further improve service composition; (2) avoiding any bias introduced by human experts when they annotate relevant services. This again will help service discovery and ultimately serve service composition accordingly. LSA-EM is proposed in the paper for clustering purpose. Our experiments show that the LSA-EM method has high performance. It not only outperforms KMeans, EM, and LSA-KMeans by a substantial margin, but also shows potential to adapt to huge applications in the future. By using SVM algorithm, we found out the correlation between the number of class-labels and classification accuracy. As for the WSDL data the optimum number of class-labels is around 35.

Acknowledgements. This work is supported by National Natural Science Foundations of China (61170192), Natural Science Foundations of CQ (CSTC 2007BB2372).

References

1. Dumais, S.T.: Latent Semantic Analysis. Annual Reiview of Information Science and Technology 38(1), 188–230 (2004)
2. Borman, S.: The Expectation Maximization Algorithm—A Short Tutorial (2009), http://www.seanborman.com/publications/EMalgorithm.pdf
3. Burges, C.J.C.: A Tutorial on Support Vector Machines for Pattern Recognition. Data Mining and Knowledge Discovery 2(2), 121–167 (1998)
4. Aversano, L., Bruno, M., et al.: Using Concept Lattices to Support Service Selection. International Journal of Web Services Research 3(4), 32–51 (2006)
5. Minh, H.Q., Niyogi, P., Yao, Y.: Mercer's Theorem, Feature Maps, and Smoothing. In: Lugosi, G., Simon, H.U. (eds.) COLT 2006. LNCS (LNAI), vol. 4005, pp. 154–168. Springer, Heidelberg (2006)
6. Hsu, C.W., Lin, C.J.: A Comparison of Methods for Multi-class Support Vector Machines. IEEE Transactions on Neural Networks 13(2), 415–425 (2002)
7. Joachims, T.: A Statistical Learning Model of Text Classification for Support Vector Machines. In: Proc. of the 24th Annual International ACM SIGIR Conference on Research and Development in Information Retrieval, pp. 128–136 (2001)
8. Lewis, D.D., Li, F., Rose, T., Yang, Y.: RCV 1: A new benchmark collection for text categorization research. Journal of Machine Learning Research 5(3), 361–397 (2004)
9. Forman, G., Cohen, I.: Learning from Little: Comparison of Classifiers Given Little Training. In: Boulicaut, J.-F., Esposito, F., Giannotti, F., Pedreschi, D. (eds.) PKDD 2004. LNCS (LNAI), vol. 3202, pp. 161–172. Springer, Heidelberg (2004)
10. Debole, F., Sebastiani, F.: An analysis of the Relative Hardness of Reuters-21578 Subsets. Journal of the American Society for Information Science and Technology 56(6), 584–596 (2004)

Link Communities Detection via Local Approach

Lei Pan, Chongjun Wang, and Junyuan Xie

National Key Laboratory for Novel Software Technology
Department of Computer Science and Technology
Nanjing University
Nanjing, China
panleipanlei@gmail.com, {chjwang,jyxie}@nju.edu.cn

Abstract. The traditional community detection algorithms were always focusing on the node community, while some recent studies have shown great advantage of link community approach which partitions links instead of nodes into communities. Here, we proposed a novel algorithm LBLC (local based link community) to detect link communities in networks based on some local information. A local link community can be detected by maximizing a local link fitness function from a seed link, which was ranked by another algorithm previously. The proposed LBLC algorithm has been tested on both synthetic and real world networks. The experimental results showed LBLC achieves meaningful link community structure.

Keywords: community detection, link community, local community.

1   Introduction

Mining community structures has become a general problem which exists in many fields including: Computer-Science, Mathematics, Physics, Biology, Sociology and so on. It has developed rapidly and been used widely in many applications: web data mining, social network analysis, criminal network mining, protein interaction network analysis, metabolic network analysis, genetic network analysis, customers relation-ship mining and user online behavior analysis, etc. After years of development and with the rise of social network web site such as Facebook and Twitter, community detection has been one of the most popular re-search areas in web data mining, and many representative algorithms have been proposed including GN Algorithm [1][2], Fast Newman (FN) Algorithm [3], Radicchi fast split Algorithm [4], and Simulated Annealing based GA Algorithm [5]. But the topic of overlapping community detection has also been given more attention recently. Many methods based on clique percolation theory have been proposed such as CPM Algorithm [6] and Clique Graph [7]. Another kind of algorithms is local based detecting methods which focus on the local information about each node and demonstrate the superiority to find overlapping communities. These algorithms include EAGLE Algorithm [8], LFM Algorithm [9], Greedy clique expansion (GCE) [10], Model based Seed Expansion

T. Li et al. (Eds.): RSKT 2012, LNAI 7414, pp. 282–291, 2012.

algorithm [11], Multi-Level Local Community Detecting [12], and some other kind like probabilistic model based methods as illustrated in [13], [14]. In some real-world networks, although a node could belong to multiple communi-ties, its internal links in different communities can be resolved easily. These links build up different personal communities or different social networks. So link commu-nity detection algorithms which use link information in networks to extract link community structure have been proposed. Line graph based method is a novel approach which converts original graphs into line graphs and then applies other community detection methods on it [15]. Yong-Yeol showed us the prevalent link community structure in many real networks [16]. More researchers started to pay attention to this aspect. Some modified node-community based methods can also be used to find improved link community. For example, map equation with random walk can be used to detect link community [17]. Considering the local information about link community, we propose a local based link commu-nity detection algorithm LBLC (local based link community) in this pa-per. It could find local link communities of the network through some given seed links selected by a ranking algorithm to obtain the global link community structure. We use some metrics to rank links and find that the edge-clustering coefficient is very effective and is also simple to use. The local community can be obtained by a greedy expansion method using a fitness function. And the fitness function we use in this algorithm has been improved to conform to the requirements of the link community. Thus, the algorithm has good performance in link com-munity structure detection and overlapping community coverage. This paper is organized as follows. In section 2, we provide problem definition and some ba-sic concepts about community detection. In section 3, we present the pro-posed LBLC algorithm. In section 4, we make some experiments in several data sets to evaluate our algorithm, and show some comparisons. In section 5, we draw a conclusion.

2   Community Detection

2.1   Problem Definition

Given a network $G = (V, E)$, where V denotes the node set, E denotes the edge set. The purpose of community detection is to identify some collections of the network. The collection is called community.

Overlapping Community. As we mentioned above, a node could belong to multiple communities in many real networks. For example, individuals have dif-ferent social network relationships in the society such as schoolmates, co-workers and families. Researchers may have many different research interests in differ-ent research areas. So a node i in the network can be shared by m^x different communities. In turn, any two communities can share n_s nodes which defined as overlap size between them [6]. We may call the community with these overlap-ping nodes the overlapping community. Due to the existence of a large portion of overlapping nodes, highly overlapping communities may have more external connections than internal ones (depends on the overlapping level of the overlapping nodes), which is quite the contrary to non-overlapping communities [16] as shown in Figure 1. The red node in Figure 1 belongs to two communities and it has more external links than internal links for the right community.

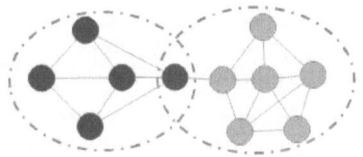

Fig. 1. An example of overlapping community

Local Community. Local communities are essentially local structures, involving the nodes belonging to the modules themselves plus at most an extended neighborhood of them [9]. Local communities are always being constructed from some starting seeds, and then being expanded to optimal size by merging its neighbor nodes so that the global community structure can be obtained. As illustrated in Figure 2, the natural overlapping community of each seed can be found by using some local property based approach. It is noted that nodes between local communities can be highly overlapped depending on the selection of the initial seeds. The merging of two similar local communities can be controlled by a duplicate measure. In Figure 2, the red nodes are initial nodes; Green nodes are first neighbors of them; Blue nodes are next layer neighbors.

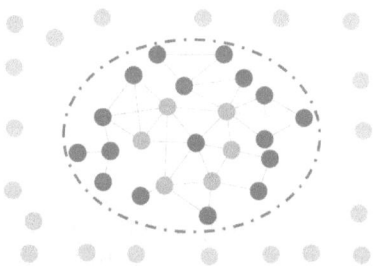

Fig. 2. An example of local community

Link Community. The traditional community detection algorithms focus on the nodes in the network, and communities found by them are constructed within a group of nodes. But in link community, the objects are links rather than nodes. Actually, the links represent relationships in social networks, and the relationships can be friends on Facebook, professional contacts on LinkedIn, dates on an online dating site, jobs or workers on employment websites, or people to follow on Twitter. As mentioned earlier, nodes can become overlapped, but their internal links in different communities can be resolved easily. Nodes can belong to multiple communities; whereas links often exist for one dominant reason (two people are in the same family, work together or have common interests). So instead of assuming that a community is a set of nodes with many links between them, link community considers a community to be a set of closely interrelated links [16]. As shown in Figure 3, the red node has three different social net-works: college mates, colleagues, families, while link community represents these links. The benefit of using the link community is to find an overlapping and natural community structure of networks. Its easy to implement because a node has multiple links belonging to multiple communities and we can cluster the links into different partitions so that the nodes become overlapped naturally. By converting link communities to node communities, the problems of detecting multiple relationships between nodes and finding the overlapping nodes are solved effectively.

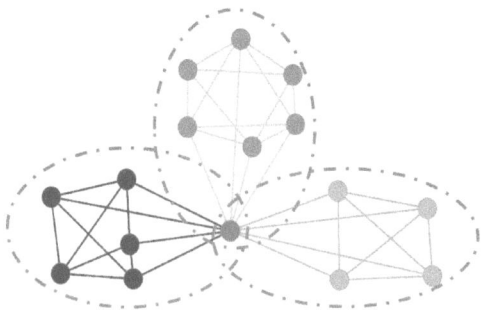

Fig. 3. An example of link community. An individual node has three social networks which means it belongs to multiple communities, connections in different colors represent different relationships.

3   LBLC Algorithm

3.1   Local Based Link Community Detection (LBLC)

Given a Network G with nodes V and links E, LBLC works by first ranking the link seeds in G, then expanding by greedily maximizing a local fitness function, and stopping expansion only when all neighbor links have negative fitness values. Finally, the LBLC finds all local link communities by iteration. In this section, we introduce some core concepts of LBLC: the selection of initial seeds, fitness function, and local information based greedy expansion.

Initial Seed. Local community detecting method needs a seed to start finding community. Because our seeds are links of the network, we want to select seed links from the dense-linked areas, so we rank the links by edge clustering coefficient [4]. The edge cluster-ing coefficient is defined as:

$$C_{i,j}^{(g)} = \frac{z_{i,j}^{(g)} + 1}{s_{i,j}^{(g)}} = \frac{z_{i,j}^{(g)} + 1}{min(k_i - 1, k_j - 1) = 0}$$

Where $z_{i,j}^{(g)}$ is the number of cyclic structures of order g the link(i,j) belongs to, and $s_{i,j}^{(g)}$ is the number of possible cyclic structures of order g that can be built given the degree of the nodes. The purpose of plus one at the numerator is to prevent the following situation: when the number of triangles is zero, $C_{i,j}^{(g)} = 0$, irrespective of k_i and k_j, or even $C_{i,j}^{(g)}$ is indeterminate that $min(k_i-1, k_j-1) = 0$.

Link Fitness Function. With an input of a subgraph S, the original fitness function [9] returns a real value which indicates how dense the links are in S and how well it corresponds to the rest of the network. We consider the links rather than nodes. Thus, we propose a lightly modified link-based fitness function as:

$$lf_S^{(i,j)} = (C_{i,j}^{(g)} + 2) \cdot (f_{S+(i,j)} - f_{S-(i,j)})$$

where $f_S = \frac{m_{in}^S}{(m_{in}^S + m_{out}^S)}$, S is a community, α is a positive real-valued parameter that can control the size of community,. m_{in}^S is the number of internal links of the community, and m_{out}^S is the external links that connect the community. The symbol $S + (i,j)(S - (i,j))$ indicates the subgraph obtained from module S with link(i,j) inside (outside) it. $lf_S^{(i,j)}$ denotes the contribution of the link to the local community. $C_{i,j}^{(g)}$ shows the links density of its area.

3.2   Description of LBLC Algorithm

In the previous sections, we have described the basic concepts. Based on them, we can outline our proposed LBLC algorithm as algorithm 1.

Algorithm1: LBLC

Input: A Network G
Output: Local link communities.
1. Rank links of the network according to their edge-clustering coefficient, and join the sorted objects into a set S.
2. Select a seed link with the largest edge-clustering coefficient value from S (the link must not exist in any other communities), then find a Local Link Community from this seed link, and remove this link from S.
3. Repeat step 2 until all links are covered.

It is notable that we dont need overlapping link communities, since the nodes of link communities already overlapped. The algorithm of finding local link community is illustrated as Algorithm 2. The computational complexity of the first step of LBLC is $O(m)$, where m is the total links of network. The time complexity of the second step is hard to calculate, as it depends on the size of the local community. The rough estimated time to build a local link community with e links is approximately $O(e^2)$ for a fixed -value. Therefore, for the whole network, the time complexity is approximately $O(m+ce^2)$, where c is the number of local communities to be found. The worst-case complexity is $O(m + m^2)$, when the community is of size comparable with m. This is not the case in general, in most situations the algorithm runs much faster and almost linearly when communities are all small enough.

Algorithm2: Finding A Local Link Community

Input: A start link seed
Output: A Local link community
1. Use the seed link, calculate fitness function values of all of its first layer neighbors and add them to the candidate set C.
2. If C is not empty, get a link with largest fitness value from C (not in any other communities)add it to the community whose seed link belongs to. Then recalculate fitness of all links in the candidate set.
3. If C is empty, we will get the next layer neighbor links of the local community. Furthermore, we will calculate their fitness value, and add these links with positive fitness into C.
4. When step 3 cannot add any more links with positive fitness, the algorithm will stop; otherwise, repeat step 2.

3.3   Evaluation of Link Community

For those data without any prior knowledge or meta-data or labeled information, the modularity [2] is often used to measure the quality of a particular division of a node-based non-overlapping network. But there is no generally accepted metric to measure the quality of link communities. The partition density [16] may be an effective meth-od to do this job.

Extend Modularity. Actually, the link community can also be seen as overlapping community. So the extended modularity [16] can also be used to evaluate the link communities by some transformation. Extended Modularity(EQ) is defined as

$$EQ = \frac{1}{2m} \sum_C \sum_{i,j \in C} \frac{1}{O_i O_j} \left[A_{ij} - \frac{k_i k_j}{2m} \right]$$

where A_{ij} is the element of adjacency matrix of the network (we only consider undirected and unweighted networks here). $A_{ij} = 1$ if there is an edge between node i and j, $A_{ij} = 0$ otherwise. $m = \frac{1}{2} \sum_{ij} A_{ij}$ is the total number of edges in the network. The degree k_i of any node i is defined as $k_i = \sum_j A_{ij}$. And O_i is the number of communities node i belongs to. It should be noted that EQ degenerate to Q when each node belongs to only one community, and EQ is equal to 0 when all nodes belong to the same community. Certainly, a high value of EQ indicates a significant overlapping community structure. Similar to modularity, the extended modularity suffers a resolution limit [18] beyond which no community can be detected.

Link Partition Density. For a network with M links, P_1, P_2, P_c is a partition of the links into C partitions. Partition P_c has m_c links and n_c nodesthen the link partition density is defined as

$$D_c = \frac{m_c - (n_c - 1)}{n_c(n_c - 1)/2 - (n_c - 1)}$$

$$D = \frac{1}{M} \sum_C m_c \cdot D_c$$

where D_c is normalizedm_c by the possible minimum and maximum numbers of links between nodes in P_c(Assume $D_c = 0$ if $n_c = 2$). The partition density D is the average of D_c. The advantage of defining D is that it does not suffer the resolution limit [18] since each term is local in C. The partition density D measures the quality of link partitions when the maximum D is 1, but D can become negative: $D = 1$ when each community is a fully connected clique, and $D = 0$ when each community is a tree.

4 Experiments

In order to quantitatively test our algorithm, we test it on simulation network and on the real-world network respectively. Our experiments are done on a single Dell PC (Intel(R) Pentiuim(R) CPU P6000 @1.87GHz processor with 2G bytes of main memory on Microsoft Windows 7 OS). Our programming environment is Java 6.0.

4.1 Computer-Generated Networks

We use two computer-generated benchmarks: Newman-benchmark[2] and LFR-benchmark [19]. Newman-benchmark consists of $n = 128$ nodes divided into four communities of 32. Average degree $\bar{k} = 16$. LFR-benchmark has been proposed to construct the real-world like benchmark. The benchmark not only produces nodes and communities with such statistical distributions, but also has overlapping and hierarchical structures. We could see the apparent community structure of the example and the overlapping nodes found by LBLC in Figure 4. But it should be noted that all these network models are based on nodes, while the links of networks are ignored.

4.2 Real-World Networks

In this part we demonstrate LBLCs ability to detect meaningful link communities in the real-world networks. The network datasets descriptions are given in Table 1.

   The experimental results are shown in Figure 5. From the figure we can see that LBLC performed better than LC algorithm in karate and polbooks network. As the football network is formed by 12 non-overlapping communities, the

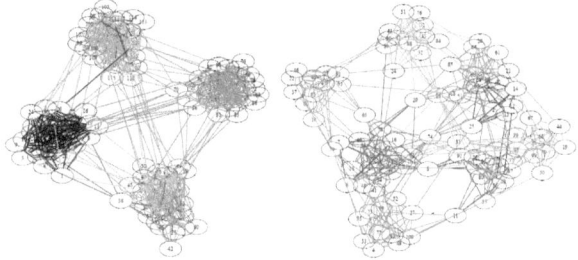

Fig. 4. Newman benchmark ($Z_{out} = 1$) and LFR-benchmark ($n = 100$, $O_n = 10$, $O_m = 2$, $k = 10$ others are default)

Table 1. Real-World Networks Used In Experiments

Networks	V(G)	E(G)	Description
karate	34	78	Zachartys karate club [20]
dolphins	62	160	Dophin social network [21]
polbooks	105	441	Books about US politics [2]
football	115	613	American College football [2]
netscience	1589	2742	Coauthorships in network[22]

hierarchy-based LC algorithm works well, but the local link communities found by LBLC Algorithm have higher EQ than by LC algorithm in the other three networks, although some D values found by LBLC algorithm may not be as well as by LC. In summary, LBLC detected reasonable community topological morphology.

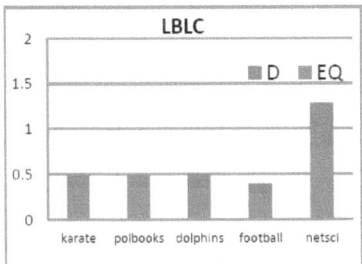
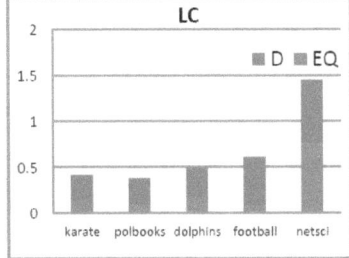

Fig. 5. Extended modularity and link density of real networks

The two parameters α and g could affect the experimental results. As shown in Figure 6, parameters α-value and g-value with their relative EQ. For different networks, different α-values produce different EQ. We achieve the maximum EQ when $\alpha = 0.1$ in the network polbooks, but in the network football this value is 0.4. The EQ is essentially higher when parameter $g = 4$ than when $g = 3$, but the complexity of computing also rises.

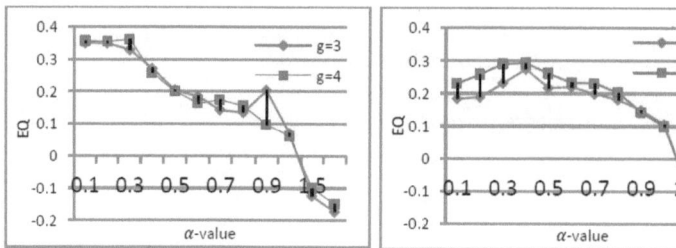

Fig. 6. The parameters(α and g) of LBLC and the results. The left use polbooks as the test dataset and the right use football.

5   Conclusion

A local based link community detecting algorithm (LBLC) is proposed in this pa-per. The LBLC focuses on the links of community rather than nodes. It first ranks all links in the network, then selects an initial seed link and makes expansion of it to obtain a natural local link community. It obtains the global link community structure of the network iteratively. Our future work will be focused on benchmarks and meas-ure functions of the link community.

Acknowledgments. This Paper was supported by the National Natural Science Foundation of China under Grant 60503021, 60721002, 60875038 and 61105069, by the Technology Support Program of Jiangsu Province under Grant BE2010180 and BE2011171, by the the Innovation Fund of Nanjing University under Grant 2011CL07.

References

1. Girvan, M., Newman, M.E.J.: Community Structure in Social and Biological Networks. Proceedings of National Academy of Science 99, 7821–7826 (2002), doi:10.1038/nature03288
2. Newman, M.E.J., Girvan, M.: Finding and Evaluating Community Structure in Networks. Physical Review E 69, 026113 (2004), doi:10.1103/PhysRevE.69.026113
3. Newman, M.E.J.: Fast Algorithm for Detecting Community Structure in Networks. Physical Review E 69, 066133 (2004), doi:10.1103/PhysRevE.69.066133
4. Radicchi, F., Castellano, C., Cecconi, F., Loreto, V., Parisi, D.: Defining and identifying communities in networks. Proceedings of the National Academy of Sciences of the United States of America 101(9), 2658–2663 (2004), doi:10.1073/pnas.0400054101
5. Guimera, R., Nunes Amaral, L.A.: Functional cartography of complex metabolic networks. Nature 433(7028), 895–900 (2005), doi:10.1038/nature03288
6. Palla, G., Derenyi, I., Farkas, I., Vicsek, T.: Uncovering the overlapping community structure of complex networks in nature and society. Nature 435(7043), 814–818 (2005), doi:10.1038/nature03607

7. Evans, T.S.: Clique graphs and overlapping communities. Journal of Statistical Mechanics: Theory and Experiment 2010(12), P12037+ (2010), doi:10.1088/1742-5468/2010/12/P12037
8. Shen, H., Cheng, X., Cai, K., Hu, M.B.: Detect overlapping and hierarchical community structure in networks. Physica A: Statistical Mechanics and its Applications 388(8), 1706–1712 (2009), doi:10.1016/j.physa.2008.12.021
9. Lancichinetti, A., Fortunato, S., Kertész, J.: Detecting the overlapping and hierarchical community structure in complex networks. New Journal of Physics 11(3), 033015 (2009), doi:10.1088/1367-2630/11/3/033015
10. Lee, C., Reid, F., McDaid, A., Hurley, N.: Detecting highly overlapping community structure by greedy clique expansion. In: SNA-KDD 2010, pp. 33–42 (February 2010)
11. McDaid, A., Hurley, N.: Detecting highly overlapping communities with Model-based Overlapping Seed Expansion
12. Havemann, F., Heinz, M., Struck, A., Glaser, J.: Identification of overlapping communities and their hierarchy by locally calculating community changing resolution levels. Journal of Statistical Mechanics: Theory and Experiment 2011(01), P01023+ (2011), doi:10.1088/1742-5468/2011/01/P01023
13. Karrer, B., Newman, M.E.J.: Stochastic blockmodels and community structure in networks. Physical Review E 83(1), 016107+ (2011), doi:10.1103/PhysRevE.83.016107
14. Ball, B., Karrer, B., Newman, M.E.J.: An efficient and principled method for detecting communities in networks, arXiv:1104.3590v1 (April 2011)
15. Evans, T.S., Lambiotte, R.: Line graphs, link partitions, and overlapping communities. Physical Review E (Statistical, Nonlinear, and Soft Matter Physics) 80(1), 016105+ (2009), doi:10.1103/PhysRevE.80.016105
16. Ahn, Y.Y., Bagrow, J.P., Lehmann, S.: Link communities reveal multi-scale complexity in networks. Nature 466, 761–764 (2010), doi:10.1038/nature09182
17. Kim, Y., Jeong, H.: The map equation for link community, arXiv:1105.0257v1 (May 2011)
18. Fortunato, S., Barthlemy, M.: Resolution limit in community detection. Proceedings of the National Academy of Sciences 104(1), 36–41 (2007), doi:10.1073/pnas.0605965104
19. Lancichinetti, A., Fortunato, S.: Community detection algorithms: A comparative analysis. Physical Review E 80(5), 056117+ (2009), doi:10.1103/PhysRevE.80.056117
20. Zachary, W.W.: An Information Flow Model for Conict and Fission in Small Groups. J. Anthropological Research 33, 452–473 (1977)
21. Lusseau, D.: The Emergent Properties of a Dolphin Social Network. Proc. Biol. Sci. 270(suppl. 2), S186–S188 (2003), doi:10.1098/rsbl.2003.0057
22. Newman, M.E.J.: Finding community structure in networks using the eigenvectors of matrices. J. Physical Review E 74(3), 036104 (2006)

Maximizing Influence Spread
in a New Propagation Model

Hongchao Yang, Chongjun Wang*, and Junyuan Xie

Department of Computer Science of Nanjing University,
State Key Laboratory for Novel Software Technology,
Nanjing, China
yanghongchao@iip.nju.edu.cn, {chjwang,jyxie}@nju.edu.cn

Abstract. Study on information propagation in social networks has a long history. The influence maximization problem has become a popular research area for many scholars. Most of algorithms to solve the problem are based on the basic greedy algorithm raised by David Kempe etc. However, these algorithms seem to be ineffective for the large-scaled networks. On seeing the bottleneck of these algorithms, some scholars raised some heuristic algorithms. However, these heuristic algorithms just consider local information of networks and cannot get good results. In this paper, we studied the procedure of information propagation in layered cascade model, a new propagation model in which we can consider the global information of networks. Based on the analysis on layered cascade model, we developed heuristic algorithms to solve influence maximization problem, which perform well in experiments.

Keywords: social network, information propagation, network model, influence maximization.

1 Introduction

Recent years study on the propagation of information in social network catches many researchers' attentions since it can be widely used in practice. When studying the propagation of information in social networks, a key problem, called "influence maximization problem", is mostly concerned by researchers. The goal of influence maximization problem is to find k initial nodes which can yield the largest number of nodes that get information when propagation is finished.

David Kempe etc. tried to find the solutions of this problem in some propagation models, such as IC (independent cascade) model and GT (general threshold) model. They proved that influence maximization problem is NP hard in these models and we can get a $(1-1/e)$-approximation answer if greedy algorithm is applied to influence maximization problem. (Note: the following text will refer to the greedy algorithm raised by David Kempe etc. as basic greedy algorithm.)

---

* Corresponding author.

T. Li et al. (Eds.): RSKT 2012, LNAI 7414, pp. 292–301, 2012.

Since the basic greedy algorithm can get a good approximation, most algorithms are various improvements of basic greedy algorithm. However, time consuming is a big problem no matter what kind of improvements are made. Any algorithm based on basic greedy algorithm must inevitably imitate the process of information propagation by a large number of times, which is #P hard.

In this paper, we design a new propagation model, from which we can learn the whole procedure of information propagation and global information of networks. Based on the model, we design heuristic algorithms which perform better than extant heuristic algorithms and can be applied in large-scale networks.

Paper Organization. Section 2 presents a new propagation model. Section 3 studies the role edges play in influence spread. In Section 4 we design an algorithm to solve influence maximization problem. Section 5 shows our experimental results and analysis of them. Conclusions are included in Section 6.

2   Layered Network and Layered Cascade Model

2.1   Layered Network

Before the discussion of layered cascade model, we must introduce layered network, whose definition is given in Definition 1,

Definition 1 (Layer). *Divide a graph G(V,E) into n subgraphs, which are denoted as $Layer_0(V_0, E_0), Layer_1(V_1, E_1) \ldots Layer_{n-1}(V_{n-1}, E_{n-1})$, where $\cup_{i=0}^{n-1} V_i = V$ and $V_i \cap V_j = \emptyset$ for any $i \neq j$. $\forall v_{ij}, v_{i'j'} \in V$, $(v_{ij}v_{i'j'}) \notin E$ if $|i - i'| > 1$, where v_{ij} denotes the jth node of the ith subgraph. Such subgraphs are layers.*

To obtain a layered network, first we select some nodes to make $Layer_0$. Then we put all neighbors of nodes in $Layer_0$ but not selected into $Layer_0$ into $Layer_1$. Similarly we put all neighbors of nodes in $Layer_{i-1}$ but not added into the upper layers into $Layer_i$ until there are no more nodes can be added. Now we would like to introduce some parameters of a layered network here.

- Up-degree(u_{ij}): the number of v_{ij}'s neighbor nodes in the *(i-1)*th layer;
- Inner-degree (s_{ij}): the number of v_{ij}'s neighbor nodes within the *i*th layer;
- Down-degree(d_{ij}): the number of v_{ij}'s neighbor nodes in the *(i+1)*th layer;
- Number of Layer-path (r_{ij}): the path of length i from any node of $Layer_0$ to v_{ij} is called a layer-path of v_{ij}. And the number of v_{ij}'s layer-paths is denoted as r_{ij}. When $i = 0$, we set r_{ij} as 1.

For nodes of two neighbor layers, there is following relationships between their numbers of layer-paths: $r_{ij} = \sum_{k=1}^{u_{ij}} r_{(i-1)k}$

2.2   Layered Cascade Model

We use layered cascade model to describe the propagation in layered networks. Restrictions for propagation in layered cascade model are listed as following.

- All the nodes in $Layer_0$ are owners of information in initial state;
- The way information transmitted between nodes is as the same as that in IC model. An information owner can transmit information to its neighbors which have not got information with possibility p. An information owner cannot try to transmit information to its neighbors for the second time whether the transmission is successful;
- Information is propagated from $Layer_0$(top layer) to $Layer_{n-1}$(bottom layer). When propagation from $Layer_{i-1}$ to $Layer_i$ is over, information begins to be propagated within $Layer_i$. When no more nodes in $Layer_i$ can get information, information begins to be propagated from $Layer_i$ to $Layer_{i+1}$;
- When no more nodes in $Layer_{n-1}$ can get information, information start to spread to upper layers in the same way until no nodes can get information.

In this paper, we only discuss the situation that when p is very small, because in LC model (or IC model) influence spread tends to be similar regardless which set of nodes own information initially.

3   Information Propagation in Network

3.1   Role of Edges in Networks

Suppose A is a node of degree d. When information is being propagated in the network, k neighbors of A get the information before A, then A will get the information with possibility $p_a = 1 - (1 - p)^k \approx kp$ in LC model. If we denote the possibilities that the d neighbors of A get information before A as p_1, p_2 ...p_d respectively, then $p_a = 1 - \prod_{i=1}^{d}(1 - p * p_i) \approx p * \sum_{i=1}^{d} p_i$.

The first expression of p_a can be explained as k edges are involved in information transmission and the possibility for A to get information is $k * p$, which equals to that each edge makes tantamount contribution to A to make it activated, and the value of contribution is quantified as p. The second expression of p_a can be explained as: edges connected with A make contribution of value p with possibility p_i. Hence, when information spread in LC model, we can understand the role of edges as following: an edge makes contribution of value p when an information owner tries to activate its neighbor via the edge no matter whether the activation is successful. Such edges which make contribution are useful edges.

When information propagation comes to an end, the ultimate number of information owners can be defined as $N = \sum_{i-1}^{|V|} p_i$, where p_i denotes the possibility that node v_i gets information. If we assume that k_i edges make contribution to v_i, then: $N = p \sum_{i=1}^{|v|} k_i = p|E_{inf}| = p(|E| - |E_{ninf}|)$. E_{inf} and E_{ninf} denotes the set of useful edges and useless edges.

In order to study the information spread, we will give a more general formulation of N. If edge e_i make contribution with possibility p_i, then:

$$N = p \sum_{i=0}^{|E|} p_i \tag{1}$$

where p_i relies on concrete procedure of information propagation, which has close relationships with original information owners and topology of network.

3.2   Information Propagation in LC Model

In LC model, most ultimate information owners get information from the propagation from $layer_0$ to $layer_{n-1}$. We are going to talk about the propagation from $layer_0$ to $layer_{n-1}$ rather than the whole procedure of propagation.

We use p_{ij} and p_{ij}^* to denote the possibility that v_{ij} becomes information owner ultimately and the possibility that v_{ij} get information by the transmission on layer-paths separately. Then, $p_{ij} \geq p_{ij}^*$. For p_{ij}^*, we have Theorem1.

Theorem 1 (Value of p_{ij}^*). *For node v_{ij} in a layered network, v_{ij} can get information by the transmission on layer-paths with possibility $p_{ij}^* = p^i * r_{ij}$.*

Proof (of theorem). Basis:: When $i = 1$, we have $p_{1j}^* = p * u_{1j} = p * r_{1j}$ by learning relationships of r_{ij} between different layers in section 2.1; Induction:: Assume that for nodes of $Layer_{i-1}$ we have $p_{(i-1)j}^* = p^{i-1} * r_{(i-1)j}$,

$$p_{ij}^* = \sum_{k=1}^{u_{ij}}(p * p^{i-1} * r_{(i-1)j_k}) = p^i \sum_{i=1}^{u_{ij}} r(i-1)j_k = p^i * r_{ij}$$

Since all nodes in $Layer_0$ are information owners, no edge makes contribution within $Layer_0$, that is: $C_0 = 0$, where C_i denotes contribution made within $Layer_i$. But each edge between $Layer_0$ and $Layer_1$ will make contribution. We use $C_{i(i+1)}$ to denote contribution made between $Layer_i$ and $Layer_{i+1}$. Then $C_{01} = p * |E_{01}| = p \sum_{j=1}^{v_1} u_{1j} = p \sum_{j=1}^{|v_1|} r_{1j}$.

When information starts to be propagated within $Layer_1$, we do not know the initial information owners of $Layer_1$ and as a result we cannot get value of contribution made by edges within $Layer_1$. However, we can deduce the range of value of contribution within $Layer_1$. In the worst case, when the nodes which get information from $Layer_0$ try to transmit information to their neighbors within $Layer_1$, none neighbor gets information. In other words, in $Layer_1$ only edges connected with nodes that get in-formation from top layer are useful. Contribution made within $Layer_1$ is minimized.

$Min(C_1) = p * |\{e_{1j1j'}|v_{1j} \ got \ information \ from \ Layer_0 \ while \ v_{1j'} \ did \ not\}|$
$= p \sum_{e_{1j1j'} \in E_0}[p_{1j}^*(1 - p_{1j'}^*) + p_{1j'}^*(1 - p_{1j}^*)] \approx p \sum_{e_{1j1j'} \in E_0}(p_{1j}^* + p_{1j'}^*)$
$= p \sum_{e_{1j1j'} \in E_0}(p * r_{1j} + p * r_{1j'}) = (\frac{p^2}{2}) * [\sum_{j=1}^{|V_1|}(s_{1j} * r_{1j}) + \sum_{j=1}^{|V_1|}(s_{1j} * r_{1j'})]$
$= p^2 \sum_{j=1}^{|V_1|}(s_{1j} * r(1j))$

In the best case, all edges except edge between nodes that get information from top layer make contribution. At this case, contribution is maximized.

$Max(C_1) = p * |\{e_{1j1j'}|One \ of \ v_{1j} \ and \ v_{1j'} \ get \ information \ from \ Layer_0\}|$
$= p \sum_{e_{1j1j'} \in E_0}[p_{1j}^*(1 - p_{1j'}^*) + p_{1j'}^*(1 - p_{1j}^*) + (1 - p_{1j}^*) * (1 - p_{1j'}^*)]$
$= p \sum_{e_{1j1j'} \in E_0}(1 - p_{1j}^* * p_{1j'}^*) \approx p * |E_1| = (\frac{p}{2}) * \sum_{j=1}^{|V_1|} s_{1j}$

For propagation from $Layer_1$ to $Layer_2$, because $e_{1j2j'}$ makes contribution with possibility p_{1j}, we get C_{12} according to formula (1): $C_{12} = p \sum_{e_{1j2j'} \in E} p_{1j}$. While $p_{1j}^* \leq p_{1j} < 1$, we have: $Min(C_{12}) = p \sum_{e_{1j2j'} \in E} p_{1j}^* = p \sum_{j=1}^{|V_1|}(d_{1j}*p_{1j}^*) = p^2 \sum_{j=1}^{|V_1|}(d_{1j}*r_{1j})$ and $Max(C_{12}) = p \sum_{e_{1j2j'} \in E} 1 = (\frac{p}{2}) * \sum_{j=1}^{|V_1|} d_{ij}$.

To make our conclusion more generalized, we have the following theorem.

Theorem 2 (Range of Contribution Made by Edges in Layered network). *For a given layered network $G(V,E)$, the range of contribution made in $Layer_i$ and range of contribution between $Layer_i$ and $Layer_{i+1}$ are:*

$$p^{i+1} \sum_{j=1}^{|V_i|}(s_{ij}*r_{ij}) \leq C_i \leq (\frac{p}{2}) * \sum_{j=1}^{|V_i|} s_{ij}; \quad p^{i+1} \sum_{j=1}^{|V_i|}(d_{ij}*r_{ij}) \leq C_{i(i+1)} \leq (\frac{p}{2}) * \sum_{j=1}^{|V_i|} d_{ij}$$

The deduction of C_i and $C_{(i(i+1))}$ is the same with that of C_1 and C_{12}.

4   Influence Maximization Problem

First we will give a formative definition of the problem.

Definition 2 (Influence Maximization Problem). *In a given network $G(V, E)$, $S(S \subseteq V, |S| = k)$ is the set of initial information owners. The number of ultimate information owners when the propagation from S ends up is denoted as $\sigma(S)$. Find set S^* to insure that $\sigma(S^*) = Max_{S \subset V}\sigma(S)$ and $|S^*| = k$.*

To solve the problem, we select proper initial information owners to involve more edges to make contribution. In a layered network, the total contribution is: $C = \sum_{i=0}^{n-1} C_i + \sum_{i=0}^{n-2} C_{i(i+1)}$. Its range is: $\sum_{i=0}^{n-1} Min(C_i) + \sum_{i=0}^{n-2} Min(C_{i(i+1)}) \leq Min(sum_{i=0}^{n-1}C_i + \sum_{i=0}^{n-2} C_{i(i+1)}) \leq C \leq Max(sum_{i=0}^{n-1}C_i + \sum_{i=0}^{n-2} C_{i(i+1)}) \leq \sum_{i=0}^{n-1} Max(C_i) + \sum_{i=0}^{n-2} Max(C_{i(i+1)})$

In order to maximize the contribution we should enlarge the lower bound of the range of total contribution. We use $F(S)$ to denote the lower bound, while $f(S)$ and $g(s)$ denote contribution made by edges within layers and between layers, we have : $F(S) = f(s) + g(s) = \sum_{i=0}^{n-1} \sum_{j=1}^{|V_i|} p^{i+1}(s_{ij} * r_{ij}) + \sum_{i=0}^{n-2} \sum_{j=1}^{|V_i|} p^{i+1}(d_{ij} * r_{ij})$

Hence, we should select proper S to maximize $F(S)$. In this paper we are going to apply greedy strategy to choose nodes one by one. Our algorithm is named as LCMB (Layered Cascade Model Based Algorithm).

In LCMB the complexity of computing $F(S+node_j)$ is $O(m)$ where m denotes the number of edges, so the complexity to select a new node in LCMB is $O(mn)$ where n denotes the number of nodes. The total complexity will be $O(kmn)$.

In complex networks nodes with larger degree tend to be more influential. Nodes with large degree are influence-maximization friendly. Since most real world networks yield power-law distribution of the node degrees, nodes of small degree take up the majority of a network, by less considering which can greatly

Table 1. Algorithms input: network $G(V, E)$, k; Output: Initial set S

Algorithm: LCMB	Algorithm: MLCMB
1.S = empty; 2.F(S) = 0; 3.for i=0 to k do 4. for every node(j) in V/S, do 5. delta = F(S+node(j))-F(S); 6. if delta > Max(delta) then 7. Max(delta) = delta; 8. Newnode = node(j); 9. end if 10. end for 11. S = S+newnode; 12.end for	1.S = {node(degmax)}; 2.Get the value of F(S); 3.Lastnode = node(degmax); 4.for i=0 to k do 5. for node(j) whose degree is larger than lastnode.degree*M in V/S, do 6. delta = F(S+node(j))-F(S); 7. if delta > Max(delta) then 8. Max(delta) = delta; 9. Newnode = node(j); 10. end if 11. end for 12. S = S+newnode; 13. Lastnode = newnode; 14.end for

promote efficiency of the algorithm. Hence, we can get the new algorithm called MLCMB (Modified Layered Cascade Model Based Algorithm). We initialize set S to contain the node of maximum degree. We just consider those whose degree is no less than the M-times ($M < 1$) of degree of the node last selected. The larger M is, the higher efficiency is, while $\sigma(S)$ may decrease. Complexity of MLCMB relies on M and structure of network. It is hard to give its complexity by mathematic means. We will further discuss the efficiency of MLCMB.

5 Experiment

We made experiments to compare different algorithms. Outcome of algorithms (number of information owners) and time cost are concerned.

5.1 Experiment Setup

Our experiments are based on three real world datasets. The first dataset, denoted as ASTRO-PH, is a collaboration network from e-print arXir. There are 18772 nodes and 396160 edges in ASTRO-PH. The second dataset, denoted as COND-MAT, also comes from e-print arXir. There are 23133 nodes and 186936 edges in COND-MAT. The third dataset, denoted as NotreDame, whose nodes represent pages from Univer-sity of Notre Dame and edges represent hyperlinks between them. There are 325729 nodes and 1497134 edges in NotreDame. All dataset can be get from Stanford's large network dataset collection (http://snap.stanford.edu/data). We set p as 0.01 and 0.005 separately to conduct experiments based on the following algorithms.

- DDIC (DegreeDiscountIC): The heuristic algorithm raised in [6].
- Degree: A basic algorithm that merely select nodes with largest degrees.
- LCMB: Our algorithm based on layered cascade model.
- MLCMB: Modified algorithm based on LCMB. M is set as 0.5.
- Random: The algorithm randomly selects some nodes as the outcome of it.
- Greedy: The basic greedy algorithm raised by David Kempe etc.

For a given outcome we simulated the process of information propagation on every network for 20000 times to get the information spread. Our simulation is run on a server with 2.59 GHz Dual-Core Intel Pentium E5300 and 2G memory.

And for each dataset, we compared the influence spread generated by different al-gorithms when the size of the initial set varies from 1 to 50. We compared time cost by different algorithms when the size of the initial set is 10. However, we did not show how spread influence various with the initial set when Greedy is applied, because the time spend by our server is much more than the time we can accept.

5.2   Experiment Results

We summarize experiment results in different datasets.

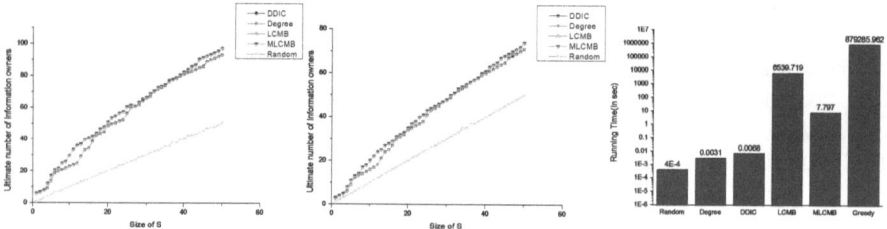

Fig. 1. Result got by running algorithms on ASTRO-PH

In all above three figures, the first two plots show influence spreads in different al-gorithms as the size of the initial set varies from 1 to 50 in three datasets when the propagation probability is 0.01 and 0.005. The third plot shows time cost by each algorithm to find a set of size 10.

As we can see from Figure 1 to Figure 3, Random as the baseline performs badly in the experiment. Almost no nodes other than nodes in the initial set get information. Degree and DDIC perform better than Random and curves illustrate the influence spreads of Degree and DDIC almost overlap with each other. LCMB and MLCMB perform better than Degree and DDIC. And influence spreads of these two algorithms are also almost the same. In ASTRO-PH, influence spread of LCMB and MLCMB is 5.36% and 4.40% higher respectively than that of Degree and DDIC when propagation probability is 0.01 and 0.005. In COND-MAT, influence spreads of these LCMB and MLCMB is 4.95% and

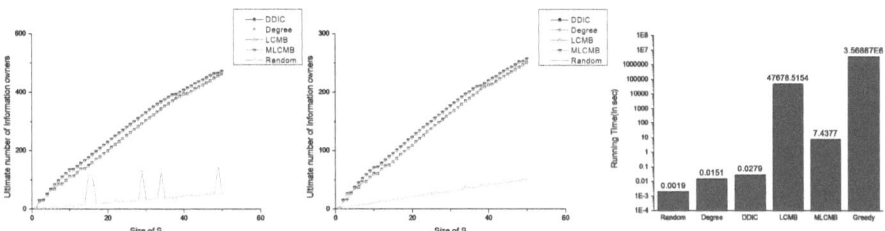

Fig. 2. Result got by running algorithms on COND-MAT

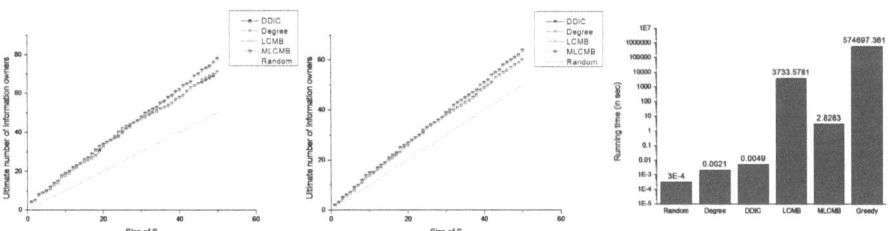

Fig. 3. Result got by running algorithms on NotreDame

3.74% higher respectively than that of Degree and DDIC when propagation probability is 0.01 and 0.005. In NotreDame, influence spread of LCMB and MLCMB is 8.09% and 7.02% higher respectively than that of Degree and DDIC when propagation probability is 0.01 and 0.005.

The third plot in all figures shows the running time of each algorithm ($p = 0.01$). We can see that Random, Degree and DDIC spend little time to get an outcome. Com-pared with above algorithms, LCMB costs much more time to get an outcome. In ASTRO-PH, the modified MLCMB is about 800 times faster than LCMB. Though still slower than algorithms like DDIC, it spends less than 10 seconds to get a result, which is acceptable. In COND-MAT, the modified MLCMB is about 1300 times faster than LCMB and spends less than 3 seconds. In NotreDame, the modified MLCMB is about 6000 times faster than LCMB.

5.3   Discussion on the Results

Several interesting phenomenon are shown by the results of experiment.

Firstly, the influence spread of Degree and DDIC are almost the same in all three networks experimented. In all of three networks, Degree and DDIC seem to perform equally to get the influence spread. DDIC always tries to select nodes with larger degrees and avoid selecting nodes whose neighbors are selected. DDIC subtract the degree of nodes whose neighbors have been selected with a calculated value, which can be viewed as "discount", and then rank the overall nodes by the gotten "discount degree". The more the neighbors of a given node are selected, the heavier the "dis-count" will be. If the "discount" is not heavy enough to change the original rank of degrees, the result of DDIC will be the same as that of Degree. Similarly, if the "dis-count" just changes the original rank of degrees a little, the result of DDIC will not be much better than that of Degree. Situations in which "discount" is not heavy enough can occur when length of route between hub nodes of the network are long.

Secondly, the influence spread of LCMB and MLCMB are almost the same and larger than that of Degree and DDIC. LCMB and MLCMB perform better than De-gree and DDIC for the nature that DDIC and Degree consider their neighbors only while LCMB and MLCMB consider contribution made by edges of the whole network. Based on LCMB, MLCMB just considers nodes with higher degrees. The de-gree bound is determined largely by the parameter of MLCMB. The smaller M is, the quicker MLCMB will be. The larger M is, the better results can be achieved. In our experiment, we get outcome of MLCMB which is no worse than that of LCMB. It indicates that the nodes' degrees we consider when M is 0.5 are big enough for us to select proper nodes. Moreover, we can further decrease the value of M to get higher speed and ensure the effectiveness of MLCMB. We believe it is meaningful to find the critical value of M which leads to the highest speed and the greatest effectiveness at the same time. However, such critical value may be different in different networks.

Thirdly, the modified algorithm MLCMB can speed up the original LCMB with different times regarding to different networks. In detailed, MLCMB is about 800 times, 1300 times and 6000 times faster than LCMB in three networks separately. When we want to choose a new node, we just consider nodes whose degrees are no less than M times of the degree of last chosen node. Most real world networks yield power-law distribution of the node degrees. And the larger the power exponent of power-law distribution is, the more the nodes with small degrees are. Hence, running time of MLCMB on such networks can be magically decreased. Because if quite a lot of nodes are nodes with small degrees in a network, when we are selecting new nodes into the initial set, they are likely to be ignored, which helps saving time. Compared with ASTRO-PH, there are a larger percentage of nodes with small degrees in NotreDame, which result to the greater speed up when MLCMB is applied to NotreDame.

6   Conclusions

In this paper we learn how information is propagated in networks and try to discover the influence spread by learning contribution made by edges in the network. And we design efficient heuristic algorithms base on our analysis to solve the influence maximization problem. Our heuristic algorithms perform better than other heuristic algorithms and run fast even applied to large-scale networks.

Several future researches are being anticipated. First, we plan to further learn the contribution made by edges and find a tighter bound of total contribu-tion, which will be helpful to get better result to solve influence maximization problem. Second, as we learned, an algorithm may perform differently in dif-ferent networks, that is, the topo-logical structure of the network can influence the result of an algorithm. We are will-ing to discover how different topological structures hamper or facilitate influence spread in networks, which is meaningful in designing effective heuristic algorithms.

Acknowledgement. This Paper was supported by the National Natural Science Foundation of China under Grant 60503021, 60721002, 60875038 and 61105069, by the Technology Support Program of Jiangsu Province under Grant BE2010180 and BE2011171, by the Innovation Fund of Nanjing University under Grant 2011CL07.

References

1. Domingos, P., Richardson, M.: Mining the Network Value of Customers. In: KDD (2001)
2. Richardson, M., Domingos, P.: Mining Knowledge-sharing Sites for Viral Marketing. In: SIGKDD (2002)
3. Kempe, D., Kleinberg, J., Tardos, E.: Maximizing the Spread of Influence through a Social Network. In: SIGKDD (2003)
4. Kempe, D., Kleinberg, J., Tardos, É.: Influential Nodes in a Diffusion Model for Social Networks. In: Caires, L., Italiano, G.F., Monteiro, L., Palamidessi, C., Yung, M. (eds.) ICALP 2005. LNCS, vol. 3580, pp. 1127–1138. Springer, Heidelberg (2005)
5. Leskovec, J., Krause, A., Guestrin, C., et al.: Cost-effective Outbreak Detection in Networks. In: KDD (2007)
6. Chen, W., Wang, Y., Yang, S.: Efficient Influence Maximization in Social Networks. In: KDD (2009)
7. Wang, Y., Cong, G., Song, G., Xie, K.: Community-based Greedy Algorithm for Mining Top-K Influential Nodes in Mobile Social Networks. In: KDD (2010)
8. Chen, W., Wang, Y., Yang, S.: Scalable Influence Maximization for Prevalent Viral Marketing in Large-Scale Social Networks. In: KDD (2010)
9. Watts, D.J., Strogatz, S.H.: Collective Dynamics of 'Small-world' Networks. Nature 393, 440 (1998)
10. Goldenberg, J., Libai, B., Muller, E.: Talk of the Network: A Complex Systems Look at the Underlying Process of Word-of-Mouth. Marketing Letters (2001)
11. Goldenberg, J., Libai, B., Muller, E.: Using Complex Systems Analysis to Advance Marketing Theory Development. Academy of Marketing Science Review (2001)
12. Goyal, A., Bonchi, F., Lakshmanan, L.V.S.: Learning Influence Probabilitie in Social Networks. In: WSDM (2010)
13. Backstrom, L., Huttenlocher, D., Kleinberg, J., Lan, X.: Group Formation in Large Social Networks: Membership, Growth, and Evolution. In: KDD (2006)
14. Apolloni, A., Channakeshava, K., Durbeck, L., et al.: A study of Information Diffusion over a Realistic Social Network Model
15. Kossinets, G., Watts, D.J.: Empirical Analysis of an Evolving Social Network. Science 311, 88 (2006)
16. Kernighan, B.W., Lin, S.: A Efficient Heuristic Procedure for Partitioning Graphs. Bell System Technical Journal 49, 291 (1970)

Semi-supervised Clustering Ensemble Based on Multi-ant Colonies Algorithm*

Yan Yang, Hongjun Wang, Chao Lin, and Jinyuan Zhang

School of Information Science and Technology,
Provincial Key Lab of Cloud Computing and Intelligent Technology,
Southwest Jiaotong University, Chengdu, 610031, P.R. China
{yyang,wanghongjun}@swjtu.edu.cn, linchao0916@126.com, lingsuch@qq.com

Abstract. Semi-supervised clustering ensemble has emerged as an important elaboration of classical clustering problem that improves quality and robustness in clustering by combining the results of different clustering components with user provided constraints. In this paper, we propose a novel semi-supervised consensus clustering algorithm based on multi-ant colonies. Our method incorporates pairwise constraints not only in each ant colony clustering process, but also in computing new similarity matrix during the multi-ant colonies ensemble. Experimental results demonstrate the effectiveness of the proposed method.

Keywords: semi-supervised clustering, clustering ensemble, multi-ant colonies clustering, pairwise constraints.

1   Introduction

Clustering problem has attracted extensive attention due to its high impact on various important applications, such us document mining, image retrieval, and bioinformatics. Basically clustering aims to group data objects into clusters such that objects in the same cluster are similar to each other while objects in different clusters are dissimilar [1]. Clustering ensembles have emerged as a powerful method for improving both the quality and the robustness of the clustering by aggregating multiple clustering solutions into a single one [2]. Strehl and Ghosh proposed three effective and efficient combiners for solving cluster ensembles based on a hypergraph model [3]. We presented an aggregated clustering approach which imitates the cooperative behavior of multi-ant colonies [4]. Ayad and Kamel developed a voting-based consensus of cluster ensembles [5]. Jia et al. introduced a selective spectral clustering ensemble [6]. Masson et al. gave an ensemble clustering in the belief functions framework [7].

However, clustering ensemble without any prior knowledge or background information is still a challenging problem. Recently, there is an emerging interest in

---

* This work is partially supported by the National Science Foundation of China (Nos. 61170111, 61003142 and 61152001) and the Fundamental Research Funds for the Central Universities (No. SWJTU11ZT08).

T. Li et al. (Eds.): RSKT 2012, LNAI 7414, pp. 302–309, 2012.

incorporating limited supervision information into clustering ensemble to obtain user-desired and more accurate partition. Duan et al. studied a semi-supervised clustering framework based on a combination of consensus-based and constrained clustering techniques [8]. Greene et al. explained constraint selection by committee in semi-supervised clustering [9]. Iqbal et al. solved semi-supervised clustering ensemble by voting [10]. Wang et al. explored a semi-supervised cluster ensemble model based on both semi-supervised learning and ensemble learning technologies using Bayesian network and EM algorithm [11].

In this paper, we focus on the constrained clustering ensemble, where the pairwise constraints provide the supervision information: a must-link (ML) constraint specifies that the pair of instances should always be assigned to the same group, and a cannot-link (CL) constraint specifies that the pair of instances should never be placed into the same group [12]. First, we consider supervision provided in the form of the ML and CL constraints on pairs of instances in each ant colony clustering process to enhance the performance of clustering component. Next, we address the problem of finding a consensus clustering as multi-ant colonies clustering problem by incorporating user provided constraints to compute new similarity matrix. Finally, we demonstrate the effectiveness of our method.

The rest of this paper is organized as follows: Section 2 discusses the ant colony clustering algorithm with pairwise constraints. Section 3 establishes the semi-supervised clustering ensemble framework. Section 4 reports the test results evaluating the performance of the proposed algorithm. Finally, Section 5 provides conclusions and future work.

2   Ant Colony Clustering with Pairwise Constraints

The ant colony clustering algorithm is inspired by the behavior of ant colonies in clustering their corpses and sorting their larvae. One of the early studies related to this domain is LF algorithm [13], and then there have been some successful algorithms [14]. In general, the ant colony clustering works as follows: first, data objects are randomly projected onto a plane with a Cartesian grid. Second, each ant chooses an object at random, and picks up or moves or drops down the object according to picking-up or dropping probability with respect to the similarity of the current object within a local region. Finally, clusters are collected from the plane.

Suppose that an ant is located at site r at time t, and finds an object o_i at that site. The local density of objects similar to type o_i at the site r is given by [14].

$$f(o_i) = max \left\{ 0, \frac{1}{s^2} \sum_{o_j \in Neigh_{s \times s(r)}} \left[1 - \frac{d(o_i, o_j)}{\alpha(1 + ((v-1)/v_{max}))} \right] \right\} \quad (1)$$

where $f(o_i)$ is a measure of the average similarity density of object o_i with the other objects o_j present in its neighborhood. $Neigh_{s \times s(r)}$ denotes a square of

$s \times s$ sites surrounding site r. $d(o_i, o_j)$ is the distance between two objects o_i and o_j in the space of attributes. α is a factor that defines the scale of similarity between objects.

The parameter v denotes the speed of the ants, and v_{max} is the maximum speed. Fast moving ants form clusters roughly on large scales, while slow ants group objects at smaller scales by placing objects with more accuracy. There are three versions of clustering components based on ants moving with different speed.

(1) v is a constant. All ants move with the same speed at any time.
(2) v is random. The speed of each ant is distributed randomly in $[1, v_{max}]$.
(3) v is randomly decreasing. The speed term starts with large value (forming clusters), then the value of the speed gradually decreases in a random manner (helping ants to cluster more accurately).

The probability conversion function is a function of $f(o_i)$ that converts the average similarity of a data object into the probability of picking-up or dropping-down for an ant. The picking-up probability for a randomly moving ant that is currently not carrying an object to pick up an object is given by

$$P_p = 1 - sigmoid(f(o_i)) \tag{2}$$

The dropping-down probability for a randomly moving loaded ant to deposit an object is given by

$$P_d = sigmoid(f(o_i)) \tag{3}$$

Where sigmoid function has a natural exponential form, only one parameter needs to be adjusted in the calculation.

According to our work before done in [15], we incorporate the pairwise constraints to guide the clustering process towards an accurate partition. We note that the smaller the similarity of a data object is, the higher the picking-up probability is and the lower the dropping-down probability is, and vice versa. So we can make use of the ML and CL to help ants picking-up or dropping-down the objects.

Let NuM be the number of ML constraints among the object o_i and the other objects o_j present in its neighborhood that can be denoted as follows:

$$NuM = \begin{cases} NuM + 1 & if(o_i, o_j) \in ML \\ 0 & otherwise \end{cases} \tag{4}$$

Let NuC be the number of CL constraints among the object o_i and the other objects o_j present in its neighborhood that can be denoted as follows:

$$NuC = \begin{cases} NuC + 1 & if(o_i, o_j) \in CL \\ 0 & otherwise \end{cases} \tag{5}$$

If NuM is greater than a given constant, it means there are many objects that must belong to the same cluster in this object's neighborhood, and then the ant drops down the object. When NuC is greater than a given constant, it implies the object is unlikely to its neighborhood, so the ant must pick it up and move it to a new position.

3   Semi-supervised Clustering Ensemble

Fig. 1 shows a schematic diagram for semi-supervised consensus clustering based on multi-ant colonies. The first phase generates a diverse collection of three clustering components using ant colony clustering algorithm with pairwise constraints under different moving speed such as constant, random, and randomly decreasing. The second phase combines three clustering components produced from the previous phase using the hypergraph model to be an aggregated similarity matrix. The third phase is the ensemble process that incorporates the ML and CL constraints to form what is called a consensus clustering.

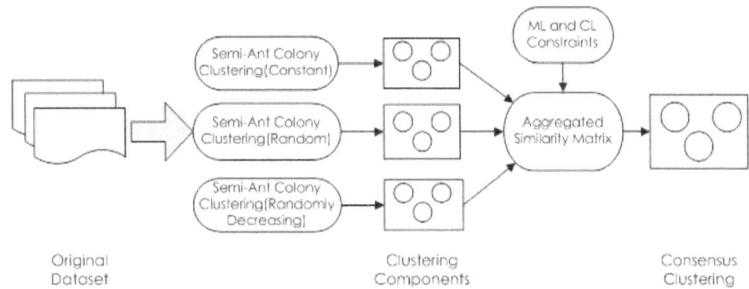

Fig. 1. Semi-supervised clustering ensemble based on multi-ant colonies

Let $\mathbf{O} = \{\mathbf{o_1}, \mathbf{o_2}, ..., \mathbf{o_n}\}$ denote a set of objects, and a clustering component of these n objects into k clusters can be represented as a label vector $\lambda \in N^n)$. Given r group clustering components with the q-th grouping $\lambda^{(q)}$ having $k^{(q)}$ clusters, a binary membership indicator matrix $H^{(q)\in I^{n \times k^{(q)}}}$ is constructed, in which each cluster is represented as a hyperedge (column). We can further define a concatenated block matrix

$$H = (H^{(1)}, ..., H^{(r)}) \tag{6}$$

as the adjacency matrix of a hypergraph with n vertices and $SIGMA_{q=1}^{r} k^{(q)}$ hyperedges. Each row of matrix $\mathbf{H}$ denotes a vertex (object) and each column specifies a hyperedge, where 1 indicates that the vertex corresponding to the row belongs to the same cluster and 0 indicates that it is not or the data is unknown. Thus, the set of clustering components have been transformed to a hypergraph with adjacency matrix $\mathbf{H}$.

Generally speaking, two objects are considered to be fully similar if they are in the same cluster, or they are fully dissimilar if they are not. So the similarity measure can be viewed as the fraction of clustering components in which two objects are in the same cluster. The aggregated similarity matrix $\mathbf{S}$ among n objects can be defined by

$$S = \frac{1}{r}HH^T \tag{7}$$

where matrix $\mathbf{H^T}$ is the transposition of matrix $\mathbf{H}$, $\mathbf{S}$ is $n \times n$ sparse matrix.

For ML and CL constraints are often represented as

$S_{ij} = 1,$ $if(o_i, o_j)$ likely to be in same class

$S_{ij} = 0,$ $if(o_i, o_j)$ unlikely to be in same class

So we make use of these limited degrees of supervision to update the value of S to improve the accuracy of clustering ensemble.

Let us illustrate clustering ensemble process using a simple example in Table 1 [9]. There are 5 objects $o_i(i = 1, 2, , 5)$ corresponding to 2 label vectors of clustering components. This clusterings are represented as the adjacency matrix $\mathbf{H}$ of hypergraph shown in Table 2, where $r = 3$, $k^{(1,2)} = 2$, the number of vertices (objects)=5, the number of hyperedges=6. By formula (7) the aggregated similarity matrix $\mathbf{S}$ is in the form of a symmetric matrix shown in Table 3. If known $ML = \{(o_3, o_4)\}$, then the modified matrix $\mathbf{S}$ is shown in Table 4.

Table 1. Label vectors

	$\lambda^{(1)}$	$\lambda^{(2)}$	$\lambda^{(3)}$
o_1	1	1	1
o_2	1	2	2
o_3	2	1	2
o_4	2	2	2
o_5	2	2	2

Table 2. The adjacency matrix of hypergraph

H	$H^{(1)}$		$H^{(2)}$		$H^{(3)}$	
v_1	1	0	1	0	1	0
v_2	1	0	0	1	0	1
v_3	0	1	1	0	0	1
v_4	0	1	0	1	0	1
v_5	0	1	0	1	0	1

Table 3. Similarity Matrix S

	o_1	o_2	o_3	o_4	o_5
o_1	1	0.33	0.33	0	0
o_2	0.33	1	0.33	0.67	0.67
o_3	0.33	0.33	1	0.67	0.67
o_4	0	0.67	0.67	1	1
o_5	0	0.67	0.67	1	1

4   Experimental Results

The proposed semi-supervised clustering ensemble algorithm was implemented in VC++6.0 and is tested on two datasets from UCI machine learning repository (http://www.ics.uci.edu/mlearn/MLRepository.html). Fisher's Iris Plants Dataset contains three classes of 50 instances each in a 4D space, and Wine Dataset consists of three classes of 178 instances in a 13D space.

Table 4. Similarity Matrix S with ML & CL constraints

	o_1	o_2	o_3	o_4	o_5
o_1	1	0	0.33	0	0
o_2	0	1	0.33	0.67	0.67
o_3	0.33	0.33	1	1	0.67
o_4	0	0.67	1	1	1
o_5	0	0.67	0.67	1	1

We evaluated the clustering ensemble algorithm (En-Ant) compared to the average ant colony clustering algorithm (Average-Ant) and Cop-Kmeans algorithm [15] with different pairwise constrains cases using F-measure in Fig. 2 [13]. We conducted 30 trials on each dataset and selected randomly ML and CL constraints. It is not involve in pairwise constraints when number of constraints is equal to 0.

It can be observed from Fig. 2 that the results of clustering ensemble are better than that of the average results before the ensemble and Cop-Kmeans. In general, it is clear that the clustering quality of all of algorithms improves with the increase of the number of pairwise constraints. Fig. 3 shows the F-measure of

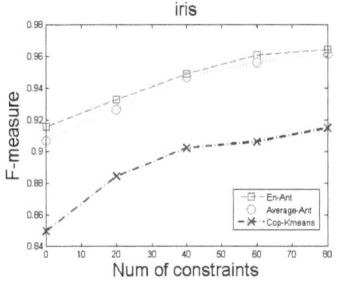
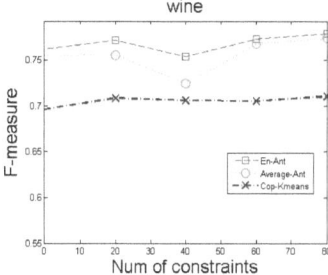

Fig. 2. F-measure of En-Ant, Aaverage-Ant and Cop-Kmeans algorithms on the Iris and Wine datasets

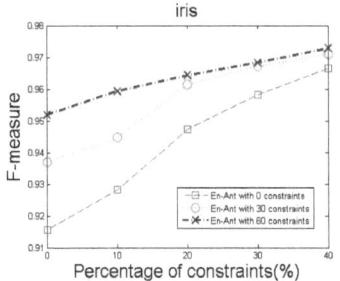
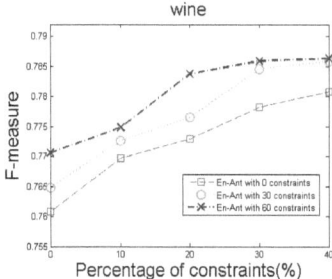

Fig. 3. F-measure of En-Ant with different initial constraints on the Iris and Wine datasets

semi-supervised multi-ant colonies ensemble under clustering components with null, 20%, and 40% ML and CL constraints incorporating different pairwise constraints in aggregated similarity matrix respectively. It is noted that the clustering components with high constraints percent has superior quality. It is also noted that the clustering performance of all algorithms enhances with the increase of the number of pairwise constraints.

5   Conclusion

Combining the information provided by a large collection of clustering components will produce a superior clustering solution. In this paper we have explored multi-ant colonies to solve the clustering problem with semi-supervised policy. Experimental results show that using pairwise constraint information improves the accuracy and stability of existing clustering algorithms. For future work, we will study consensus model for massive dataset.

References

1. Han, J., Kamber, M.: Data Mining: Concepts and Techniques, 2nd edn. Morgan Kaufmann (2006)
2. Topchy, A., Jain, A.K., Punch, W.: Clustering ensembles: models of consensus and weak partitions. IEEE Trans. on Pattern Analysis and Machine Intelligence 27(12), 1866–1881 (2005)
3. Strehl, A., Ghosh, J.: Cluster ensembles–a knowledge reuse framework for combining multiple partitions. Machine Learning Research 3, 583–617 (2002)
4. Yang, Y., Kamel, M.S.: An aggregated clustering approach using multi-ant colonies algorithms. Pattern Recognition 39(7), 1278–1289 (2006)
5. Ayad, H.G., Kamel, M.S.: On voting-based consensus of cluster ensembles. Pattern Recognition 43(5), 1943–1953 (2010)
6. Jia, J.H., Xiao, X., Liu, B.X., et al.: Bagging-Based Spectral Clustering Ensemble Selection. Pattern Recognition Letters 32(10), 1456–1467 (2011)
7. Masson, M.H., Denoeux, T.: Ensemble clustering in the belief functions framework. International Journal of Approximate Reasoning 52(1), 92–109 (2011)
8. Duan, C., Cleland-Huang, J., Mobasher, B.: A consensus based approach to constrained clustering of software requirements. In: Proceeding of the 17th ACM Conference on Information and Knowledge Management (CIKM 2008), pp. 1073–1082 (2008)
9. Greene, D., Cunningham, P.: Constraint Selection by Committee: An Ensemble Approach to Identifying Informative Constraints for Semi-supervised Clustering. In: Kok, J.N., Koronacki, J., Lopez de Mantaras, R., Matwin, S., Mladenič, D., Skowron, A. (eds.) ECML 2007. LNCS (LNAI), vol. 4701, pp. 140–151. Springer, Heidelberg (2007)
10. Iqbal, A.M., Moh'd, A., Khan, Z.A.: Semi-supervised clustering ensemble by voting. In: Proceeding of the International Conference on Information and Communication Systems (ICICS 2009), pp. 1–5 (2009)
11. Wang, H.J., et al.: Semi-Supervised Cluster Ensemble Model Based on Bayesian Network. Journal of Software 21(11), 2814–2825 (2010)

12. Wagstaff, K., Cardie, C.: Clustering with instance-level constraints. In: Proceeding of the International Conference on Machine Learning, pp. 1103–1110 (2000)
13. Yang, Y., Kamel, M.: Clustering ensemble using swarm intelligence. In: Proceeding of IEEE Swarm Intelligence Symposium, pp. 65–71 (2003)
14. Lumer, E., Faieta, B.: Diversity and adaptation in populations of clustering ants. In: Proceeding of the Third International Conference on Simulation of Adaptive Behavior: From Animals to Animats, vol. 3, pp. 499–508 (1994)
15. Yang, Y., Chen, J., Tan, W.: Enhancing ant-based clustering using pairwise constraints. In: Proceeding of the 4th International Conference on Intelligent Systems & Knowledge Engineering (ISKE 2009), pp. 76–81 (2009)

Semi-supervised Hierarchical Co-clustering*

Feifei Huang[1], Yan Yang[1], Tao Li[2], Jinyuan Zhang[1],
Tonny Rutayisire[1], and Amjad Mahmood[1]

[1] School of Information Science and Technology,
Provincial Key Lab of Cloud Computing and Intelligent Technology,
Southwest Jiaotong University, Chengdu, 610031, P.R. China
[2] School of Computer Science, Florida International University,
Miami, FL 33199, USA
feifei-huang521@163.com, yyang@swjtu.edu.cn, taoli@cs.fiu.edu,
lingsuch@qq.com, rutantonio14@yahoo.com, amjad.pu@gmail.com

Abstract. Hierarchical co-clustering aims at generating dendrograms for the rows and columns of the input data matrix. The limitation of using simple hierarchical co-clustering for document clustering is that it has a lot of feature terms and documents, and it also ignores the semantic relations between feature terms. In this paper a semi-supervised clustering algorithm is proposed for hierarchical co-clustering. In the first step feature terms are clustered using a little supervised information. In the second step, the feature terms are merged as new feature attributes. And in the last step, the documents and merged feature terms are clustered using hierarchical co-clustering algorithm. Semantic information is used to measure the similarity during the hierarchical co-clustering process. Experimental results show that the proposed algorithm is effective and efficient.

Keywords: document clustering, co-clustering, semi-supervised clustering, hierarchical clustering.

1 Introduction

Document clustering is a fundamental and effective tool for efficient organization, summarization, navigation and retrieval of massive amount of documents[1]. In general, the problem of document clustering is described as follows: given a set of documents, group them into different clusters where documents in the same cluster are similar to each other and documents in different clusters are dissimilar to each other[2].

Hierarchical co-clustering in document clustering uses both documents and feature terms for clustering[3]. For massive documents and feature terms, using hierarchical co-clustering directly, the time complexity will be increased, and

* This work is partially supported by the National Science Foundation of China (Nos. 61170111, 61003142 and 61152001) and the Fundamental Research Funds for the Central Universities (No. SWJTU11ZT08).

at the same time, the accuracy might be reduced. Note that hierarchical co-clustering only considers the term weight for document, but ignores the semantic relations between feature terms[4]. So, finding feature term clusters effectively, and combining them effectively, may be able to reduce the feature dimensions and improve the efficiency, and also improve the clustering accuracy[5].

In this paper, first of all, we perform clustering on the feature terms by using semi-supervised clustering algorithm, and then combine the feature terms in the same cluster as a single feature attribute. After that the hierarchical co-clustering is used to cluster documents and feature terms simultaneously. During the hierarchical co-clustering process, the semantic information is used to supplement a cooperative matrix.

The rest of the paper is organized as follows: Section 2 provides a detailed description of how we used a semi-supervised clustering approach to cluster feature terms and then combining the feature terms in the same cluster to be a feature attribute of a document. In section 3, latent semantic information is used to measure the Similarity between Documents (SD) and the Similarity between Feature Sets (SFS) which results into the cooperative matrix. In section 4, we introduce the improved hierarchical co-clustering approach and explain in details its underlying idea, where as section 5 describes the experimental procedure, datasets, evaluation methods and results. Finally, section 6 summarizes the work in this paper, features our conclusions and gives a snapshot of future work.

2   Feature Clustering

The general model used by document clustering is a vector space model matrix[6], shown as formula(1), where each row d_i represents a document, each column T_j represents a feature term, w_{ij} is the weight of feature term T_j in document d_i. The greater the weight is, the more frequently the feature term T_j appears in the ith document, and less frequently in other documents [6].

$$W_{m \times n} = \begin{array}{c} \\ d_1 \\ d_2 \\ \vdots \\ d_m \end{array} \begin{pmatrix} T_1 & T_2 & \cdots & T_n \\ w_{11} & w_{12} & \cdots & w_{1n} \\ w_{21} & w_{22} & \cdots & w_{2n} \\ \vdots & \vdots & \vdots & \vdots \\ w_{m1} & w_{m2} & \cdots & cw_{mn} \end{pmatrix} \tag{1}$$

Semi-supervised feature terms clustering uses priori knowledge to guide clustering for those feature terms which have relationship among each other. the pair-wise constraints are used as priori information. First of all, the pair-wise constraint sets, containing Must-link sets (M) and Cannot-link sets (C), are found out in feature term set appeared frequently. After that the constraint sets can be expanded by using the K nearest set method, and clustering according to the partition results of the constraint set. Finally the feature terms in the same cluster are combined as a single feature attribute[7].

Wagstaff proposed Must-link and Cannot-link [8], and analyzed the nature of the two constraints. M and C have the properties of symmetry and transitivity [8], as shown in formulas (2) and (3).

$$\begin{cases} (x_i, x_j) \in M \Leftrightarrow (x_j, x_i) \in M \\ (x_i, x_j) \in C \Leftrightarrow (x_j, x_i) \in C \end{cases} \tag{2}$$

$$\begin{cases} (x_i, x_j) \in M \& (x_j, x_k) \in M \Leftrightarrow (x_i, x_k) \in M \\ (x_i, x_j) \in M \& (x_j, x_k) \in C \Leftrightarrow (x_i, x_k) \in C \end{cases} \tag{3}$$

In many cases, the expanded information is not enough, so M and C need to be expanded again by using the K nearest set method. The advantages of the K nearest set method is that the distribution information of the objects is be found by a minimal overhead, then a large Pair-wise constraint set is be constructed.

The K nearest set method to expand M is that if an object appears when the distance between the object and one of a Must-link pair is very short, and the distance between the object and another of the Must-link pair is shorter than the distance between the two of the pair, and the relation between the object and any of the two of the Must-link pair are both not belong to set C, then the relation between the object and one of the Must-link pair belongs to M [8].

Similarly, the K nearest set method to expand C is that if an object appears when the distance between the object and one of a Cannot-link pair is very short, and the distance between the object and another of the Cannot-link pair is farer than the distance between the two of the pair, and the relation between the object and another of the Cannot-link pair are both not belong to set M, then the relation between the object and one of the Must-link pair belongs to C [8].

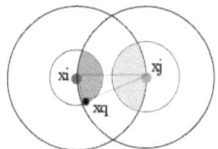   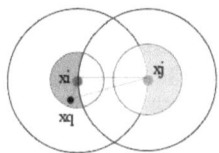

Fig. 1. Expand M with K nearest Set   **Fig. 2.** Expand C with K nearest Set

As shown in Fig.1, $(x_i, x_j) \in M$, $\mathrm{dis}(x_i, x_j)$ is the distance between x_i and x_j, if x_q is one member of the K nearest set for x_i, if $\mathrm{dis}(x_q, x_j) \leq \mathrm{dis}(x_i, x_j)$, and $(x_i, x_q) \notin C$, $(x_q, x_j) \notin C$, then$(x_q, x_j) \in M$.

As shown in Fig.2, $(x_i, x_j) \in C$, $\mathrm{dis}(x_i, x_j)$ is the distance between x_i and x_j, if x_q is one member of the K nearest set for x_i, if $\mathrm{dis}(x_q, x_j) \geq \mathrm{dis}(x_i, x_j)$, and $(x_q, x_j) \notin M$, then$(x_q, x_j) \in C$.

According to the K nearest set method above, the steps to extend constraints set are:

1 Extend M and C by symmetry and transitivity.
2 Extend M with K nearest set method.

3 Extend C with K nearest set method.

4 Loop until convergence.

When convergence, M has reflexive, symmetry and transitivity properties, so Must-link is also a partition, and is a result of clustering. Result of clustering can be represented as A_1, A_2, , A_k, in total, there are k classes.

After feature clustering, all those feature terms in each cluster are combined to be a new attribute, which is the sum of the weights of feature terms in the same cluster as shown in formula (4). cw_{ip} in formula (4) as the new standardized attribute value after combined feature terms, is the sum value of the weights in formula (1), represents the weight of the pth feature terms cluster in the ith document.

$$cw_{ip} = \frac{\sum\limits_{t_j \in A_p} w_{ij}}{\max\limits_{i<m,p<k} cw_{ip}} \qquad i \leq p \leq k \qquad (4)$$

After feature terms are combined, the new vector space model matrix generated as formula (5). t_1, t_2, $\cdots$, t_k represent the 1st, 2nd, $\cdots$, k^{th}, feature term clusters which are no longer represented by feature terms, but feature sets.

$$W_{m \times k} = \begin{array}{c} \\ d_1 \\ d_2 \\ \vdots \\ d_m \end{array} \begin{array}{cccc} t_1 & t_2 & \cdots & t_k \\ \left(\begin{array}{cccc} cw_{11} & cw_{12} & \cdots & cw_{1k} \\ cw_{21} & cw_{22} & \cdots & cw_{2k} \\ \vdots & \vdots & \vdots & \vdots \\ cw_{m1} & cw_{m2} & \cdots & cw_{mk} \end{array} \right) \end{array} \qquad (5)$$

3   Similarity Measurement

The value of the weights can not uniquely identify the degree of closeness between them. By the latent semantic information, finding out SD and SFS will obtain better cluster results.

Semantic information used here means that the higher the similarity between two feature sets is, the more times they occurred in the same document, and the closer the relation between the documents in which the two feature sets both occurred. Also, the higher the similarity between the documents is, the more times they occurred the same feature sets are in the two documents at the same time, and the closer the relation between these feature sets which occurred in the two documents at the same time [9].

First step is calculating similarity rel1 (t_i, t_j) for the 1st iteration between feature set t_i and t_j, calculated as formula (6)[9], $sim(d_p, d_q)$ represents the angle cosine between document d_p and document d_q, $CommD$ represents the document set both t_i and t_j occurred at the same time. Then standardized the matrix is calculated as formula (7).

$$rel^1(t_i, t_j) = \begin{cases} 1 \, , & \text{if } t_i = t_j \\ \sum\limits_{d_p, d_q \in CommD\&d_p \neq d_q} sim(d_p, d_q) \, , & \text{otherwise} \end{cases} \quad (6)$$

$$r^1(t_i, t_j) = \begin{cases} 1 \, , & \text{if } t_i = t_j \\ \dfrac{rel^1(t_i, t_j)}{\max\limits_{g,h<r,g \neq h} rel^1(t_g, t_h)} \, , & \text{otherwise} \end{cases} \quad (7)$$

Method for calculating similarity rel1 (d_i, d_j) for the 1st iteration between document d_i and document d_j is shown in formula (8)[9], $sim(t_p, t_q)$ represents the angle cosine between feature set t_p and t_q, $CommT$ represents the feature set which are both occurred in document d_i and d_j at the same time, then standardize the matrix as formula (9).

$$rel^1(d_i, d_j) = \begin{cases} 1 \, , & \text{if } d_i = d_j \\ \sum\limits_{t_p, t_q \in CommT\&t_p \neq t_q} sim(t_p, t_q) \, , & \text{otherwise} \end{cases} \quad (8)$$

$$r^1(d_i, d_j) = \begin{cases} 1 \, , & \text{if } d_i = d_j \\ \dfrac{rel^1(d_i, d_j)}{\max\limits_{g,h<r,g \neq h} rel^1(d_g, d_h)} \, , & \text{otherwise} \end{cases} \quad (9)$$

Method for calculating the f^{st} iteration similarity relf (t_i, t_j) between feature set t_i and t_j, calculated as formula (10). Then standardize the formula as before.

$$rel^f(t_i, t_j) = \begin{cases} 1 \, , & \text{if } t_i = t_j \\ \sum\limits_{d_p, d_q \in CommD\&d_p \neq d_q} r^{f-1}(d_p, d_q) \, , & \text{otherwise} \end{cases} \quad (10)$$

Method for calculating the f^{st} iteration similarity relf (d_i, d_j) between document d_i and document d_j is shown in formula (11). Then standardize the formula as before.

$$rel^f(d_i, d_j) = \begin{cases} 1 \, , & \text{if } d_i = d_j \\ \sum\limits_{t_p, t_q \in CommT\&t_p \neq t_q} r^{f-1}(t_p, t_q) \, , & \text{otherwise} \end{cases} \quad (11)$$

In each iteration, SFS is calculated with the last SD, and then SD is updated by the latest SFS. So after iteration, SFS and SD are closer to the true value. After many times of iteration, until $th^f \leq th$, as shown in formula (12), the final SFS (r (t_i, t_j)) and the final SD (r (d_i, d_j)) are generated.

$$th^f = \sum_{i,j<k} |r^f(t_i, t_j) - r^{f-1}(t_i, t_j)| + \sum_{p,q<m} |r^f(d_p, d_q) - r^{f-1}(d_p, d_q)| \quad (12)$$

A cooperative matrix with similarity of the two types of objects which contains documents and features, shown as formula (13), is constructed. In this cooperative matrix, compared to the weight matrix of original feature sets and documents (shown in formula (5)), SD and SFS are merged to improve the similarity matrix. So the two kinds of objects are considered as the same objects to be clustered by the hierarchical algorithm simultaneously.

$$
r_{m \times k} = \begin{array}{c} \\ t_1 \\ t_2 \\ t_k \\ \\ \vdots \\ d_1 \\ d_2 \\ \\ \vdots \\ d_m \end{array}
\begin{array}{c}
t_1 \quad\quad t_2 \quad\quad \cdots \quad\quad t_k \quad\quad d_1 \quad\quad d_2 \quad\quad \cdots \quad\quad d_m
\end{array}
\left(
\begin{array}{cccccccc}
r(t_1,t_1) & r(t_1,t_2) & \cdots & r(t_1,t_k) & cw_{11} & cw_{21} & \cdots & cw_{m1} \\
r(t_2,t_1) & r(t_2,t_2) & \cdots & r(t_2,t_k) & cw_{12} & cw_{22} & \cdots & cw_{m2} \\
r(t_k,t_1) & r(t_k,t_2) & \cdots & r(t_k,t_k) & cw_{1k} & cw_{2k} & \cdots & cw_{mk} \\
\vdots & \vdots & \vdots & \vdots & \vdots & \vdots & \vdots & \vdots \\
cw_{11} & cw_{12} & \cdots & cw_{1k} & r(d_1,d_1) & r(d_1,d_2) & \cdots & r(d_1,d_m) \\
cw_{21} & cw_{22} & \cdots & cw_{2k} & r(d_2,d_1) & r(d_2,d_2) & \cdots & r(d_2,d_m) \\
\vdots & \vdots & \vdots & \vdots & \vdots & \vdots & \vdots & \vdots \\
cw_{m1} & cw_{m2} & \cdots & cw_{mk} & r(d_m,d_1) & r(d_m,d_2) & \cdots & r(d_m,d_m)
\end{array}
\right)
$$

(13)

4   Hierarchical Co-clustering

Hierarchical co-clustering clusters the documents and feature sets at the same time. In this paper, the documents and feature sets as leaf nodes are clustered hierarchically from bottom to top with the similarity matrix shown in the 3rd section.

For example, the document title and feature sets are shown in Table 1 and Table 2 respectively.

Table 1. Documents Title

d_1	Swarm Intelligence Research
d_2	Ant colony algorithm
d_3	Ants in Animal World
... ...	
... ...	

Table 2. Feature Sets

t_1	Swarm Intelligence
t_2	Ant
t_3	Animal
t_4	Ant colony
... ...	

Traditional hierarchical clustering considers only one kind of objects, i.e "Document", as expressed in Fig 3. However, in hierarchical co-clustering, there are two kinds of objects, "Documents" and "Feature sets". The process of hierarchical co-clustering is shown in Fig.4, in which, clustering results in each clustering step contains a set of documents (represented by blue nodes) and feature sets(represented by orange nodes).

Hierarchical co-clustering makes use of the similarity between documents and feature sets, SD and SFS. It gets better results than the traditional hierarchical clustering, and generates the topic in each cluster at the same time, because the feature sets are the topic terms in each cluster of the result.

5   Experiments Evaluation

In this paper, experiments are carried out on data from the Reuters-21578 collection, which is a standard document-clustering corpus composed of 21578 news

Fig. 3. Traditional hierarchical clustering process   **Fig. 4.** Hierarchical co-clustering process

articles in 1987 [10]. The experiment is performed 10 times, where in each run, total of 1000 documents of 7 classes, 125 in each class, are selected randomly. The experimental steps are as follows.

5.1 Data Preprocessing

Cleaning of a document is to get rid of unwanted elements, like tags, stop-words, and stemming of words. The textual contents are extracted ignoring the textual structure and organization [11].

The most commonly used document representation is vector space, shown as formula (1). A well-known approach for computing term weight w_{ij} is the TF-IDF-weighting. As shown in formula (14), tf_{ij} is the number of occurrences of the term j in a document i. the df_i is the document frequency, that is the number of documents in which the term j occurs at least once. And N is the total number of documents[12].

$$w^{ij} = tf_{ij} \lg \frac{N}{df_j} \tag{14}$$

In order to reduce the dimension of the feature vector, the terms whose document frequency is less than 0.5% or greater than 90%, are removed. Only a small number of $numw$ terms with the highest weights in each document are selected as indexing terms[13].

5.2 Experiment Process

First of all, the Frequent Term Set (FTS) is sampled from the 1000 documents, 100 feature terms with the largest df. And the Must-link and Cannot-link are found out from FTS. After that the constraint sets are extended, and the semi-supervised clustering for the feature terms is performed, shown in the 2rd section.

Then, the similarity matrix, containing the weight between feature sets and documents, SD and SFS, is calculated by the method shown in section 3.

Based on the similarity matrix, hierarchical co-clustering is performed. When the number of clusters is equal to 7, the algorithm terminates.

5.3   Experiment Results

In this paper, F-measure and NMI [9] metrics are used as the evaluation measure for the clustering results. The analysis is shown in Fig.5 and Fig.6. When calculating the similarity matrix, if the threshold th is set to different values, F-measure and NMI are also changed along with th, Comparing the results, when th is set to be 0.008, the best clustering results are obtained. The threshold th is obtained by experiment. If the data set changes, The threshold th is different.

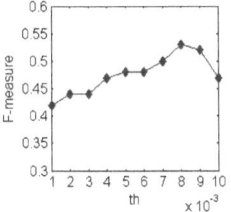   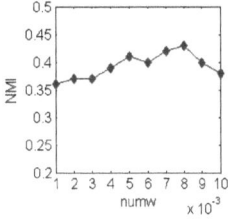

Fig. 5. F-measure changing along with th **Fig. 6.** NMI changing along with th

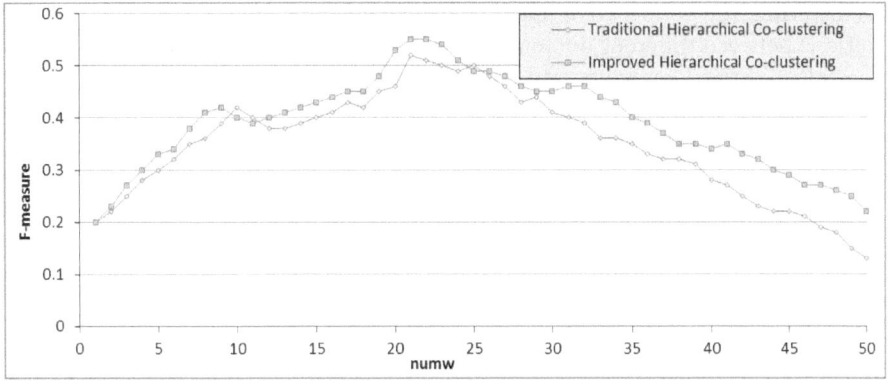

Fig. 7. F-measure comparison

Comparing the result of the traditional hierarchical co-clustering with the result of improved algorithm introduced in this paper, F-measure and NMI comparisons are shown in Fig.7 & 8 respectively. The experiment is repeated for different values of $numw$, keeping th fixed to 0.008. It is observed that result of the hierarchical co-clustering with similarity matrix after semi-supervised clustering is better than the traditional hierarchical co-clustering. The best clustering results are obtained at $numw = 22$.

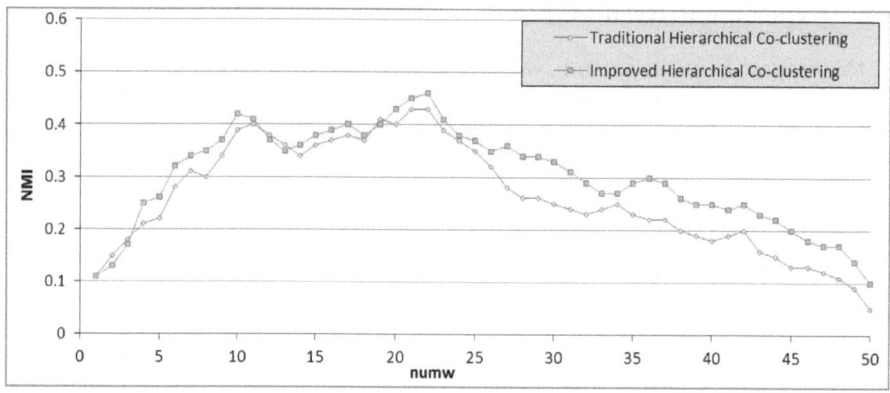

Fig. 8. NMI comparison

6   Conclusion

In this paper, an improved hierarchical co-clustering approach is proposed. First, using semi-supervised clustering to cluster the feature terms, the feature terms in the same cluster are combined. As a result, not only the dimension of vector space is reduced, but also the efficiency is improved. Second, before the hierarchical co-clustering, the cooperative matrix is constructed by finding out the semantic relations between different documents and different feature sets. At last, the two types of objects (documents and feature sets) are clustered with the method of hierarchical co-clustering. The experiments carried out show that the clustering is improved by using this approach.

The first direction of the future work is to identify the topics for document clusters in the results of hierarchical co-clustering. It is necessary to find out the feature term sets which are more representative to describe the each document cluster. Another interesting future direction is to optimize the method of calculating the weight of feature terms in preprocess to further improve the clustering.

References

1. Song, Y., Pan, S., Liu, S., Wei, F., Zhou, M., Qian, W.: Constrained Text Co-Clustering with Supervised and Unsupervised Constraints. Knowledge and Data Engineering PP(99), 1–2 (2012)
2. Wu, J.S., Lai, J.H., Wang, C.H.: A Novel Co-clustering Method with Intra-similarities. In: Data Mining Workshops, vol. 12(1), pp. 300–306 (2011)
3. Li, J.X., Shao, B., Li, T., Ogihara, M.: Hierarchical Co-Clustering: A New Way to Organize the Music Data. Multimedia 14(2), 1–2 (2011)
4. Ye, Y.M., Li, X.T., Wu, B., Li, Y.: Feature Weighting Information-Theoretic Co-Clustering for Document Clustering. In: Proceedings of the 2nd International Conference on Computer Science and its Applications, pp. 1–6 (2009)

5. Zhong, J., Liu, L., Liang, C.: Semi-supervised Text Clustering Based on the Pairwise Constraints. Computer Engineering 37(13), 183–186 (2011) (in Chinese)
6. Yang, Y., Kamel, M., Jin, F.: Topic Discovery from Document Using Ant-Based Clustering Combination. In: Zhang, Y., Tanaka, K., Yu, J.X., Wang, S., Li, M. (eds.) APWeb 2005. LNCS, vol. 3399, pp. 100–108. Springer, Heidelberg (2005)
7. Affk, W., Cardie, C.: Clustering with instance level constraints. In: Proceedings of the 17th International Conference on Machine Learning, pp. 1103–1110 (2000)
8. Pan, J., Kong, F., Wang, R.: Semi-supervised Clustering Used Weighted Pair-wise Constraints Projection. Zhejiang University Journal (Engineering Science) 45(4), 934–940 (2011) (in Chinese)
9. Wang, M., Fu, J.: Two Stages of Text Clustering Based On the Co-clustering. Pattern Recognition and Artificial Intelligence 22(6), 848–853 (2009) (in Chinese)
10. Lewis, D.D.: Reuters 21578 Text Categorization Test Collection [EB/OL], http://www.daviddlewis.com/resources/testcollections/reuters21578/
11. Deodhar, M., Ghosh, J.: A Framework for Simultaneous Co-clustering and Learning from Complex Data. Knowledge Discovery in Databases 4(3), 250–259 (2007)
12. Li, J.X., Li, T.: HCC: A Hierarchical Co-Clustering Algorithm. In: Proceedings of Special Interest Group on Information Retrieval, pp. 861–862 (2010)
13. Li, X.Y., Zeng, L.P., Shi, H.J.: Text Clustering Using the Similarity Words. Computer Engineering and Design 30(8), 1966–1968 (2009) (in Chinese)

A Granular Computing Perspective on Image Organization within an Image Retrieval Context

Orland Hoeber[1,2] and Minglun Gong[1]

[1] Department of Computer Science
Memorial University of Newfoundland
St. John's, NL A1B 3X5
Canada
{hoeber,gong}@mun.ca
[2] Department of Computer Science
University of Regina
Regina, SK S4S 0A1
Canada
orland.hoeber@uregina.ca

Abstract. The field of granular computing deals with representing, organizing, and processing information based on different levels of abstraction or aggregation. In the domain of image search, an increasingly common approach is to organize and aggregate the retrieved images within multi-level structures. In this paper, we will explore some of the core principles of granular computing within the context of image retrieval, discussing how our hierarchical approach to image clustering supports the searchers' decision-making tasks within the context of image retrieval.

Keywords: granular computing, image retrieval, image organization, multi-resolution SOM, interaction, decision-making.

1 Introduction

Granular computing deals with structuring information associated with a problem at different levels of abstraction (granulation). At a high level of abstraction, information is aggregated resulting in a small number of large granules that represent general information. At a low level of abstraction, the degree of aggregation is much lower, resulting in many smaller granules that represent much more specific information. The advantage of problem solving within a granular framework is that it allows irrelevant details to be ignored and the problem to be addressed at the most appropriate level of abstraction. Viewing the problem from a high level of granularity leads to approximate solutions; delving deeper into lower levels leads to successive increases in precision [21].

Bargeila and Pedrycz [2] have suggested that there are three fundamental elements required to engage in granular computing. A granular computing framework must support multiple levels of information granularity, allow for the encoding and decoding of information between these levels of granularity, and

T. Li et al. (Eds.): RSKT 2012, LNAI 7414, pp. 320–328, 2012.
© Springer-Verlag Berlin Heidelberg 2012

support non-homogeneous computation and analysis methods depending on the level of granularity. Yao [21] reinforced these fundamental elements through the specification of three principles of granular computing: *multilevel granularity*, *granularity conversion*, and *focused effort*.

Granular computing has also been discussed from three complementary perspectives: philosophy, methodology, and computation [21–23]. The philosophical perspective deals with structured thinking (e.g., the human thought process of decomposition and integration). The methodological perspective deals with structured problem solving (e.g., the classical "divide and conquer" approach). The computational perspective deals with structured information processing (e.g., representing the granular information in structures that lend themselves to computation, and the actual process of computation on these structures). The common theme across all of these perspectives is the fundamental notion of multilevel granular decomposition and hierarchical organization.

In this paper, we will discuss how our recent work to develop a visual and interactive interface that supports Web image search tasks [7, 17] follows a granular computing paradigm. We will use Yao's principles as the basis for not only describing how images are organized, but also how searchers can interact within this interface and the support it provides for decision making in the context of image retrieval. This discussion will primarily follow the methodological perspective on granular computing, focusing on how these systems support the human decision-making processes required in image retrieval.

2 Fundamentals of Image Retrieval

At a conceptual level, image retrieval is not very different from other types of information retrieval (e.g., document retrieval). However, at a practical level, there are significant differences that make it more difficult from an information-centric perspective, but also more effective from a human-centric perspective.

Many image retrieval approaches used on the Web are based on keyword search algorithms [8]. The image index is generated based on terms that are used in relation to the images (e.g., tags used in links, words used on the same page as the image, etc.). Searchers provide a textual description of their image needs, and the underlying search engine matches these terms to the index, returning a ranked set of corresponding images. This approach can work well if the images are accurately and fully described within the index, and if the searcher is able to provide a complete and precise description of what it is they are seeking. Unfortunately, these constraints are seldom met.

The issue of creating a robust index of images on the Web is especially challenging, given the sheer number of images as well as the difficulty with automatically determining accurate index terms. Some have attempted to address this problem from the perspective of content-based image retrieval [13], wherein the searcher provides one or more images or sketches that describe what they are seeking, and matches are made based on visual similarity. While appealing, such approaches are often not effective due to the differences in how humans and computers evaluate image similarity [4].

Another approach is to assume that other researchers will be able to continue to improve the information-centric process for matching queries to image indexes, and instead focus on the human-centric aspects of image retrieval. Many of the top search engines provide their image search results in a simple scrollable grid of images. This naïve approach does little to take advantage of the powerful visual capabilities of the human mind [20].

An increasingly popular approach is to support similarity-based image browsing [14, 19], wherein images are organized based on visual similarity and the searcher's task becomes one of browsing and exploring within this image space. Methods such as these take advantage of the searcher's ability to easily identify the relevance of images with just a glance. They can be applied to the entire document collection, or to a subset extracted as a result of a user-supplied query. Our research follows this stream, focusing on providing granular support at multiple levels of abstraction for the human-centric decision-making activities that are necessary for effective Web image retrieval.

3   Granular Organization of Images

Due to the ambiguity that is common in Web image search queries [1], the relationships among the set of retrieved images can be rather complicated. Some images may be related to one another because they convey similar meaning, whereas others may be visually similar. These different methods for determining the relationships between images provide different frames of reference for the creation of granular worlds that provide abstract views of the image collection.

Considering the semantic and visual similarity of images, there are three different approaches that can be followed for generating a granular framework for the organization of images: (1) using a semantic frame of reference, images may organized into granular structures based on the meanings or conceptual features contained within the images; (2) using a visual frame of reference, granules may be defined based on the appearance of the images; and (3) combining the semantic and visual frames of reference, granular structures may be generated that simultaneously group images based both on their meanings and their appearance. Given that searchers are often not only interested in subject matter but also appearance and aesthetics when searching for images, our research follows the last of these options.

3.1   Feature Vector Generation

To achieve our goal of using both semantic and visual information to organize the images, we extract two feature vectors for each image in the collection. The semantic feature vector captures the conceptual meaning of the image, while the visual feature vector describes the appearance and visual characteristics of the image.

To compute the semantic feature vectors, we assume that all images in the collection are associated with one or more tags that describe their contents.

This assumption holds for photo collections such as Flickr. For untagged image collections, automatic annotation techniques may be employed such as the one proposed by Li and Wang [9]. Another alternative is to derive tags for the images through a query expansion process, wherein the tags for an image are deduced based on its source expanded query [7, 18].

In order to determine the semantic relationship between any pair of images based on these tags, we must have access to some external knowledge base that understands the semantic relationships between the tags. Wikipedia is well-suited to this task given the fact that it includes many hyperlinked articles on people, places, and things, matching the conceptual subjects of many images. Using Wikipedia, we can compute the relatedness $R(i, j)$ between two images i and j using the average relatedness between their tags:

$$R(i, j) = \frac{\sum_{s \in T(i), t \in T(j)} WLM(s, t)}{\|T(i)\| \times \|T(j)\|}$$

Here, $T(\cdot)$ denotes a set of tags associated with an image, $\| \cdot \|$ represents the number of elements in a set, and $WLM(\cdot, \cdot)$ is the Wikipedia Link-based Measure [10] that computes the semantic relationship between two tags based on the hyperlink structure between their associated articles in Wikipedia.

Given a search results set that contains N images, the above process generates an $N \times N$ table, in which each entry stores the semantic relatedness of two corresponding images. To convert this table into a set of semantic feature vectors, classical multi-dimensional scaling (MDS) is applied [3]. MDS assigns each image i a high-dimensional vector $S(i)$ such that for any pair of images i and j, the distance between $S(i)$ and $S(j)$ is similar to the value of $R(i, j)$.

Determining the visual feature vector is a much simpler operation, given the extensive work that has been devoted to this problem within the domain of computer vision. Many different approaches have been studied that can be applied to this work [15]. One simple yet effective method is to use a colour histogram. The 3D colour space is quantized into a set of bins, and for each image i, we count the number of pixels that belong in each bin. These counts are then normalized, resulting in a visual feature vector $V(i)$.

3.2   Similarity-Based Image Organization

To organize the image collection, we use our previously proposed approach [16, 17], which places images on a 2D canvas with the proximity between images indicating the similarities among them. A near-optimal location for each image is obtained by training a Self Organizing Map (SOM) using feature vectors associated with the image set. Once the SOM is trained, an image is placed at the location of its Best Matching Unit (BMU) in the SOM. Since the SOM can effectively map high dimensional space to a 2D canvas in a topology preserving manner, images with similar feature vectors are placed together on the canvas.

As previously noted, our goal in organizing the images is to combine the semantic and visual frames of reference in the production of a granular structure.

One option is to use the semantic and visual feature vectors independently within the SOM, determining approximate locations based on semantics first and then more specific locations based on visual similarity. However, such an organization might appear rather disjoint at the transitions between semantic features due to this independence. A more flexible approach is to use these vectors simultaneously within the SOM. That is, a hybrid feature $H(i)$ vector is generated as a weighted concatenation of the semantic and visual feature vectors:

$$H(i) = (\alpha S(i), \beta V(i))$$

When $\alpha \gg \beta$, the approach places semantically close images in the same area of the canvas; at the same time the visual features will also be used to provide a seemingly continuous visual organization. Furthermore, the values of α and β can be controlled by the end user, allowing the searcher to control the degree to which each frame of reference impacts the granularization of the images.

3.3   Hierarchical Structure of Image Space

When the number of retrieved images N is large, it is impractical to display all of these images within the limited resolution of a computer screen at the same time. While traditional search engines address this problem by organizing the image set in a scrollable grid covering multiple pages, this approach is not feasible for our work. Since the images are organized based on semantic and visual features, there is no guarantee that the more relevant images will be placed at the top of the organizational structure. Our solution is to generate a hierarchical granular structure using a multi-resolution SOM. Such a structure provides a high-level overview of the search results set, and allows the searcher to interactively adjust the level of granularity via zoom operations. The details of this interaction will be discussed in the following section.

The process of generating the multi-resolution SOM structure is as follows. After the image set is organized using the SOM, the multi-resolution SOM structure is constructed in a bottom-up approach, with the resolution of the space reduced by half along both dimensions with each step. Hence, a granule at level n corresponds to four granules in level $n - 1$. The image associated with a high-level granule is selected from those associated with the granules is subsumes, where the selection criteria is based on how close the feature vector of an image is to the average feature vector of the group. Hence at a given level of granularity, users only see images that best represent the collection of images contained within the granule, providing searchers with multi-level overviews of the image organization, depending on the level of the SOM that is being shown.

This approach for organizing the search result images in a hierarchy that preserves their conceptual and visual similarity follows the principle of multi-level granularity. Rather than organizing the images once and providing these in a singe image space that the searchers must scroll through, our approach generates a hierarchical granular structure based on both the semantic and visual features of the images. By displaying the images associated with the multi-resolution

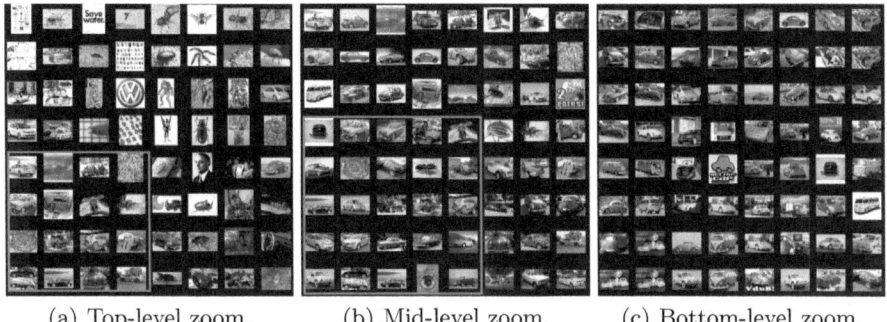

(a) Top-level zoom.   (b) Mid-level zoom.   (c) Bottom-level zoom.

Fig. 1. Zooming from a high level of granularity (a) into lower levels of granularity (b and c) results in a more specific focus on the semantic and visual features that the searcher is seeking. The red box illustrates the region of zoom from the top-level zoom to the mid-level zoom, and from the mid-level zoom to the bottom-level zoom.

SOM at a high level of granularity, the searchers are provided with a general overview of the image set. Reducing the level of granularity results in a reduction in the abstraction of the image space, providing images that are more descriptive of the region of the image space that they occupy (see Figure 1).

4   Granular Interaction Mechanisms

User interaction is a fundamental element of any software system that is designed to support and enhance user-guided tasks or activities [12]. It is also extremely important for systems that seek to take advantage of human visual processing capabilities [20], since such systems are not only designed to visually represent information, but also to allow users to directly interact with this information.

The primary goal for the interaction mechanisms within our system is to allow the searcher to explore within the image space with pan and zoom operations. As the searcher identifies a region of interest within the high-level overview of the image space, they can zoom into this area (using the normal mouse-wheel or two-finger trackpad drag interaction). Since the image sizes remain constant, this zoom operation dynamically creates more space between the images. Once sufficient space is created, the images contained within the next lower level of the multi-resolution SOM are shown. This process continues as the searcher zooms deeper and deeper into the image space, and also works in the reverse direction for zooming out. At the bottom level of the hierarchy, further zooming results in an increase in the image sizes.

This method of interaction follows Shneiderman's [11] popular Visual Information Seeking Mantra for supporting interaction within visual software: "overview first, zoom and filter, then details-on-demand". The zoom operation is an example of Yi et al.'s [24] interaction category of "abstract/elaborate", where the fundamental goal is to show the user more or less detail depending on the direction of the zoom.

Within the context of granular computing, zooming conforms to the principle of granularity conversion. As the searcher zooms into a region of interest, the granular structure is traversed to a lower-level of granularity, increasing its specificity. As a result, the images that are shown become more specific to the regions of the image space they occupy. Zooming out reverses this process, showing the image collection at a higher-level of granularity and increasing the generality of the information that is shown (i.e., the images are more general representations of the regions they cover). Conceptually, this zoom process represents a movement up or down within the information pyramid [2] produced by the hierarchical granular structure within our system.

5   Support for Intelligent Decision-Making

One of the fundamental features of a Web information retrieval support system is to enhance and promote the searcher's decision-making processes as they evaluate and explore the search results [5]. This holds true regardless of whether the searcher is seeking documents, images, or other types of information.

Our system supports the searcher in making intelligent decisions regarding the images they are seeking at different levels of granularity. When viewing the images at a high level in the multi-resolution SOM (see Figure 1(a)), the searchers are able to make course-grained decisions with respect to which region they should explore in more detail (i.e., zoom into). As the searcher delves deeper into the multi-resolution SOM (see Figure 1(b)), the granularity decreases and the granules become more specific, allowing searchers to make fine-grained decisions regarding the regions in which they feel relevant images may be present. When the searcher arrives at the bottom level in the multi-resolution SOM (see Figure 1(c)), the images are no longer aggregated into granules.

Within each step of traversing the multi-resolution organization of the image space, the number of images that are shown is relatively small (in comparison to the total number of images that are retrieved). This reduction in information supports a more intelligent approach to decision making: rather than considering each image individually for relevance, the searcher considers the representative images at various levels of granularity while traversing the granular structure. As a result, a smaller number of incremental decisions regarding the relevance of images within a given region of the image space are made as the searcher zoom (and perhaps pans) within the image space. At the lowest level of granularity, a small collection of semantically and visually similar images are finally shown, and the searcher can then focus on making a specific relevance decision for each image.

These decision-making processes supported by our system conform to the principle of focused effort. That is, decisions are made at the given level of granularity in order to solve the problem of finding relevant images. At a high level of granularity, the searcher is able to arrive at an approximate solution regarding the region of space that contains the images they are seeking. As they zoom into the image space and are shown the images at a lower level of granularity, they are able to make more precise decisions. When they finally reach the lowest level of granularity, the image space is effectively filtered to show a very focused set of images, upon which low-level relevance decisions can then be made.

6   Conclusion

In this paper, we have outlined how our approach to interactive image organization and exploration follows a model of granular computing, supporting the methodological perspective of structured problem solving that the human mind embraces. The images are organized at multiple levels of abstraction and the searcher is supported in traversing these levels using a zoom operation. At a given level of abstraction, the searcher able to make decisions of increasing specificity as regions of the image space that contain relevant images are sought. This is in contrast to the common approach in image search of providing a scrollable grid of images, which requires that the searcher make relevance decisions of images individually.

Our future work in this domain includes devising and evaluating additional frames of reference that generate alternative granular worlds for resolving the tasks associated with image retrieval. We also wish to evaluate the benefits of simultaneously showing different granular structures to the searcher, as well as make further enhancements to the interface with the goal of supporting the human-centred aspects of image retrieval on the Web [6].

References

1. André, P., Cutrell, E., Tan, D.S., Smith, G.: Designing Novel Image Search Interfaces by Understanding Unique Characteristics and Usage. In: Gross, T., Gulliksen, J., Kotzé, P., Oestreicher, L., Palanque, P., Prates, R.O., Winckler, M. (eds.) INTERACT 2009, Part II. LNCS, vol. 5727, pp. 340–353. Springer, Heidelberg (2009)
2. Bargiela, A., Pedrycz, W.: Granular Computing: An Introduction. Springer (2002)
3. Cox, T.F., Cox, M.A.A.: Multidimensional Scaling, 2nd edn. Chapman and Hall/CRC (2000)
4. Datta, R., Joshi, D., Li, J., Wang, J.Z.: Image retrieval: Ideas, influences, and trends of the new age. ACM Computing Surveys 40(2), 1–60 (2008)
5. Hoeber, O.: Web information retrieval support systems: The future of Web search. In: Proceedings of the IEEE/WIC/ACM International Conference on Web Intelligence - Workshops (International Workshop on Web Information Retrieval Support Systems), pp. 29–32 (2008)
6. Hoeber, O.: Human-centred Web search. In: Jouis, C., Biskri, I., Ganascia, J.G., Roux, M. (eds.) Next Generation Search Engines: Advanced Models for Information Retrieval, pp. 217–238. IGI Global (2012)
7. Hoque, E., Hoeber, O., Strong, G., Gong, M.: Combining conceptual query expansion and visual search results exploration for Web image retrieval. Journal of Ambient Intelligence and Humanized Computing (in press)
8. Kherfi, M.L., Ziou, D., Bernardi, A.: Image retrieval from the World Wide Web: Issues, techniques, and systems. ACM Computing Surveys 36(1), 35–67 (2004)

9. Li, J., Wang, J.Z.: Real-time computerized annotation of pictures. IEEE Transactions on Pattern Analysis and Machine Intelligence 30(6), 985–1002 (2008)
10. Milne, D., Witten, I.H.: Learning to link with Wikipedia. In: Proceedings of the ACM Conference on Information and Knowledge Management, pp. 509–518 (2008)
11. Shneiderman, B.: The eyes have it: a task by data type taxonomy for information visualizations. In: Proceedings of IEEE Symposium on Visual Languages, pp. 336–343 (1996)
12. Shneiderman, B., Plaisant, C.: Designing the User Interface: Strategies for Effective Human-Computer Interaction, 5th edn. Addison Wesley (2009)
13. Smeulders, A.W.M., Worring, M., Santini, S., Gupta, A., Jain, R.: Content-based image retrieval at the end of the early years. IEEE Transactions on Pattern Analysis and Machine Intelligence 22(12), 1349–1380 (2000)
14. Snavely, N., Seitz, S.M., Szeliski, R.: Photo tourism: Exploring photo collections in 3d. In: Proceedings of the ACM International Conference on Computer Graphics and Interactive Techniques, pp. 835–846 (2006)
15. Strong, G., Gong, M.: Organizing and browsing photos using different feature vectors and their evaluations. In: Proceedings of the ACM International Conference on Image and Video Retrieval, pp. 3:1–3:8 (2009)
16. Strong, G., Gong, M.: Similarity-based image organization and browsing using multi-resolution self-organizing map. Image and Vision Computing 29(11), 774–786 (2011)
17. Strong, G., Hoeber, O., Gong, M.: Visual Image Browsing and Exploration (Vibe): User Evaluations of Image Search Tasks. In: An, A., Lingras, P., Petty, S., Huang, R. (eds.) AMT 2010. LNCS, vol. 6335, pp. 424–435. Springer, Heidelberg (2010)
18. Strong, G., Hoque, E., Gong, M., Hoeber, O.: Organizing and browsing image search results based on conceptual and visual similarities. In: Proceedings of the International Symposium on Visual Computing, pp. 481–490 (2010)
19. Torres, R.S., Silva, C.G., Medeiros, C.B., Rocha, H.V.: Visual structures for image browsing. In: Proceedings of the International Conference on Information and Knowledge Management, pp. 49–55 (2003)
20. Ware, C.: Information Visualization: Perception for Design, 2nd edn. Morgan Kaufmann (2004)
21. Yao, Y.: The Art of Granular Computing. In: Kryszkiewicz, M., Peters, J.F., Rybiński, H., Skowron, A. (eds.) RSEISP 2007. LNCS (LNAI), vol. 4585, pp. 101–112. Springer, Heidelberg (2007)
22. Yao, Y.: Granular computing: past, present, and future. In: Proeceedings of the IEEE International Conference on Granular Computing, pp. 80–85 (2008)
23. Yao, Y.: Human-inspired granular computing. In: Yao, J. (ed.) Novel Developments in Granular Computing: Applications for Advanced Human Reasoning and Soft Computation, pp. 1–15. IGI Global (2010)
24. Yi, J.S., Ah Kang, Y., Stasko, J.T., Jacko, J.A.: Toward a deeper understanding of the role of interaction in information visualization. IEEE Transactions on Visualization and Computer Graphics 13(6), 1224–1231 (2007)

Granular Computing
in Opinion Target Extraction

Jianyang Wu, Haipeng Wang, Xiaojun Xiang, and Lin Shang

State Key Laboratory for Novel Software Technology,
Department of Computer Science and Technology,
Nanjing University, Nanjing 210093, China
wu.wujy@163.com, wanghaipeng007@gmail.com, xxjune_2006@163.com,
shanglin@nju.edu.cn

Abstract. In opinion analysis, opinion target extraction is the basic component. Dealing with the opinion target extraction on short comments, we adopt two-dimensional vectors to represent words, which are composed of the frequency of the word and the proportion the word companying with subjective words respectively. Neural network is adopted in the opinion target extraction process, and the method is tested in different granularity sizes, the result of experiments shows the effectiveness of the method.

Keywords: Granularity, Opinion target, Neural network.

1 Introduction

Opinion targets show some distribution characteristics in comments. In machine learning, opinion target extraction mainly relies on the distribution characteristics. These characteristics exist in the co- occurrence information between the paragraphs in the same document or between documents. For example, topic model is extracted by the co-occurrence of words on document level. However, Lun-Wei Ku etc. [1,2] use co-occurrence information between words based on both paragraphs in the same document and documents. Tengfei Ma etc. [8] extract explicit and implicit opinion targets from news comments by using Centering Theory. The approach uses global information in news articles as well as contextual information in adjacent sentences of comments. Kim and Hovy [9] use semantic role labeling as an intermediate step to label an opinion holder and topic using FrameNet data. They decompose the task into three phases: identifying an opinion-bearing word, labeling semantic roles related to the word in the sentence, and then finding the holder and the topic of the opinion word among the labeled semantic roles. Ruppenhofer etc. [10] argue that while automatic semantic role labeling systems (ASRL) have an important contribution to sources and targets extraction, they cannot solve the problem for all cases. Based on the experience of manually annotating opinions, sources, and targets in various genres, they present linguistic phenomena that require knowledge beyond that

T. Li et al. (Eds.): RSKT 2012, LNAI 7414, pp. 329–335, 2012.

of ASRL systems. They address issues about targets and sources extraction and inferred opinions.

However, these opinion target extraction methods are applied on long blog and news comments, which may show the different dependence on paragraph length comparing with the short ones. For short comments, methods based on fine-grained measures may generate the different orders of targets and may reduce the dependence on the paragraph length. Thus we propose the method using various granularity to extract opinion targets.

We adopt a two-layer BP neural network with the hyperbolic tangent function as the activation function and using iterative method to extract opinion target. We compute the attribute values of words in various granularity sizes and then compare opinion target extraction results in various granularity sizes. Experiments show that a reasonable granularity obtains satisfactory result.

This paper is organized as follows: section 2 introduces the representation method of words in low granularity. Then we introduce the algorithm structure and procedure in section 3. We conduct a series of experiments in section 4. A brief conclusion of the whole work is given in section 5.

2   Preliminaries

Opinion targets refer to the object described by the emotion expressed in a paragraph or a sentence of comments, specifically perform as the object modified by the evaluation words in the sentence.

In general, we use a noun or noun phrase as an opinion target. Yi[3], Bing Liu[4], Lin HF[5], Bin Qin etc.[6], Oren Etzioni etc.[7] regard nouns and noun phrases as the candidate targets in advance during the preprocess of the opinion targets extraction, then begin the opinion targets extraction process.

In papers[3-7], it is shown that the frequent item mining algorithm will work if opinion targets appear frequently in the comments, Oren Etzioni, et al. [7] use a threshold value to obtain the candidate opinion targets, i.e., nouns or noun phrases with frequency less than the threshold can't be the opinion targets.

Generally, we have the following knowledge.

1. Opinion targets and the emotional words appear together at a high proportion. The opinion targets are mainly used for evaluation and description, in the semantic unit of the consumers' emotional expression, the opinion targets and emotional words appear together at a high proportion. In expression of a semantic unit, we need two factors to give our view of some opinion target, those are emotional words and opinion targets. Therefore, in the same semantic unit, it is a higher proportion together with opinion targets and emotional words.

2. Low frequency words may also be emotional words. In papers [3-7], we can see that it is the biggest problem during the frequent items mining for opinion target extraction: for mining the low frequency opinion target directly. Although these algorithms have used other methods to get low-frequency opinion targets, the results sound not good. Based on the priori knowledge above, we know that whether a word is an opinion target not only relates to the category of itself,

but also the frequency accompanying with emotional words. In this paper, we adopted a two-dimensional vector to represent a word, one is the proportion of the word accompanying with emotional words, the other is the normalized value of the word frequency. By vector calculating, we can decide whether the word an opinion target.

In a formal written expression, the sentence is the basic unit of language, which consists of words and phrases to express a complete meaning, such as telling others a thing, asking a question, expressing some emotion etc.. It is always ending up with period, ellipsis, question mark or excla- mation mark. However, in a network comment, the user or consumers rarely who use the correct punctuation in the middle of the paragraph as the end of the sentence, usually occupy the majority. It is shown that except the end of the sentence only 3748 cases in total 14408 phone reviews are ending with period, i.e. 10,624 of them don't have a period as an end. By adjusting the punctuation set of the end of the semantic unit to change the granularity of the opinion comments. In this paper we design method to present results of extracting opinion target in a different granularity.

3   Algorithm

In our method, we use a two-dimensional vector to represent a word. By calculating the term vector, we can decide whether the word belongs to opinion targets.

We use formula $y = f(\varphi)$ to represent the relationship between words and its features, where is for the input of the word vector, y is for the score to determine whether the word is an opinion target.

A neural network is employed to compute the word score. The neural network contains one hidden layer, one input layer and one output layer. The input layer has three neurons, which are $word_{wf}$, $word_{occur}$ and constants input. The output layer has one neuron. We set the output of the training data 1 when it is opinion target, otherwise, the value is 0.The neural network structure is shown in Figure.1.

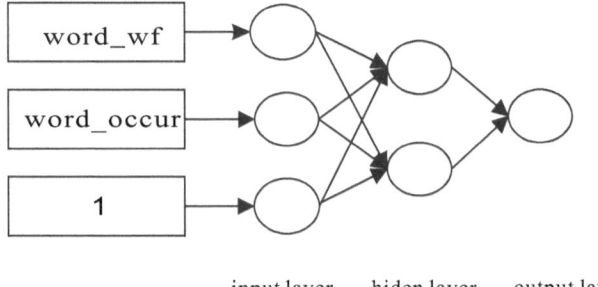

input layer   hiden layer   output layer

Fig. 1. Neural network structure for opinion target extraction

BP neural network is used with the hyperbolic tangent function tanh() as the activation function, the initial weights between the input layer and hidden layer are random numbers between $[-0.2, 0.2]$, the initial weights between the hidden layer and output layer are random numbers between $[-2.0, 2.0]$. The two-dimensional vector is composed of the word frequency ($word_{wf}$) and the proportion the word accompanying with subjective words $word_{occur}$ respectively.

The algorithm main procedure as follows: Generate a training set, initial opinion targets in the training set with score 1, non-opinion targets with score 0. Then, a three-layers BP neural network is adopted with the hyperbolic tangent function as the activation function, normalize the word vector as input, use iterative method to extract the opinion targets. The followings give the details of the methods.

3.1   The Algorithm to Compute the Features of the Words

First we divide the text content according to the punctuation set of the semantic units. Different punctuation sets lead to different divisions, so that we can get the words different features values. The algorithm is shown in the Algorithm.1.

Algorithm 1. Algorithm to get the words features: $word_{wf}$ and $word_{wf}$.

Input:
   The text database information, punctuation set consists of semantic units.
Output:
   The features values: $word_{wf}$ and $word_{wf}$.
1: Define $segReview(cur, punc_{set})$; This function returns $review_{sent}$ and $review_{doc}$, where $review_{sent}$ is the form $[doc_1, doc_2, doc_3, ...]$, in which doc_i is the form $[sent_1, sent_2, sent_3, ...]$, $sent_j$ is $[word_1, word_2, word_3, ...]$. $review_{doc}$ is the form $[doc_1, doc_2, doc_3, ...]$, in which doc_i is $[word_1, word_2, word_3, ...]$.
2: Define $getWf(review_{sent})$; This function computes the $word_{wf}$ value of input words.
3: Define $getOccur(review_{sent}, word_{wf})$; This function computes the $word_{wf}$ value according to $word_{wf}$ and $review_{sent}$.
4: Define $normalize(dict, list)$; This function normalizes the dictionary data, eliminates the effect of the words in the list and returns dict.
5: Define $init(cur, punc_{set})$;
6: **return** $word_{wf}, word_{wf}$.

3.2   Extracting the Opinion Target with Neural Network

Then we design the ComputeScore function, which is an iterative method to obtain the opinion target. The algorithm is presented in the Algorithm.2.

4   Experiment

4.1   Data Set

We obtain network comments about mobile phones from "Zhongguancun online" by web crawler. The comments are classified as commendatory and derogatory

Algorithm 2. Opinion target extraction algorithm using neural network.

Input:

Opinion target list $list_0$ and non-opinion target list $list_1$.

Output:

Opinion target list.

1: Define $computeScore(list_0, list_1, n_{list}, word_{wf}, word_{occur})$; Define a static variable tag to control the iteration times: $tag = 0$.

2: Generate a 3 neurons input layer, 2 neurons hidden layer, 1 neuron output layer neural network: $n = bp.NN(3, 2, 1)$.

3: For word in $list_0$, Construct the words input vector, the output is 0, like the form $[[x_1, x_2, x_3], [0]]$, and add it to the matrix $matrix_0$. Same to $list_1$, merge the two matrices, and add them to the matrix according to the last value crossly.

4: Train the neural network: $n.train(matrix)$, forward in n_list and word not in $list_0$ and word not in $list_1$. When the input vector is composed of the words not in the training set for candidate opinion target, the output is 0: $value = n.test([[x_1, x_2, x_3], [0]])$. Compute and save the words and the words' scores.

5: Give 0.5 as the cut-off point between evaluation and non-opinion targets. The words of scores greater than 0.5 are divided into opinion targets, less than 0.5 into non-opinion targets. Select six of the highest scores in the opinion targets to add into $list_1$ and six of the lowest scores in the non-opinion targets to add into $list_0$.

6: **return** n_{list}.

according to the "merit" comments and "weakness" comments. But we don't use the commendatory and derogatory information in the experiment.

4.2   Experiment Result

The deviation results of the different punctuation sets are described as Table.1.

Table 1. Deviation results of different punctuation sets

Iteration Times \ Set	$punc_{set_1}$	$punc_{set_2}$	$punc_{set_3}$	$punc_{set_4}$	$punc_{set_5}$	$punc_{set_6}$
1	5.38637	0.965414	0.563252	0.563087	0.685464	0.912328
2	5.38637	0.884616	0.576632	0.576542	1.067238	1.067348
3	5.38637	0.888758	0.584629	0.584556	1.080655	1.090348
4	5.38637	0.892061	0.590518	0.590405	1.064112	1.083865
5	5.38637	0.894947	0.595188	0.595105	0.199994	1.085128
6	5.38637	0.899201	0.600211	0.600119	0.199995	0.199999
7	5.38637	0.903410	0.604813	0.604819	0.199996	0.199999
8	5.38637	0.906918	0.609220	0.609103	0.199997	0.199999
9	5.38637	0.910436	0.612511	0.612503	0.199996	0.1999997
10	5.38637	0.913653	0.615803	0.615812	0.199993	0.199999

Here we list the former 10 targets and the final 10 targets for the different punctuation sets. The results are described as Table 2. (The training terms is expressed as A(B,C), in which A stands for Chinese word, B stands for the same word in English, C stands for the Hanyu Pinyin of the word.)

Table 2. Opinion Targets for the Different Punctuation Sets

Set Iteration Times	$punc_{set_1}$	$punc_{set_2}$	$punc_{set_3}$	$punc_{set_4}$	$punc_{set_5}$	$punc_{set_6}$
1		软件 (software, ruanjian)	软件 (software, ruanjian)	软件 (software, ruanjian)	卡 (card, ka)	卡 (card, ka)
2		效果 (effect, xiaoguo)	效果 (effect, xiaoguo)	效果 (effect, xiaoguo)	效果 (effect, xiaoguo)	键盘 (keyboard, jianpan)
3		卡 (card, ka)	卡 (card, ka)	卡 (card, ka)	键盘 (keyboard, jianpan)	时间 (time, shijian)
4		时间 (time, shijian)	时间 (time, shijian)	时间 (time, shiajin)	时间 (time, shijian)	效果 (effect, xiaoguo)
5		键盘 (keyboard, jianpan)	键盘 (keyboard, jianpan)	键盘 (keyboard, jianpan)	软件 (software, ruanjian)	音乐 (music, yinyue)
6		速度 (speed, sudu)	速度 (speed, sudu)	速度 (speed, sudu)	音乐 (music, yinyue)	软件 (software, ruanjain)
7		音乐 (music, yinyue)	音乐 (music, yinyue)	音乐 (music, yinyue)	款 (style, kuan)	款 (style, kuan)
8		款 (style, kuan)	款 (style, kuan)	款 (style, kuan)	耳机 (earphone, erji)	耳机 (earphone, erji)

Because of the attribute values are different for every term on different granularity, the experiment results show that based on different granularity the extracted sequences are different for every term. But in general, the extracted terms are similar. According to the statistics of the database, beside the punctuation, there are 432295 words and 23323 paragraphs in the comments, so on average every paragraph contains 18.5 words; every comment article contains 30 words. No matter what the granularity is, each paragraph will be considered as boundary of calculating attribute value though it contains small amount of words. The differences among the attribute values of the terms on different granularity are small, thus the final results among all the granularity are similar. It's obvious that, the method with any punctuation set to partition the comments can only use paragraph as semantic unit which will put the wf values of most terms close to 1. Because $wf = wf_1$ occur and most paragraph contains sentiment words, if we use the paragraph as semantic unit, the value of wf is equal to the value of occur with a high possibility. So it leads that the differences of the attribute values among the words are quite small, and the trained Neural Network doesn't work very well.

5 Conclusion

This paper presents a new method of opinion target extraction on short comments. We adopt a two-dimensional vector to represent every word, which is composed of the word frequency and the proportion the word accompanying with subjective words respectively. We compute the attribute values of words in various granularity sizes to compare opinion target extraction results. Experiments show that a reasonable granularity obtains satisfactory result.

Acknowledgement. We really appreciate National Natural Science Foundation project (61170180), National Natural Science Foundation key project (61035003) and Jiangsu Province Natural Science Foundation key research project (BK2011005).

References

1. Ku, L.-W., Li, L.-Y., Wu, T.-H., Chen, H.-H.: Major topic detection and its application to opinion summarization. In: SIGIR 2005, pp. 627–628 (2005)
2. Ku, L.-W., Liang, Y.-T., Chen, H.-H.: Opinion Extraction, Summarization and Tracking in News and Blog Corpora. In: AAAI 2006 (2006)
3. Yi, J., Nasukawa, T., Bunescu, R.: Sentiment analyzer: extracting sentiments about a given topic using natural language processing techniques. In: Wu, X.D., Tuzhilin, A. (eds.) Proc. of the IEEE Int'l Conf. on Data Mining (ICDM), pp. 427–434 (2003)
4. Hu, M., Liu, B.: Mining opinion features in customer reviews. In: Hendler, J.A. (ed.) Proc. of the AAAI 2004, pp. 755–760. AAAI Press, Menlo Park (2004)
5. Ni, M.S., Lin, H.F.: Mining product reviews based on association rule and polar analysis. In: Zhu, Q.M., et al. (eds.) Proc. of the NCIRCS 2007, pp. 628–634 (2007)
6. Liu, H.Y., Zhao, Y.Y., Qin, B., Liu, T.: Target extraction and sentiment classification. Journal of Chinese Information Processing 24(1), 84–88 (2010)
7. Popescu, A.M., Etzioni, O.: Extracting product features and opinions from reviews. In: Mooney, R.J. (ed.) Proc. of the HLT/EMNLP 2005, pp. 339–346. ACL, Morristown (2005)
8. Ma, T., Wan, X.: Opinion Target Extraction in Chinese News Comments. In: Proceedings of the 23rd International Conference on Computational Linguistics (COLING 2010), pp. 782–790 (2010)
9. Kim, S.-M., Hovy, E.: Extracting Opinions, Opinion Holders, and Topics Expressed in Online News Media Text. In: SST 2006 Proceedings of the Workshop on Sentiment and Subjectivity in Text, pp. 1–8 (2006)
10. Ruppenhofer, J., Somasundaran, S., Wiebe, J.: Finding the Sources and Targets of Subjective Expressions. In: The Sixth International Conference on Language Resources and Evaluation, LREC 2008 (2008)

Granular Covering Selection Methods Dependent on the Granule Size

Piotr Artiemjew

Department of Mathematics and Computer Science
University of Warmia and Mazury
Olsztyn, Poland
artem@matman.uwm.edu.pl

Abstract. In today's increasingly computerized world, the amount of electronic information is rising at an incredible pace. The huge number of internet resources, in particular diverse data bases, forces users to use some methods of acceleration to get the awaited result in a reasonable time. There are a lot of problems which require acceleration, for instance, the speed of search engines, classification modules, statistical tools, marketing tools, and many more. There are a lot of approaches, which try to overcome problems through the speed of information processing. One of the main groups is based on usage of software methods as acceleration, optimization, and approximation tools. An exemplary paradigm which, among others, gives the tools for approximation of a huge amount of information, is the granular rough computing paradigm. This is a sub-paradigm of the rough set theory which was proposed by Professor Pawlak in 1982. It provides tools which are useful in widely understood classification problems, and in lowering the amount of information with the preservation of important knowledge.

In this paper we focus our attention on the granular methods developed recently by Professor Polkowski. An important element of these methods is the covering search method. We show and experimentally check three of our methods dependent on the number of objects inside granules.

Keywords: granular computing, rough sets, decision systems, covering selection.

1 Introduction

In this paper we describe granular covering finding methods based on covering by granules of a minimal, average, and maximal number of objects. The general idea of granulation of knowledge based on granular computing methods was proposed by Professor Polkowski in [4–7].

This paper is organized as follows: in sect. 2, we provide basic information of used granular computing method - see [4–6]. In sect. 3, we describe in detail our ideas of covering finding methods. Sect. 4 reports an experimental session performed on exemplary data from UCI repository.

Firstly we describe the basic facts of granular rough computing.

T. Li et al. (Eds.): RSKT 2012, LNAI 7414, pp. 336–341, 2012.

2   Granular Rough Computing

The maintenance of internal knowledge in the smaller modules of information is possible by the use of granules of knowledge. The granules are the collections of objects which are indiscernible w.r.t. the chosen similarity measure. The theoretical forms of granules considered in this paper proposed by Professor Polkowski in - [4–6] - are defined below. For the decision system (U, A, d), where U is the universe of objects, A is the set of conditional attributes, and $d \notin A$ is a decision attribute, the granule of the central object $u \in U$ of the fixed radius $r_{gran} \in \{\frac{0}{card\{A\}}, \frac{1}{card\{A\}}, ..., \frac{card\{A\}}{card\{A\}}\}$ is defined in the following way:

$$g_{r_{gran}}(u) = \{v \in U; \frac{card\{IND(u, v)\}}{card\{A\}} \geq r_{gran}\}$$

where $IND(u, v) = \{a \in A; a(u) = a(v)\}$.

The computation of granules in the degree r_{gran} can be accelerated by using the following defined indiscernibility matrix, $\mu_A = [c_{ij}]_{card\{U\} \times card\{U\}}$, where

$$c_{ij} = \begin{cases} 1, & \text{if } \frac{IND(u_i, u_j)}{card\{A\}} \geq r_{gran} \\ 0, & \text{otherwise} \end{cases}$$

It is obvious that to compute granules of decision systems we only need to build the triangular part of the indiscernibility matrix.

In the next section we show our approach to covering finding methods, which will be used in the granular mechanism proposed in [4–6].

3   Covering Finding Methods

For the sake of clarity we will start our consideration from the hypothetical indiscernibility table, which is formed from the original decision system (U, A, d) for a granular radius r_{gran}, where $U = \{u_1, u_2, u_3, u_4, u_5, u_6\}$ is the set of objects consisting of $card\{A\}$ conditional attributes, and containing decision attribute $d \notin A$, where $d \in D = \{1, 2\}$. Decision concepts X, Y are defined as follows, $X = \{u_1, u_2\}$ and $Y = \{u_3, u_4, u_5, u_6\}$. The indiscernibility table created for granular radius r_{gran} is shown in Table 1. Based on the fixed indescernible matrix, the granules of decision system (U, A, d) look as follows, $g_{r_{gran}}(u_1) = \{u_1, u_2, u_3\}$, $g_{r_{gran}}(u_2) = \{u_1, u_2, u_3\}$, $g_{r_{gran}}(u_3) = \{u_1, u_2, u_3, u_4\}$, $g_{r_{gran}}(u_4) = \{u_3, u_4, u_5\}$, $g_{r_{gran}}(u_5) = \{u_4, u_5\}$, $g_{r_{gran}}(u_6) = \{u_6\}$.

In the next subsection we present basic methods of covering search, which are dependent on the size of chosen granules. The chosen granules should represent the original decision system completely, and should include important information from all objects of the original decision system. The original decision system is said to be covered if the covering U_{cover} is equal U. The granule can be assigned to the covering if at least one new object is transferred.

The covering finding methods based on the selection of the shortest size granules are described in the next subsection.

Table 1. The indiscernibility table in degree r_{gran}, 1 - objects are indiscernible in degree r_{gran}, 0 - objects are discernible in degree r_{gran}

	u_1	u_2	u_3	u_4	u_5	u_6
u_1	1	1	1	0	0	0
u_2	1	1	1	0	0	0
u_3	1	1	1	1	0	0
u_4	0	0	1	1	1	0
u_5	0	0	0	1	1	0
u_6	0	0	0	0	0	1

3.1   Covering by Granules with Minimal Size

The first method of covering finding is based on choice of granules, according to increasing cardinality. If in the same step we have granules with the same size, this problem is resolved hierarchically. We choose granules from the smallest index to the highest. For our exemplary granule data set the covering works in a deterministic way.

$Step1$: $g_{r_{gran}}(u_6) \rightarrow U_{cover}$, $U_{cover} = \{u_6\}$,

$Step2$: $g_{r_{gran}}(u_5) \rightarrow U_{cover}$, $U_{cover} = \{u_5, u_6\}$,

$Step3$: $g_{r_{gran}}(u_1) \rightarrow U_{cover}$, $U_{cover} = \{u_1, u_2, u_3, u_5, u_6\}$,

$Step4$: $g_{r_{gran}}(u_2) \nrightarrow U_{cover}$, nothing new - the covering doesn't change,

$Step5$: $g_{r_{gran}}(u_4) \rightarrow U_{cover}$, $U_{cover} = U$, the original decision system has been covered,

Finally $U_{cover} = \{g_{r_{gran}}(u_6), g_{r_{gran}}(u_5), g_{r_{gran}}(u_1), g_{r_{gran}}(u_4)\}$.

3.2   Covering by Granules with Average Size

To select the granules with an average size (average number of objects), we have to define the average size. By average size we understand the arithmetic mean of size of all granules rounded to the nearest integer value. If the fraction part is equal $\frac{1}{2}$ the value is rounded up to the nearest integer value. If our average value is equal φ, the granules can have size $1, 2, \ldots, \varphi - 1, \varphi, \varphi + 1, \ldots, card\{U\}$. An exemplary policy of covering search is the following:

if $\varphi - 1 < card\{U\} - \varphi$, then granules are chosen in the following order of their size,

$$\varphi, \varphi - 1, \varphi + 1, \varphi - 2, \varphi + 2, \ldots, 1, 2\varphi - 1, 2\varphi, \ldots, card\{U\},$$

in case $\varphi - 1 > card\{U\} - \varphi$, the order of choice is the following:

$$\varphi, \varphi - 1, \varphi + 1, \varphi - 2, \varphi + 2, \ldots, 2\varphi - card\{U\}, card\{U\}, 2\varphi - card\{U\} - 1, \ldots, 1,$$

whereas in case of equality $\varphi - 1 = card\{U\} - \varphi$, selection is as follows:

$$\varphi, \varphi - 1, \varphi + 1, \varphi - 2, \varphi + 2, \ldots, 1, card\{U\}.$$

If we have more than one granule of the same size, we have a tie, which is resolved hierarchically, as in sect. 3.1. To clarify our approach we consider the following example:

In our case, the granules contain the following numbers of objects: $card\{g_{r_{gran}}(u_1)\} = 3$, $card\{g_{r_{gran}}(u_2)\} = 3$, $card\{g_{r_{gran}}(u_3)\} = 4$, $card\{g_{r_{gran}}(u_4)\} = 3$, $card\{g_{r_{gran}}(u_5)\} = 2$, $card\{g_{r_{gran}}(u_6)\} = 1$, hence the arithmetic mean is the following:

$$\frac{3+3+4+3+2+1}{6} = \frac{16}{6} = 2\frac{2}{3}$$

After rounding up we get $\varphi = 3$, we know that $card\{U\} = 6$. Our example fulfils the property $\varphi - 1 < card\{U\} - \varphi$, it is $3 - 1 < 6 - 3$, $2 < 3$ hence during granule selection we have the following order:

$$\varphi, \varphi - 1, \varphi + 1, \varphi - 2, \varphi + 2, \ldots, 1, 2\varphi - 1, 2\varphi, \ldots, card\{U\};$$

in our case we choose granules which contain the following number of objects,

$$3, 2, 4, 1, 5, 6.$$

The described methods form the covering in the following way:

$Step1:$ $g_{r_{gran}}(u_1) \rightarrow U_{cover}$, $U_{cover} = \{u_1, u_2, u_3\}$,
$Step2:$ $g_{r_{gran}}(u_2) \not\rightarrow U_{cover}$, nothing new - the covering doesn't change,
$Step3:$ $g_{r_{gran}}(u_4) \rightarrow U_{cover}$, $U_{cover} = \{u_1, u_2, u_3, u_4, u_5\}$,
$Step4:$ $g_{r_{gran}}(u_5) \not\rightarrow U_{cover}$, nothing new - the covering doesn't change,
$Step5:$ $g_{r_{gran}}(u_3) \not\rightarrow U_{cover}$, nothing new - the covering doesn't change,
$Step6:$ $g_{r_{gran}}(u_6) \rightarrow U_{cover}$, $U_{cover} = U$, the original decision system is covered,
We obtained the following covering $U_{cover} = \{g_{r_{gran}}(u_1), g_{r_{gran}}(u_4), g_{r_{gran}}(u_6)\}$.

3.3   Covering by Granules of Maximal Size

The inverse method to minimal size selection works in the following way:
$Step1:$ $g_{r_{gran}}(u_3) \rightarrow U_{cover}$, $U_{cover} = \{u_1, u_2, u_3, u_4\}$,
$Step2:$ $g_{r_{gran}}(u_1) \not\rightarrow U_{cover}$, nothing new - the covering doesn't change,
$Step3:$ $g_{r_{gran}}(u_2) \not\rightarrow U_{cover}$, nothing new - the covering doesn't change,
$Step4:$ $g_{r_{gran}}(u_4) \rightarrow U_{cover}$, $U_{cover} = \{u_1, u_2, u_3, u_4, u_5\}$,
$Step5:$ $g_{r_{gran}}(u_5) \not\rightarrow U_{cover}$, nothing new - the covering doesn't change,
$Step6:$ $g_{r_{gran}}(u_6) \rightarrow U_{cover}$, $U_{cover} = U$, the original decision system has been covered,
Finally $U_{cover} = \{g_{r_{gran}}(u_6), g_{r_{gran}}(u_5), g_{r_{gran}}(u_1), g_{r_{gran}}(u_4)\}$.
We have obtained granular covering in the form $U_{cover} = \{g_{r_{gran}}(u_3), g_{r_{gran}}(u_4), g_{r_{gran}}(u_6)\}$

4   Experimental Session

We have carried out experiments with data from UCI repository. The missing values were complete by the most common values in decision concepts. A short description is as follows: the original data were split into five sub-datasets each time and used in the five time CV-5 method. For exemplary fold, the chosen training set consisted of four parts of original data and the test set consisted of the fifth part. The training data set is granulated and used for classification of test data by means of the one nearest neighbor classifier. The granular decision systems are formed from covering granules with use of majority voting strategy. The distance between objects in 1-NN classifier is computed by modified Canberra metric, the modification consists of use normalization by maximal and minimal values of attributes instead of sum of descriptors. The experiments were performed for three methods of covering finding. The mean values of accuracy for Australian Credit data set from five Cross Validation 5 method, for considered covering finding methods is presented in the Table 2. Analogous result with information of granular decision systems' size is in the Table 3.

Table 2. 5 times Cross Validation 5 method; The result of accuracy for all tests with Australian Credit data set; r_{gran} = Granulation radius

r_{gran}	$accCovMin$	SD	$accCovMax$	SD	$accCovAverage$	SD
0	0.444928	0	0.444927	0	0.444928	0
0.071428	0.444928	0	0.444927	0	0.444928	0
0.142857	0.444928	0	0.444927	0	0.444928	0
0.214286	0.821739	0.008696	0.444927	0	0.452174	0.011594
0.285714	0.813624	0.004752	0.444927	0	0.747537	0.044521
0.357143	0.840291	0.006608	0.704058	0.014723	0.812753	0.006029
0.428571	0.842899	0.003594	0.774493	0.010665	0.823768	0.008928
0.5	0.835362	0.003014	0.833333	0.002898	0.823768	0.004753
0.571429	0.839711	0.005565	0.834202	0.003361	0.841159	0.005334
0.642857	0.822029	0.002665	0.825217	0.005912	0.83913	0.011594
0.714286	0.809564	0.005681	0.809276	0.005333	0.805217	0.00742
0.785714	0.805507	0.008347	0.80145	0.005217	0.806376	0.001274
0.857143	0.800581	0.007419	0.795361	0.004405	0.802318	0.001622
0.928571	0.80145	0.008116	0.796522	0.004753	0.802899	0.00116
1	0.801449	0.006377	0.796522	0.005332	0.803189	0.000463

Table 3. 5 x Cross Validation 5 test; Granular decision system size for Australian Credit data set; r_{gran} = Granulation radius

r_{gran}	$GSsizeCovMin$	SD	$GSsizeCovMax$	SD	$GSsizeCovAverage$	SD
0	1	0	1	0	1	0
0.071428	2.36	0.064	1	0	2	0
0.142857	5.12	0.256	1	0	2.48	0.255999
0.214286	4.76	0.128	2.72	0.127999	2.92	0.144
0.285714	9.4	0.08	8.36	0.272	5.28	0.623999
0.357143	31.08	0.336	22.96	0.144	10.44	0.383999
0.428571	74.36	1.248	51.68	0.895999	23.88	0.735999
0.5	125.6	0.96	101.64	0.847999	61.6	0.88
0.571429	205.52	1.216	162.4	1.2	145.84	1.312
0.642857	341.64	0.991998	291.24	0.927996	304.08	1.744
0.714286	466.44	0.672003	446.12	0.655999	467.24	0.911999
0.785714	534.08	0.33601	532.28	0.176	533.96	0.288
0.857143	546.28	0.144006	546.2	0.080004	546.28	0.144006
0.928571	547.84	0.128015	547.88	0.144006	547.8	0.080004
1	552	0	552	0	552	0

5   Conclusions

The results of these experiments show an interesting dependence between the quality of classification, size of granular decision systems, and the methods of covering finding. First of all, it has turned out that the average method gives for smaller granular radii the smallest granular decision systems with maintenance of knowledge from the original decision system in average quality; but close to the best method. The best method was the minimal algorithm, which worked best for smaller granular decision systems. For a higher number of granular objects all methods were comparable. This means that with the rise of the granular radius of granular decision system, the size of the granules converged into one, and in the case of such small granules, the method of covering finding has a smaller influence on the quality of granular decision systems. In the future we are planning to compare other families of covering finding methods, particularly methods based on randomized procedures.

Acknowledgements. The research has been supported by grant 1309-802 from Ministry of Science and Higher Education of the Republic of Poland.

References

1. Artiemjew, P.: Rough Mereological Classifiers Obtained from Weak Variants of Rough Inclusions. In: Wang, G., Li, T., Grzymala-Busse, J.W., Miao, D., Skowron, A., Yao, Y. (eds.) RSKT 2008. LNCS (LNAI), vol. 5009, pp. 229–236. Springer, Heidelberg (2008)
2. Artiemjew, P.: On Classification of Data by Means of Rough Mereological Granules of Objects and Rules. In: Wang, G., Li, T., Grzymala-Busse, J.W., Miao, D., Skowron, A., Yao, Y. (eds.) RSKT 2008. LNCS (LNAI), vol. 5009, pp. 221–228. Springer, Heidelberg (2008)
3. Pawlak, Z.: Rough sets. Int. J. Computer and Information Sci. 11, 341–356 (1982)
4. Polkowski, L.: Formal granular calculi based on rough inclusions (a feature talk). In: Proceedings 2006 IEEE Int. Conference on Granular Computing GrC 2006, pp. 57–62. IEEE Press (2006)
5. Polkowski, L.: The paradigm of granular rough computing. In: Proceedings ICCI 2007. 6th IEEE Intern. Conf. on Cognitive Informatics, pp. 145–163. IEEE Computer Society, Los Alamitos (2007)
6. Polkowski, L.: A Unified Approach to Granulation of Knowledge and Granular Computing Based on Rough Mereology: A Survey. In: Pedrycz, W., Skowron, A., Kreinovich, V. (eds.) Handbook of Granular Computing, pp. 375–401. John Wiley & Sons, New York (2008)
7. Polkowski, L.: Granulation of Knowledge: Similarity Based Approach in Information and Decision Systems. In: Meyers, R. (ed.) Encyclopedia of Complexity and System Sciences, article 00788. Springer (2009)

Rough Set Approximations in Incomplete Multi-scale Information Systems

Shen-Ming Gu, Xiao-Hui Sun, and Wei-Zhi Wu

School of Mathematics, Physics and Information Science,
Zhejiang Ocean University, Zhoushan, Zhejiang 316000, China
{gsm,wuwz}@zjou.edu.cn, sxh4006@126.com

Abstract. With the view point of granular computing, there are different granules at different levels of scale in data sets having hierarchical scale structures. In this paper, the notion of incomplete multi-scale information systems is first introduced. Using tolerance relation, the lower and upper approximations in incomplete multi-scale information systems are then defined and their properties are examined.

Keywords: Granular computing, Incomplete information system, Multi-scale information system, Rough set, Tolerance relation.

1 Introduction

Granular computing (GrC) is a basic approach for knowledge representation and data mining. The purpose of GrC is to seek for an approximation scheme which can effectively solve a complex problem, albeit not in the most precise way [22]. Basic ingredients of GrC are subsets, classes, and clusters of a universe. The topic of fuzzy information granulation was first proposed and discussed by Zadeh [27] in 1979. A general framework of GrC was presented by Zadeh [28] in the context of fuzzy set theory. Ever since the introduction of the concept of "GrC", we have witnessed a rapid development and a fast growing interest in the topic [1, 2, 12, 16, 18, 23–26].

Various methods of GrC concentrating on concrete models in specific contexts have been proposed over the years. Rough set theory is perhaps one of the most advanced areas that popularize GrC [8, 9, 12, 16, 18, 23–26]. It was originally proposed by Pawlak [14] as a formal tool for modelling and processing incomplete information. Rough set models enable us to precisely define and analyze many notions of GrC. For example, Yao [25] proposed a partition model of GrC. The model is constructed by granulating a finite universe of discourse through a family of pairwise disjoint subsets under an equivalence relation. The partition model is actually important and is based on the Pawlak approximation space [13, 15, 20]. Qian et al. [17, 18] proposed the concept of multi-granulation rough sets. It is different from Pawlak's rough sets since the former is constructed on the basis of a family of the binary relations instead of a single one.

Most applications based on rough set theory can fall into the attribute-value representation model, called an information system [21]. Usually, in an information system, each object at each attribute can only take on one value, we call such

T. Li et al. (Eds.): RSKT 2012, LNAI 7414, pp. 342–350, 2012.

an information system a single scale information system. However, people can observe objects or deal with data hierarchically structured into different scales [11]. Hence, the multi-scale information system is a new and interesting topic in the theory of GrC. In many real-life multi-scale information systems, an object can take on as many values as they are scales under the same attribute. For example, maps can be hierarchically organized into different scales, from large to small and vice versa. The political subdivision of China at the top level has 34 provinces, autonomous regions, and directly governed city regions. Under each province, there are many prefecture-level cities. And, under each prefecture-level city, there are several counties, so on and so forth down the hierarchy. With respect to different scales, a point in space may be located in a province, or in a prefecture-level city, or in a county, etc. Another example is that the examination results of mathematics for students can be recorded as natural numbers between 0 to 100, it can also be graded as "Excellent", "Good", "Moderate", "Bad", and "Unacceptable". Sometimes, if needed, it might be graded into two values, "Pass" and "Fail". Therefore, how to discover knowledge in multi-scale information systems is of particular importance in data mining. Recently, Wu and Leung [22] proposed a formal approach to granular computing in data sets having hierarchical scale structures measured at different levels of granulations. Gu [7] introduced an approach to rule acquisition in consistent multi-scale decision systems. Gu [6] proposed some algorithms for knowledge acquisition in inconsistent multi-scale decision systems.

This paper mainly focuses on the knowledge approximation in incomplete multi-scale information systems under different levels of granulations. The organization of this paper is as follows. In the next section, we give some basic notions related to Pawlak rough sets. The concept of incomplete multi-scale information system is explored in Section 3. In Section 4, we propose the rough set approximations in incomplete multi-scale information systems, and examine some properties. We then conclude the paper with a summary in Section 5.

2   Classical Rough Sets and Information Systems

2.1   Pawlak Rough Sets

Let U be a nonempty finite set of objects called the universe of discourse, and $R \subseteq U \times U$ an equivalence relation on U, then the pair (U, R) is called a Pawlak approximation space. The equivalence relation R and the approximation space (U, R) may be regarded as the available information or knowledge about the objects under consideration. For two elements $x, y \in U$, if $(x, y) \in R$, we say that x and y are indistinguishable. The equivalence relation R partitions U into disjoint subsets U/R, called a quotient set of U. Elements of U/R are called the elementary sets. The empty set $\emptyset$ and the union of one or more elementary sets are called definable, observable, measurable, or composed sets.

Given a approximation space (U, R), for $X \subseteq U$, the lower approximation and upper approximation of X with respect to (w.r.t.) (U, R), denoted by $\underline{R}(X)$ and $\overline{R}(X)$, are respectively defined as follows:

$$\underline{R}(X) = \cup\{[x]_R : [x]_R \subseteq X\}, \tag{1}$$

$$\overline{R}(X) = \cup\{[x]_R : [x]_R \cap X \neq \emptyset\}, \tag{2}$$

where $[x]_R = \{y \in U : (x, y) \in R\}$ is the R equivalence class containing x. The lower approximation $\underline{R}(X)$ is the union of equivalence classes which are subsets of X. The upper approximation $\overline{R}(X)$ is the union of equivalence classes which have a nonempty intersection with X. The pair $(\underline{R}(X), \overline{R}(X))$ is called the Pawlak rough set of X w.r.t. (U, R).

2.2   Information Systems

An information system is an ordered pair $S = (U, AT)$, where $U = \{x_1, x_2, \ldots, x_n\}$ is a non-empty, finite set of objects called the universe of discourse and $AT = \{a_1, a_2, \ldots, a_m\}$ is a non-empty, finite set of attributes, such that $a : U \to V_a$ for any $a \in AT$, i.e., $a(x) \in V_a$, $x \in U$, where $V_a = \{a(x) : x \in U\}$ is called the domain of a.

When the precise values of some of the attributes in an information system are not known, then such a system is called an incomplete information system and is still denoted without confusion by $S = (U, AT)$. In this case, a special symbol "$*$" is used to indicate that the value of the attribute is unknown, that is, if $a(x) = *$, then we say that the value of x is unknown on the attribute a.

For an incomplete information system, Slowinski et al. [4, 19] defined rough approximations based on similarity and fuzzy similarity relations. Greco et al. [3, 5] discussed rough set approach to multi-criteria decision problems.

Given an incomplete information system $S = (U, AT)$ and $A \subseteq AT$, if all unknown values are assumed to be compared with any other values in the domain of the corresponding attributes, Kryszkiewicz [10] defined a tolerance relation $\mathcal{T}(A)$ on U as follows:

$$\mathcal{T}(A) = \{(x, y) \in U \times U : \forall a \in A, a(x) = a(y) \\ \vee a(x) = * \vee a(y) = *\}. \tag{3}$$

Obviously, tolerance relation $\mathcal{T}(A)$ is reflexive and symmetric, while it is not necessarily transitive.

Let S be an incomplete information system, $A \subseteq AT$, and $X \subseteq U$, the lower approximation and upper approximation of X with respect to $\mathcal{T}(A)$ are defined as follows:

$$\underline{A}_{\mathcal{T}}(X) = \{x \in U : \mathcal{T}_A(x) \subseteq X\}, \tag{4}$$

$$\overline{A}_{\mathcal{T}}(X) = \{x \in U : \mathcal{T}_A(x) \cap X \neq \emptyset\}, \tag{5}$$

where $\mathcal{T}_A(x)$ is the tolerance class of x such that $\mathcal{T}_A(x) = \{y \in U : (x, y) \in \mathcal{T}(A)\}$.

3 Multi-scale Information Systems

3.1 Complete Multi-scale Information Systems

A multi-scale information system is a tuple $S = (U, AT)$, where $U = \{x_1, x_2, \ldots, x_n\}$ is a non-empty, finite set of objects called the universe of discourse, $AT = \{a_1, a_2, \ldots, a_m\}$ is a non-empty, finite set of attributes, and each $a_j \in AT$ is a multi-scale attribute, i.e., for the same object in U, attribute a_j can take on different values at different scales. We assume that all the attributes have the same number I of levels of granulations. Hence a multi-scale information system can be represented as a system $(U, \{a_j^k : k = 1, 2, \ldots, I, j = 1, 2, \ldots, m\})$, where $a_j^k : U \rightarrow V_j^k$ is a surjective function and V_j^k is the domain of the k-th scale attribute a_j^k. For $1 \leq k \leq I - 1$, there exists a surjective function $g_j^{k,k+1} : V_j^k \rightarrow V_j^{k+1}$, such that $a_j^{k+1} = g_j^{k,k+1} \circ a_j^k$, i.e.

$$a_j^{k+1}(x) = g_j^{k,k+1}(a_j^k(x)) \tag{6}$$

where $g_j^{k,k+1}$ is called a granular information transformation function.

For $k \in \{1, 2, \ldots, I\}$, we denote $AT^k = \{a_j^k : j = 1, 2, \ldots, m\}$. Then a multi-scale information system $S = (U, AT)$ can be decomposed into I information systems $S^k = (U, AT^k), k = 1, 2, \ldots, I$.

3.2 Incomplete Multi-scale Information Systems

An incomplete multi-scale information system is a system $(U, \{a_j^k : k = 1, 2, \ldots, I, j = 1, 2, \ldots, m\})$, in which some attribute domain $V_{a_j^k}$ may contain the symbol "*". i.e., there exist $x \in U$ and $j \in \{1, 2, \ldots, m\}$, such that $a_j^k(x) = *$. Therefore, we define a new granular information transformation function $\xi_j^{k,k+1}$ as:

$$a_j^{k+1}(x) = \xi_j^{k,k+1}(a_j^k(x)) = \begin{cases} *, & \text{if } a_j^k(x) = *, \\ g_j^{k,k+1}(a_j^k(x)), & \text{otherwise.} \end{cases} \tag{7}$$

where $1 \leq k \leq I - 1$. Thus, an incomplete multi-scale information system is associated with I incomplete information systems $S^k = (U, AT^k), k = 1, 2, \ldots, I$.

Example 1. Tables 1-3 depict an example of an incomplete multi-scale information system $(U, \{a_j^k : k = 1, 2, 3, j = 1, 2, 3, 4\})$, where $U = \{x_1, x_2, \ldots, x_{10}\}$. $AT^k = \{a_1^k, a_2^k, a_3^k\}, k = 1, 2, 3$. The first level of granulation is (U, AT^1), the second level of granulation is (U, AT^2), and the third level of granulation is (U, AT^3). The granular information transformation functions $g_j^{k,k+1}$ are defined as follows:

$$a_j^2(x) = g_j^{1,2}(a_j^1(x)) = \begin{cases} E, & \text{if } 90 \leq a_j^1(x) \leq 100, \\ G, & \text{if } 80 \leq a_j^1(x) < 90, \\ M, & \text{if } 70 \leq a_j^1(x) < 80, \\ B, & \text{if } 60 \leq a_j^1(x) < 70, \\ U, & \text{otherwise.} \end{cases} \tag{8}$$

$$a_j^3(x) = g_j^{2,3}(a_j^2(x)) = \begin{cases} F, \text{ if } a_j^2(x) = U, \\ P, \text{ otherwise.} \end{cases} \tag{9}$$

where $j = 1, 2, 3, 4$.

Table 1. The first level of granulation

U	a_1^1	a_2^1	a_3^1	a_4^1
x_1	95	91	93	94
x_2	93	90	90	88
x_3	88	83	87	81
x_4	88	83	87	81
x_5	82	85	85	78
x_6	75	76	78	55
x_7	72	71	*	71
x_8	61	*	57	63
x_9	57	58	65	67
x_{10}	*	67	73	*

Table 2. The second level of granulation

U	a_1^2	a_2^2	a_3^2	a_4^2
x_1	E	E	E	E
x_2	E	E	E	G
x_3	G	G	G	G
x_4	G	G	G	G
x_5	G	G	G	M
x_6	M	M	M	U
x_7	M	M	*	M
x_8	B	*	U	B
x_9	U	U	B	B
x_{10}	*	B	M	*

Table 3. The third level of granulation

U	a_1^3	a_2^3	a_3^3	a_4^3
x_1	P	P	P	P
x_2	P	P	P	P
x_3	P	P	P	P
x_4	P	P	P	P
x_5	P	P	P	P
x_6	P	P	P	F
x_7	P	P	*	P
x_8	P	*	F	P
x_9	F	F	P	P
x_{10}	*	P	P	*

4   Properties of Rough Set Approximations

Proposition 1. *For an incomplete multi-scale information system* $(U, \{a_j^k : k = 1, 2, \ldots, I, j = 1, 2, \ldots, m\})$, $1 \leq k \leq I - 1$, *and* $A^k \subseteq AT^k$, *then we have* $\mathcal{T}(A^k) \subseteq \mathcal{T}(A^{k+1})$.

Proof.

$$\mathcal{T}(A^k)$$
$$= \{(x, y) \in U \times U : \forall a_j^k \in A^k, a_j^k(x) = a_j^k(y) \vee a_j^k(x) = * \vee a_j^k(y) = *\}$$
$$\subseteq \{(x, y) \in U \times U : \forall a_j^k \in A^k, \xi(a_j^k(x)) = \xi(a_j^k(y)) \vee \xi(a_j^k(x)) = * \vee \xi(a_j^k(y)) = *\}$$
$$= \{(x, y) \in U \times U : \forall a_j^{k+1} \in A^{k+1}, a_j^{k+1}(x) = a_j^{k+1}(y) \vee a_j^{k+1}(x) = * \vee a_j^{k+1}(y) = *\}$$
$$= \mathcal{T}(A^{k+1}).$$

Therefore $\mathcal{T}(A^k) \subseteq \mathcal{T}(A^{k+1})$.

Proposition 2. *For an incomplete multi-scale information system* $(U, \{a_j^k : k = 1, 2, \ldots, I, j = 1, 2, \ldots, m\})$, $1 \leq k \leq I - 1$, *and* $A^k \subseteq AT^k$, *then* $\mathcal{T}_{A^k}(x) \subseteq \mathcal{T}_{A^{k+1}}(x)$ *for all* $x \in U$.

Proof. $\forall x \in U$,

$$\mathcal{T}_{A^k}(x) = \{y \in U : (x, y) \in \mathcal{T}(A^k)\}$$
$$\subseteq \{y \in U : (x, y) \in \mathcal{T}(A^{k+1})\}$$
$$= \mathcal{T}_{A^{k+1}}(x).$$

Therefore $\mathcal{T}_{A^k}(x) \subseteq \mathcal{T}_{A^{k+1}}(x)$.

Example 2. For the incomplete multi-scale information system $(U, \{a_j^k : k = 1, 2, 3, j = 1, 2, 3, 4\})$ in Example 1. Let $A^k = \{a_1^k\}$, $k = 1, 2, 3$, then the tolerance classes can be obtained as follows:

$\mathcal{T}_{A^1}(x_1) = \{x_1, x_{10}\}, \quad \mathcal{T}_{A^2}(x_1) = \{x_1, x_2, x_{10}\}, \quad \mathcal{T}_{A^3}(x_1) = U - \{x_9\};$
$\mathcal{T}_{A^1}(x_2) = \{x_2, x_{10}\}, \quad \mathcal{T}_{A^2}(x_2) = \{x_1, x_2, x_{10}\}, \quad \mathcal{T}_{A^3}(x_2) = U - \{x_9\};$
$\mathcal{T}_{A^1}(x_3) = \{x_3, x_4, x_{10}\}, \mathcal{T}_{A^2}(x_3) = \{x_3, x_4, x_5, x_{10}\}, \mathcal{T}_{A^3}(x_3) = U - \{x_9\};$
$\mathcal{T}_{A^1}(x_4) = \{x_3, x_4, x_{10}\}, \mathcal{T}_{A^2}(x_4) = \{x_3, x_4, x_5, x_{10}\}, \mathcal{T}_{A^3}(x_4) = U - \{x_9\};$
$\mathcal{T}_{A^1}(x_5) = \{x_5, x_{10}\}, \quad \mathcal{T}_{A^2}(x_5) = \{x_3, x_4, x_5, x_{10}\}, \mathcal{T}_{A^3}(x_5) = U - \{x_9\};$
$\mathcal{T}_{A^1}(x_6) = \{x_6, x_7, x_{10}\}, \mathcal{T}_{A^2}(x_6) = \{x_6, x_7, x_{10}\}, \quad \mathcal{T}_{A^3}(x_6) = U - \{x_9\};$
$\mathcal{T}_{A^1}(x_7) = \{x_6, x_7, x_{10}\}, \mathcal{T}_{A^2}(x_7) = \{x_6, x_7, x_{10}\}, \quad \mathcal{T}_{A^3}(x_7) = U - \{x_9\};$
$\mathcal{T}_{A^1}(x_8) = \{x_8, x_{10}\}, \quad \mathcal{T}_{A^2}(x_8) = \{x_8, x_{10}\}, \quad \mathcal{T}_{A^3}(x_8) = U - \{x_9\};$
$\mathcal{T}_{A^1}(x_9) = \{x_9, x_{10}\}, \quad \mathcal{T}_{A^2}(x_9) = \{x_9, x_{10}\}, \quad \mathcal{T}_{A^3}(x_9) = \{x_9, x_{10}\};$
$\mathcal{T}_{A^1}(x_{10}) = U, \quad \mathcal{T}_{A^2}(x_{10}) = U, \quad \mathcal{T}_{A^3}(x_{10}) = U.$

Obviously, $\mathcal{T}_{A^1}(x_i) \subseteq \mathcal{T}_{A^2}(x_i) \subseteq \mathcal{T}_{A^3}(x_i)$, $i = 1, 2, \ldots, 10$.

In an incomplete multi-scale information system $(U, \{a_j^k : k = 1, 2, \ldots, I, j = 1, 2, \ldots, m\})$, $1 \leq k \leq I$, S^k is an incomplete information system, in which

$A^k \subseteq AT^k$, for $X \subseteq U$, the lower approximation and upper approximation of X with respect to $\mathcal{T}(A^k)$ are defined as follows:

$$\underline{A^k}_{\mathcal{T}}(X) = \{x \in U : \mathcal{T}_{A^k}(x) \subseteq X\}, \tag{10}$$

$$\overline{A^k}_{\mathcal{T}}(X) = \{x \in U : \mathcal{T}_{A^k}(x) \cap X \neq \emptyset\}, \tag{11}$$

where $\mathcal{T}_{A^k}(x)$ is the tolerance class of x w.r.t. $\mathcal{T}(A^k)$, i.e., $\mathcal{T}_{A^k}(x) = \{y \in U : (x, y) \in \mathcal{T}(A^k)\}$.

Proposition 3. *For an incomplete multi-scale information system* $(U, \{a_j^k : k = 1, 2, \ldots, I, j = 1, 2, \ldots, m\})$, $1 \leq k \leq I - 1$, $A^k \subseteq AT^k$, *and* $X \subseteq U$, *then the lower approximation and upper approximation of* X *satisfy following properties:*

$$\underline{A^k}_{\mathcal{T}}(X) \supseteq \underline{A^{k+1}}_{\mathcal{T}}(X) \tag{12}$$

$$\overline{A^k}_{\mathcal{T}}(X) \subseteq \overline{A^{k+1}}_{\mathcal{T}}(X) \tag{13}$$

Proof. (12) $\forall x \in U$,

$$\begin{aligned}
x \in \underline{A^{k+1}}_{\mathcal{T}}(X) &\Longleftrightarrow \mathcal{T}_{A^{k+1}}(x) \subseteq X \\
&\Longrightarrow \mathcal{T}_{A^k}(x) \subseteq X \\
&\Longleftrightarrow x \in \underline{A^k}_{\mathcal{T}}(X).
\end{aligned}$$

Therefore $\underline{A^k}_{\mathcal{T}}(X) \supseteq \underline{A^{k+1}}_{\mathcal{T}}(X)$.
(13) $\forall x \in U$,

$$\begin{aligned}
x \in \overline{A^k}_{\mathcal{T}}(X) &\Longleftrightarrow \mathcal{T}_{A^k}(x) \cap X \neq \emptyset \\
&\Longrightarrow \mathcal{T}_{A^{k+1}}(x) \cap X \neq \emptyset \\
&\Longleftrightarrow x \in \overline{A^{k+1}}_{\mathcal{T}}(X).
\end{aligned}$$

Therefore $\overline{A^k}_{\mathcal{T}}(X) \subseteq \overline{A^{k+1}}_{\mathcal{T}}(X)$.

Example 3. In Example 1, let $A^k = \{a_1^k\}$, $k = 1, 2, 3$, and $X = \{x_1, x_2, x_3, x_5, x_{10}\}$, then $\underline{A^1}_{\mathcal{T}}(X) = \{x_1, x_2, x_5, x_{10}\}$, $\underline{A^2}_{\mathcal{T}}(X) = \{x_1, x_2, x_{10}\}$, $\underline{A^3}_{\mathcal{T}}(X) = \emptyset$; $\overline{A^1}_{\mathcal{T}}(X) = U$, $\overline{A^2}_{\mathcal{T}}(X) = U$, $\overline{A^3}_{\mathcal{T}}(X) = U$.
Obviously,

$$\underline{A^1}_{\mathcal{T}}(X) \supseteq \underline{A^2}_{\mathcal{T}}(X) \supseteq \underline{A^3}_{\mathcal{T}}(X),$$
$$\overline{A^1}_{\mathcal{T}}(X) \subseteq \overline{A^2}_{\mathcal{T}}(X) \subseteq \overline{A^3}_{\mathcal{T}}(X).$$

5   Conclusion

In this paper, we have developed a new knowledge representation system called an incomplete multi-scale information system. An incomplete multi-scale information system can be used to represent a data set having hierarchical scale structures measured at different levels of granulations in which some values of the attributes are not known. By using tolerance relations to construct the tolerance classes, we have introduced rough set approximations in incomplete multi-scale information systems and examined some basic properties. Our future work will focus on new approaches to knowledge acquisition in incomplete multi-scale information systems.

Acknowledgments. This work was supported by grants from the National Natural Science Foundation of China (Nos. 61075120, 11071284, and 61173181), the Zhejiang Provincial Natural Science Foundation in China (No. LZ12F03002), and the Scientific Research Project of Science and Technology Department of Zhejiang in China (No. 2008C13068).

References

1. Bargiela, A., Pedrycz, W.: Granular Computing: An Introduction. Kluwer Academic Publishers, Boston (2002)
2. Bargiela, A., Pedrycz, W.: Toward a theory of granular computing for human-centered information processing. IEEE Transactions on Fuzzy Systems 16, 320–330 (2008)
3. Dembczynski, K., Greco, S., Slowinski, R.: Rough set approach to multiple criteria classification with imprecise evaluations and assignments. European J. Operational Research 198(2), 626–636 (2009)
4. Greco, S., Matarazzo, B., Slowinski, R.: Rough set processing of vague information using fuzzy similarity relations. In: Calude, C.S., Paun, G. (eds.) Finite Versus Infinite – Contributions to an Eternal Dilemma, pp. 149–173. Springer, London (2000)
5. Greco, S., Matarazzo, B., Słowiński, R.: Handling Missing Values in Rough Set Analysis of Multi-attribute and Multi-criteria Decision Problems. In: Zhong, N., Skowron, A., Ohsuga, S. (eds.) RSFDGrC 1999. LNCS (LNAI), vol. 1711, pp. 146–157. Springer, Heidelberg (1999)
6. Gu, S.M., Wu, W.Z.: Knowledge Acquisition in Inconsistent Multi-scale Decision Systems. In: Yao, J., Ramanna, S., Wang, G., Suraj, Z. (eds.) RSKT 2011. LNCS, vol. 6954, pp. 669–678. Springer, Heidelberg (2011)
7. Gu, S.M., Wu, W.Z., Zheng, Y.: Rule acquisition in consistent multi-scale decision systems. In: 8th International conference on Fuzzy System and Knowledge Discovery, pp. 390–393. IEEE Computer Society, Los Alamitos (2011)
8. Hu, Q.H., Liu, J.F., Yu, D.R.: Mixed feature selection based on granulation and approximation. Knowledge-Based Systems 21, 294–304 (2008)
9. Inuiguchi, M., Hirano, S., Tsumoto, S.: Rough Set Theory and Granular Computing. Springer, Heidelberg (2002)
10. Kryszkiewicz, M.: Rough set approach to incomplete information systems. Inf. Sci. 112, 39–49 (1998)
11. Leung, Y., Zhang, J.S., Xu, Z.B.: Clustering by scale-space filtering. IEEE Transactions on Pattern Analysis and Machine Intelligence 22, 1396–1410 (2000)
12. Lin, T.Y., Yao, Y.Y., Zadeh, L.A.: Data Mining, Rough Sets and Granular Computing. Physica-Verlag, Heidelberg (2002)

13. Ma, J.-M., Zhang, W.-X., Wu, W.-Z., Li, T.-J.: Granular Computing Based on a Generalized Approximation Space. In: Yao, J., Lingras, P., Wu, W.-Z., Szczuka, M.S., Cercone, N.J., Ślęzak, D. (eds.) RSKT 2007. LNCS (LNAI), vol. 4481, pp. 93–100. Springer, Heidelberg (2007)
14. Pawlak, Z.: Rough Sets: Theoretical Aspects of Reasoning about Data. Kluwer Academic Publishers, Boston (1991)
15. Polkowski, L., Skowron, A.: Rough mereology: A new paradigm for approximate reasoning. International Journal of Approximate Reasoning 15, 333–365 (1996)
16. Qian, Y.H., Liang, J.Y., Dang, C.Y.: Knowledge structure, knowledge granulation and knowledge distance in a knowledge base. International Journal of Approximate Reasoning 50, 174–188 (2009)
17. Qian, Y.H., Liang, J.Y., Wei, W.: Pessimistic rough decision. In: Second International Workshop on Rough Sets Theory, vol. 50, pp. 440–449 (2010)
18. Qian, Y.H., Liang, J.Y., Yao, Y.Y., Dang, C.Y.: MGRS: A multi-granulation rough set. Information Sciences 180, 949–970 (2010)
19. Slowinski, R., Vanderpooten, D.: A generalized definition of rough approximations based on similarity. IEEE Transactions on Knowledge and Data Engineering 12(2), 331–336 (2000)
20. Wu, W.Z.: Rough Set Approximations Based on Granular Labels. In: Sakai, H., Chakraborty, M.K., Hassanien, A.E., Ślęzak, D., Zhu, W. (eds.) RSFDGrC 2009. LNCS (LNAI), vol. 5908, pp. 93–100. Springer, Heidelberg (2009)
21. Wu, W.Z.: Attribute Granules in Formal Contexts. In: An, A., Stefanowski, J., Ramanna, S., Butz, C.J., Pedrycz, W., Wang, G. (eds.) RSFDGrC 2007. LNCS (LNAI), vol. 4482, pp. 395–402. Springer, Heidelberg (2007)
22. Wu, W.Z., Leung, Y.: Theory and applications of granular labelled partitions in multi-scale decision tables. Information Sciences 181, 3878–3897 (2011)
23. Yao, Y.Y.: Stratified rough sets and granular computing. In: Dave, R.N., Sudkamp, T. (eds.) 18th International Conference of the North American Fuzzy Information Processing Society, pp. 800–804. IEEE Press, New York (1999)
24. Yao, Y.Y.: Information granulation and rough set approximation. International Journal of Intelligent Systems 16, 87–104 (2001)
25. Yao, Y.Y.: A Partition Model of Granular Computing. In: Peters, J.F., Skowron, A., Grzymała-Busse, J.W., Kostek, B.z., Świniarski, R.W., Szczuka, M.S. (eds.) Transactions on Rough Sets I. LNCS, vol. 3100, pp. 232–253. Springer, Heidelberg (2004)
26. Yao, Y.Y., Liau, C.J., Zhong, N.: Granular Computing Based on Rough Sets, Quotient Space Theory, and Belief Functions. In: Zhong, N., Raś, Z.W., Tsumoto, S., Suzuki, E. (eds.) ISMIS 2003. LNCS (LNAI), vol. 2871, pp. 152–159. Springer, Heidelberg (2003)
27. Zadeh, L.A.: Fuzzy sets and information granularity. In: Gupta, N., Ragade, R., Yager, R.R. (eds.) Advances in Fuzzy Set Theory and Applications, pp. 3–18. North-Holland, Amsterdam (1979)
28. Zadeh, L.A.: Towards a theory of fuzzy information granulation and its centrality in human reasoning and fuzzy logic. Fuzzy Sets and Systems 90, 111–127 (1997)

Three Granular Structure Models in Graphs

Guang Chen[1] and Ning Zhong[1,2]

[1] International WIC Institute, Beijing University of Technology
Beijing, 100124, P.R. China
`cg@emails.bjut.edu.cn`
[2] Department of Life Science and Informatics, Maebashi Institute of Technology
Maebashi-City, 371-0816, Japan
`zhong@maebashi-it.ac.jp`

Abstract. The granular structures emphasize a multilevel and multi-view understanding of problems. This paper gives a study on how to granulate a graph, and how to extract the granular structures in the graph. There are three kinds of objects in the graph, vertices, edges and the combinations of vertices and edges. Differing from previous researches on graph clustering which focused on the classification of vertices, we study three granular structure models for the three kinds of objects in the graph.

Keywords: Granular Structures, Graph, Vertex-oriented granulation, Edge-oriented granulation, Combination-oriented granulation.

1 Purpose of the Study

Granular computing concerns a particular human-centric paradigm of problem solving by means of multiple levels of granularity and its applications in machines [14]. Many researchers have made significant progress on concrete models and methods of granular computing [1, 6–9, 11, 17–20].The triarchic theory of granular computing [15] offers a conceptual framework of granular computing by weaving together three powerful ideas: structured thinking, structured problem solving, and structured information processing. It emphasizes the exploitation of useful structures known as granular structures characterized by multilevel and multiview. The multi-level requires that a granular structure consists of a family of integrative levels with different granularity [15]; the multiview requires that it should be necessary to consider a family of multilevel structures with each representing a different view. A single hierarchical granular structure provides a multilevel understanding and representation of a problem or a system. But it typically captures one particular aspect and therefore offers one view. By constructing a family of hierarchies, it is possible to obtain multiple different views. Granular structures are a family of complementary hierarchies working together for a complete and comprehensive multiview understanding and representation.

The study of 'structures' has existed in many disciplines and fields. In the study of these fields, structures have different names and its composition has

T. Li et al. (Eds.): RSKT 2012, LNAI 7414, pp. 351–358, 2012.

different contents, but they have the same and uniform characteristics and features in fact. Table 1 shows the structures in different fields. As can be seen from the table, the structures in each fields have its own name and content, but they all have common features, such as, they all have basic component, which is named granule in our study; they all have the features of multi-level and multi-view. Each level in the structures contains a number of granules, it is a special granularity understanding of the problem. Each single structure typically captures one particular aspect and therefore offers one view of the problem. By constructing a family of structures, it is possible to obtain multiple different views of the problem. So the study of granular structures has two unique tasks. One is to extract high-level commonalities of different disciplines and to synthesize their results into an integrated whole by ignoring low-level details. The other is to make explicit ideas hidden in discipline-specific discussions in order to arrive at a set of discipline-independent principles. Through the study of granular structures, we attempt to achieve the following goals:

To make implicit principles explicit,
To make invisible principles visible,
To make domain-specific principles domain-independent,
To make subconscious effects conscious.

Many real world problems depicted and analyzed by graphs, have achieved good impacts. In this paper, we study how to extract the granular structures for graphs, and how to represent granular structures by graphs. The granular structures are an abstract and summary of the graph; it can be seen as a epitome of the graph. We can achieve a well-rounded and manoeuvrable understanding of the graph by granular structures. By graph based granular structures, we can make implicit principles in graphs explicit; make invisible principles in graphs visible; make subconscious effects in graphs conscious.

Table 1. General characters of structures in some different subjects

Subject	granules	granular structures
General systems theory	Unit	System structure
Cluster analysis	Cluster	Relations among clusters
Problem solving	Sub problem	Structure of problems
Social networks	Community	Social structures
Rough sets	Equivalence class	Layered rough sets
Quotient space theory	Quotient set	Quotient structure
Formal concept analysis	Formal concept	Concept lattice
Brain informatics	Neuron, brain region	Brain structure
Concept formation and learning	Concept	Relations among concepts
Writing	Word, sentence, chapter	Article structure

2   Related Studies

Some researchers have considered the relation of granular computing and graph. Stell [12] discussed the granularity for graphs and hypergraphs, and its relation with rough set theory and mathematical morphology. His work denoted that handling spatial data at several levels of detail the granulation of graphs was an important topic, and demonstrated that there were several quite different kinds of granulation for graphs. Wong and Wu [13] used the idea of granularity and hypergraph method to investigate database scheme. The database scheme was presented by hypergraph. Chen, et al. [3] studied how to use hypergraph to present granules and granular structures, and concreted a hypergraph model for granular computing. Previous researches explored how to construct granules or group clusters for graph in different domains. There was less discussion in how to construct granular structure for graphs, how to present the relationships and mappings among multiple levels, how to transform among different granularities, and how to solve problem under the granular structures.

Many researchers have studied clustering on graphs [2, 5, 10]. Graph clustering consists in grouping the vertices of the networks into different clusters according to various different criteria, i.e., edge structure of the graph, vertices of the same type, etc. Commonly, there should be many edges within each cluster and relatively few between the clusters.

The study of graph clustering focused on the clustering of the vertices, and had less consideration on the clustering of the edges and combination of vertices and edges. But in many problems we need to granulate the edges and the combination of vertices and edges. So our work study three kinds of granulation: vertex-oriented granulation, edge-oriented granulation and combination-oriented granulation, for three kinds of objects in graphs.

3   Granules and Granular Structures

Granular computing exploits structures in terms of granules, levels, and hierarchies based on multilevel and multiview representations [14, 15, 17]. A granule is a group of elements which have similar characteristics and can be seen as a whole unit. A granule plays two distinct roles. It may be an element of another granule and be a part of forming the other granule. It may also consist of a family of granules and be considered to be a whole. All granules in a level may collectively show a certain structure. This is the internal structure of a granulated view. Granules in a level, although may be relatively independent, are somehow related to a certain degree. A hierarchy represents one view of a problem with multiple levels of granularity, and a hierarchical structure is a understanding of the problem in a specific perspective. So granular computing exploits multi-level and multi-view representations in problem solving with granular structure.

4   Three Granular Structure Models in Graphs

There are three kinds of information in graphs: vertices, edges and the combination of vertices and edges. Previous researches about the clustering for graphs has focused on the classification of vertices, but in many problems, for example, intelligence traffic problems and protein structure analysis, we need to granulate the edges and the combination of vertices and edges. So we propose three granular structure models in graphs to study the granulation of vertices, edges and the combination of vertices and edges.

4.1   Vertex-Oriented Granular Structure Model in Graphs

The research objects of the vertex-oriented granular structure model are the vertices in graphs. This model studies the relation of these vertices and the structure of these vertices.

For many problems, we have to granulate the vertices and extract the structure of the granules. For example, in the field of electronic commerce, we always need to group the customers who have similar buying habits, we can use vertices to represent the customers, use edges to represent the comparability of the buying habits, and granulate the customers using vertex-oriented model.

Previous studies of graph clustering almost belong to vertex-oriented model. In this type of model, edges are only used to describe the relation of the vertices. The granulation of vertices is based on the information of edges and the results of the granulation only concern the vertices. Fig. 1 shows that the granules in this model only contain vertices.

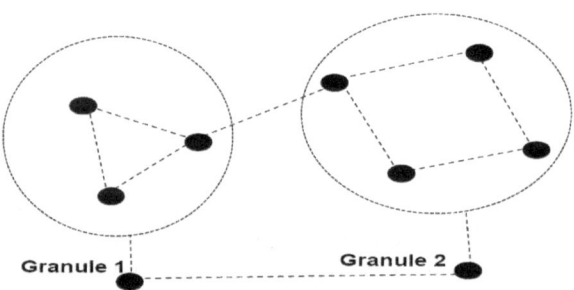

Fig. 1. Granulation of vertices

There is a detailed discussion of this model in our paper [4], so we overleap the granular methods for this model and focus on the discussion of the second model.

4.2   Edge-Oriented Granular Structure Model in Graphs

Differing from the first model, the research objects of the edge-oriented granular structure model are the edges in graphs. This model studies the relation of the edges and the structure of these edges. In many problems, we have to group the edges and explore the structure of these granules.

For example, in intelligence traffic control problem, the traditional single point signal control can't meet a requirement of urban traffic control, and regional traffic signals control can promote urban road network's efficiency more efficiently. We need to group the roads to a traffic region according to their relations, and have a unified control for the whole roads in a traffic region. We can use edges to represent the roads, then granulate the edges into a traffic region. Fig. 2 shows some granules in a road net, each granule is composed of edges.

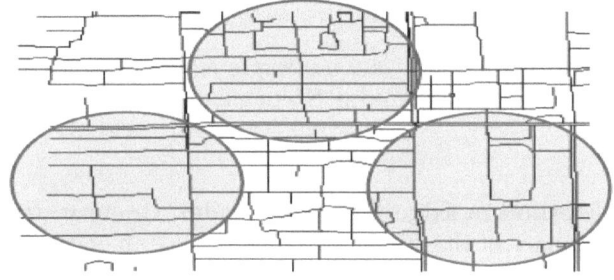

Fig. 2. Granules in road net

In edge-oriented model, edges are not used to describe the relation of the vertices, and we need to granulate the edges based on the relations of the edges, but not to granulate the vertices. Previous researches lacked this type of model. So we study it as the second model of granular structures in graphs. Fig. 3 shows that the granules in this model only contain edges.

Granular Structures in Edge-Oriented Model

Definition 1. *Let $G = (V, E)$ be a graph, V is defined as a set of vertices and E a set of edges. A element $g \in 2^E$ is called a* **Granule** *in edge-oriented Granular Structure Model of graph G, where 2^E is the power set of E.*

Definition 2. *Let $G = (V, E)$ be a graph, V is defined as a set of vertices and E a set of edges, 2^E is the power set of E. $l = \{g_i \mid 1 \leq i \leq m, g_i \in 2^V\}$ is a subset of 2^E. We call l a* **Level** *in edge-oriented Granular Structure Model of graph G, if it satisfies $\cup\{g_i \mid 1 \leq i \leq m\} = E$.*

Definition 3. *Consider two levels l_1 and l_2, if $|l_2| \leq |l_1|$, then we call that l_1 has a finer granularity than l_2, and l_2 has a coarser granularity than l_1, shorter*

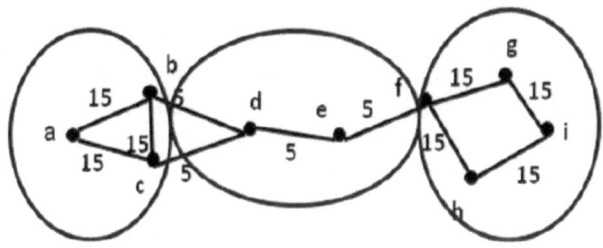

Fig. 3. Granulation of edges

form l_1 is finer than l_2, denoted as $l_2 \leq l_1$, $\leq$ is called finer relation between levels .

Definition 4. *Suppose $G \subseteq 2^V$ is a nonempty family of granules of the levels. The poset $(G, \leq)$ is called a* **Granular Structure***, where $\leq$ is the finer relation between levels.*

Granulation Method in Edge-Oriented Model. Granular structures are an abstract and summary of the graph. It can be seen as an epitome of the graph. Through the granular structures, the comprehension of the graph is clear at a glance. The granular structures have both glancing and particular information of the graph. When we want to have a finer or coarser understanding of the graph, we only need to switch among the levels in structures.

Algorithm 1. Connected-graph-based Algorithm

---

1. Construct the undirected weighted graph
2. Set thresholds
3. FOR each vertex (each threshold can produce a level)
4.   Compare the weight of edges
5.   Select the edges satisfied the threshold
6.   Then cut the connection between edge and vertex
7. END FOR
8. Each sub-connected-graph is a granule
9. FOR each level except the last
10.   Mapping it to the next one
11. END FOR

A typical method is based on the connected graph method. An indirected and weighted graph is used in this method. The algorithm for this model is described as Algorithm 1: for an undirected weighted graph, set some thresholds; For each threshold, select the edges exceed it, then each sub-connected-graph of the forest is a granule, and a level is produced. For different threshold, we can get different levels. When the given threshold changed, a new level will be constructed. For example, in Fig. 3, the edges $< a, b >, < a, c >, < b, c >$ form a granule, and edges $< h, g >, < g, i >, < h, i >, < f, h >$ form another granule.

4.3 Combination-Oriented Granular Structure Model in Graphs

The granules in this model not only contain the vertices or edges, they are the special combinations of some vertices and edges. The combination must be handle as a aptotic whole, can't be break up. We often encounter such a kind of problems, in which the vertices and edges in a graph are regarded as special whole, for example, as shown in Fig.4, in protein structures prediction, each amino acid is such a combination, and we should determine the combinations for protein structures analysis. We consider such a combination as a granule.

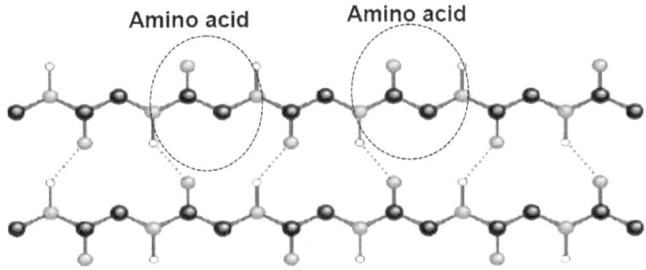

Fig. 4. Amino acid in protein structures

5   Conclusion

The triarchic theory emphasizes the exploitation of useful structures known as granular structures characterized by multilevel and multiview. This paper study the granular structures in graphs. For three kinds of objects in graphs, we have three granular structure models: vertex-oriented model, edge-oriented model and combination-oriented model. The models may make the implicit and invisible structures in graphs explicit and visible.

Acknowledgments. This work is partially supported by National Natural Science Foundation of China (No.60905027) and Beijing Natural Science Foundation (No.4102007).

References

1. Bargiela, A., Pedrycz, W.: Granular Computing: An Introduction. Kluwer Academic Publishers, Boston (2002)
2. Brandes, U., Gaertler, M., Wagner, D.: Experiments on Graph Clustering Algorithms. In: Di Battista, G., Zwick, U. (eds.) ESA 2003. LNCS, vol. 2832, pp. 568–579. Springer, Heidelberg (2003)
3. Chen, G., Zhong, N., Yao, Y.Y.: Hypergraph Model of Granular Computing. In: 2008 IEEE International Conference on Granular Computing, pp. 80–85. IEEE Press, New York (2008)
4. Chen, G., Zhong, N.: Granular Structures in Graphs. In: Yao, J., Ramanna, S., Wang, G., Suraj, Z. (eds.) RSKT 2011. LNCS, vol. 6954, pp. 649–658. Springer, Heidelberg (2011)
5. Flake, G.W., Tarjan, R.E., Tsioutsiouliklis, K.: Graph Clustering and Minimum Cut Trees. Internet Mathematics 1(1), 385–408 (2004)
6. Hobbs, J.R.: Granularity. In: 9th International Joint Conference on Artificial Intelligence, pp. 432–435 (1985)
7. Luo, J., Yao, Y.: Granular State Space Search. In: Butz, C., Lingras, P. (eds.) Canadian AI 2011. LNCS (LNAI), vol. 6657, pp. 285–290. Springer, Heidelberg (2011)
8. Pawlak, Z.: Rough Sets. International Journal of Computer and Information Sciences 11, 341–356 (1982)
9. Polkowski, L.: A Model of Granular Computing with Applications: Granules from Rough Inclusions in Information Systems. In: IEEE International Conference on Granular Computing, pp. 9–16 (2006)
10. Schaeffer, S.E.: Graph Clustering. Computer Science Review 1(1), 27–64 (2007)
11. Skowron, A., Stepaniuk, J.: Information Granules: Towards Foundations of Granular Computing. International Journal of Intelligent Systems 16, 57–85 (2001)
12. Stell, A.J.: Granulation for Graphs. In: Freksa, C., Mark, D.M. (eds.) COSIT 1999. LNCS, vol. 1661, pp. 417–432. Springer, Heidelberg (1999)
13. Wong, S.K.M., Wu, D.: Automated Mining of Granular Database Scheme. In: IEEE International Conference on Fuzzy Systems, vol. 1, pp. 690–694 (2002)
14. Yao, Y.Y.: A Partition Model of Granular Computing. In: Peters, J.F., Skowron, A., Grzymała-Busse, J.W., Kostek, B.z., Świniarski, R.W., Szczuka, M.S. (eds.) Transactions on Rough Sets I. LNCS, vol. 3100, pp. 232–253. Springer, Heidelberg (2004)
15. Yao, Y.Y.: The Art of Granular Computing. In: Kryszkiewicz, M., Peters, J.F., Rybiński, H., Skowron, A. (eds.) RSEISP 2007. LNCS (LNAI), vol. 4585, pp. 101–112. Springer, Heidelberg (2007)
16. Yao, Y., Miao, D., Zhang, N., Xu, F.: Set-Theoretic Models of Granular Structures. In: Yu, J., Greco, S., Lingras, P., Wang, G., Skowron, A. (eds.) RSKT 2010. LNCS (LNAI), vol. 6401, pp. 94–101. Springer, Heidelberg (2010)
17. Yao, Y.Y., Zhong, N.: Granular Computing. Wiley Encyclopedia of Computer Science and Engineering 3, 1446–1453 (2009)
18. Zadeh, L.A.: Towards a Theory of Fuzzy Information Granulation and Its Centrality in Human Reasoning and Fuzzy Logic. Fuzzy Sets and Systems 90, 111–127 (1997)
19. Zhang, L., Zhang, B.: The Quotient Space Theory of Problem Solving. Fundamenta Informatcae 59, 287–298 (2004)
20. Zhu, W., Wang, F.Y.: On Three Types of Covering-based Rough Sets. IEEE Transactions on Knowledge and Data Engineering 19(8), 1131–1144 (2007)

A Competition Strategy
to Cost-Sensitive Decision Trees

Fan Min and William Zhu

Lab of Granular Computing,
Zhangzhou Normal University, Zhangzhou 363000, China
minfanphd@163.com, williamfengzhu@gmail.com

Abstract. Learning from data with test cost and misclassification cost has been a hot topic in data mining. Many algorithms have been proposed to induce decision trees for this purpose. This paper studies a number of such algorithms and presents a competition strategy to obtain trees with lower cost. First, we generate a population of decision trees using λ-ID3 and EG2 algorithms through considering information gain and test cost. λ-ID3 is a generalization of three existing algorithms, namely ID3, IDX, and CS-ID3. EG2 is another parameterized algorithm, and its parameter range is extended in this work. Second, we post-prune these trees by considering the tradeoff between the test cost and the misclassification cost. Finally, we select the best decision tree for classification. Experimental results on the mushroom dataset with various cost settings indicate: 1) there does not exist an optimal parameter for λ-ID3 or EG2; 2) the competition strategy is effective in selecting an appropriate decision tree; and 3) post-pruning can help decreasing the average cost effectively.

Keywords: Cost-sensitive learning, decision tree, heuristic function, competition strategy.

1 Introduction

Cost-sensitive learning is one of the most challenging problems in data mining research [1]. It has attracted much research interests from the areas of decision trees [2,3], ANN [4,5], Bayesian networks [6], Rough sets [7,8,9,10], etc. There are various types of costs [11]. Test cost and misclassification cost are more often addressed [12]. The test cost is the measurement cost of determining the value of an attribute a exhibited by an object [2,12]. The misclassification cost is the penalty of deciding that an object belongs to class J when its real class is K [3].

Learning from data with both test cost and misclassification cost is especially interesting. The ICET algorithm proposed by Turney [2] might be the first decision tree approach to this issue, which was later addressed by Ling et al. [13]. ICET uses a genetic algorithm [14] to evolve a population of biases for a decision tree induction algorithm. ICET is based on the EG2 [15] algorithm, which has a parameter $\omega \in [0,1]$ to control the heuristic function. In the rough set society, there are also attribute reduction ([16,17,18]) or classification ([19,20]) approaches to data with both costs.

T. Li et al. (Eds.): RSKT 2012, LNAI 7414, pp. 359–368, 2012.
© Springer-Verlag Berlin Heidelberg 2012

In this paper, we propose a new approach named CC-ID3 (Competition strategy from Cost-sensitive ID3) to obtain low cost decision trees. Firstly, we introduce a parameter λ to the heuristic function of the IDX algorithm [21] and present the λ-ID3 algorithm. λ-ID3 coincides with ID3 [22], IDX [21], and CS-ID3 [23] while λ is 0, -1, and -0.5, respectively. We also point out the upper bound of ω in the EG2 is rather low. Then we construct a population of decision trees using different λ values for λ-ID3, and different ω values for EG2. These trees are post-pruned to keep the tradeoff between the test cost and the misclassification cost. Finally these trees compete with each other on the training set, and the one with the least cost is selected for the classification purpose. Although a number of trees are built, only one is selected for classification. This is the major difference between our strategy and bagging [24].

The mushroom dataset [25] is selected for the experimentation. We use various settings on both test cost and misclassification cost to simulate different situations of applications. Experiments are undertaken using an open source software called COSER (cost-sensitive rough sets) [26]. Experimental results show that: 1) no λ (or ω) always beats others on different test cost settings, hence it is not good to use a fixed λ (or ω) as did in IDX and CS-ID3; 2) decision trees with low cost in the training set generally tend to have low cost in the testing set, therefore the competition strategy is effective in selecting an appropriate decision tree; and 3) post-pruning can help decreasing the average cost effectively especially when misclassification costs are not too high compared with test costs.

2   Related Works

In this section, the concept of decision systems with test cost and misclassification cost is revisited. Then the computation of average cost of a decision tree is presented. Finally a number of existing test-cost-sensitive decision tree induction algorithms are reviewed.

2.1   The Data Model

Decision system is a fundamental data model in data mining and machine learning. We consider decision systems with test cost and misclassification cost.

Definition 1. *A decision system with test cost and misclassification cost (DS-TM) S is the 7-tuple [17] :*

$$S = (U, C, d, V, I, tc, mc), \tag{1}$$

where U is a finite set of objects called the universe, C is the set of conditional attributes, d is the decision attribute, $V = \{V_a | a \in C \cup \{d\}\}$ where V_a is the set of values for each $a \in C \cup \{d\}$, $I = \{I_a | a \in C \cup \{d\}\}$ where $I_a : U \to V_a$ is an information function for each $a \in C \cup \{d\}$, $tc : C \to \mathbb{R}^+ \cup \{0\}$ is the test cost function, and $mc : k \times k \to \mathbb{R}^+ \cup \{0\}$ is the misclassification cost function, where $\mathbb{R}^+$ is the set of positive real numbers, and $k = |I_d|$.

Terms *conditional attribute, attribute* and *test* are already employed in the literature, and these have the same meaning throughout this paper. $U, C, d, \{V_a\}$ and $\{I_a\}$ can be displayed in a classical decision table.

The test cost function is stored in a vector $tc = [tc(a_1), tc(a_2), \ldots, tc(a_{|C|})]$. We adopt the test cost independent model [7] to define the cost of a test set. That is, the test cost of $B \subseteq C$ is given by $tc(B) = \sum_{a \in B} tc(a)$. The misclassification cost function can be represented by a matrix $mc = \{mc_{k \times k}\}$, where $mc_{i,j}$ is the cost of classifying an object of the i-th class into the j-th class. Similar to Turney's work [2], we require that $m_{i,i} \equiv 0$. However, this requirement could be easily loosed if necessary.

2.2   Computation of the Average Cost

The computation of the average cost is as follows. Let T be a decision tree, U be the testing dataset, and $x \in U$. x follows a path from the root of T to a leaf. Let the set of attributes on the path be $A(x)$. The test cost of x is $tc(A(x))$. Let the real class label of x be $C(x)$ and x is classified as $T(x)$. The misclassification cost of x is $mc(C(x), T(x))$. Therefore the total cost of x is $tc(A(x)) + mc(C(x), T(x))$. Finally, the average cost of T on U is

$$T_c(U) = \sum_{x \in U} (tc(A(x)) + mc(C(x), T(x)))/|U|. \tag{2}$$

2.3   Related Algorithms

Now we review a number of existing decision tree induction algorithms: ID3 [22], EG2 [15], IDX [21] and CS-ID3 [23]. We begin with their general structure and focus on their heuristic functions. These functions are all based on the information gain due to its stability in generating similar results on the training and testing sets. Test cost is considered in heuristic functions except in the ID3 algorithm. Therefore EG2, CS-ID3 and IDX are essentially test-cost-sensitive.

The constructions of decision trees generally share a common scenario as follows [22]. Let the training set be U. If U is empty or contains only objects of one class, the simplest decision tree is just a leaf labelled with the class. Otherwise any $a \in C$ can split U into a family of subsets $\{U_1, U_2, \ldots, U_{|V_a|}\}$. According to the heuristic function, we can select the *best* attribute to split U and repeat the process in each subset. This divide-and-conquer approach terminates until no more nodes can be produced. For any leaf corresponding to an object subset of two or more classes, the class label that minimizes the misclassification cost is assigned.

ID3 [22] is the most popular decision tree induction algorithm for symbolic data. The heuristic function is the information gain, which is denoted by

$$g(a) = \Delta I_a \tag{3}$$

for $a \in C$. Naturally, the attribute maximizing $g(a)$ is selected. Since the computation of ΔI_a is fundamental to data miners, we will not list the detail here.

EG2 [15] employs the following heuristic function:

$$ICF(a) = \frac{2^{\Delta I_a} - 1}{(tc(a) + 1)^\omega}, \text{ where } 0 \le \omega \le 1. \tag{4}$$

ω is the parameter to adjust the impact of the test cost. If $\omega = 0$, the test cost is ignored and maximizing $ICF(a)$ is equivalent to maximizing ΔI_a. In other words, EG2 coincides with ID3. If $\omega = 1$, the heuristic function is strongly biased by test cost.

It is reasonable that $\omega \ge 0$ to ensure that expensive attributes are never biased. However, there does not exist a good explanation why $\omega \le 1$. Can the heuristic function be even stronger biased by the test cost? The answer is yes, and we will show the reason through experimentation in Section 4.

IDX [21] employs the following heuristic function::

$$IDX(a) = \frac{\Delta I_a}{tc(a)}. \tag{5}$$

Finally, CS-ID3 [23] employs the following heuristic function:

$$CS(a) = \frac{(\Delta I_a)^2}{tc(a)}. \tag{6}$$

One may observe that $ICF(a)$, $IDX(a)$ and $CS(a)$ are quite similar in that they give penalty to expensive attributes. However their forms are different, or the degrees of penalty are different.

3   New Algorithms

In this section, we first present λ-ID3 as a generalization of ID3, IDX, and CS-IDX. Then we discuss the post-pruning technique. Finally we propose the competition strategy.

3.1 λ-ID3

We present the following heuristic function:

$$f(a)^\lambda = \Delta I_a tc(a)^\lambda, \text{ where } \lambda \le 0. \tag{7}$$

Using this function, we can follow the same routine of ID3 to obtain a new algorithm called λ-ID3. One observes that

$$f(a)^\lambda = \begin{cases} g(a), & \lambda = 0; \\ \sqrt{CS(a)}, & \lambda = -0.5; \\ IDX(a), & \lambda = -1. \end{cases} \tag{8}$$

In other words, λ-ID3 coincides with ID3, CS-ID3, and IDX when λ is 0, -0.5, and -1, respectively.

3.2   The Post-prune Technique

We employ the following post-prune technique. After the decision tree is built, we traverse it in postorder. Denote the current non-leaf node by N, the subtree rooted at N be T_N, and the set of instances reaching N be $U_N \subseteq U$. We try to assign a class label to N that minimizes the average misclassification cost. The cost is denoted by $c(U_N)$. If this cost is no more than the average cost of the subtree rooted by N, namely $c(U_N) \leq T_{Nc}(U_N)$, the subtree does not help decreasing the cost. In this case N is set to a leaf.

Pre-prune is another important technique. In the simplest case, it only considers the effectiveness of the current test. If no test can help decreasing the average cost, then the node is made a leaf. Unfortunately, this technique may run in local optimal and loses subtrees that create less cost. Therefore we will not employ it in our algorithm.

3.3   The Competition Strategy

CC-ID3 uses a competition strategy to select the decision tree with the best performance on the training set. It works as follows:

Step 1. A number of decision trees are produced using different algorithms and/or different parameter values.

Step 2. These decision trees are post-pruned.

Step 3. The decision tree with the least cost on the training set is selected for classification.

This strategy is similar to bagging [24] in that both make use of a number of classifiers. However they are essentially different. The bagging approach produces a number of classifiers which can be employed simultaneously for classification. However in CC-ID3 only one decision tree should be selected. This is because different decision trees require different attributes, and the test cost is increased accordingly.

CC-ID3 is also similar to ICET [2] in that both adjust parameters of the decision tree induction algorithm. The mechanism of CC-ID3 is simpler because it produces a number of decision trees and select one directly. In contrast, ICET employs a genetic algorithm to evolve biases and produce decision trees in some generations. Moreover, CC-ID3 only uses different parameters, while ICET treats test costs as biases and tries to adjust them. CC-ID3 seldom generate more than 40 decision trees while ICET often need 1,000. Therefore CC-ID3 is also more efficient. Unfortunately, in this work CC-ID3 cannot be compared with ICET because the former deals with symbolic data while the latter deals with numerical ones. The comparison will be made in our further works.

4   Experiments

The main purpose of our experiments is to answer the following questions.

1. Are there optimal settings of λ and/or ω?
2. Can CC-ID3 improve the performance over any particular decision tree?

3. Which algorithm is more appropriate as the basis of CC-ID3?
4. Can post-pruning improve the performance of the decision tree?

Experiments are undertaken on the mushroom dataset, where $|U| = 8124$ and $|C| = 22$. This dataset is selected mainly because the size is large enough to test the performance of the algorithm. Moreover, its distribution is stable under different data sampling.

Each time we use 60% of the dataset as the training data, and the remaining part as the testing data. The test cost is set to a random integer in $[1, 10]$. The misclassification cost matrix is set to $\begin{bmatrix} 0 & 500 \\ 50 & 0 \end{bmatrix}$.

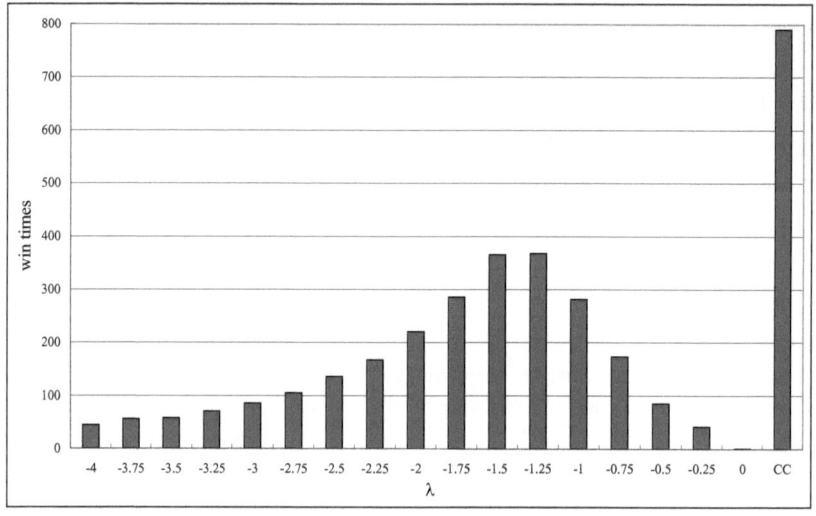

Fig. 1. The win times for different λ and CC-ID3

First we study the influence of λ to λ-ID3. For each cost setting we run λ-ID3 by setting $\lambda = 0, -0.25, -0.5, \ldots, -4$. If the algorithm performs the best on the testing set with a certain λ, we say respective λ *wins*. We use 100 test cost setting to obtain the number of win times for each λ values. Results are depicted in Fig. 1, from where we observe the following.

1. $\lambda = -1.25$ is the best setting. In other words, ID3 ($\lambda = 0$), CS-ID3 ($\lambda = -0.5$) and IDX ($\lambda = -1$) all perform not good.
2. With the best λ we obtain the best result in less than 40% of the time. Therefore it is not a good choice to fix the λ setting.
3. CC-ID3 is an effective approach to improve the performance of λ-ID3. In about 80% of the time it produces the best result.

Second we study the influence of ω to EG2. We let $\omega = 0, 0.25, 0.5, \ldots, 4$. Results are depicted in Fig. 2, where we observe the following.

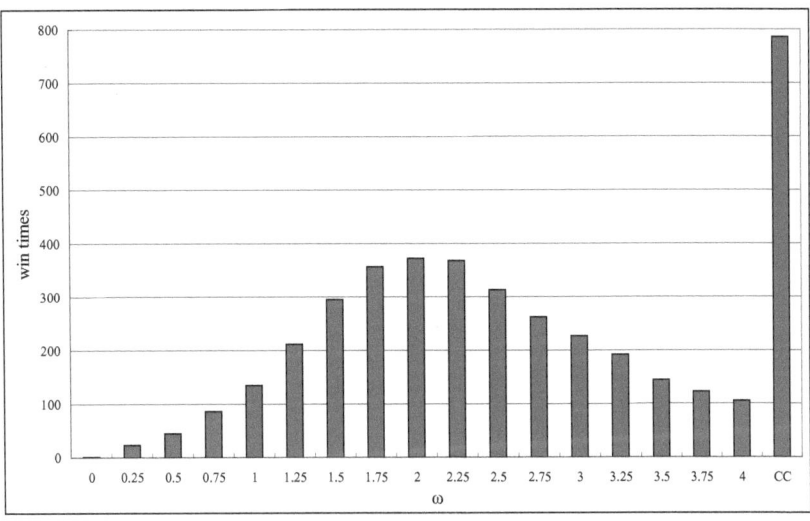

Fig. 2. The win time for different ω and CC-ID3

1. $\omega = 2$ is the best setting. Therefore $\omega \le 1$ is not a reasonable setting for EG2.
2. Even if we know the best setting of ω, in only about 40% of the time we obtain the best result.
3. CC-ID3 is an effective approach to improve the performance of EG2. In about 80% of the time it produces the best result.

Table 1. Average cost for different approaches (1,000 test cost settings)

ID3	λ-ID3 $\lambda = -1.5$	EG2 $\omega = 2.0$	CC-ID3 on λ-ID3	CC-ID3 on EG2	CC-ID3 on both algorithms
8.35315	5.04883	4.97522	4.93822	4.89279	4.87866

Third we study which algorithm is more appropriate as the basis of CC-ID3. We use the same settings mentioned above. We are interested in results of CC-ID3 based on λ-ID3, EG2, and both. The average cost is listed in Table 1. Results show that both λ-ID3 and EG2 with the best settings significantly outperforms ID3, which does not take into consideration the test cost. EG2 is more appropriate than λ-ID3 as the basis of CC-ID3, but the advantage is not significant. The best choice is to employ both algorithms to produce candidates.

Finally we study the effectiveness of the post-prune technique. As discussed in Section 3.2, post-prune always incur higher misclassification cost and lower

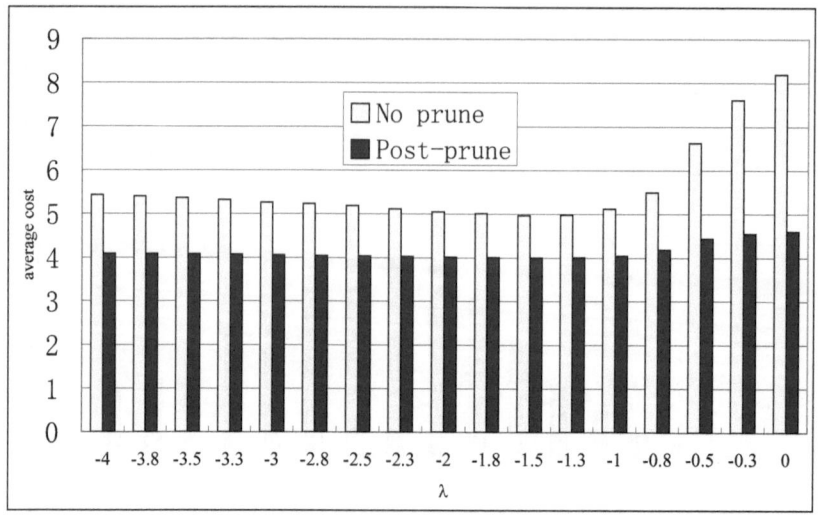

Fig. 3. The effect of the post-prune technique

test cost. When misclassification costs are very high compared with test costs, post-prune is essentially not executed since it always produce high cost on the training set. In contrast, if misclassification costs are comparable with test costs, post-prune is very effective on the training set in producing lower cost. We let the misclassification cost matrix be $\begin{bmatrix} 0 & 100 \\ 10 & 0 \end{bmatrix}$ and compare the performance of the CC-ID3 algorithm with and without the post-prune technique. Naturally, only results on the testing set is illustrated since these on the training set are not interesting. Results are depicted in Fig. 3. Here we observe that post-prune is significant in decreasing the average cost.

5   Conclusions and Further Works

In this paper, we have proposed the competition strategy for cost-sensitive decision tree selection. This strategy is simple because it does not change the structure of any candidate decision tree. It is efficient because not too many candidate decision trees are needed. It is effective in selecting a good decision tree for classification. The contributions of the work also include the design of λ-ID3, the parameter range extension of EG2, and the post-prune technique. Note that we only tested λ-ID3 and EG2, which have similar performance on the training and testing sets. Other decision tree induction algorithm may be inappropriate for the competition strategy. In the future we will design algorithms based on C4.5 [27] to deal with numerical data and compare the performance with ICET [2]. A more interesting issue is to introduce waiting time to the data model and build time-cost-sensitive decision trees.

Acknowledgements. This work is in part supported by National Science Foundation of China under Grant No. 61170128, the Natural Science Foundation of Fujian Province, China under Grant No. 2011J01374, State key laboratory of management and control for complex systems open project under Grant No. 20110106, and the Education Department of Fujian Province under Grant No. JA11176.

References

1. Yang, Q., Wu, X.: 10 challenging problems in data mining research. International Journal of Information Technology and Decision Making 5(4), 597–604 (2006)
2. Turney, P.D.: Cost-sensitive classification: Empirical evaluation of a hybrid genetic decision tree induction algorithm. Journal of Artificial Intelligence Research 2, 369–409 (1995)
3. Fan, W., Stolfo, S., Zhang, J., Chan, P.: Adacost: Misclassification cost-sensitive boosting. In: Proceedings of the 16th International Conference on Machine Learning, pp. 97–105 (1999)
4. Zhou, Z., Liu, X.: Training cost-sensitive neural networks with methods addressing the class imbalance problem. IEEE Transactions on Knowledge and Data Engineering 18(1), 63–77 (2006)
5. Kukar, M., Kononenko, I.: Cost-sensitive learning with neural networks. In: Proceedings of the 13th European Conference on Artificial Intelligence, pp. 445–449 (1998)
6. Chai, X.Y., Deng, L., Yang, Q., Ling, C.X.: Test-cost sensitive Naïve Bayes classification. In: Proceedings of the 5th International Conference on Data Mining, pp. 51–58 (2004)
7. Min, F., Liu, Q.: A hierarchical model for test-cost-sensitive decision systems. Information Sciences 179, 2442–2452 (2009)
8. Min, F., He, H., Qian, Y., Zhu, W.: Test-cost-sensitive attribute reduction. Information Sciences 181, 4928–4942 (2011)
9. Zhu, W., Wang, F.: Reduction and axiomization of covering generalized rough sets. Information Sciences 152(1), 217–230 (2003)
10. Zhu, W.: Generalized rough sets based on relations. Information Sciences 177(22), 4997–5011 (2007)
11. Turney, P.D.: Types of cost in inductive concept learning. In: Proceedings of the Workshop on Cost-Sensitive Learning at the 17th ICML, pp. 1–7 (2000)
12. Hunt, E.B., Marin, J., Stone, P.J. (eds.): Experiments in induction. Academic Press, New York (1966)
13. Ling, C.X., Sheng, V.S., Yang, Q.: Test strategies for cost-sensitive decision trees. IEEE Transactions on Knowledge and Data Engineering 18(8), 1055–1067 (2006)
14. Grefenstette, J.J.: Optimization of control parameters for genetic algorithms. IEEE Transactions on Systems, Man, and Cybernetics, 122–128 (1986)
15. Núñez, M.: The use of background knowledge in decision tree induction. Machine Learning 6, 231–250 (1991)
16. Yao, Y.Y., Zhao, Y.: Attribute reduction in decision-theoretic rough set models. Information Sciences 178(17), 3356–3373 (2008)
17. Min, F., Zhu, W.: Minimal Cost Attribute Reduction through Backtracking. In: Kim, T.-H., Adeli, H., Cuzzocrea, A., Arslan, T., Zhang, Y., Ma, J., Chung, K.-I., Mariyam, S., Song, X. (eds.) DTA/BSBT 2011. CCIS, vol. 258, pp. 100–107. Springer, Heidelberg (2011)

18. Jia, X., Li, W., Shang, L., Chen, J.: An Optimization Viewpoint of Decision-Theoretic Rough Set Model. In: Yao, J., Ramanna, S., Wang, G., Suraj, Z. (eds.) RSKT 2011. LNCS, vol. 6954, pp. 457–465. Springer, Heidelberg (2011)
19. Liu, D., Yao, Y.Y., Li, T.R.: Three-way investment decisions with decision-theoretic Rough sets. International Journal of Computational Intelligence Systems 4, 66–74 (2011)
20. Li, H., Zhou, X.: Risk decision making based on decision-theoretic rough set: a three-way view decision model. International Journal of Computational Intelligence Systems 4(1), 1–11 (2011)
21. Norton, S.: Generating better decision trees. In: Proceedings of the 11th International Joint Conference on Artificial Intelligence, pp. 800–805 (1989)
22. Quinlan, J.R.: Induction of decision trees. Machine Learning 1, 81–106 (1986)
23. Tan, M.: Cost-sensitive learning of classification knowledge and its applications in robotics. Machine Learning 13, 7–33 (1993)
24. Breiman, L.: Bagging predictors. Machine Learning 24(2), 123–140 (1996)
25. Blake, C.L., Merz, C.J.: UCI repository of machine learning databases (1998), http://www.ics.uci.edu/~mlearn/mlrepository.html
26. Min, F., Zhu, W., Zhao, H., Pan, G.: Coser: Cost-senstive rough sets (2011), http://grc.fjzs.edu.cn/~fmin/coser/
27. Quinlan, J.R. (ed.): C4.5 Programs for Machine Learning. Morgan kaufmann Publisher, San Mateo (1993)

An Information-Theoretic Interpretation of Thresholds in Probabilistic Rough Sets

Xiaofei Deng and Yiyu Yao

Department of Computer Science, University of Regina
Regina, Saskatchewan, Canada S4S 0A2
{deng200x,yyao}@cs.uregina.ca

Abstract. In a probabilistic rough set model, the positive, negative and boundary regions are associated with classification errors or uncertainty. The uncertainty is controlled by a pair of thresholds defining the three regions. The problem of searching for optimal thresholds can be formulated as the minimization of uncertainty induced by the three regions. By using Shannon entropy as a measure of uncertainty, we present an information-theoretic approach to the interpretation and determination of thresholds.

1 Introduction

Probabilistic rough sets are an extension of Pawlak [9] rough sets by introducing a pair of thresholds for defining probabilistic rough sets approximations, or equivalently, three probabilistic regions [11,12,16,17]. A key issue is the interpretation and estimation of thresholds. This can be approached from the viewpoint of trade-off between error and applicability (or generality) of rough set classification. The Pawlak positive and negative regions do not have classification error, but are only applicable to a small set of objects in practical applications. On the other hand, probabilistic positive and negative regions have some classification errors, but are applicable to more objects. The pair of thresholds controls the level of trade-off. It maybe interpreted and determined based on various notions and measures related to classification errors and applicability.

A pair of thresholds in a decision-theoretic rough set model (DTRSM) [6,11,16,17], an example of probabilistic rough set models, is determined by minimizing overall classification cost. Several more recent attempts include a game-theoretic framework [1,2,3], a multi-view decision model [5], a model based on an optimization viewpoint [4], and a method using probabilistic model criteria [7,8]. This paper is a new contribution along the same line. We propose an information-theoretic interpretation of thresholds. By minimizing Shannon entropy of probabilistic rough set three regions, one can determine a pair of thresholds. The proposed model and interpretation are closely related to machine learning methods based on information-theoretic measures, for example, ID3 algorithm [10].

T. Li et al. (Eds.): RSKT 2012, LNAI 7414, pp. 369–378, 2012.

2   An Overview of Probabilistic Rough Sets

In this section, we briefly examine three fundamental issues in a probabilistic rough set model, namely, three regions defined by a pair of thresholds, interpretations of the thresholds, and determination of the thresholds.

2.1   Three Probabilistic Regions Defined by a Pair of Thresholds

Suppose U is a finite set of objects called the universe and $E \subseteq U \times U$ is an equivalence relation on U, i.e., E is reflexive, symmetric and transitive. The equivalence class of E containing an object $x \in U$ is given by $[x]_E = [x] = \{y \in U \mid xEy\}$. The family of all equivalence classes, $U/E = \{[x]_E \mid x \in U\}$, is called the quotient set of the universe and is also called a partition of U. For a subset $C \subseteq U$, $Pr(C|[x])$ denotes the conditional probability of an object in C given that the object is in $[x]$. The main results of probabilistic rough sets, introduced and studied by Yao et al. [16,17] in a decision theoretic rough set model, are parameterized approximations. For a pair of thresholds (α, β) with $0 \leq \beta < \alpha \leq 1$, the (α, β)-probabilistic lower and upper approximations of C are expressed by:

$$\underline{apr}_{(\alpha,\beta)}(C) = \bigcup\{[x] \in U/E \mid Pr(C|[x]) \geq \alpha\},$$
$$\overline{apr}_{(\alpha,\beta)}(C) = \bigcup\{[x] \in U/E \mid Pr(C|[x]) > \beta\}. \tag{1}$$

According to the lower and upper approximations, we can define the following probabilistic positive, negative and boundary regions:

$$\begin{aligned} \mathrm{POS}_{(\alpha,\beta)}(C) &= \underline{apr}_{(\alpha,\beta)}(C) \\ &= \{x \in U \mid Pr(C|[x]) \geq \alpha\}, \\ \mathrm{NEG}_{(\alpha,\beta)}(C) &= (\overline{apr}_{(\alpha,\beta)}(C))^c \\ &= \{x \in U \mid Pr(C|[x]) \leq \beta\}, \\ \mathrm{BND}_{(\alpha,\beta)}(C) &= (\mathrm{POS}_{(\alpha,\beta)}(C) \cup \mathrm{NEG}_{(\alpha,\beta)}(C))^c \\ &= \{x \in U \mid \beta < Pr(C|[x]) < \alpha\}, \end{aligned} \tag{2}$$

where $(\overline{apr}_{(\alpha,\beta)}(C))^c = U - \overline{apr}_{(\alpha,\beta)}(C)$ denotes the complement of $\overline{apr}_{(\alpha,\beta)}(C)$. The three probabilistic regions are pair-wise disjoint and their union is the entire universe U. As some of them maybe empty, they do not necessarily form a partition of U. The three probabilistic regions can be interpreted based on the notions of three-way decisions and ternary classifications [13,14].

2.2   Thresholds and Classification Errors and Costs

Suppose C is a set of instances of a concept, the equivalence class $[x]$ contains objects that have the same description as x. The conditional probability can be viewed as a measure of the degree of confidence that an object with the same description as x is in C. For an object y with the same description as x, we accept it to be in C if our confidence is at or above α level, namely, $Pr(C|[x]) \geq \alpha$.

For an equivalence class $[x]$, by accepting all its members to be in C, we may make incorrect acceptance errors. Since $[x] \cap C^c$ is the set of elements not in C, the rate of incorrect acceptance errors is given by [14]:

$$IAE([x], C) = \frac{|[x] \cap C^c|}{|[x]|} = 1 - Pr(C|[x]),$$

where $|\cdot|$ denotes the cardinality of a set. By the condition $Pr(C|[x]) \geq \alpha$, we have $IAE([x], C) \leq 1 - \alpha$, namely, the rate of incorrect acceptance errors is at or below $1 - \alpha$. For the positive region, its rate of incorrect acceptance errors is,

$$IAE(\text{POS}_{(\alpha,\beta)}(C), C) = \frac{|\text{POS}_{(\alpha,\beta)}(C) \cap C^c|}{|\text{POS}_{(\alpha,\beta)}(C)|}$$

$$= \sum_{\substack{[x] \in U/E \\ Pr(C|[x]) \geq \alpha}} \frac{|[x] \cap C^c|}{|\text{POS}_{(\alpha,\beta)}(C)|}$$

$$= \sum_{\substack{[x] \in U/E \\ Pr(C|[x]) \geq \alpha}} \frac{|[x]|}{|\text{POS}_{(\alpha,\beta)}(C)|} \frac{|[x] \cap C^c|}{|[x]|}$$

$$= \sum_{\substack{[x] \in U/E \\ Pr(C|[x]) \geq \alpha}} \frac{|[x]|}{|\text{POS}_{(\alpha,\beta)}(C)|} IAE([x], C), \qquad (\leq 1 - \alpha).$$

The quantity $|[x]|/|\text{POS}_{(\alpha,\beta)}(C)|$ may be interpreted as the probability of an equivalence class $[x]$ with respect to the probabilistic positive region. Therefore, $IAE(\text{POS}_{(\alpha,\beta)}(C), C)$ is the expected rate of incorrect acceptance errors of all equivalence classes in the positive region, which is also bounded by $1 - \alpha$.

We reject an object to be in C if our confidence of the object being in C is at or below β level, namely, $Pr(C|[x]) \leq \beta$. This introduces incorrect rejection errors $IRE([x], C) = |[x] \cap C|/|[x]| \leq \beta$. The incorrect rejection error of the negative region is defined by [14]:

$$IRE(\text{NEG}_{(\alpha,\beta)}(C), C) = \frac{|\text{NEG}_{(\alpha,\beta)}(C) \cap C|}{|\text{NEG}_{(\alpha,\beta)}(C)|}$$

$$= \sum_{\substack{[x] \in U/E \\ Pr(C|[x]) \leq \beta}} \frac{|[x]|}{|\text{NEG}_{(\alpha,\beta)}(C)|} IRE([x], C), \qquad (\leq \beta).$$

Again, $IRE(\text{NEG}_{(\alpha,\beta)}(C), C)$ is bounded by β.

When the confidence is too low to warrant an acceptance and, at the same time, too high to warrant a rejection, we choose a third option of noncommitment. In sequential decision-making, the third choice may be understood as a decision of deferment [13,16,17]. For the boundary region, two new types of errors are introduced, namely, noncommitment for positives and noncommitment for negatives. They are defined, respectively, by [14]:

$$NPE([x], C) = \frac{|[x] \cap C|}{|[x]|},$$

$$NPE(\text{BND}_{(\alpha,\beta)}(C), C) = \frac{|\text{BND}_{(\alpha,\beta)}(C) \cap C|}{|\text{BND}_{(\alpha,\beta)}(C)|};$$

$$NNE([x], C) = \frac{|[x] \cap C^c|}{|[x]|},$$

$$NNE(\text{BND}_{(\alpha,\beta)}(C), C) = \frac{|\text{BND}_{(\alpha,\beta)}(C) \cap C^c|}{|\text{BND}_{(\alpha,\beta)}(C)|}.$$

It can be verified that $\beta < NPE < \alpha$ and $1 - \alpha < NNE < 1 - \beta$.

Pawlak rough set model is a special case of probabilistic rough set model when $\alpha = 1$ and $\beta = 0$. This means that decisions of acceptance or rejection are made without any error, i.e., $IAE(\text{POS}_{(\alpha,\beta)}(C), C) = 0$ and $IRE(\text{NEG}_{(\alpha,\beta)}(C), C) = 0$. Whenever we are not sure, we make a decision of noncommitment. This leads to larger values of $NPE(\text{BND}_{(\alpha,\beta)}(C), C)$ and $NNE(\text{BND}_{(\alpha,\beta)}(C), C)$. The Pawlak boundary region may be too large due to intolerance to incorrect acceptance and incorrect rejection. In contrast, by allowing certain levels of error, a probabilistic model may have a smaller boundary region. The sizes of the three regions are controlled by the pair of thresholds (α, β).

Based on the interpretation of $1 - \alpha$ and β, a user can simply select a suitable pair of thresholds (α, β) according to the tolerance of errors. However, this interpretation does not explicitly take into consideration of different costs usually associated with different decisions. A cost-sensitive interpretation is given in the decision theoretic rough set model [16,17].

2.3   Determination of Thresholds

The thresholds are related to classification errors, benefits or risks of the resulting three regions of three-way decisions. It suggests that an optimal pair of thresholds must induce three regions with a minimum total cost or a total maximum benefit. Determining the thresholds may be formulated as an optimization problem [15].

Consider a pair of thresholds (α, β). Let $R_P(\alpha, \beta)$, $R_N(\alpha, \beta)$ and $R_B(\alpha, \beta)$ denote the risks of the positive, negative, and boundary regions, respectively. It is reasonable to require that a pair of threshold (α, β) should minimize the following overall risks:

$$R(\alpha, \beta) = R_P(\alpha, \beta) + R_N(\alpha, \beta) + R_B(\alpha, \beta). \tag{3}$$

That is, finding a pair of thresholds can be formulated as the following optimization problem:

$$\arg \min_{(\alpha,\beta)} R(\alpha, \beta). \tag{4}$$

There may exist more than one pair of thresholds that minimizes the overall risks $R(\alpha, \beta)$. Depending on particular formulations and interpretations of risks, one may obtain concrete probabilistic rough set models.

Based on the well-established Bayesian decision theory, the decision-theoretic rough set model [16,17] determines a pair of thresholds by minimizing the overall risk or cost of a ternary classification. Given a loss, risk or cost function for correct and incorrect acceptance, rejection, or noncommitment, one can calculate the values of the pair of thresholds. By imposing different sets of conditions on the cost function, one can obtain many sub-classes of probabilistic models. Two examples are a symmetric model with $\alpha + \beta = 1$ and a majority-oriented model with $0 \leq \beta < 0.5 \leq \alpha \leq 1$. Herbert and Yao [3] introduce game theory into decision-theoretic rough sets and propose a game-theoretic rough set (GTRS) model. The problem of determining the thresholds is formulated as a game consisting of two players competing for two properties of ternary classification, say, accuracy and precision. By gradually modifying various decision costs, the GTRS model searches for a cost function that defines the thresholds. Li and Zhou [5] consider a special class of functions and divide users into three groups, namely, optimistic, pessimistic, and neutral. They examine the relationships of different thresholds and different groups of users. Jia et al. [4] investigate the problem of learning thresholds from data by minimizing a cost function.

3   An Information-Theoretic Interpretation

In a majority-oriented probabilistic rough set model with $0 \leq \beta < 0.5 \leq \alpha < 1$, the distribution of objects in the positive region is biased towards C, and the distribution of objects in the negative region is biased towards C^c. One is more certain about the classification of objects in the positive and negative regions than objects in the boundary region. In the Pawlak model (i.e., $\beta = 0$ and $\alpha = 1$), one has full certainty with respect to the positive and negative regions. However, uncertainty of the boundary region is not considered. By allowing some levels of uncertainty in the probabilistic positive and negative regions, we can determine a pair of thresholds that minimizes the overall uncertainty of the three regions.

3.1   Shannon Entropy of a Partition

Suppose $\pi = \{b_1, b_2, \ldots, b_n\}$ is a partition of a universe U, namely, $\cup_{i=1}^{n} b_i = U$ and $b_i \cap b_j = \emptyset$ for $i \neq j$. One can associate a probability distribution:

$$P_\pi = \left(\frac{|b_1|}{|U|}, \frac{|b_2|}{|U|}, \ldots, \frac{|b_n|}{|U|} \right), \tag{5}$$

with the partition π, where $Pr(b_i) = |b_i|/|U|$ denotes the probability of the block b_i. With respect to P_π, the Shannon entropy is defined by

$$H(\pi) = H(P_\pi) = -\sum_{i=1}^{n} Pr(b_i) \log Pr(b_i) = -\sum_{i=1}^{n} \frac{|b_i|}{|U|} \log \frac{|b_i|}{|U|}. \tag{6}$$

The Shannon entropy may be viewed as a measure of uncertainty of a partition π. Assume that blocks of a partition represent a set of classes. Based on the probability distribution P_π, one wants to predict the class of an arbitrary object. In one extreme case, the sizes of all blocks are the same and P_π is a uniform distribution, one is most uncertain in predicting a class (i.e., a block) of an object; the information entropy reaches the maximum value $H(\pi) = \log n$. In another extreme case, there is only one block, one is certain about the class of object; the entropy reaches the minimum value 0. In general, Shannon entropy measures the degree of uncertainty in such a prediction.

3.2　Uncertainties of Three Probabilistic Regions

A pair of thresholds (α, β) with $0 \le \beta < 0.5 \le \alpha \le 1$ produces the following three regions with respect to $\emptyset \ne C \subseteq U$:

$$\pi_{(\alpha,\beta)} = \{\text{POS}_{(\alpha,\beta)}(C), \text{NEG}_{(\alpha,\beta)}(C), \text{BND}_{(\alpha,\beta)}(C)\}. \tag{7}$$

We compute the uncertainty of $\pi_C = \{C, C^c\}$ by using Shannon entropy,

$$
\begin{aligned}
H(\pi_C|\text{POS}_{(\alpha,\beta)}(C)) &= -Pr(C|\text{POS}_{(\alpha,\beta)}(C))\log Pr(C|\text{POS}_{(\alpha,\beta)}(C)) \\
&\quad -Pr(C^c|\text{POS}_{(\alpha,\beta)}(C))\log Pr(C^c|\text{POS}_{(\alpha,\beta)}(C)), \\
H(\pi_C|\text{NEG}_{(\alpha,\beta)}(C)) &= -Pr(C|\text{NEG}_{(\alpha,\beta)}(C))\log Pr(C|\text{NEG}_{(\alpha,\beta)}(C)) \\
&\quad -Pr(C^c|\text{NEG}_{(\alpha,\beta)}(C))\log Pr(C^c|\text{NEG}_{(\alpha,\beta)}(C)), \\
H(\pi_C|\text{BND}_{(\alpha,\beta)}(C)) &= -Pr(C|\text{BND}_{(\alpha,\beta)}(C))\log Pr(C|\text{BND}_{(\alpha,\beta)}(C)) \\
&\quad -Pr(C^c|\text{BND}_{(\alpha,\beta)}(C))\log Pr(C^c|\text{BND}_{(\alpha,\beta)}(C)), \tag{8}
\end{aligned}
$$

where $Pr(C|\text{POS}_{(\alpha,\beta)}(C))$ denotes the conditional probability of an object x in C given that the object is in the positive probabilistic region $\text{POS}_{(\alpha,\beta)}(C)$ and so on. The conditional probabilities can be estimated by:

$$
\begin{aligned}
Pr(C|\text{POS}_{(\alpha,\beta)}(C)) &= \frac{|C \cap \text{POS}_{(\alpha,\beta)}(C)|}{|\text{POS}_{(\alpha,\beta)}(C)|}, \\
Pr(C|\text{NEG}_{(\alpha,\beta)}(C)) &= \frac{|C \cap \text{NEG}_{(\alpha,\beta)}(C)|}{|\text{NEG}_{(\alpha,\beta)}(C)|}, \\
Pr(C|\text{BND}_{(\alpha,\beta)}(C)) &= \frac{|C \cap \text{BND}_{(\alpha,\beta)}(C)|}{|\text{BND}_{(\alpha,\beta)}(C)|}. \tag{9}
\end{aligned}
$$

Conditional probabilities of C^c can be similarly estimated.

The overall uncertainty of three regions can be computed as their average uncertainty, which is called the conditional entropy of π_C given $\pi_{(\alpha,\beta)}$, namely,

$$
\begin{aligned}
H(\pi_C|\pi_{(\alpha,\beta)}) &= Pr(\text{POS}_{(\alpha,\beta)}(C))H(\pi_C|\text{POS}_{(\alpha,\beta)}(C)) \\
&\quad +Pr(\text{NEG}_{(\alpha,\beta)}(C))H(\pi_C|\text{NEG}_{(\alpha,\beta)}(C)) \\
&\quad +Pr(\text{BND}_{(\alpha,\beta)}(C))H(\pi_C|\text{BND}_{(\alpha,\beta)}(C)). \tag{10}
\end{aligned}
$$

The probabilities of the three regions are given by:

$$Pr(\text{POS}_{(\alpha,\beta)}(C)) = \frac{|\text{POS}_{(\alpha,\beta)}(C)|}{|U|},$$

$$Pr(\text{NEG}_{(\alpha,\beta)}(C)) = \frac{|\text{NEG}_{(\alpha,\beta)}(C)|}{|U|},$$

$$Pr(\text{BND}_{(\alpha,\beta)}(C)) = \frac{|\text{BND}_{(\alpha,\beta)}(C)|}{|U|}. \tag{11}$$

The conditional entropy takes into consideration of all uncertainties in three regions. In some sense, it provides another motivation for introducing probabilistic rough sets. Although Pawlak rough sets model has the minimum uncertainties of 0 for both positive and negative regions, it may not have the minimum uncertainties for three regions. A probabilistic model attempts to make a trade-off among uncertainties of three regions.

3.3   Determining Optimal Thresholds by Entropy Minimization

Consider the partition $\pi_C = \{C, C^c\}$ induced by $\emptyset \neq C \subseteq U$, its Shannon entropy is defined by

$$H(\pi_C) = -\frac{|C|}{|U|}\log\frac{|C|}{|U|} - \frac{|C^c|}{|U|}\log\frac{|C^c|}{|U|}. \tag{12}$$

The quantity $H(\pi_C) - H(\pi_C|\pi_{(\alpha,\beta)})$ may be viewed as a reduction of uncertainty produced by probabilistic rough set classification. Since $H(\pi_C)$ is independent of the pair of thresholds (α, β), we can minimize $H(\pi_C|\pi_{(\alpha,\beta)})$ to obtain the maximum uncertainty reduction. Thus, the problem of finding an optimal pair of thresholds is formulated as the following optimization problem:

$$\arg\min_{(\alpha,\beta)} H(\pi_C|\pi_{(\alpha,\beta)}). \tag{13}$$

One can search the space of all possible thresholds to obtain a pair of optimal thresholds. Our formulation is closely related to some machine learning algorithm, say, decision tree construction method used in ID3 class [10]. In ID3, one selects an attribute based on entropy minimization. In our method, we select a pair of thresholds based on entropy minimization. In general, one may use other measures of uncertainty to determine the thresholds.

4   An Example

The main ideas of the proposed information-theoretic interpretation are illustrated by a simple example. Table 1 summarizes probabilistic information about a concept C with respect to a partition of 15 equivalence classes, where X_i denotes an equivalence class. For the convenience of computing the three probabilistic regions, we list the equivalence classes in a decreasing order according to the conditional probabilities $Pr(C|X_i)$. The possible values of two thresholds α and β

are given by conditional probabilities. According to the condition $0 \leq \beta < 0.5 \leq \alpha \leq 1$ of a majority-oriented model, we have the following domains of α and β:

$$D_\alpha = \{\alpha_1, \alpha_2, \alpha_3, \alpha_4, \alpha_5\} = \{1.0, 0.9, 0.8, 0.6, 0.5\},$$
$$D_\beta = \{\beta_1, \beta_2, \beta_3, \beta_4\} = \{0, 0.1, 0.2, 0.4\}.$$

They produce $5 \times 4 = 20$ pairs of thresholds.

Table 1. Probabilistic information of a concept C

	X_1	X_2	X_3	X_4	X_5	X_6	X_7	X_8
$Pr(X_i)$	0.0177	0.1285	0.0137	0.1352	0.0580	0.0069	0.0498	0.1070
$Pr(C\|X_i)$	1.0	1.0	1.0	1.0	0.9	0.8	0.8	0.6

	X_9	X_{10}	X_{11}	X_{12}	X_{13}	X_{14}	X_{15}
$Pr(X_i)$	0.1155	0.0792	0.0998	0.1299	0.0080	0.0441	0.0067
$Pr(C\|X_i)$	0.5	0.4	0.4	0.2	0.1	0.0	0.0

For the pair of thresholds $(\alpha_1, \beta_1) = (1, 0)$, we have the Pawlak ternary classification. The three regions are $\text{POS}_{(1,0)}(C) = \bigcup\{X_1, X_2, X_3, X_4\}$, $\text{BND}_{(1,0)}(C) = \bigcup\{X_5, \ldots, X_{13}\}$ and $\text{NEG}_{(1,0)}(C) = \bigcup\{X_{14}, X_{15}\}$. The probability of the boundary region is

$$Pr(\text{BND}_{(1,0)}(C)) = \sum_{i=5}^{13} Pr(X_i)$$
$$= 0.058 + 0.0069 + 0.0498 + \cdots + 0.1299 + 0.008 = 0.6541.$$

Similarly, we have $Pr(\text{POS}_{(1,0)}(C)) = 0.2951$ and $Pr(\text{NEG}_{(1,0)}(C)) = 0.0508$. For the positive and negative regions, we have $Pr(C|\text{POS}_{(1,0)}(C)) = 1$ and $Pr(C|\text{NEG}_{(1,0)}(C)) = 0$, respectively. The uncertainties of both regions are 0, that is, $H(\pi_C|\text{POS}_{(1,0)}(C)) = 0$ and $H(\pi_C|\text{NEG}_{(1,0)}(C)) = 0$. For the boundary region, the conditional probability of C can be computed by

$$Pr(C|\text{BND}_{(1,0)}(C)) = \frac{\sum_{i=5}^{13} Pr(C|X_i)Pr(X_i)}{\sum_{i=5}^{13} Pr(X_i)}$$
$$= \frac{0.058 * 0.9 + 0.0069 * 0.8 + \cdots + 0.1299 * 0.2 + 0.008 * 0.1}{0.058 + 0.0069 + \cdots + 0.1299 + 0.008} = 0.4860.$$

The Shannon entropy of the boundary region, $H(\pi_C|\text{BND}_{(1,0)}(C))$, is given by

$$H(\pi_C|\text{BND}_{(1,0)}(C))$$
$$= -0.4860 * \log 0.4860 - (1 - 0.4860) * \log(1 - 0.4860)$$
$$= 0.9994,$$

where base 2 is used. The overall uncertainties of the three probabilistic regions are given by

$$H(\pi_C|\pi_{(1,0)}) = 0.2951 * 0 + 0.6541 * 0.9994 + 0.0508 * 0 = 0.6537.$$

By following the same procedure one can compute uncertainties for ternary classifications defined by two additional pairs of thresholds $(\alpha_2, \beta_2) = (0.9, 0.1)$ and $(\alpha_3, \beta_3) = (0.8, 0.2)$. The results are summarized in Table 2, in which each cell contains each of the three regions, its probability and its entropy.

Table 2. Uncertainties induced by three ternary classifications

(α, β)	POS	BND	NEG	$H(\pi_C \mid \pi_{(\alpha,\beta)})$
$(1, 0)$	$\cup\{X_1, X_2, X_3, X_4\}$	$\cup\{X_5, \ldots, X_{13}\}$	$\cup\{X_{14}, X_{15}\}$	
	$0.2951 * 0$	$0.6541 * 0.9994$	$0.0508 * 0$	0.6537
$(0.9, 0.1)$	$\cup\{X_1, \ldots, X_5\}$	$\cup\{X_6, \ldots, X_{12}\}$	$\cup\{X_{13}, X_{14}, X_{15}\}$	
	$0.3531 * 0.1209$	$0.5881 * 0.9929$	$0.0588 * 0.1038$	0.6327
$(0.8, 0.2)$	$\cup\{X_1, \ldots, X_7\}$	$\cup\{X_8, \ldots, X_{11}\}$	$\cup\{X_{12}, \ldots, X_{15}\}$	
	$0.4098 * 0.2506$	$0.4015 * 0.9991$	$0.1887 * 0.5892$	0.6150

It can be observed that although Pawlak positive and negative regions have the minimum uncertainty 0, the boundary region has a larger uncertainty. In contrast, probabilistic positive and negative regions have larger uncertainty, but the boundary region has a lower uncertainty. A pair of thresholds controls the trade-off among uncertainties of the three regions.

One can compute the corresponding conditional entropy values for all possible pairs of thresholds and store them in a matrix form. For Table 1, the results are given by the following 5×4 matrix:

$$
\begin{array}{c}
\\
\alpha_1 = 1.0 \\
\alpha_2 = \mathbf{0.9} \\
\alpha_3 = 0.8 \\
\alpha_4 = 0.6 \\
\alpha_5 = 0.5
\end{array}
\begin{array}{cccc}
\beta_1 = 0 & \beta_2 = 0.1 & \beta_3 = \mathbf{0.2} & \beta_4 = 0.4 \\
\left(0.6537 \right. & 0.6520 & 0.6213 & 0.6228 \\
0.6337 & 0.6327 & \mathbf{0.6115} & 0.6220 \\
0.6290 & 0.6286 & 0.6150 & 0.6318 \\
0.6756 & 0.6758 & 0.6701 & 0.6912 \\
0.7216 & 0.7225 & 0.7234 & \left. 0.7465 \right)
\end{array}
$$

We can find a pair of thresholds with minimum conditional entropy by searching a minimum value in the matrix. For this example, we have $\alpha = 0.9$ and $\beta = 0.2$.

5   Conclusion

An information-theoretic interpretation is proposed and examined. The problem of searching for an optimal pair of thresholds is formulated as the minimization of Shannon entropy of ternary classification produced by probabilistic positive, negative and boundary regions. Complementary to existing interpretations, the new interpretation enables us to compute the thresholds by using only data in an information table. As future work, we will investigate algorithms that can find an optimal, or close to optimal, pair of thresholds quickly. It should also be noted that the proposed framework is also applicable if other measures of uncertainty are used.

Acknowledgements. This work is partially supported by a Discovery Grant from NSERC Canada.

References

1. Azam, N., Yao, J.: Multiple Criteria Decision Analysis with Game-theoretic Rough Sets. In: Li, T., Nguyen, H.S., Wang, G., Grzymala-Busse, J.W., Janicki, R., Hassanien, A.E., Yu, H. (eds.) RSKT 2012. LNCS (LNAI), vol. 7414, pp. 400–409. Springer, Heidelberg (2012)
2. Herbert, J.P., Yao, J.T.: Learning optimal parameters in decision-theoretic rough sets. In: Wen, P., Li, Y., Polkowski, L., Yao, Y., Tsumoto, S., Wang, G. (eds.) RSKT 2009. LNCS, vol. 5589, pp. 610–617. Springer, Heidelberg (2009)
3. Herbert, J.P., Yao, J.T.: Game-theoretic rough sets. Fundamenta Informaticae 108, 267–286 (2011)
4. Jia, X., Li, W., Shang, L., Chen, J.: An Optimization Viewpoint of Decision-Theoretic Rough Set Model. In: Yao, J., Ramanna, S., Wang, G., Suraj, Z. (eds.) RSKT 2011. LNCS, vol. 6954, pp. 457–465. Springer, Heidelberg (2011)
5. Li, H.X., Zhou, X.Z.: Risk decision making based on decision-theoretic rough set: A three-way view decision model. International Journal of Computational Intelligence Systems 4, 1–11 (2011)
6. Li, H.X., Zhou, X.Z., Li, T.R., Wang, G.Y., Miao, D.Q., Yao, Y.Y. (eds.): Decision-Theoretic Rough Sets Theory and Recent Research. Science Press, Beijing (2011) (in Chinese)
7. Liu, D., Li, T.R., Ruan, D.: Probabilistic model criteria with decision-theoretic rough sets. Information Science 181, 3709–3722 (2011)
8. Liu, D., Li, T.R., Li, H.X.: A multiple-category classification approach with decision-theoretic rough sets. Fundamenta Informaticae 115, 173–188 (2012)
9. Pawlak, Z.: Rough sets. International Journal of Computer and Information Sciences 11, 341–356 (1982)
10. Quinlan, J.R.: Introduction of decision trees. Machine Learning 1, 81–106 (1986)
11. Yao, Y.Y.: Probabilistic approaches to rough sets. Expert Systems 20, 287–297 (2003)
12. Yao, Y.Y.: Probabilistic rough set approximations. International Journal of Approximate Reasoning 49, 255–271 (2008)
13. Yao, Y.Y.: Three-way decisions with probabilistic rough sets. Information Sciences 180, 341–353 (2010)
14. Yao, Y.Y.: The superiority of three-way decisions in probabilistic rough set models. Information Science 181, 1080–1096 (2011)
15. Yao, Y.Y.: An Outline of a Theory of Three-Way Decisions. In: Yao, J., Yang, Y., Slowinski, R., Greco, S., Li, H., Mitra, S., Polkowski, L. (eds.) RSCTC 2012. LNCS (LNAI), vol. 7413, pp. 1–17. Springer, Heidelberg (2012)
16. Yao, Y.Y., Wong, S.K.M.: A decision theoretic framework for approximating concepts. International Journal of Man-machine Studies 37, 793–809 (1992)
17. Yao, Y.Y., Wong, S.K.M., Lingras, P.J.: A decision-theoretic rough set model. In: Ras, Z.W., Zemankova, M., Emrich, M.L. (eds.) Methodologies for Intelligent Systems 5, pp. 17–24. North-Holland, New York (1990)

Cost-Sensitive Classification
Based on Decision-Theoretic Rough Set Model

Huaxiong Li[1,2], Xianzhong Zhou[1], Jiabao Zhao[1], and Bing Huang[3]

[1] School of Management and Engineering, Nanjing University,
Nanjing, Jiangsu, 210093, P.R. China
[2] State Key Laboratory for Novel Software Technology, Nanjing University,
Nanjing, Jiangsu, 210093, P.R. China
[3] School of Information Science, Nanjing Audit University,
Nanjing, Jiangsu, 211815, P.R. China
{huaxiongli,zhouxz,jbzhao}@nju.edu.cn, hbhuangbing@126.com

Abstract. A framework of cost-sensitive classification based on decision-theoretic rough set model is proposed to determine the local minimum total cost classification and the local optimal test attributes set. Based on the proposed classification strategy, a cost-sensitive classification algorithm CSDTRS is presented. CSDTRS focuses on searching for an optimal test attributes set with minimum total cost including both misclassification cost and test cost, and then determine the classification based on the optimal test attributes set. A heuristic function for evaluating the attribute is presented to determine which attribute should be added in the optimal test attributes set. Experiments on four UCI data sets are performed to show the effectiveness of the proposed classification algorithm.

Keywords: decision-theoretic rough set, cost-sensitive, misclassification cost, test cost, attribute selection.

1 Introduction

A fundamental application of rough set theory is to induce rules from lower and upper approximations and to make classification based on the induced rules. In Pawlak rough set model [16], the lower and upper approximations are defined by two extreme cases of the relationships between an equivalence class and the set to be approximated. For lower approximation, the equivalence class must be completely included in the set, and for upper approximation, the equivalence class has a non-empty intersection with the set, while the actual degree of membership is not taken into consideration. This makes the Pawlak rough set intolerant to classification error and very sensitive to the accuracy of input data [18]. We may regard the Pawlak rough set as an accuracy-sensitive classification method.

However in practice, cost-sensitive classification instead of accuracy-sensitive classification may be a better choice for real world data analysis. The reasons are that: (1) Due to the uncertainty and noise interference in the process of data

T. Li et al. (Eds.): RSKT 2012, LNAI 7414, pp. 379–388, 2012.
© Springer-Verlag Berlin Heidelberg 2012

acquisition and transfer, it is impossible to get sufficiently high classification accuracy, thus the accuracy-sensitive classification can not be fully realized, and it seems that admitting some extent of the misclassification may lead to a better utilization of the properties of the data set. (2) On the other hand, in many real-world data analysis, the costs of different misclassification are often unequal, and the practical goal of the classification may be to minimize the misclassification cost instead of simply minimizing the misclassification rate. Therefore, it is necessary to admit some extent misclassification and introduce cost-sensitive analysis in rough set model.

Decision-theoretic rough set model (DTRS) [18–23], proposed by Yao in the early 1990s, provides a solution on the problems mentioned above. In recent years, DTRS has received much attention, and many examples of theoretical research and applications of the DTRS are frequently mentioned in literatures [2, 4–13, 17, 24, 25]. By introducing membership functions of set inclusion with statistical information, DTRS model allows a certain level of classification in the process of data analysis. Moreover, in DTRS model, the classification label of an instance is determined by minimizing the expected misclassification cost instead of simply maximizing the likelihood probability, therefore DTRS is suitable for cost-sensitive classification. In this paper, we propose a sequential decision strategy for cost-sensitive classification based on DTRS model. In the framework of the proposed cost-sensitive classification method, the misclassification cost and test cost are both concerned, and the attributes used for classification are sequentially selected based on the efficiency of reducing the total cost so that the optimal classification with minimum misclassification cost and test cost are determined.

2   Decision-Theoretic Rough Set Model

This section reviews some basic notions of DTRS model [18, 20, 22, 23], which presents a theoretical basis for the proposed classification method.

To simplify the description, we consider on a binary classification problem. The set of labels is given by $\Omega = \{X, \neg X\}$ indicating that each instance is either labeled by X or labeled by $\neg X$. The set of actions is given by $\mathcal{A} = \{a_P, a_N, a_B\}$, representing the three actions in classifying an instance, i.e., respectively deciding $POS(X)$ (classify the instance to X), deciding $NEG(X)$ (classify the instance to $\neg X$), and deciding $BND(X)$ (delay the classification decision due to dilemma). Table 1 presents a general classification cost matrix. The cost λ_{ij} forms a matrix denoted as $(\lambda_{ij})_{2\times3}$, where $i \in \{P, B, N\}$, and $j \in \{P, N\}$.

In general, the costs of making the right decision are always less than that of making the wrong decision, therefore we have $\lambda_{PP} \leq \lambda_{BP} \leq \lambda_{NP}$ and $\lambda_{NN} \leq \lambda_{BN} \leq \lambda_{PN}$. Moreover, a reasonable assumption is that the costs of making the right decision are equal to zero, then we get a simplified classification cost matrix as presented in Table 2.

Table 1. Classification cost matrix

Class labels	$POS(X)$	$BND(X)$	$NEG(X)$
X	λ_{PP}	λ_{BP}	λ_{NP}
$\neg X$	λ_{PN}	λ_{BN}	λ_{NN}

Table 2. Simplified classification cost matrix

Class labels	$POS(X)$	$BND(X)$	$NEG(X)$
X	0	λ_{BP}	λ_{NP}
$\neg X$	λ_{PN}	λ_{BN}	0

Then the expected classification cost $R(a_i|[x]_\Re)$ for taking the individual actions can be expressed as:

$$R(a_P|[x]_\Re) = \lambda_{PP}P(X|[x]_\Re) + \lambda_{PN}P(\neg X|[x]_\Re) = \lambda_{PN}P(\neg X|[x]_\Re);$$
$$R(a_N|[x]_\Re) = \lambda_{NP}P(X|[x]_\Re) + \lambda_{NN}P(\neg X|[x]_\Re) = \lambda_{NP}P(X|[x]_\Re);$$
$$R(a_B|[x]_\Re) = \lambda_{BP}P(X|[x]_\Re) + \lambda_{BN}P(\neg X|[x]_\Re), \tag{1}$$

where $[x]_\Re$ denotes the equivalence class of x under equivalence relation $\Re$. The Bayesian decision procedure leads to the following minimum-cost decision rules:

If $R(a_P|[x]_\Re) \leq R(a_N|[x]_\Re)$ and $R(a_P|[x]_\Re) \leq R(a_B|[x]_\Re)$, decide $POS(X)$;
If $R(a_N|[x]_\Re) \leq R(a_P|[x]_\Re)$ and $R(a_N|[x]_\Re) \leq R(a_B|[x]_\Re)$, decide $NEG(X)$;
If $R(a_B|[x]_\Re) \leq R(a_P|[x]_\Re)$ and $R(a_B|[x]_\Re) \leq R(a_N|[x]_\Re)$, decide $BND(X)$.(2)

Therefore, in order to minimizing the classification cost, the optimal classification function $\phi^*([x]_\Re)$ should be:

$$\phi^*([x]_\Re) = \arg \min_{\mathcal{D}\in\{a_P,a_N,a_B\}} R(\mathcal{D}|[x]_\Re) \tag{3}$$

3   Cost-Sensitive Classification Based on DTRS

By introducing the Bayesian decision theory, the DTRS model provides a classification result with minimum classification cost. Although the classification result may not be always correct, it will be the most appropriate decision with minimum risks. Normally, the classification cost will decrease with the increasing of the attributes used for classification since the classification accuracy may be improved when there are more attributes can be used for classification.

However, different attributes may have different test costs. It is practically useful to find a set of attributes with a minimal test cost. The classification cost performance measure is too simplistic for many practical scenario, in which one must balance the classification cost and the test cost of acquiring the data with regard to a given attribute set. Therefore, we consider on both of two costs: one is classification cost, and the other is test cost. As far as we know, in rough set research, there are few works focus on both of classification cost and test cost. Some literatures concern singly on classification cost while other literatures concern singly on test cost. For example, Jia et al. discuss the minimum decision cost attribute reduction in DTRS model which considering on the classification cost [4]. Min et al. propose a test-cost-sensitive attribute reduction for rough set model which focuses on the test cost [14, 15]. It is necessary to combine both of the two cost and present a classification strategy with minimum total cost. In [3], Herbert and Yao present a game-theoretic rough sets model which may be used to study trade-off between the two types of cost. In [1], Chai et al. consider both classification cost and test cost and propose a test-cost sensitive naive Bayes classification method. However, it is used for cost-sensitive learning from incomplete data, in which many attribute values are missing. We will introduce the classification strategy to DTRS model and propose a cost-sensitive classification strategy which minimizes the total cost of the classification including both classification cost and test cost.

First of all, we denote the data set as a decision information table [23]: $S = (U, At = C \cup D, \{V_a \mid a \in At\}, \{I_a \mid a \in At\})$. We associate two types of evaluation functions $\mathcal{C}$ and $\mathcal{T}$ to an attribute set which respectively denote the classification cost and the test cost for the given attribute set, i.e., $\mathcal{C} : U \times 2^C \to \mathbb{R}^+ \cup \{0\}, \mathcal{T} : 2^C \to \mathbb{R}^+ \cup \{0\}$. For a designated attribute set $B \subseteq C$, the classification cost are computed by the expected cost of three types of classification decision including a_P, a_B, and a_N, which are formalized as:

$$\mathcal{C}(x, B) = \begin{cases} \lambda_{PP} P(X|[x]_B) + \lambda_{PN} P(\neg X|[x]_B), & \phi^*([x]_B) = a_P, \\ \lambda_{BP} P(X|[x]_B) + \lambda_{BN} P(\neg X|[x]_B), & \phi^*([x]_B) = a_B, \\ \lambda_{NP} P(X|[x]_B) + \lambda_{NN} P(\neg X|[x]_B), & \phi^*([x]_B) = a_N. \end{cases} \tag{4}$$

Based on the assumption that $\lambda_{ii} = 0$, the classification function $\mathcal{C}(B)$ can be simplified as:

$$\mathcal{C}(x, B) = \begin{cases} \lambda_{PN} P(\neg X|[x]_B), & \phi^*([x]_B) = a_P, \\ \lambda_{BP} P(X|[x]_B) + \lambda_{BN} P(\neg X|[x]_B), & \phi^*([x]_B) = a_B, \\ \lambda_{NP} P(X|[x]_B), & \phi^*([x]_B) = a_N. \end{cases} \tag{5}$$

On the other hand, the test cost of a designated condition attribute set $B \subseteq C$ is computed by the summation cost of all test attributes in B, and all instances in U have the same test cost on a given attribute, therefore we have:

$$\mathcal{T}(B) = \sum_{c_i \in B} T(c_i), \tag{6}$$

where $T(c_i)$ is the test cost of a single attribute c_i. The test cost of a single attribute can be acquired according to practical scenario. By combining the classification cost and test cost, we associate a pair of instance and attribute set (x, B) with a cost function called total cost function $\mathcal{F}(x, B)$, which are defined by the summation of the two types of the costs:

$$\mathcal{F}(x, B) = \mathcal{C}(x, B) + \mathcal{T}(B), \tag{7}$$

The objective of the minimum cost classification is to find an attribute subset $B \subseteq C$, which minimizes the total cost $\mathcal{F}(x, B)$, i.e., the optimal attribute sets B^* selected out for classification are computed by: $B^* = \arg\min_{B \in 2^C} \mathcal{F}(x, B)$, and the classification result based on B^* will output the predicted label $\phi^*([x]_{B^*})$. Normally, to find a classification strategy with global optimal total cost is computationally difficult, since all the subsets of attributes should be tested in the process of classification. In this paper, we consider to find a local approximation solutions.

Suppose $x \in U$ is a test instance, and consider a test attribute set $B \subseteq C$. All instances in $[x]_B$ will be assigned by the same classification label since the instances have completely same attribute values on B. Therefore, we may represent the classification label of x with regard to test attribute set B by $\phi^*([x]_B)$. The classification cost $\mathcal{C}(x, B)$ and test cost $\mathcal{T}(B)$ for classifying $[x]_B$ as $a_j \in \{P, B, N\}$ are respectively computed according to formula (5),(6), and the total cost of the classification are computed by formula (7). The test attribute set B varies from a single attribute $\{c_i\}$ to the entire attribute set $\{c_1, c_2, \ldots, c_{|C|}\}$.

To acquire a minor total cost, we should add the candidate attribute used for test in a right sequence strategy so that the total cost may decrease. Considering a given attribute set B, whether or not for a new attribute $\overline{b_i} \in C - B$ should be added to test attribute set B depends on the contribution of $\overline{b_i}$ for reducing the total cost. In other words, $\overline{b_i}$ should be added to test attribute set B if $\mathcal{F}(x, B) - \mathcal{F}(x, B \cup \{\overline{b_i}\}) > 0$, otherwise $\overline{b_i}$ is unnecessary to be added in B. The difference of the two total cost may also be used to determine which attribute should be taken for the further test. Therefore, we take the difference as a heuristic significance function for the sequential test strategy:

$$sig(x, B, \overline{b_i}) = \mathcal{F}(x, B) - \mathcal{F}(x, B \cup \{\overline{b_i}\}) \tag{8}$$

The test process starts with a single attribute $c_i \in C$ used for test, and the total cost for classifying an instance x using c_i is computed based on formula (5),(6), and (7). The optimal single attribute c_i^* is selected with the minimum total cost. Then the significance of the remaining attributes will be respectively computed according to formula (8), deciding whether a new attribute should be added in and if so, which attribute should be selected. The sequential process will terminate when all of the remaining attributes have no contribution to decrease the total cost or the entire attribute sets have been tested. The detail of the sequential classification algorithm is presented in Figure 1.

ALGORITHM: Cost-Sensitive classification based on DTRS (CSDTRS)
INPUT: A decision table $S = (U, C \cup D\}$, a cost matrix λ_{ij},
           test cost function $T(c_i)$, an instance x for classification;
OUTPUT: A classification label, an attribute subset for test B, total cost TC.
PROCESS:

$\quad\quad B \leftarrow \arg\min_{c_i \in C} \mathcal{F}(x, \{c_i\}); \; \overline{B} \leftarrow C - B;$
$\quad\quad TC = \mathcal{C}(x, B) + \mathcal{T}(B);$
$\quad\quad$ **While** $\overline{B} \neq \emptyset$ **Do**
$\quad\quad\quad\quad$ **For** any $\overline{b}_i \in \overline{B}$
$\quad\quad\quad\quad\quad\quad$ Compute $TC' = \mathcal{C}(x, B \cup \{\overline{b}_i\}) + \mathcal{T}(B \cup \{\overline{b}_i\});$
$\quad\quad\quad\quad\quad\quad$ Compute $sig(x, B, \overline{b}_i) = TC - TC';$
$\quad\quad\quad\quad$ **End of For**
$\quad\quad\quad\quad \overline{b}_i^{\,*} = \arg\max_{\overline{b}_i \in \overline{B}} sig(x, B, \overline{b}_i);$
$\quad\quad\quad\quad$ **If** $sig(x, B, \overline{b}_i^{\,*}) \leq 0;$
$\quad\quad\quad\quad\quad\quad$ Break;
$\quad\quad\quad\quad$ **End of If**
$\quad\quad\quad\quad B \leftarrow B \cup \{\overline{b}_i^{\,*}\};$
$\quad\quad\quad\quad \overline{B} \leftarrow \overline{B} - \{\overline{b}_i^{\,*}\};$
$\quad\quad\quad\quad TC \leftarrow TC';$
$\quad\quad$ **End of While**
$\quad\quad$ Compute $\phi^*([x]_B)$ by formula (5);
OUTPUT: $\phi^*([x]_B), B, TC.$

Fig. 1. The CSDTRS Algorithm

4 Experimental Analysis

This section presents an experimental evaluation of the proposed cost-sensitive
sequential classification algorithm. Experiments are performed on four UCI data
sets listed in Table 3. In the four data sets, Mushroom, Breast-cancer-wisconsin,
and Hepatitis contain missing values, and we delete those instances with missing
values in data sets Mushroom, Breast-cancer-wisconsin, and fill in missing values
with most common values for data set Hepatitis so that all data are complete
data. In addition, considering that the proposed algorithm is designed to deal
with binary classification problem, we convert the four classes in the data set
Car to two classes by merging the class labeled with "good" and "vgood" into
the class labeled with "acc" so that all data sets in Table 3 have two classes.

Table 3. Experimental data sets from UCI machine learning repository

ID	Data	Classes	Attributes	Raw size	New size
1	Mushroom	2	22	8124	5644
2	Breast-cancer-wisconsin	2	9	699	683
3	Hepatitis	2	12	155	155
4	Car	2	6	1728	1728

In the experiments, the misclassification costs are unbalanced. We assume that the cost of misclassifying $NEG(X)$ as $POS(X)$ is greater than that of misclassifying $POS(X)$ as $NEG(X)$, and the costs of the delayed classification are assigned in between two misclassification costs. The classification cost matrix $(\lambda_{ij})_{2\times3}$ used for experiment is presented in Table 4. In UCI Machine Learning Repository, the

Table 4. Classification cost matrix for experiments

Class labels	$POS(X)$	$BND(X)$	$NEG(X)$
X	0	1	2
$\neg X$	10	1	0

selected four data sets are not assigned with any test cost information. For experiment test purpose, we randomly assign 3 test costs to each attribute in the four data sets (test cost 1-3). The random test costs are subjected to normal distribution $\mathcal{N}(\mu, \sigma)$, where $\mu = 0.2$ and $\sigma = 0.1$. For each data set associated

Table 5. CSDTRS performance (on test cost 1)

Data Sets	All att	CSDTRS att (Average)	Reduction rate(att)	TC of all att	TC of CSDTRS	Reduction rate(TC)
Mushroom	22	1.0500	95.23%	4.4065	0.0675	98.47%
Breast	9	1.0500	88.33%	1.6199	0.2196	86.44%
Hepatitis	12	1.0400	91.33%	2.8409	0.9206	67.59%
Car	6	1.4900	75.17%	1.2747	0.5304	58.39%

Table 6. CSDTRS performance (on test cost 2)

Data Sets	All att	CSDTRS att (Average)	Reduction rate(att)	TC of all att	TC of CSDTRS	Reduction rate(TC)
Mushroom	22	1.3700	93.77%	4.2418	0.1843	95.66%
Breast	9	1.8100	79.89%	1.3256	0.1349	89.82%
Hepatitis	12	1.0200	91.50%	3.2418	1.0028	69.07%
Car	6	1.4200	76.33%	1.4845	0.7833	47.23%

with a certain test cost, we randomly select 100 instances as test data (For reproducibility, the previous 100 instances in the data sets are selected). Each instance in the data set is classified according to the proposed CSDTRS algorithm, and the optimal test attributes sets are also outputted. Then the average numbers of the test attributes (CSDTRS att (average)) per instance are computed, which are presented to compare with the total number of the attributes (All att). In order to present a contrastive analysis, the total classification costs of CSDTRS(TC of CSDTRS) and the total classification costs based on all attributes (TC of all att)

Table 7. CSDTRS performance (on test cost 3)

Data Sets	All att	CSDTRS att (Average)	Reduction rate(att)	TC of all att	TC of CSDTRS	Reduction rate(TC)
Mushroom	22	1.3600	**93.82%**	4.2410	**0.2507**	**94.09%**
Breast	9	1.5400	**82.89%**	1.5513	**0.1392**	**91.03%**
Hepatitis	12	1.0800	**91.00%**	3.4083	**0.9995**	**70.67%**
Car	6	1.2100	**79.83%**	1.0331	**0.4124**	**60.08%**

are respectively calculated. All experiment results with regard to three test cost setting are presented in Table 5-7.

The results reported in the tables reveal two interesting regularities. First, the CSDTRS always takes a very less number of attributes as test attributes when compared to the entire condition attribute sets, e.g., in Table 5, the average numbers of the test attributes per instance for Mushroom data set is about 1.05, while the sum of the condition attributes is 22, which indicates that about 95.23% attributes have no contribution to reduce the total cost, thus we denote the rate as reduction rate (att) in Table 5-7. Secondly, the total cost using CSDTRS is far less than that of classification based on entire attribute set. For example, in Table 5, the total cost per instance based on entire attribute set is about 4.4065, while the total cost per instance for CSDTRS is only 0.0675, which reduce the total cost by 98.47% (reduction rate (TC)). Similar results can be found in Table 6-7. The reason is that CSDTRS only takes a part of attributes as test attributes, which have high contribution to reduce the total cost of classification. Although the misclassification cost of CSDTRS may be higher, the test cost may be dramatically lower than that of classification based on entire attribute set, therefore the CSDTRS performs with a very low total cost of classification.

5   Conclusion

Traditional data analysis usually focuses on consistent misclassification cost, which aims to minimize the classification error. However, in real world data analysis, cost-sensitive classification instead of accuracy-sensitive classification may be a better choice since the misclassification cost in practice are often imbalanced. In this paper, a cost-sensitive classification algorithm CSDTRS is proposed to find local optimal classification with minimum total cost. The proposed CSDTRS is designed to search for an optimal test attributes set with minimum total cost including both misclassification cost and test cost, and then determine the classification based on the optimal test attributes set. A heuristic function to evaluate the attribute is presented to determine which attribute should be added in the optimal test attributes set. Experiments on four UCI data sets are performed to validate the effectiveness of the proposed classification algorithm. The study not only presents a new view of decision-theoretic rough set model, but also provides a new solution on cost-sensitive classification.

Acknowledgments. This research is supported by the National Natural Science Foundation of China under grant No. 70971062,70971063,61170105, and the Natural Science Foundation of Jiangsu, China (BK2011564).

References

1. Chai, X., Deng, L., Yang, Q.: Test-cost sensitive naive bayes classification. In: Proceedings of ICDM 2004, pp. 51–58. IEEE Computer Society, Washington (2004)
2. Herbert, J.P., Yao, J.: Game-Theoretic Risk Analysis in Decision-Theoretic Rough Sets. In: Wang, G., Li, T., Grzymala-Busse, J.W., Miao, D., Skowron, A., Yao, Y. (eds.) RSKT 2008. LNCS (LNAI), vol. 5009, pp. 132–139. Springer, Heidelberg (2008)
3. Herbert, J.P., Yao, J.T.: Game-theoretic rough sets. Fundamenta Informaticae 3-4, 267–286 (2011)
4. Jia, X.Y., Shang, L., Chen, J.J.: Attribute reduction based on minimum decision cost. Journal of Frontiers of Computer Science and Technology 5, 155–160 (2011) (in Chinese)
5. Li, H.X., Liu, D., Zhou, X.Z.: Survey on decision-theoretic rough set model. Journal of Chongqing University of Posts and Telecommunications (Natural Science Edition) 22, 624–630 (2010) (in Chinese)
6. Li, H., Zhou, X., Zhao, J., Liu, D.: Attribute Reduction in Decision-Theoretic Rough Set Model: A Further Investigation. In: Yao, J., Ramanna, S., Wang, G., Suraj, Z. (eds.) RSKT 2011. LNCS, vol. 6954, pp. 466–475. Springer, Heidelberg (2011)
7. Li, H.X., Zhou, X.Z.: Risk decision making based on decision-theoretic rough set: A multi-view decision model. International Journal of Computational Intelligence Systems 4, 1–11 (2011)
8. Li, H.X., Zhou, X.Z., Li, T.R., Wang, G.Y., Miao, D.Q., Yao, Y.Y. (eds.): Decision-Theoretic Rough Sets Theory and Recent Research. Science Press, Beijing (2011) (in Chinese)
9. Li, W., Miao, D.Q., Wang, W.L., Zhang, N.: Hierarchical rough decision theoretic framework for text classification. In: Proceedings of ICCI 2010, pp. 484–489. IEEE Press (2010)
10. Liu, D., Li, H.X., Zhou, X.Z.: Two decades' research on decision-theoretic rough sets. In: Proceedings of ICCI 2010, pp. 968–973. IEEE Press (2010)
11. Liu, D., Li, T., Hu, P., Li, H.: Multiple-Category Classification with Decision-Theoretic Rough Sets. In: Yu, J., Greco, S., Lingras, P., Wang, G., Skowron, A. (eds.) RSKT 2010. LNCS (LNAI), vol. 6401, pp. 703–710. Springer, Heidelberg (2010)
12. Liu, D., Li, T.R., Li, H.X.: A multiple-category classification approach with decision-theoretic rough sets. Fundamenta Informaticae 115, 173–188 (2012)
13. Liu, D., Li, T.R., Ruan, D.: Probabilistic model criteria with decision-theoretic rough sets. Information Sciences 181, 3709–3722 (2011)
14. Min, F., Liu, Q.H.: A hierarchical model for test-cost-sensitive decision systems. Information Sciences 179, 2442–2452 (2009)
15. Min, F., He, H.P., Qian, Y.H., Zhu, W.: Test-cost-sensitive attribute reduction. Information Sciences 181, 4928–4942 (2011)
16. Pawlak, Z.: Rough sets. International Journal of Computer and Information Science 11, 341–356 (1982)

17. Yao, J., Herbert, J.P.: Web-Based Support Systems with Rough Set Analysis. In: Kryszkiewicz, M., Peters, J.F., Rybiński, H., Skowron, A. (eds.) RSEISP 2007. LNCS (LNAI), vol. 4585, pp. 360–370. Springer, Heidelberg (2007)
18. Yao, Y.: Decision-Theoretic Rough Set Models. In: Yao, J., Lingras, P., Wu, W.-Z., Szczuka, M.S., Cercone, N.J., Ślęzak, D. (eds.) RSKT 2007. LNCS (LNAI), vol. 4481, pp. 1–12. Springer, Heidelberg (2007)
19. Yao, Y.Y.: The superiority of three-way decision in probabilistic rough set models. Information Sciences 181, 1080–1096 (2011)
20. Yao, Y.Y.: Three-way decisions with probabilistic rough sets. Information Sciences 180, 341–353 (2010)
21. Yao, Y.Y.: Two semantic issues in a probabilistic rough set model. Fundamenta Informaticae 108, 249–265 (2011)
22. Yao, Y.Y., Wong, S.K.M., Lingras, P.: A decision-theoretic rough set model. Methodologies for Intelligent Systems 5, 17–24 (1990)
23. Yao, Y.Y., Zhao, Y.: Attribute reduction in decision-teoretic rough set models. Information Sciences 178, 3356–3373 (2008)
24. Yu, H., Chu, S., Yang, D.: Autonomous Knowledge-Oriented Clustering Using Decision-Theoretic Rough Set Theory. In: Yu, J., Greco, S., Lingras, P., Wang, G., Skowron, A. (eds.) RSKT 2010. LNCS (LNAI), vol. 6401, pp. 687–694. Springer, Heidelberg (2010)
25. Zhou, X., Li, H.: A Multi-View Decision Model Based on Decision-Theoretic Rough Set. In: Wen, P., Li, Y., Polkowski, L., Yao, Y., Tsumoto, S., Wang, G. (eds.) RSKT 2009. LNCS, vol. 5589, pp. 650–657. Springer, Heidelberg (2009)

Decision-Theoretic Rough Sets
with Probabilistic Distribution

Dun Liu[1], Tianrui Li[2], and Decui Liang[1]

[1] School of Economics and Management, Southwest Jiaotong University
Chengdu 610031, P.R. China
newton83@163.com, decuiliang@126.com
[2] School of Information Science and Technology, Southwest Jiaotong University
Chengdu 610031, P.R. China
trli@swjtu.edu.cn

Abstract. In the previous decision-theoretic rough sets model (DTRS), its loss function values are precise. This paper extends the precise values of loss functions to a more realistic stochastic environment. Considering all loss functions in DTRS model obey a certain of probabilistic distribution, the extension of decision-theoretic rough set models under uniform distribution and normal distribution are proposed in this paper. An empirical study validates the reasonability and effectiveness of the proposed approach.

Keywords: Decision-theoretic rough sets, uniform distribution, normal distribution, stochastic.

1 Introduction

Rough Set Theory (RST) was proposed by Pawlak in the early 1980s [13]. It generalizes the classical set theory so as to handle incomplete, inexact and insufficient information. In RST, two crisp subsets of a universe, lower and upper approximations, were defined to characterize a crisp subset of the universe. The two approximations divide the universe into three pairwise disjoint regions: the positive region, boundary region and negative region. With respect to the three regions, Yao [23–25, 27] introduced the notion of three-way decisions, consisting of the positive, boundary and negative rules. Positive rules, negative rules and boundary rules correspond to making decisions of acceptance, rejection and deferring a definite decision, respectively. The ideas of three-way decisions have in fact been applied to many fields, both in theoretical analysis [4–7, 9–11] and applications, including medical clinic [12, 28], products inspecting process [16], environmental management [3], data packs selection [14], E-learning [1], email spam filtering [29, 30] and oil investment [8, 17].

By considering the tolerance of errors in the three-way decisions, Yao et al. introduced Bayesian decision procedure to propose decision-theoretic rough sets (DTRS) [18, 19, 21], and the pair of thresholds can be directly calculated by minimizing the decision cost with Bayesian theory, which gives a brief semantics

T. Li et al. (Eds.): RSKT 2012, LNAI 7414, pp. 389–398, 2012.
© Springer-Verlag Berlin Heidelberg 2012

explanation in practical applications with minimum decision risks [9, 24–26]. The loss functions in aforementioned studies are assumed as precise numerical values. Generally, people prefer precise information because precise information gives them a greater sense of security and confidence in their ability to predict unknown outcomes in their environment [2]. However, the decision makers may hardly estimate the precise loss function values in real decision procedure, and the uncertain information (including stochastic, vague or rough information) sometimes serves people better than precise information does. As one important descriptions of uncertainty, stochastic number is used to indicate that a particular subject is seen from point of view of randomness, which refers to systems whose behavior is intrinsically non-deterministic, sporadic and categorically not intermittent. In mathematics, specifically in probability theory, we say a random variable is stochastic means that it obeys one certain discrete probability distribution or continuous probability distribution. In this paper, the stochastic numbers are appropriate instead of the precise numbers in realistic decision problems. In this paper, we consider the loss functions obey a certain of probabilistic distribution, and mainly discuss two probabilistic distribution, namely, uniform distribution and normal distribution, and focus on discussing the stochastic extension of decision-theoretic rough sets model.

The remainder of this paper is organized as follows: Section 2 provides the basic concepts of probabilistic distribution and DTRS. the extension of decision-theoretic rough set models under uniform distribution and normal distribution are discussed in Section 3. Then, a case study of PPP project investment problem is given to illustrate our approach in Section 4. Section 5 concludes the paper and outlines the future work.

2   Preliminaries

Basic concepts, notations and results of the probabilistic distribution and DTRS are briefly reviewed in this section [19, 20, 22].

2.1   Probabilistic Distribution

Probability distribution is a function that describes the probability of a random variable taking certain values. For a more precise definition one needs to distinguish between discrete and continuous random variables. In the discrete case, one can easily assign a probability to each possible value. In contrast, when a random variable takes values from a continuum, probabilities are nonzero only if they refer to finite intervals [15]. In this paper, we discuss two continuous probability distribution, namely uniform distribution and normal distribution as follows.

In probability theory, the uniform distribution is a family of probability distributions such that for each member of the family, all intervals of the same length on the distribution's support are equally probable. The support is defined by the two parameters, a and b, which are its minimum and maximum values.

The distribution is often abbreviated $U(a, b)$. The probability density function of the uniform distribution is:

$$f(x) = \begin{cases} \frac{1}{b-a} & a \leq x \leq b \\ 0 & \text{others} \end{cases}$$

Suppose $X \sim U(a, b)$, the expected value and variances are denoted as $E(X) = \frac{a+b}{2}$ and $V(X) = \frac{b^2 - a^2}{12}$.

Furthermore, the normal (or Gaussian) distribution is a continuous probability distribution that has a bell-shaped probability density function, known as the Gaussian function or informally the bell curve:

$$f(x; \mu, \sigma^2) = \frac{1}{\sigma\sqrt{2\pi}} e^{-\frac{(x-\mu)^2}{2\sigma^2}}.$$

where parameter μ is the mean or expectation (location of the peak) and σ^2 is the variance. σ is known as the standard deviation. The distribution with $\mu = 0$ and $\sigma^2 = 1$ is called the standard normal. A normal distribution is often used as a first approximation to describe real-valued random variables that cluster around a single mean value.

2.2   Decision-Theoretic Rough Sets

For the Bayesian decision procedure, the DTRS model is composed of 2 states and 3 actions. The set of states is given by $\Omega = \{C, \neg C\}$ indicating that an object is in C and not in C, respectively. The set of actions is given by $\mathcal{A} = \{a_P, a_B, a_N\}$, where a_P, a_B, and a_N represent the three actions in classifying an object x, namely, deciding $x \in \text{POS}(C)$, deciding x should be further investigated $x \in \text{BND}(C)$, and deciding $x \in \text{NEG}(C)$, respectively. The loss function λ regarding the risk or cost of actions in different states is given by the 3×2 matrix:

	C (P)	$\neg C$ (N)
a_P	λ_{PP}	λ_{PN}
a_B	λ_{BP}	λ_{BN}
a_N	λ_{NP}	λ_{NN}

In the matrix, λ_{PP}, λ_{BP} and λ_{NP} denote the losses incurred for taking actions of a_P, a_B and a_N, respectively, when an object belongs to C. Similarly, λ_{PN}, λ_{BN} and λ_{NN} denote the losses incurred for taking the same actions when the object belongs to $\neg C$. $Pr(C|[x])$ is the conditional probability of an object x belonging to C given that the object is described by its equivalence class $[x]$. For an object x, the expected loss $R(a_i|[x])$ associated with taking the individual actions can be expressed as:

$$R(a_P|[x]) = \lambda_{PP} Pr(C|[x]) + \lambda_{PN} Pr(\neg C|[x]),$$
$$R(a_B|[x]) = \lambda_{BP} Pr(C|[x]) + \lambda_{BN} Pr(\neg C|[x]),$$
$$R(a_N|[x]) = \lambda_{NP} Pr(C|[x]) + \lambda_{NN} Pr(\neg C|[x]).$$

The Bayesian decision procedure suggests the following minimum-cost decision rules:

(P) If $R(a_P|[x]) \leq R(a_B|[x])$ and $R(a_P|[x]) \leq R(a_N|[x])$, decide $x \in POS(C)$;
(B) If $R(a_B|[x]) \leq R(a_P|[x])$ and $R(a_B|[x]) \leq R(a_N|[x])$, decide $x \in BND(C)$;
(N) If $R(a_N|[x]) \leq R(a_P|[x])$ and $R(a_N|[x]) \leq R(a_B|[x])$, decide $x \in NEG(C)$.

Since $Pr(C|[x]) + Pr(\neg C|[x]) = 1$, we simplify the rules based only on the probability $Pr(C|[x])$ and the loss function. By considering a reasonable kind of loss functions with $\lambda_{PP} \leq \lambda_{BP} < \lambda_{NP}$ and $\lambda_{NN} \leq \lambda_{BN} < \lambda_{PN}$, the decision rules (P)-(N) can be expressed concisely as:

(P) If $Pr(C|[x]) \geq \alpha$ and $Pr(C|[x]) \geq \gamma$, decide $x \in POS(C)$;
(B) If $Pr(C|[x]) \leq \alpha$ and $Pr(C|[x]) \geq \beta$, decide $x \in BND(C)$;
(N) If $Pr(C|[x]) \leq \beta$ and $Pr(C|[x]) \leq \gamma$, decide $x \in NEG(C)$.

The thresholds values α, β, γ are given by: $\alpha = \frac{(\lambda_{PN}-\lambda_{BN})}{(\lambda_{PN}-\lambda_{BN})+(\lambda_{BP}-\lambda_{PP})}$, $\beta = \frac{(\lambda_{BN}-\lambda_{NN})}{(\lambda_{BN}-\lambda_{NN})+(\lambda_{NP}-\lambda_{BP})}$ and $\gamma = \frac{(\lambda_{PN}-\lambda_{NN})}{(\lambda_{PN}-\lambda_{NN})+(\lambda_{NP}-\lambda_{PP})}$.

In addition, as a well-defined boundary region, the conditions of rule (B) suggest that $\alpha > \beta$, that is, $\frac{(\lambda_{BP}-\lambda_{PP})}{(\lambda_{PN}-\lambda_{BN})} < \frac{(\lambda_{NP}-\lambda_{BP})}{(\lambda_{BN}-\lambda_{NN})}$. It implies $0 \leq \beta < \gamma < \alpha \leq 1$. In this case, after tie-breaking, the following simplified rules are obtained:

(P1)   If $Pr(C|[x]) \geq \alpha$, decide $x \in POS(C)$;
(B1)   If $\beta < Pr(C|[x]) < \alpha$, decide $x \in BND(C)$;
(N1)   If $Pr(C|[x]) \leq \beta$, decide $x \in NEG(C)$.

3   Extension of Decision-Theoretic Rough Set Models

Following our discussions in Section 2, we mainly discuss the two scenarios that the loss functions obey uniform distribution and normal distribution, namely, $\lambda \sim U(a,b)$ and $\lambda \sim N(\mu, \sigma^2)$, respectively.

3.1   Decision-Theoretic Rough Sets with Uniform Distribution

Suppose $\lambda_{PP} \sim U(a_{PP}, b_{PP})$, $\lambda_{BP} \sim U(a_{BP}, b_{BP})$, $\lambda_{NP} \sim U(a_{NP}, b_{NP})$; $\lambda_{NN} \sim U(a_{NN}, b_{NN})$, $\lambda_{BN} \sim U(a_{BN}, b_{BN})$, $\lambda_{PN} \sim U(a_{PN}, b_{PN})$. According to Section 2.1, we choose the expected value $E(\lambda)$ as the measure to estimate λ. Considered the conditions of $a_{PP} \leq a_{BP} < a_{NP}$, $a_{NN} \leq a_{BN} < a_{PN}$; $b_{PP} \leq b_{BP} < b_{NP}$, $b_{NN} \leq b_{BN} < b_{PN}$, we have $\frac{a_{PP}+b_{PP}}{2} \leq \frac{a_{BP}+b_{BP}}{2} < \frac{a_{NP}+b_{NP}}{2}$ and $\frac{a_{NN}+b_{NN}}{2} \leq \frac{a_{BN}+b_{BN}}{2} < \frac{a_{PN}+b_{PN}}{2}$. Therefore, we directly utilize $E(\lambda)$ to instead of λ and get the $\overline{\alpha}$, $\overline{\beta}$, $\overline{\gamma}$ as:

$$\overline{\alpha} = \frac{E(\lambda_{PN}) - E(\lambda_{BN})}{(E(\lambda_{PN}) - E(\lambda_{BN})) + (E(\lambda_{BP}) - E(\lambda_{PP}))}$$

$$= \frac{(a_{PN} + b_{PN}) - (a_{BN} + b_{BN})}{[(a_{PN} + b_{PN}) - (a_{BN} + b_{BN})] + [(a_{BP} + b_{BP}) - (a_{PP} + b_{PP})]},$$

$$\overline{\beta} = \frac{E(\lambda_{BN}) - E(\lambda_{NN}))}{E(\lambda_{BN}) - E(\lambda_{NN})) + E(\lambda_{NP}) - E(\lambda_{BP}))}$$

$$= \frac{(a_{BN} + b_{BN}) - (a_{NN} + b_{NN})}{[(a_{BN} + b_{BN}) - (a_{NN} + b_{NN})] + [(a_{NP} + b_{NP}) - [a_{BP} + b_{BP})]},$$

$$\overline{\gamma} = \frac{E(\lambda_{PN}) - E(\lambda_{NN})}{(E(\lambda_{PN}) - E(\lambda_{NN})) + (E(\lambda_{NP}) - E(\lambda_{PP}))}$$

$$= \frac{(a_{PN} + b_{PN}) - (a_{NN} + b_{NN})}{[(a_{PN} + b_{PN}) - (a_{NN} + b_{NN})] + [(a_{NP} + b_{NP}) - (a_{PP} + b_{PP})]}.$$

3.2 Decision-Theoretic Rough Sets with Normal Distribution

Suppose $\lambda_{PP} \sim N(\mu_{PP}, \sigma_{PP}^2)$, $\lambda_{BP} \sim N(\mu_{BP}, \sigma_{BP}^2)$, $\lambda_{NP} \sim N(\mu_{NP}, \sigma_{NP}^2)$; $\lambda_{NN} \sim N(\mu_{NN}, \sigma_{NN}^2)$, $\lambda_{BN} \sim N(\mu_{BN}, \sigma_{BN}^2)$, $\lambda_{PN} \sim N(\mu_{PN}, \sigma_{PN}^2)$. In statistics, the 68-95-99.7 rule, states that for a normal distribution, nearly all values lie within 3 standard deviations of the mean. About 68% of values drawn from a normal distribution are within one standard deviation σ away from the mean; about 95% of the values lie within two standard deviations; and about 99.7% are within three standard deviations. Here, we introduce confidence intervals to our study, and three confidence intervals of λ can be expressed as: $Pr(\mu - \sigma \leq \lambda \leq \mu + \sigma) \approx 0.6827$; $Pr(\mu - 2\sigma \leq \lambda \leq \mu + 2\sigma) \approx 0.9545$; $Pr(\mu - 3\sigma \leq \lambda \leq \mu + 3\sigma) \approx 0.9973$.

Therefore, we can utilize the intervals $[\mu - n\sigma, \mu + n\sigma]$ $(n = 1, 2, 3)$ to instead of λ. Note that, in this scenario, λ is not a number, but an interval. For simplicity, we only calculate the minimum and maximum values for the three parameters as:

$$\alpha^{min} = \frac{(\mu_{PN} - n\sigma_{PN}) - (\mu_{BN} + n\sigma_{BN})}{[(\mu_{PN} + n\sigma_{PN}) - (\mu_{BN} - n\sigma_{BN})] + [(\mu_{BP} + n\sigma_{BP}) - (\mu_{PP} - n\sigma_{PP})]},$$

$$\alpha^{max} = \frac{(\mu_{PN} + n\sigma_{PN}) - (\mu_{BN} - n\sigma_{BN})}{[(\mu_{PN} - n\sigma_{PN}) - (\mu_{BN} + n\sigma_{BN})] + [(\mu_{BP} - n\sigma_{BP}) - (\mu_{PP} + n\sigma_{PP})]};$$

$$\beta^{min} = \frac{(\mu_{BN} - n\sigma_{BN}) - (\mu_{NN} + n\sigma_{NN})}{[(\mu_{BN} + n\sigma_{BN}) - (\mu_{NN} - n\sigma_{NN})] + [(\mu_{NP} + n\sigma_{NP}) - (\mu_{BP} - n\sigma_{BP})]},$$

$$\beta^{max} = \frac{(\mu_{BN} + n\sigma_{BN}) - (\mu_{NN} - n\sigma_{NN})}{[(\mu_{BN} - n\sigma_{BN}) - (\mu_{NN} + n\sigma_{NN})] + [(\mu_{NP} - n\sigma_{NP}) - (\mu_{BP} + n\sigma_{BP})]};$$

$$\gamma^{min} = \frac{(\mu_{PN} - n\sigma_{PN}) - (\mu_{NN} + n\sigma_{NN})}{[(\mu_{PN} + n\sigma_{PN}) - (\mu_{NN} - n\sigma_{NN})] + [(\mu_{NP} + n\sigma_{NP}) - (\mu_{PP} - n\sigma_{PP})]},$$

$$\gamma^{max} = \frac{(\mu_{PN} + n\sigma_{PN}) - (\mu_{NN} - n\sigma_{NN})}{[(\mu_{PN} - n\sigma_{PN}) - (\mu_{NN} + n\sigma_{NN})] + [(\mu_{NP} - n\sigma_{NP}) - (\mu_{PP} + n\sigma_{PP})]}.$$

where, $n = 1, 2, 3$ correspond to the three different degrees of confidence. In DTRS model, it requires all $\lambda_{\bullet\bullet} \geq 0$, that is, $\mu_{\bullet\bullet} - n\sigma_{\bullet\bullet} \geq 0$. Furthermore, the values of α^{min}, β^{min} and γ^{min} may less than 0, as well, the values of α^{max}, β^{max} and γ^{max} may more than 1. Here, we should constrain all α, β and γ in the interval $[0, 1]$. Therefore, the lower and upper bound for the three parameters can be calculate as:

$$\alpha \in [\alpha^{Low}, \alpha^{Upp}] = [max(0, \alpha^{min}), min(1, \alpha^{max})];$$
$$\beta \in [\beta^{Low}, \beta^{Upp}] = [max(0, \beta^{min}), min(1, \beta^{max})];$$
$$\gamma \in [\gamma^{Low}, \gamma^{Upp}] = [max(0, \gamma^{min}), min(1, \gamma^{max})].$$

Specially, we consider two special cases: (1). $\lambda_1 = \mu_{\bullet\bullet} - n\sigma_{\bullet\bullet}$, (2). $\lambda_2 = \mu_{\bullet\bullet} + n\sigma_{\bullet\bullet}$. We can calculate the corresponding values of α^1, α^2, β^1 and β^2 as follows:

$$\alpha^1 = \frac{(\mu_{PN} - n\sigma_{PN}) - (\mu_{BN} - n\sigma_{BN})}{[(\mu_{PN} - n\sigma_{PN}) - (\mu_{BN} - n\sigma_{BN})] + [(\mu_{BP} - n\sigma_{BP}) - (\mu_{PP} - n\sigma_{PP})]},$$

$$\beta^1 = \frac{(\mu_{BN} - n\sigma_{BN}) - (\mu_{NN} - n\sigma_{NN})}{[(\mu_{BN} - n\sigma_{BN}) - (\mu_{NN} - n\sigma_{NN})] + [(\mu_{NP} - n\sigma_{NP}) - (\mu_{BP} - n\sigma_{BP})]};$$

$$\alpha^2 = \frac{(\mu_{PN} + n\sigma_{PN}) - (\mu_{BN} + n\sigma_{BN})}{[(\mu_{PN} + n\sigma_{PN}) - (\mu_{BN} + n\sigma_{BN})] + [(\mu_{BP} + n\sigma_{BP}) - (\mu_{PP} + n\sigma_{PP})]},$$

$$\beta^2 = \frac{(\mu_{BN} + n\sigma_{BN}) - (\mu_{NN} + n\sigma_{NN})}{[(\mu_{BN} + n\sigma_{BN}) - (\mu_{NN} + n\sigma_{NN})] + [(\mu_{NP} + n\sigma_{NP}) - (\mu_{BP} + n\sigma_{BP})]}.$$

Obviously, we can get $\alpha^1, \alpha^2 \in [\alpha^{Low}, \alpha^{Upp}]$ and $\beta^1, \beta^2 \in [\beta^{Low}, \beta^{Upp}]$.

4   An Illustration

In this section, we illustrate the extended model by an example of decision in Public-Private Partnerships (PPP) project investment. A PPP project is funded and operated through a partnership of government and one or more private sectors according to their contract. Through a PPP project, the government can reduces financial expenditure and sufficiently allocates the limited recourse, and the private sectors can benefit from the PPP project by using their technology, fund and professional knowledge. However, because of the implementation of PPP project, there exists many risk and uncertain factors.

During the risk assessment of PPP project investment, we have two states $\Omega = \{C, \neg C\}$ indicating that the project is a good project and bad project, respectively. With respect to the three-way decision, the set of actions is given by $\mathcal{A} = \{a_P, a_B, a_N\}$, where a_P, a_B, and a_N represent investment, need further analysis and do not investment, respectively. There are 6 parameters in the model.

Table 1. The costs of 6 types of PPP projects

X	$\lambda_{\bullet\bullet} \sim U(a,b)$											
	a_{PP}	b_{PP}	a_{BP}	b_{BP}	a_{NP}	b_{NP}	a_{PN}	b_{PN}	a_{BN}	b_{BN}	a_{NN}	b_{NN}
x_1	0.5	1.5	1	3	2	6	2.5	7.5	1.5	4.5	1	3
x_2	2	4	4	6	7	9	3	5	0.25	0.75	0	0
x_3	0.5	1.5	1	2	2	4	4	6	0.5	1.5	0	0
x_4	3	5	6	8	10	12	5	9	3	6	1	3
x_5	1	3	2	4	3	5	2	4	0.5	1.5	0	1
x_6	0	2	1	3	3	4	4	6	2	4	1	3

X	$\lambda_{\bullet\bullet} \sim N(\mu,\sigma^2)$											
	μ_{PP}	σ_{PP}	μ_{BP}	σ_{BP}	μ_{NP}	σ_{NP}	μ_{PN}	σ_{PN}	μ_{BN}	σ_{BN}	μ_{NN}	σ_{NN}
x_1	1	0.1	2	0.2	4	0.25	5	0.3	3	0.15	2	0.05
x_2	3	0.2	5	0.4	8	0.6	4	0.3	0.5	0.05	0	0
x_3	1	0.05	1.5	0.1	3	0.2	5	0.5	1	0.05	0	0
x_4	4	0.2	7	0.3	11	0.5	7	0.3	4.5	0.2	2	0.1
x_5	2	0.1	3	0.1	4	0.1	3	0.2	1	0.1	0.5	0
x_6	1	0.05	2	0.1	3.5	0.15	5	0.2	3	0.1	2	0.05

λ_{PP}, λ_{BP} and λ_{NP} denote the costs incurred for taking actions of investment, need further analysis and do not investment when a PPP project is good; λ_{PN}, λ_{BN} and λ_{NN} denote the costs incurred for taking actions of investment, need further analysis and do not investment when a PPP project is bad. Here, we consider two scenarios: (1). $\lambda_{\bullet\bullet}$ obey uniform distribution; (2). $\lambda_{\bullet\bullet}$ obey normal distribution. Table 1 shows the costs for six types of PPP projects.

Based on Table 1 and Formula in Section 3.1 and 3.2, the three thresholds for each PPP project can calculate. Here, we consider $\lambda_{\bullet\bullet} \sim U(a,b)$ and $\lambda_{\bullet\bullet} \sim N(\mu,\sigma^2)$ ($n = 1, 2, 3$), Table 2 lists the calculating results of thresholds α, β and γ for each PPP investment project.

In Table 2, it can be seen that different pairs of thresholds are obtained for different types of PPP projects. For each $x_i \in X$ ($i = 1, 2, \cdots, 6$, $n = 1, 2, 3$), we have $\alpha^1, \alpha^2 \in [\alpha^{Low}, \alpha^{Upp}]$ and $\beta^1, \beta^2 \in [\beta^{Low}, \beta^{Upp}]$. Suppose now that $Pr(X|x_i) = 0.67$ for all $x_i \in X$. By using rules (P1), (B1) and (N1), we compare the thresholds in Table 2 and the conditional probability $Pr(X|x_i)$ for each x_i, then list the decision regions in Table 3 by considering different α, β and γ.

To sum up, given a PPP project x, we can simply compare the relation between the conditional probability $Pr(X|x)$ and the thresholds (α, β) to make an optimal decision, the three-way decision approach provides a reasonable way to solve the practical PPP project investment in applications.

Table 2. The values of α, β and γ in 6 types of PPP projects

X	Uniform Distribution			Normal Distribution (n=1)					
	$\overline{\alpha}$	$\overline{\beta}$	$\overline{\gamma}$	α^{Low}	α^{Upp}	β^{Low}	β^{Upp}	γ^{Low}	γ^{Upp}
x_1	0.6667	0.3333	0.5000	0.4133	1.0000	0.2051	0.5106	0.3955	0.6321
x_2	0.6364	0.1429	0.4444	0.4884	0.8462	0.0874	0.2245	0.3663	0.5443
x_3	0.8889	0.4000	0.7143	0.6635	1.0000	0.3115	0.4884	0.5806	0.8800
x_4	0.4545	0.3846	0.4167	0.3077	0.6667	0.2716	0.5185	0.3511	0.4954
x_5	0.6667	0.3333	0.5556	0.4857	0.9200	0.2105	0.5000	0.4694	0.6585
x_6	0.6667	0.4000	0.5455	0.4928	0.9020	0.2787	0.5476	0.4622	0.6436

X	Normal Distribution (n=2)						Normal Distribution (n=3)					
	α^{Low}	α^{Upp}	β^{Low}	β^{Upp}	γ^{Low}	γ^{Upp}	α^{Low}	α^{Upp}	β^{Low}	β^{Upp}	γ^{Low}	γ^{Upp}
x_1	0.2444	1.0000	0.1395	0.8235	0.3108	0.8043	0.1238	1.0000	0.0808	1.0000	0.2407	1.0000
x_2	0.3784	1.0000	0.0714	0.4286	0.3036	0.6765	0.2934	1.0000	0.0526	1.0000	0.2520	0.8596
x_3	0.4915	1.0000	0.2813	0.6111	0.4706	1.0000	0.3561	1.0000	0.2394	0.7931	0.3784	1.0000
x_4	0.2000	1.0000	0.2184	0.7209	0.2958	0.5918	0.1176	1.0000	0.1633	1.0000	0.2484	0.7126
x_5	0.3500	1.0000	0.1429	0.7778	0.3962	0.7838	0.2444	1.0000	0.0833	1.0000	0.3333	0.9394
x_6	0.3590	1.0000	0.2121	0.7647	0.3906	0.7609	0.2529	1.0000	0.1486	1.0000	0.3285	0.9036

X	Normal Distribution (n=1)			Normal Distribution (n=1)		
	α^1	β^1	γ^1	α^2	β^2	γ^2
x_1	0.6727	0.3158	0.4911	0.6615	0.3492	0.5078
x_2	0.6436	0.1385	0.4458	0.6303	0.1467	0.4433
x_3	0.8875	0.4043	0.7087	0.8900	0.3962	0.7190
x_4	0.4528	0.3871	0.4174	0.4561	0.3824	0.4160
x_5	0.6552	0.2857	0.5349	0.6774	0.3750	0.5745
x_6	0.6667	0.3958	0.5429	0.6667	0.4038	0.5478

X	Normal Distribution (n=2)			Normal Distribution (n=2)		
	α^1	β^1	γ^1	α^2	β^2	γ^2
x_1	0.6800	0.2963	0.4808	0.6571	0.3636	0.5147
x_2	0.6522	0.1333	0.4474	0.6250	0.1500	0.4423
x_3	0.8857	0.4091	0.7018	0.8909	0.3929	0.7229
x_4	0.4510	0.3898	0.4182	0.4576	0.3803	0.4154
x_5	0.6429	0.2308	0.5122	0.6875	0.4118	0.5918
x_6	0.6667	0.3913	0.5400	0.6667	0.4074	0.5500

X	Normal Distribution (n=3)			Normal Distribution (n=3)		
	α^1	β^1	γ^1	α^2	β^2	γ^2
x_1	0.6889	0.2745	0.4688	0.6533	0.3768	0.5208
x_2	0.6627	0.1273	0.4493	0.6204	0.1529	0.4414
x_3	0.8833	0.4146	0.6931	0.8917	0.3898	0.7263
x_4	0.4490	0.3929	0.4190	0.4590	0.3784	0.4148
x_5	0.6296	0.1667	0.4872	0.6970	0.4444	0.6078
x_6	0.6667	0.3864	0.5368	0.6667	0.4107	0.5520

Table 3. The decision region for each PPP projects when $Pr(X|x_i) = 0.67$

(α, β)	$POS(X)$	$BND(X)$	$NEG(X)$
$(\overline{\alpha}, \overline{\beta})$	$\{x_1, x_2, x_4, x_5, x_6\}$	$\{x_3\}$	$\varnothing$
(α^1, β^1) $(n = 1)$	$\{x_2, x_4, x_5, x_6\}$	$\{x_1, x_3\}$	$\varnothing$
(α^2, β^2) $(n = 1)$	$\{x_1, x_2, x_4, x_6\}$	$\{x_3, x_5\}$	$\varnothing$
(α^1, β^1) $(n = 2)$	$\{x_2, x_4, x_5, x_6\}$	$\{x_1, x_3\}$	$\varnothing$
(α^2, β^2) $(n = 2)$	$\{x_1, x_2, x_4, x_6\}$	$\{x_3, x_5\}$	$\varnothing$
(α^1, β^1) $(n = 3)$	$\{x_2, x_4, x_5, x_6\}$	$\{x_1, x_3\}$	$\varnothing$
(α^2, β^2) $(n = 3)$	$\{x_1, x_2, x_4, x_6\}$	$\{x_3, x_5\}$	$\varnothing$

5   Conclusions

In this paper, two extension of DTRS models are proposed by considering the loss functions obey uniform distribution and normal distribution, respectively. The probabilistic distribution elicitation of stochastic number provides a solution to obtain loss function values for DTRS model. An example of PPP project investment is utilized to illustrate our approach. Our future research work will focus on developing new DTRS models in interval and fuzzy environments.

Acknowledgements. This work is supported by the National Science Foundation of China (Nos. 61175047, 60873108, 61100117, 71090402/G1), the Youth Social Science Foundation of the Chinese Education Commission (11YJC630127) and the Fundamental Research Funds for the central Universities of China (Nos. SWJTU12CX117, SWJTU11ZT08).

References

1. Abbas, A.R., Juan, L.: Supporting E-Learning System with Modified Bayesian Rough Set Model. In: Yu, W., He, H., Zhang, N. (eds.) ISNN 2009, Part II. LNCS (LNAI), vol. 5552, pp. 192–200. Springer, Heidelberg (2009)
2. Camerer, C., Weber, M.: Recent developments in modeling preferences: uncertainty and ambiguity. Journal of Risk and Uncertainty 5, 325–370 (1992)
3. Goudey, R.: Do statistical inferences allowing three alternative decision give better feedback for environmentally precautionary decision-making. Journal of Environmental Management 85, 338–344 (2007)
4. Herbert, J.P., Yao, J.T.: Game-theoretic rough sets. Fundamenta Informaticae 108, 267–286 (2011)
5. Li, H.X., Zhou, X.Z.: Risk decision making based on decision-theoretic rough set: a three-way view decision model. International Journal of Computational Intelligence Systems 4, 1–11 (2011)
6. Liu, D., Li, H.X., Zhou, X.Z.: Two decades' research on decision-theoretic rough sets. In: Proceeding of ICCI 2010, pp. 968–973 (2010)
7. Liu, D., Li, T., Hu, P., Li, H.: Multiple-Category Classification with Decision-Theoretic Rough Sets. In: Yu, J., Greco, S., Lingras, P., Wang, G., Skowron, A. (eds.) RSKT 2010. LNCS (LNAI), vol. 6401, pp. 703–710. Springer, Heidelberg (2010)
8. Liu, D., Yao, Y.Y., Li, T.R.: Three-way investment decisions with decision-theoretic rough sets. International Journal of Computational Intelligence Systems 4, 66–74 (2011)
9. Liu, D., Li, T.R., Ruan, D.: Probabilistic model criteria with decision-theoretic rough sets. Information Sciences 181, 3709–3722 (2011)

10. Liu, D., Li, T., Liang, D.: A New Discriminant Analysis Approach under Decision-Theoretic Rough Sets. In: Yao, J., Ramanna, S., Wang, G., Suraj, Z. (eds.) RSKT 2011. LNCS, vol. 6954, pp. 476–485. Springer, Heidelberg (2011)
11. Liu, D., Li, T.R., Li, H.X.: A multiple-category classification approach with decision-theoretic rough sets. Fundamenta Informaticae 115, 173–188 (2012)
12. Pauker, S.G., Kassirer, J.P.: The threshold approach to clinical decision making. The New England Journal of Medicine 302, 1109–1117 (1980)
13. Pawlak, Z.: Rough sets. International Journal of Computer and Information Science 11, 341–356 (1982)
14. Ślęzak, D., Wróblewski, J., Eastwood, V., Synak, P.: Brighthouse: an analytic data warehouse for ad-hoc queries. In: Proceedings of the VLDB Endowment, vol. 1, pp. 1337–1345 (2008)
15. Probabilistic Distribution,
    http://en.wikipedia.org/wiki/Probability_distribution
16. Woodward, P.W., Naylor, J.C.: An application of bayesian methods in SPC. The Statistician 42, 461–469 (1993)
17. Xie, G., Yue, W., Wang, S.Y., Lai, K.K.: Dynamic risk management in petroleum project investment based on a variable precision rough set model. Technological Forecasting & Social Change 77, 891–901 (2010)
18. Yao, Y.Y., Wong, S.K.M., Lingras, P.: A decision-theoretic rough set model. In: The 5th International Symposium on Methodologies for Intelligent Systems, pp. 17–25 (1990)
19. Yao, Y.Y., Wong, S.K.M.: A decision theoretic framework for approximating concepts. International Journal of Man-machine Studies 37, 793–809 (1992)
20. Yao, Y.Y.: Probabilistic approaches to rough sets. Expert Systems 20, 287–297 (2003)
21. Yao, Y.Y.: Decision-Theoretic Rough Set Models. In: Yao, J., Lingras, P., Wu, W.-Z., Szczuka, M.S., Cercone, N.J., Ślęzak, D. (eds.) RSKT 2007. LNCS (LNAI), vol. 4481, pp. 1–12. Springer, Heidelberg (2007)
22. Yao, Y.Y.: Probabilistic rough set approximations. International Journal of Approximate Reasoning 49, 255–271 (2008)
23. Yao, Y.Y.: Three-Way Decision: An Interpretation of Rules in Rough Set Theory. In: Wen, P., Li, Y., Polkowski, L., Yao, Y., Tsumoto, S., Wang, G. (eds.) RSKT 2009. LNCS, vol. 5589, pp. 642–649. Springer, Heidelberg (2009)
24. Yao, Y.Y.: Three-way decisions with probabilistic rough sets. Information Sciences 180, 341–353 (2010)
25. Yao, Y.Y.: The superiority of three-way decision in probabilistic rough set models. Information Sciences 181, 1080–1096 (2011)
26. Yao, Y.Y.: Two semantic issues in a probabilistic rough set model. Fundamenta Informaticae 108, 249–265 (2011)
27. Yao, Y.Y.: Three-way decisions using rough sets. In: Peters, G., et al. (eds.) Rough Sets: Selected Methods and Applications in Management and Engineering. Advanced Information and Knowledge Processing, pp. 79–93 (2012)
28. Yao, J., Herbert, J.P.: Web-Based Support Systems with Rough Set Analysis. In: Kryszkiewicz, M., Peters, J.F., Rybiński, H., Skowron, A. (eds.) RSEISP 2007. LNCS (LNAI), vol. 4585, pp. 360–370. Springer, Heidelberg (2007)
29. Zhao, W.Q., Zhu, Y.L.: An email classification scheme based on decision-theoretic rough set theory and analysis of email security. In: Proceeding of 2005 IEEE Region 10 TENCON (2005), doi:10.1109/TENCON.2005.301121.
30. Zhou, B., Yao, Y., Luo, J.: A Three-Way Decision Approach to Email Spam Filtering. In: Farzindar, A., Keselj, V. (eds.) Canadian AI 2010. LNCS (LNAI), vol. 6085, pp. 28–39. Springer, Heidelberg (2010)

Multiple Criteria Decision Analysis with Game-Theoretic Rough Sets

Nouman Azam and JingTao Yao

Department of Computer Science, University of Regina,
Regina, Saskatchewan, Canada S4S 0A2
{azam200n,jtyao}@cs.uregina.ca

Abstract. Multiple criteria decision analysis plays an important role
in many real life problems found in business, economics, management,
governmental and political disputes. The game-theoretic rough set model
(GTRS) is a recent extension to rough set theory for intelligent decision
making observed with game-theoretic formulation. In this article, we
extend GTRS for formulating and analyzing multiple criteria decision
making problems in rough sets. Basic concepts of the model are defined,
reviewed and analyzed in the context of multiple criteria. Applicability
of GTRS is demonstrated by considering different examples, including
multiple criteria effective classification, rule mining and feature selection.

1 Introduction

We are involved in decision making all the time. Majority of these decisions are
based on multiple factors or criteria. Decision makings in such situations are
often aimed at finding a proper balance or tradeoff among a group of considered
criteria. Such a tradeoff may be reached with an analysis of conflict or cooperation among the involved decision criteria. We review, discuss and analyze the
game-theoretic rough set (GTRS) model proposed by Herbert and Yao [6] for
decision making problems when multiple criteria are involved.

In Pawlak rough sets, the universe of objects is partitioned into positive,
negative and boundary regions. The inclusion of an object in positive or negative region is associated with strict conditions which usually result in a larger
boundary. This significantly limits the practical applicability of the model [13].
Probabilistic rough set approaches aim to decrease the boundary region by re-
laxing the extreme conditions used in accepting objects for positive or negative regions [10, 11]. They have been examined in many different forms, such
as decision-theoretic rough set model [12], Bayesian rough set model [9], game-
theoretic rough set model [6] and parameterized rough set model [4]. The extensive results in these studies helped in increasing our understanding and
applicability of rough set theory. The GTRS model interprets the configuration
of probabilistic regions and parameters as a decision making problem among
multiple criteria observed with a game-theoretic process [6]. The importance of
this model is that it allows for simultaneous consideration of multiple criteria in
analyzing and making intelligent decisions with rough sets.

T. Li et al. (Eds.): RSKT 2012, LNAI 7414, pp. 399–408, 2012.

The research on GTRS model has been focused on analyzing specific aspects of the rough set model, e.g. configuration of classification ability or determination of optimal parameters in probabilistic rough sets. There are further problems in rough sets which may be explored with multiple criteria, e.g. multiple criteria based rule mining and feature selection. We present a GTRS based framework for multiple criteria decision making in this article.

2 Overview of Game-Theoretic Rough Sets

Yao [13] proposed the probabilistic rough sets by introducing a pair of thresholds (α, β) for defining three regions. Moreover, the three regions are interpreted by three-way decisions [14, 15]. The decision of acceptance, rejection or deferment for an object in C based on $[x]$ is made as follows,

$$
\begin{array}{lll}
\text{Acceptance:} & \text{if } P(C|[x]) \geq \alpha, \\
\text{Rejection:} & \text{if } P(C|[x]) \leq \beta, \text{ and} \\
\text{Deferment:} & \text{if } \beta < P(C|[x]) < \alpha.
\end{array}
$$

Pawlak rough set model may be viewed as a special case of probabilistic rough set model with extreme conditions of $(\alpha, \beta) = (1, 0)$. The Pawlak model offers insufficient levels of acceptance and rejection and often results in a high level of deferment decisions [14]. This means that definite decisions in the form of acceptance or rejection may be obtained for only a few objects. Another extreme case is when $\alpha = \beta$, which results in a two-way decision model with high levels of acceptance and rejection but no deferment. This means that a definite decision is obtained for all objects, regardless of the amount of available information for making a certain decision.

The determination of a suitable (α, β) pair is one of the key challenges in probabilistic rough sets for obtaining effective decisions [6]. The GTRS model meets this challenge by providing a mechanism for determining a balanced threshold pair that leads to a moderate levels of acceptance, rejection and deferment decisions.

The thresholds in probabilistic rough sets are related to different criteria like classification errors, benefits, cost or risks associated with three-way decisions [3]. The GTRS exploits the relationship between multiple criteria and probabilistic thresholds to formulate a method for obtaining effective three-way decisions. Such a method enables the investigation as to how the probability thresholds may change in order to improve the performance of rough sets from different aspects [5]. For instance, considering the properties of accuracy and generality of the rough set model. These properties may be investigated with different measures as discussed in reference [16]. A more accurate model tends to be lesser general. On the other hand, a general model may not be very accurate. Changing the threshold levels in order to increase one measure may decrease the other. The configuration of thresholds may be realized as a competitive game among these properties.

Table 1. Accuracy versus generality

		Generality	
		$t_1 = h_1(\alpha, \beta)$	$t_2 = h_2(\alpha, \beta)$
Accuracy	$s_1 = f_1(\alpha, \beta)$	$(\alpha_1, \beta_1) \Longrightarrow (a_1, g_1)$	$(\alpha_2, \beta_2) \Longrightarrow (a_2, g_2)$
	$s_2 = f_2(\alpha, \beta)$	$(\alpha_3, \beta_3) \Longrightarrow (a_3, g_3)$	$(\alpha_4, \beta_4) \Longrightarrow (a_4, g_4)$

Table 1 may be used to represent this game where rows correspond to accuracy and columns to generality. By considering different strategies or actions as possible functions of (α, β) (that configures the thresholds in some order), a game may be formulated so that each strategy pair, e.g. $<s_1, t_1>$ leads to a particular threshold pair (α_1, β_1). The values of measures calculated with a pair, say (α_1, β_1), are represented as (a_1, g_1), which highlight the utilities or performance levels of respective measures corresponding to a particular threshold pair. By considering different choices in the form of possible (α, β) pair values, a rough set model may be obtained with game-theoretic analysis that represents a certain level of balance or tradeoff between the two properties. Therefore, the GTRS may help in eventually determining a suitable threshold pair that an entity or criterion should consider for achieving a defined performance level.

The above example highlights the capabilities of GTRS in rough set based decision making and analysis. The model provides a tradeoff mechanism among multiple entities for reaching a possible solution through consideration of different choices or preferences. The basic components in GTRS formulation include information about different criteria considered as players in a game, the strategies or available actions for each player and utilities or payoff functions for each action. Each player competes in a game to configure the probabilistic thresholds by choosing among the actions that benefits their own interests. After analyzing consequences of their opponents actions in the context of their personal interests, a mutually agreed and a balanced solution is obtained.

3   The GTRS Based Framework for Multiple Criteria Decision Analysis

In multiple criteria decision making problems, an effective solution may be investigated through different mechanisms incorporating the concepts, such as conflict resolution and analysis, cooperation and tradeoff analysis or objective minimization. A decision making problem is realized in probabilistic rough sets when different decisions are evaluated in classifying objects for obtaining effective rough set regions. This decision making problem may be formulated with multiple criteria by considering different properties of the rough set regions and classification [3, 7]. The simultaneous consideration of multiple aspects in different decision making problems of rough sets may increase our understanding of the theory by obtaining more interesting results.

A major issue remains as how to create a framework that combines multiple aspects for analyzing and obtaining effective decisions. The incorporation of a game-theoretic mechanism may be useful in investigating such a framework. We now present a general formulation framework with the model that may be utilized in modeling multiple criteria decision making problems in rough sets.

3.1   Multiple Criteria as Players in a Game

The players in GTRS formulation reflect multiple influential factors or criteria that highlight various aspects of a particular decision making problem. The selection of players depends on objectives we seek to achieve within rough sets. For instance, if we are interested in mining effective rules, the measures for evaluating rules may be considered as players. Alternatively, we may want to select features efficiently, this will require the measures for evaluating feature's importance as players in the game. The players provide various competitive or complimentary aspects for analyzing and achieving the same objectives. For analyzing the classification ability of rough sets, the measures of accuracy and generality may provide two different aspects. Individually, each criterion may seek to maximize its own benefits by competing or cooperating with others, while collectively they are incorporated in a framework for accomplishing the overall purpose of the game. Typical criteria may include measures, parameters and variables that affect the decision making process. We will present a few examples to demonstrate the usage of GTRS model in the following sections.

Example of Cooperation among Players: In rough sets based feature selection, multiple measures of a feature's importance may be utilized as possible players of a game. A cooperation among these players may be recognized when we want to obtain a mutual decision on a feature's significance or importance level by accommodating the opinions of all considered measures. Examples of such scenario were considered in [1, 2]. Reference [1] considered the measures of conditional entropy and dependency for evaluating a particular feature's importance. These measures may be defined for a particular attribute A with respect to a decision attribute D as,

$$ConditionalEntropy(D|A) = -\sum_{A_i \in A} P(A_i) \sum_{D_k \in D} P(D_k|A_i) log P(D_k|A_i), \quad (1)$$

$$Dependency(D|A) = \frac{\|POS_A(D)\|}{\|U\|}. \quad (2)$$

A mechanism was introduced with game-theoretic formulation that incorporates the actions of these measures in determining a feature's importance [1].

Example of Competition among Players: A competitive environment among players may be realized when we want to obtain a rough set model that is general and accurate at the same time. The two measures may be defined as [16],

$$Accuracy(A) = \frac{apr(A)}{\overline{apr}(A)}, \quad Generality([x]) = \frac{|[x]|}{|U|}. \tag{3}$$

where $apr(A)$ and $\overline{apr}(A)$ are the rough set based lower and upper approximations for a set A, respectively and $[x]$ represents a particular equivalence class. The two measures may be considered as complimentary to certain extend as discussed in Section 2. A competition among these measures may be considered within a game-theoretic environment for finding a suitable tradeoff point between these measures.

A similar scenario may be realized in mining effective rules. Suppose we are interested in rules belonging to probabilistic positive region of a concept (or category) C, referred to as positive rule set and denoted as PRS_C [16]. Considering the case where mining of rules that are highly accurate and widely applicable are of interest. These criteria may be evaluated with measures of confidence and coverage, defined as,

$$Confidence(PRS_C) = \frac{\text{Number of correctly classified objects by } PRS_C}{\text{Number of classified objects by } PRS_C}, \tag{4}$$

$$Coverage(PRS_C) = \frac{\text{Number of correctly classified objects by } PRS_C}{\text{Number of objects in } C}. \tag{5}$$

The confidence highlights the strength or accuracy of the rules in the PRS_C. The coverage on the other hand determines the generality of the rules in PRS_C. A competitive game environment may be used to analyze these measures.

3.2   Strategies in Multiple Criteria Analysis

Each player should have a set of possible strategies for its effective participation in a game. We may observe multiple variables in decision making problems that affect decision making in different ways. For example, changing the values of loss functions may result in different regions that lead to different decisions on classification of objects [13]. A performance increase or decrease may be observed by a particular criterion in response to these variable changes. This intuitively means that a particular criterion may increase its benefits by selecting an appropriate change in certain variables. Such variables and their associated changes may be realized as actions.

Examples of Strategies for Players: Let us consider strategies in a game between the measures of accuracy and generality in the context of probabilistic rough sets. From Equation 3, we note that the two measures are affected if changes are made to different rough set regions. In probabilistic rough sets, the (α, β) pair defines three regions, various changes in the thresholds may be considered as strategies. This suggests that the strategy set in this case would contain the threshold pairs $\{(\alpha_1, \beta_1), (\alpha_2, \beta_2), (\alpha_3, \beta_3), ...\}$, where each (α_i, β_i) pair represents an available strategy.

3.3 Payoff Functions for Analyzing Strategies

In GTRS formulation, the payoffs are determined as functions of various criteria involved in a decision making problem. For criterion c_i, the payoff for taking action a_j, represented as $u_{c_i}(a_j)$, is calculated as a function of c_i's benefits under action a_j. When measures are considered as criteria for evaluating different decisions, the payoff function $u_{c_i}(a_j)$ may be calculated as a value of measure m_i under action a_j, denoted as $m_i(a_j)$. The gains or benefits achieved by a measure (which is represented in this case as a value of a measure) in response to different actions are treated as payoffs.

Examples of Payoff Function: In the example considered in Section 3.2, the payoff for *Accuracy* in performing action $a_j = (\alpha_j, \beta_j)$ may be represented as,

$$u_{Accuracy(A)}(a_j) = \frac{\underline{apr}_{(\alpha_j, \beta_j)}(A)}{\overline{apr}_{(\alpha_j, \beta_j)}(A)}. \tag{6}$$

where $\underline{apr}_{(\alpha_j, \beta_j)}(A)$ and $\overline{apr}_{(\alpha_j, \beta_j)}(A)$ reprsents the probabilistic lower and upper approximations with respect to threshold pair (α_j, β_j).

3.4 Implementing Competition for Effective Solutions

We need to express the game as a competition or corporation among the measures. This is facilitated with a payoff table. A payoff table lists all possible actions with their respective payoffs. Table 2 presents a payoff table for a two player game. The rows represent the actions of player p_1 and columns player p_2. Each entry in the table represents a payoff pair $<u_{p_1(a_i|a_j)}, u_{p_2(a_j|a_i)}>$ corresponding to action a_i of p_1 and a_j of p_2. The players may select from n possible actions. Once the payoffs are calculated with suitable payoff functions, the table may be examined for a possible solution.

In order to examine the outcome of the game, GTRS finds an equilibrium, e.g. Nash equilibrium within the payoff table [8]. This suggests that none of the players can be benefited by changing its strategy, given the opponent's chosen action. In the above two player game, the pair $<u^*_{p_1(a_i|a_1)}, u^*_{p_2(a_j|a_2)}>$ is an equilibrium if for any action a_k, $(k \neq i, j)$,

$$u^*_{p_1(a_i|a_1)} \geq u_{p_1(a_k|a_1)} \text{ and } u^*_{p_2(a_j|a_2)} \geq u_{p_2(a_k|a_2)}. \tag{7}$$

The pair $<u^*_{p_1(a_i|a_1)}, u^*_{p_2(a_j|a_2)}>$ is an optimal solution for determining the actions of measures.

Table 2. The payoff table

		p_2		
		a_1	.....	a_n
p_1	a_1	$u_{p_1}(a_1\|a_1), u_{p_2}(a_1\|a_1)$	.....	$u_{p_1}(a_1\|a_n), u_{p_2}(a_n\|a_1)$
	.....	.....	.....	.....
	a_n	$u_{p_1}(a_n\|a_1), u_{p_2}(a_1\|a_n)$	.....	$u_{p_1}(a_n\|a_n), u_{p_2}(a_n\|a_n)$

4 A Confidence versus Coverage Game Example

The main ideas of the presented framework are illustrated by considering the example game of finding effective rules discussed in Section 3. The confidence and coverage of the rules belonging to PRS_C may be calculated with Equation 4 and 5. The PRS_C of the Pawlak rough set model have a confidence equal to 1 [16]. This means that we can make 100% accurate predictions with the PRS_C. By weakening the requirement of confidence being 1 in the Pawlak positive rule set, one expects to increase the coverage of PRS_C. This represents a situation where an inverse relationship holds between the two measures. Suitable tradeoff analysis may help in exploring an efficient solution in this case.

Table 3. Probabilistic information of a concept C

	X_1	X_2	X_3	X_4	X_5	X_6	X_7	X_8
$Pr(X_i)$	0.0277	0.0985	0.1322	0.0167	0.0680	0.0169	0.0598	0.0970
$Pr(C\|X_i)$	1.0	1.0	0.98	0.95	0.93	0.90	0.82	0.71
	X_9	X_{10}	X_{11}	X_{12}	X_{13}	X_{14}	X_{15}	
$Pr(X_i)$	0.1150	0.0797	0.0998	0.1190	0.0189	0.0420	0.0088	
$Pr(C\|X_i)$	0.63	0.55	0.34	0.21	0.13	0.0	0.0	

Let us consider the mining of effective rules with an example. Suppose Table 3 represents probabilistic information about a category or concept C with respect to a partition consisting of 15 equivalence classes. A particular equivalence class is represented as X_i, where $i = 1, 2, 3, ..., 15$. For convenience of computations, the equivalence classes are listed in decreasing order of their conditional probabilities $P(C|X_i)$. The equivalence classes may be used to generate rules. For instance, a positive rule corresponding to an equivalence class X_i is defined as,

$$X_i \rightarrow C, \text{ where } X_i \subseteq POS_{(\alpha, \beta)}(C), \tag{8}$$

where $POS_{(\alpha,\beta)}$ represents probabilistic positive region corresponding to threshold pair (α, β) [13]. Let $U = 500$ be total number of objects, then,

$$\text{objects classified by the rule } X_i \to C = P(X_i) \times U, \tag{9}$$

$$\text{objects correctly classified by the rule } X_i \to C = P(C \cap X_i) \times U =$$

$$\{P(C \cap X_i)/P(X_i)\} \times P(X_i) \times U = P(C|X_i) \times P(X_i) \times U, \text{ and} \tag{10}$$

$$\text{number of objects in category } C = \sum_{i=1}^{15} P(X_i) \times P(C|X_i) \times U \tag{11}$$

Equations 9-11 may be used to calculate measures of confidence and coverage defined in Equations 4-5. For instance, in case of probabilistic thresholds $(\alpha, \beta) = (1, 0)$, we have the Pawlak positive region $POS_{(1,0)} = \bigcup\{X_1, X_2\}$. The PRS_C in this case will contain rules generated from X_1 and X_2. The measure of confidence and coverage may be calculated as,

$$\begin{aligned} Confidence(PRS_C) &= \frac{\sum_{i=1}^{2} P(C|X_i) \times P(X_i) \times U}{\sum_{i=1}^{2} P(X_i) \times U} \\ &= \frac{(1.0 \times 0.0277 + 1 \times 0.0985) \times 500}{(0.0277 + 0.0985) \times 500} = 1.0 \end{aligned} \tag{12}$$

$$\begin{aligned} Coverage(PRS_C) &= \frac{\sum_{i=1}^{2} P(C|X_i) \times P(X_i) \times U}{\sum_{i=1}^{15} P(X_i) \times P(C|X_i) \times U} \\ &= \frac{(1.0 \times 0.0277 + 1 \times 0.0985) \times 500}{0.6456 \times 500} = 0.1955 \end{aligned} \tag{13}$$

This means that with the Pawlak model we can generate stronger positive rules that are 100% accurate but may be applicable to only 19.55% of the cases. Let us calculate the considered measures corresponding to different values of α. For instance, the pair $(Confidence(PRS_C), Coverage(PRS_C))$ containing values of considered measures, corresponding to α values of $\{0.9, 0.8, 0.7, 0.6, 0.5\}$ may be calculated as $(0.9724, 0.5422)$, $(0.9507, 0.6182)$, $(0.9055, 0.7248)$, $(0.8554, 0.8370)$ and $(0.8212, 0.9049)$, respectively. We may note that a change in α produces conflicting responses among the measures. The GTRS model may allow us to explore a possible solution among the measures in this case.

Let us utilize the formulation framework introduced in Section 3 for mining effective rules. Considering the measures of *Confidence* and *Coverage* as players in game with strategies for players formulated as possible decreases to threshold α. Suppose three actions are considered, namely, $\alpha_1 = $ no decrease, $\alpha_2 = $ a moderate decrease and $\alpha_3 = $ an aggressive decrease to threshold α. We consider an aggressive decrease as 10% and a moderate decrease as 5%. The α value corresponding to a particular action pair $<\alpha_i, \alpha_j>$ is calculated as an average, i.e. $\alpha = \{\alpha_i + \alpha_j\}/2$. Table 4 shows the payoff table corresponding to different action pairs with a starting values of thresholds $(\alpha, \beta) = (1, 0)$. The cell in bold

Table 4. Payoff table for the example game

		$Coverage(PRS_C)$		
		α_1	α_2	α_3
	α_1	$1.0, 0.1955$	$0.9898, 0.3961$	**0.9874, 0.4207**
$Confidence(PRS_C)$	α_2	$0.9898, 0.3961$	$0.9874, 0.4207$	$0.9760, 0.5186$
	α_3	$0.9874, 0.4207$	$0.9760, 0.5186$	$0.9724, 0.5422$

represents equilibrium which corresponds to action pair $<\alpha_1, \alpha_3>$. This means that none of the players can achieve a higher payoff, given the other players chosen action. For instance, changing the action of $Coverage$ from α_3 to α_2 and α_1 will result in a payoff decrease from 0.4207 to 0.3961 and 0.1955, respectively.

Let us update the value of α as an average of the values in pair $<\alpha_1, \alpha_3>$ that corresponds to an equilibrium in Table 4. The game my be repeated several times based on updated value of α until the measures fall in some predefined acceptable range. For instance, the next round may be played with an updated value of $\alpha = 0.95$. This example suggests that the GTRS based formulation may be introduced for effective modeling of multiple criteria decision making problems in rough sets.

5   Conclusion

Multiple criteria analysis plays an important role in group decision making. Game-theoretic rough sets is a new model in rough sets for analyzing and making intelligent decisions observed with a game-theoretic process. The model was utilized in probabilistic rough sets for configuring the overall classification ability in order to obtain effective region sizes and three-way decisions. In this article, we extended the model for analyzing multiple criteria decision making problems in rough sets with simultaneous consideration of multiple influential factors. In particular, we present a framework which may be utilized in modeling further decision making problems in rough sets. The proposed framework may extend the current applicability of the model to interesting real world applications. Possible application areas may include feature selection, rule mining and multi-agent decision making.

Acknowledgments. This work was partially supported by a Discovery Grant from NSERC Canada and the University of Regina FGSR Dean's Scholarship Program.

References

[1] Azam, N., Yao, J.T.: Classifying Attributes with Game-Theoretic Rough Sets. In: Watada, J., Watanabe, T., Phillips-Wren, G., Howlett, R.J., Jain, L.C. (eds.) Intelligent Decision Technologies. SIST, vol. 15, pp. 175–184. Springer, Heidelberg (2012)

[2] Azam, N., Yao, J.T.: Game-theoretic Rough Sets for Feature Selection. In: Skowron, A., Suraj, Z. (eds.) Rough Sets and Intelligent Systems - Professor Zdzisław Pawlak in Memoriam. ISRL, vol. 43, pp. 61–78. Springer, Heidelberg (2012)

[3] Deng, X.F., Yao, Y.Y.: An Information-Theoretic Interpretation of Thresholds in Probabilistic Rough Sets. In: Li, T., Nguyen, H.S., Wang, G., Grzymala-Busse, J., Janicki, R., Hassanien, A.E., Yu, H. (eds.) RSKT 2012. LNCS (LNAI), vol. 7414, pp. 369–378. Springer, Heidelberg (2012)

[4] Greco, S., Matarazzo, B., Slowinski, R.: Parameterized rough set model using rough membership and bayesian confirmation measures. International Journal of Approximate Reasoning 49(2), 285–300 (2008)

[5] Herbert, J.P., Yao, J.T.: Analysis of Data-Driven Parameters in Game-Theoretic Rough Sets. In: Yao, J., Ramanna, S., Wang, G., Suraj, Z. (eds.) RSKT 2011. LNCS, vol. 6954, pp. 447–456. Springer, Heidelberg (2011)

[6] Herbert, J.P., Yao, J.T.: Game-theoretic rough sets. Fundamenta Informaticae 108(3-4), 267–286 (2011)

[7] Jia, X.Y., Li, W.W., Shang, L., Chen, J.J.: An Optimization Viewpoint of Decision-Theoretic Rough Set Model. In: Yao, J., Ramanna, S., Wang, G., Suraj, Z. (eds.) RSKT 2011. LNCS, vol. 6954, pp. 457–465. Springer, Heidelberg (2011)

[8] Neumann, J.V., Morgenstern, O.: Theory of Games and Economic Behavior. Princeton University Press (1944)

[9] Slezak, D., Ziarko, W.: The investigation of the bayesian rough set model. International Journal of Approximate Reasoning 40(1-2), 81–91 (2005)

[10] Yao, J.T., Yao, Y.Y., Ziarko, W.: Probabilistic rough sets: Approximations, decision-makings, and applications. International Journal of Approximate Reasoning 49(2), 253–254 (2008)

[11] Yao, Y.Y.: Probabilistic approaches to rough sets. Expert Systems 20(5), 287–297 (2003)

[12] Yao, Y.Y.: Decision-Theoretic Rough Set Models. In: Yao, J., Lingras, P., Wu, W.-Z., Szczuka, M.S., Cercone, N.J., Ślęzak, D. (eds.) RSKT 2007. LNCS (LNAI), vol. 4481, pp. 1–12. Springer, Heidelberg (2007)

[13] Yao, Y.Y.: Probabilistic rough set approximations. International Journal of Approximate Reasoning 49(2), 255–271 (2008)

[14] Yao, Y.Y.: The superiority of three-way decisions in probabilistic rough set models. Information Sciences 181(6), 1080–1096 (2011)

[15] Yao, Y.Y.: An Outline of a Theory of Three-Way Decisions. In: Yao, J.T., Yang, Y., Slowinski, R., Greco, S., Li, H., Mitra, S., Polkowski, L. (eds.) RSCTC 2012. LNCS (LNAI), vol. 7413, pp. 1–17. Springer, Heidelberg (2012)

[16] Yao, Y.Y., Zhao, Y.: Attribute reduction in decision-theoretic rough set models. Information Sciences 178(17), 3356–3373 (2008)

Granularity Analysis of Fuzzy Soft Set

Dengbao Yao[1], Cuicui Wang[1], Junjun Mao[1,2,3], and Yanping Zhang[2]

[1] School of Mathematical Sciences, Anhui University, Hefei 230039, P.R. China
[2] Key Laboratory of Intelligent Computing and
Signal Processing of Ministry of Education,
Anhui University, Hefei 230039, P.R. China
[3] Department of Computer Science, University of Houston,
Houston, TX 77204–3010, USA
maojunjunuh@gmail.com

Abstract. Fuzzy soft set (FSs) is a novel mathematical tool, in order to describe and measure uncertain information of FSs perfectly, granularity analysis based on covering about FSs is originally discussed in this paper. Firstly, the α-dominance relation between any two objects is built by constructing the possibility degree or the weighted possibility degree after standardization, thus α-dominance class and α-covering approximation space could be generated on this relation. Secondly, granular capacity is proposed to measure the granular information of variable precision through introducing concepts of the description set and the indistinguishability set, Finally, an illustrative example is analysed to reflect dynamic changes of uncertain information under different granular structure.

Keywords: FSs, α-dominance relation, possibility degree, granular capacity, dynamic.

1 Introduction

In order to overcome inherent limitations of other uncertain theories(include probability theory, fuzzy set theory [1], rough set theory [2], etc.), Molodtsov [3] originated the soft set theory in 1999, which is a general mathematical tool for dealing with uncertain information. Maji, et al. [4,5] discussed many operations of soft set and firstly applied soft set theory in decision making. In fact, soft set can be combined with other uncertain theories, such as Roy and Maji [6] originally extended the soft set to fuzzy soft set (FSs) and presented a theoretic approach to cope with decision making problems. However, more or less there inevitably exists advantages and limitations in the present decision approaches. Kong, et al. [7] gave a counterexample to explain that one of algorithms in [6] is incorrect. Feng, et al. [8] and Ali [9] studied the relation among the soft set, fuzzy set and rough set, respectively, and proposed a tentative approach for decision-making. Majumda and Samanta constructed similarity measures in fuzzy soft set in [10]. Feng, et al. [11] gave deeper insights into decision making on fuzzy soft set, and firstly proposed an adjustable approach by using level soft set. Similar to [11], Jiang, et al. [12] extended this adjustable approach to intuitionistic fuzzy soft set.

T. Li et al. (Eds.): RSKT 2012, LNAI 7414, pp. 409–415, 2012.

Nowadays, few people have made granularity analysis for FSs, the main reason may be that it's hard to find an equivalence relation or preference relation to generate granular structure. Inspired by [13], which investigated some problems of granular computing based on covering models, we try to construct the α-dominance classes by introducing the possibility degree between any two objects in universe, and these classes institute a covering of universe. Based on the covering, we obtain the α-covering approximation space, thus the granular structure of the space could be measured by calculating granular capacity.

2   Preliminaries

Throughout this paper, let U denote a non-empty finite set, called initial universal set and E is a set of parameters on U, which is often the set of attributes, characters, etc.

Definition 2.1.([3]) (Soft set) Let $P(U)$ denote all subset of U, a pair (F, E) is called a soft set over U, where $F : E \to P(U)$ is a mapping from parameter set E to $P(U)$, namely, for $\forall e \in E, x \in U, F(e)(x) \in \{0, 1\}$.

Definition 2.2.([6]) (FSs) Let $F(U)$ denote the set of all fuzzy subset of U , a pair (F, E) is called a fuzzy soft set over $F(U)$, where $F : E \to F(U)$ is a mapping from parameter set E to $F(U)$, namely, for $\forall e \in E, F(e) = \{< x, \mu_{F(e)}(x) >: x \in U\} \in F(U)$.

In fact, a FSs over U is a parameterized family of subsets of U. Namely, for $\forall e \in E, F(e)$ could be considered as the set of e-elements of FSs (F, E). According to characters of the FSs, every FSs could be represented in form of a matrix, so a FSs (F, E) can be described as $(F, E) = [\mu_{ij}]_{m \times n} = [F(e_j)(x_i)]_{m \times n}$.

3 α-Dominance Class on the Possibility Degree of FSs

3.1   Standardized Methods of the Fuzzy Soft Set

Since different attributes have different dimensions, so we need conduct a pretreatment for every FSs. Suppose $(F, E) \triangleq [\mu_{ij}]_{m \times n}$ is a FSs, and $E = \{e_1, e_2, \ldots, e_n\}$ is the set of attributes, $U = \{x_1, x_2, \ldots, x_m\}$ is the object set. Let matrix $R = [r_{ij}]_{m \times n}$ be the standardization of (F, E), where r_{ij} satisfies $r_{ij} = \dfrac{\mu_{ij}}{\sum_{i=1}^{m} \mu_{ij}}(e_j$

is benefit-type) or $\dfrac{\mu_{ij}^{-1}}{\sum_{i=1}^{m} \mu_{ij}^{-1}}(e_j$ is cost-type).

3.2   The Possibility Degree of FSs

Definition 3.1. (Possibility degree) Suppose $R = [r_{ij}]_{m \times n}$ is the standardization of (F, E), the possibility degree of the object x_i is superior to x_k is defined by

$$P(x_i \geq x_k) = \frac{1}{n} \sum_{j=1}^{n} \{I(r_{ij} > r_{kj}) + \frac{1}{2} I(r_{ij} = r_{kj})\}.$$

where $i, k = 1, 2, \ldots, m$. $I(\cdot)$ is an indicator function.

In fact, the coefficient $1/n$ may be viewed as attribute weights of E. Inspired by this notion, the weighted possibility degree can be defined as follow.

Definition 3.2. (Weighted possibility degree) Suppose $R = [r_{ij}]_{m \times n}$, (F, E), $E = \{e_1, e_2, \ldots, e_n\}$ is attributes set, $\omega = \{\omega_1, \omega_2, \ldots, \omega_n\}$, $\sum_{j=1}^{n} \omega_j = 1, 0 \leq \omega_j \leq 1, j = 1, 2, \ldots, n$ are attribute weights. The weighted possibility degree of the object x_i is superior to x_k is defined by

$$P_\omega(x_i \geq x_k) = \sum_{j=1}^{n} \omega_j \{I(r_{ij} > r_{kj}) + \frac{1}{2}I(r_{ij} = r_{kj})\}.$$

Proposition 3.1. $P(x_i \geq x_k)$ and $P_\omega(x_i \geq x_k)$ satisfy following properties.
(1) $0 \leq P(x_i \geq x_k) \leq 1, 0 \leq P_\omega(x_i \geq x_k) \leq 1$;
(2) $P(x_i \geq x_k) = \frac{1}{2}, P_\omega(x_i \geq x_k) = \frac{1}{2}$;
(3) $P(x_i \geq x_k) + P(x_i \leq x_k) = 1, P_\omega(x_i \geq x_k) + P_\omega(x_i \leq x_k) = 1$.

Proof. The proof could be obtained easily according to definition 3.1 and 3.2.

3.3 α-Dominance Class and α-Covering Approximation Space

According to section 3.2, the (or weighted) possibility degree between any two objects of U could be constructed, then we can obtain a (or weighted) possibility degree matrix $P = [P_{ki}]_{m \times m} \triangleq [P(x_i \geq x_k)]_{m \times m}$ (or $P = [P_{ki}^\omega]_{m \times m} \triangleq [P_\omega(x_i \geq x_k)]_{m \times m}$). For convenience, we take the possibility degree for example in following discussion. In fact, the possibility degree measures the dominance relation between two objects, and every dominance relation could generate dominance class for a given universe.

Definition 3.3. Suppose P is the possibility degree matrix of (F, E), $\alpha(\in [0, 1])$ is a constant, for every $x_k \in U$, the α-dominance class $[x_k]^{\geq \alpha}$ of the object x_k can be defined by $[x_k]^{\geq \alpha} = \{x_i \in U : P(x_i \geq x_k) \geq \alpha\}$.

Obviously, every α-dominance class $[x_k]^{\geq \alpha}(k = 1, 2, \ldots, m)$ is a subset of U, and all α-dominance classes of objects constitute a covering of U, namely $U = \cup_{k=1}^{m}[x_k]^{\geq \alpha}$, we call $C_\alpha = \{[x_k]^{\geq \alpha} : x_k \in U\}_{k=1}^{m}$ is α-covering of U, and (U, C_α) constitute a α-covering approximation space.

4 Variable Precision Granularity Analysis Based on α-Covering Approximation Space

Actually, there usually exist some overlaps in the granulation of practical problems, traditional partition model cannot deal with them, while the covering model may play an essential role in some respects. In α-covering approximation space, objects of U construct particles by α-dominance relation. For a element in U, granularity level may change with adjusting α, this suggests uncertainty is decided by the α-dominance relation in covering granular space, and granular structure changes along with α. So we construct some measure methods to describe these dynamic granular structure.

Definition 4.1. Let (U, C_α) be a α-covering approximation space, $\alpha \in [0,1]$, for $\forall x \in U$, the description set of x in(U, C_α) is defined by $Ind_{C_\alpha}(x) = \cap\{K : K \in C_\alpha \wedge x \in K\}$. The indistinguishability set of x in $(U, C_\alpha)$is defined by $Ind_{C_\alpha}(x) = \cap\{K : K \in Des_{C_\alpha}(x)\}$.

Definition 4.2. Let (U, C_α) be a α-covering approximation space, $\alpha \in [0,1]$, where U is non-empty finite objects set called universe, $C_\alpha = \{[x_k]^{\geq \alpha} : x_k \in U\}_{k=1}^m$ is a α-covering of U,the granular capacity of C_α is defined by

$$M(C_\alpha) = \begin{cases} 0 & |[x_k]^{\geq \alpha}| = 1, k = 1, 2, \ldots, m \\ \sum_{x_k \in U} |Ind_{C_\alpha}(x_k)|/|U|^2 & otherwise \end{cases}.$$

When the granularity of α-covering C_α is the thickest, namely $C_\alpha = \{U\}$, it is maximum value 1 of the α-covering's granular capacity, this shows that it contains the most uncertain information. When the granularity of C_α is the thinnest, namely $|[x_k]^{\geq \alpha}| = 1, k = 1, 2, \ldots, m$, it is minimum value 0 of α-covering's granular capacity, and suggests the uncertainty of universe U is the weakest.

Proposition 4.1. Let (U, C_α) and (U, C_β) be two covering approximation spaces, $\alpha, \beta \in [0,1]$, if for $\forall x \in U, Ind_{C_\alpha}(x) = Ind_{C_\beta}(x)$, then $M(C_\alpha) = M(C_\beta)$.

Definition 4.3. Let (U, C_α) and (U, C_β)be two covering approximation spaces, $\alpha, \beta \in [0,1]$, if for $\forall K \in C_\alpha$,there exist $S \in C_\beta$ make $K \subset S$ be true, and for $\forall S \in C_\beta$, there exist $K \in C_\alpha$ make $K \subset S$ be true, then we call covering C_α is thinner than C_β, denoted by $C_\alpha \leq C_\beta$.

Proposition 4.2. Let C_α and C_β be two coverings, if $\alpha > \beta$, then $C_\alpha \leq C_\beta$.

Proposition 4.3. Let (U, C_α) and (U, C_β) be two covering approximation spaces, $\alpha, \beta \in [0, 1]$, if $\alpha \geq \beta$, then $M(C_\alpha) \leq M(C_\beta)$.

Proof. By $\alpha \geq \beta$, we have $C_\alpha \leq C_\beta$. For $\forall K \in C_\alpha, \exists S \in C_\beta, s.t. K \subset S$, and $\forall S \in C_\beta, \exists K \in C_\alpha, s.t. K \subset S$. Hence $\forall x_k \in U$, we have $Des_{C_\alpha}(x_k) \subset Des_{C_\beta}(x_k)$. So $Ind_{C_\alpha}(x_k) \subset Ind_{C_\beta}(x_k)$, then $|Ind_{C_\alpha}(x_k)| \leq |Ind_{C_\beta}(x_k)|$, namely $\sum_{x_k \in U} |Ind_{C_\alpha}(x_k)|/|U|^2 \leq \sum_{x_k \in U} |Ind_{C_\beta}(x_k)|/|U|^2$, therefore $M(C_\alpha) \leq M(C_\beta)$.

Proposition 4.3 indicates that by adjusting values of α, we could obtain changes of granularity under variable precision, and make dynamic analysis for uncertain information.

5   Illustrative Examples

Let (F, E) be a FSs over U, where $E = \{e_1, e_2, \ldots, e_6\}$is attributes set, $U = \{x_1, x_2, \ldots, x_5\}$is objects set (Table 1).

$$P = \begin{pmatrix} 1/2 & 1/3 & 7/12 & 5/12 & 1/3 \\ 2/3 & 1/2 & 1/2 & 7/12 & 7/12 \\ 5/12 & 1/2 & 1/2 & 1/2 & 2/3 \\ 7/12 & 5/12 & 1/2 & 1/2 & 2/3 \\ 2/3 & 5/12 & 1/3 & 1/3 & 1/2 \end{pmatrix}$$

For the sake of convenience, suppose all attributes in are benefit-type, after standardization, we obtain the matrix $R = [r_{ij}]_{m \times n}$ (Table 2).

According to definition 3.1, by computing the possibility degree between any two objects in U, we obtain the possibility degree matrix P.

Table 1. The value of (F, E)

	e_1	e_2	e_3	e_4	e_5	e_6
x_1	0.5000	0.6000	0.3000	0.6000	0.7000	0.9000
x_2	0.7000	0.5000	0.7000	0.4000	0.5000	0.6000
x_3	0.5000	0.2000	0.6000	0.9000	0.8000	0.7000
x_4	0.7000	0.3000	0.7000	0.6000	0.5000	0.6000
x_5	0.6000	0.5000	0.3000	0.7000	0.7000	0.8000

Table 2. The value of R

	e_1	e_2	e_3	e_4	e_5	e_6
x_1	0.1667	0.2857	0.1154	0.1875	0.2187	0.2500
x_2	0.2333	0.2381	0.2692	0.1250	0.1563	0.1667
x_3	0.1667	0.0952	0.2308	0.2813	0.2500	0.1944
x_4	0.2333	0.1429	0.2692	0.1875	0.1563	0.1667
x_5	0.2000	0.2381	0.1154	0.2187	0.2187	0.2222

When $\alpha = 0.4$, we could calculate 0.4-dominance class and the description set of every object(Table 3).

Table 3. 0.4-dominance class and the description set

0.4-dominance class	the description set
$[x_1]^{\geq 0.4} : \{x_1, x_3, x_4\}$	$Des_{C_{0.4}}(x_1) : [x_1]^{\geq 0.4}, [x_2]^{\geq 0.4}, [x_3]^{\geq 0.4}, [x_4]^{\geq 0.4}, [x_5]^{\geq 0.4}$
$[x_2]^{\geq 0.4} : \{x_1, x_2, x_3, x_4, x_5\}$	$Des_{C_{0.4}}(x_2) : [x_2]^{\geq 0.4}, [x_3]^{\geq 0.4}, [x_4]^{\geq 0.4}, [x_5]^{\geq 0.4}$
$[x_3]^{\geq 0.4} : \{x_1, x_2, x_3, x_4, x_5\}$	$Des_{C_{0.4}}(x_3) : [x_1]^{\geq 0.4}, [x_2]^{\geq 0.4}, [x_3]^{\geq 0.4}, [x_4]^{\geq 0.4}$
$[x_4]^{\geq 0.4} : \{x_1, x_2, x_3, x_4, x_5\}$	$Des_{C_{0.4}}(x_4) : [x_1]^{\geq 0.4}, [x_2]^{\geq 0.4}, [x_3]^{\geq 0.4}, [x_4]^{\geq 0.4}$
$[x_5]^{\geq 0.4} : \{x_1, x_2, x_5\}$	$Des_{C_{0.4}}(x_5) : [x_2]^{\geq 0.4}, [x_3]^{\geq 0.4}, [x_4]^{\geq 0.4}, [x_5]^{\geq 0.4}$

By definition 4.1, calculate the indistinguishability set of every object as follows:

$Ind_{C_{0.4}}(x_1) = \{x_1\}, Ind_{C_{0.4}}(x_2) = \{x_2, x_3, x_5\}, Ind_{C_{0.4}}(x_3) = \{x_2, x_3, x_4\},$
$Ind_{C_{0.4}}(x_4) = \{x_1, x_3, x_4\}, Ind_{C_{0.4}}(x_5) = \{x_1, x_2, x_5\}.$

According to definition 4.4, $M(C_{0.4}) = (1 + 3 + 3 + 3 + 3)/36 = 13/36$.

Similar to above discussion, when $\alpha = 0.5$, calculate 0.5-dominance class and the description set (Table 4).

Table 4. 0.5-dominance class and the description set

0.5-dominance class	the description set
$[x_1]^{\geq 0.5} : \{x_1, x_3\}$	$Des_{C_{0.5}}(x_1) : [x_1]^{\geq 0.5}, [x_2]^{\geq 0.5}, [x_4]^{\geq 0.5}, [x_5]^{\geq 0.5}$
$[x_2]^{\geq 0.5} : \{x_1, x_2, x_3, x_4, x_5\}$	$Des_{C_{0.5}}(x_2) : [x_2]^{\geq 0.5}, [x_3]^{\geq 0.5}$
$[x_3]^{\geq 0.5} : \{x_2, x_3, x_4, x_5\}$	$Des_{C_{0.5}}(x_3) : [x_1]^{\geq 0.5}, [x_2]^{\geq 0.5}, [x_3]^{\geq 0.5}, [x_4]^{\geq 0.5}$
$[x_4]^{\geq 0.5} : \{x_1, x_3, x_4, x_5\}$	$Des_{C_{0.5}}(x_4) : [x_1]^{\geq 0.5}, [x_2]^{\geq 0.5}, [x_3]^{\geq 0.5}, [x_4]^{\geq 0.5}$
$[x_5]^{\geq 0.5} : \{x_1, x_5\}$	$Des_{C_{0.5}}(x_5) : [x_2]^{\geq 0.5}, [x_3]^{\geq 0.5}, [x_4]^{\geq 0.5}, [x_5]^{\geq 0.5}$

Calculate the indistinguishability set as follows:
$Ind_{C_{0.5}}(x_1) = \{x_1\}, Ind_{C_{0.5}}(x_2) = \{x_2, x_3, x_4, x_5\}, Ind_{C_{0.5}}(x_3) = \{x_3\},$
$Ind_{C_{0.5}}(x_4) = \{x_3, x_4, x_5\}, Ind_{C_{0.5}}(x_5) = \{x_5\}.$ Thus $M(C_{0.5}) = (1+4+1+3+1)/36 = 5/18.$

Since $0.4 < 0.5$, the result is $M(C_{0.4}) < M(C_{0.5})$, the conclusion of proposition 4.3 has been checked. On the other hand, it suggests that the uncertain information of covering $C_{0.4}$ is less than $C_{0.5}$, that is to say, the granularity of the space $(U, C_{0.4})$ is thinner than $(U, C_{0.5})$.

6   Conclusions

Granular computing problems of FSs have been investigated in this paper. In order to build α-dominance relation, we present a new measure called the(or weighted) possibility degree among objects in universe, and this relation could generate α-dominance class and α-covering approximation space. In order to observe granular changes under different precision (by adjusting values of α), the concept of granular capacity is proposed. The example shows that the granular capacity becomes smaller along with the granularity of covering becomes thinner.

Acknowledgments. The work is supported by the NNSF of China (No. 61175046), and the Academic Innovation Team of Anhui University (No. KJTD001B) and the Project of Graduate Academic Innovation of Anhui University (Nos. yfc100017, yfc100018).

References

1. Zadeh, L.A.: Fuzzy sets. Inform. Cont. 8, 338–353 (1965)
2. Pawlak, Z.: Rough sets. Int. J. Inform. Comput. Sci. 11, 341–356 (1982)
3. Molodtsov, D.: Soft set theory-First results. Comput. Math. Appl. 37, 19–31 (1999)
4. Maji, P.K., Biswas, R., Roy, A.R.: Soft sets theory. Comput. Math. Appl. 45, 555–562 (2003)
5. Maji, P.K., Roy, A.R.: Application of soft sets in a decision making problem. Comput. Math. Appl. 44, 1077–1083 (2002)
6. Roy, A.R., Maji, P.K.: A fuzzy soft set theoretic approach to decision making problems. J. Comput. Appl. Math. 203, 412–418 (2007)

7. Kong, Z., Gao, L.Q., Wang, L.F.: Comment on a fuzzy soft set theoretic approach to decision making problems. J. Comput. Appl. Math. 223, 540–542 (2009)
8. Feng, F., Li, C.X., et al.: Soft sets combined with fuzzy sets and rough sets: A tentative approach. Soft. Comput. 14, 899–911 (2010)
9. Ali, M.I.: A note on soft sets, rough soft sets and fuzzy soft sets. Appl. Soft. Comput. 11, 3329–3332 (2011)
10. Majumdar, P., Samanta, S.K.: On similarity measures of fuzzy soft sets. Int. J. Adv. Soft. Comput. Appl. 2, 1–8 (2011)
11. Feng, F., Jun, Y.B., et al.: An adjustable approach to fuzzy soft set based decision making. J. Comput. Appl. Math. 234, 10–20 (2010)
12. Jiang, Y.C., Tang, Y., Chen, Q.M.: An adjustable approach to intuitionistic fuzzy soft sets based decision making. Appl. Math. Model 35, 824–836 (2011)
13. She, Y.H., Wang, G.J.: Covering model of granular computing. J. Softw. 21, 2782–2789 (2011)

Granular Approach in Knowledge Discovery[*]

Real Time Blockage Management in Fire Service

Adam Krasuski[1,2], Karol Kreński[1], Piotr Wasilewski[2], and Stanisław Łazowy[1]

[1] The Main School of Fire Service
ul. Słowackiego 52/54, 01-629 Warsaw, Poland
[2] Institute of Mathematics, University of Warsaw
ul. Banacha 2, 02-097 Warsaw, Poland
{krasuski,krenski,lazowy}@inf.sgsp.edu.pl,
piotr@mimuw.edu.pl

Abstract. This article is focused on the recognition and prediction of blockages in the fire stations using granular computing approach. Blockage refers to the situation when all fire units are out and a new incident occurs. The core of the method is an estimation of the expected return times for the fire brigades based on the granularisation of source data. This estimation, along with some other considerations allows for evaluation of the probability of the blockage.

Keywords: granular computing, probability estimating, regression, fire service.

1 Introduction

Relocation of the fire&rescue units among the fire stations in real time requires quick decisions based on an accurate prediction of future events. The wrong decisions may result not only in economic losses but also in a large number of casualties. Inappropriate management of the units deployment could also disorganise the rescue system on higher level and decrease its potential. All those factors make real time units relocation difficult. There is a need for computer systems to support the decision makers.

This article is focused on handling of the blockages in the fire stations using granular computing approach. The main contribution of the paper is the proposal of an effective method for managing the dynamic deployment of the fire&rescue units across the fire stations. The core of our method is estimating of the expected return times of the absent fire brigades which, along with the distributions of the emergencies against the hours of the day and the distribution of the emergencies per day, allows for the evaluation of the probability of the blockage. The method also allows to predict the return times of the given units and consequently to decide whether the support from other fire stations is necessary.

[*] The research presented in this paper has been supported by the individual research project awarded to the third author and realized within Homing Plus programme, edition 3/2011, of Foundation for Polish Science, co-financed from European Union, Regional Development Fund and by the grant 2011/01/D/ST6/06981 from the Polish National Science Centre.

T. Li et al. (Eds.): RSKT 2012, LNAI 7414, pp. 416–421, 2012.
© Springer-Verlag Berlin Heidelberg 2012

2   Granular Computing Approach in Solving the Problem

We approach the problem of blockage recognition and prediction in Fire Service using the paradigm of granular computing (GrC) [1]. Within this paradigm, the world is described in terms of the structured knowledge being a result of perception. In a hierarchical process of perception the acquired evidences are arranged together in forms of information granules due to their similarity, indiscernibility or functionality [2]. Granules can be organized in hierarchical networks [3]. They can have different complexity in form of the patterns, classifiers, parts of patterns, coalitions of patterns or their parts, coalitions of classifiers. Granules from higher levels of hierarchy are the abstracted or combined granules from the lower levels of the hierarchy. Granules can be used to create or to approximate complex concepts providing the basis for knowledge formation. Roughly speaking granular computing explores multiple levels of granularity in problem solving. The levels of granularity may be interpreted as the levels of abstraction, detail, complexity, and control in specific contexts.

Basically, our approach to the problem of blockage in the fire units is as follows: a large collection of detailed data from Incident Data Reporting System (IDRS) is selected. Then the attributes describing the incidents are quantized, combined or generalized creating different views of the data. These views allow for creating granules of similar incidents which is technically done by clustering.

A respective operating time distribution is attached to each of the granules. The aim is to obtain the probability distribution of time needed to handle each distinct (in granule sense) emergency situation. Once the probability distribution is found, it becomes possible to estimate in real time when the gone units are expected to return to the fire station. This can be further extended to monitor for blockage probability when the fire station gets short on the units reserves.

The research is based on Polish IDRS containing 0.26 million of reports from Warsaw agglomeration.

3   Description of the Method

3.1   Granules Generation

The data from the IDRS were first cleaned and preprocessed using the standard and dedicated methods as described in [4]. The number of attributes that described the incidents was reduced on the basis of expert judgement.

The attributes values were quantized, combined or generalized. For example, different time windows are useful for observing how the different threats, which Fire Service tries to deal with are distributed for the given area. This particular problem may be examined in hourly basis as well as season basis and provide both useful and distinct results. The clustering resulted in granules of similar incidents.

Having the clusters defined, it became possible to fit the distribution of the operating time for each cluster. One of *log-normal, gamma, weibull* were fitted to each cluster, whichever fitted best. As a result the distribution of the operating time of each cluster was obtained.

Each of the granules were labeled with an artificial but a meaningful name (three most representative terms chosen by the TF-IDF index [5]). There was also a metadata containing operating time distribution attached to each granule.

3.2   The Blockage Prediction

The blockage occurs when all units are out and a new emergency arises. It is possible to estimate the probability of such risk in the real time and perhaps to counteract it by requesting units from other fire stations. The proposed model is the composition of three independent elements: a) probability that none of the units will return within critical time interval, b) probability that a new emergency will occur in a given time of the day, c) probability that a given number of emergencies for a day will occur.

a) Probability That None of the Units will Return within Time Interval. Let us consider an example scenario: There are A, B and C fire engines in the fire station. The units A, B and C are sent to the emergencies at the respective times: 00:00, 00:20 and 01:40. The question arises how likely it is for any of the engines to return before a new emergency (blockage threat).

By providing the characteristics of A, B, C emergencies it is possible to obtain the expected return times of the units by matching them to the respective clusters. The hypothetical units A, B, C should be returning according to the distributions depicted on Figure 1.

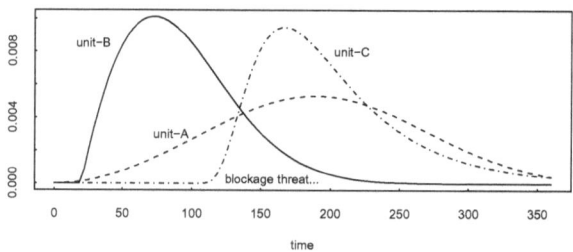

Fig. 1. All units gone creating blockage

The blockage threat starts with the C unit leaving the fire station. The other end of the blockage interval needs to be calculated by taking into account the decreasing probability that none of the units will return in any next moment in time (formula (1), Figure 2(a)). The plot starts from the value 100 which is 01:40 (unit C leaves the fire station).

$$P_N(t) = (1 - F_A(t))(1 - F_B(t))(1 - F_C(t)) \tag{1}$$

where $P_N(t)$ is the probability that none of the units will return before time t and the $F(t)$ are the Cumulative Distribution Functions for the events that the respective units will return.

b) Probability That a New Emergency within a Given Time of the Day will Occur.
Based on EWID data, the histogram of how often emergencies occur in a 24-hours period was obtained (Figure 2(b)). The sine function $F_D(t) = I - Asin(\omega t + \psi)$ was chosen for the model. The regression resulted in obtaining the following function:

$$F_D(t) = 0.042 - 0.031 sin(0.273t - 1.56) \tag{2}$$

The probability $P_D(t)$ that a new emergency will occur in a given time of the day can be calculated as follows:

$$P_D(t) = \int_{t_0}^{t} F_D(t) \tag{3}$$

c) Probability That a Number of Emergencies per Day will Occur. Another factor to be taken into account is the number of emergencies per day. The log-normal distribution was a best fit (according to MLE) to EWID data (Figure 2(c)). This probability should be regarded as a chance for occurring *n or more emergencies* rather than a fixed number of *n emergencies* and is expressed by the formula 4.

$$P_E(n) = 1 - F_E(n) \tag{4}$$

where $P_E(n)$ is the probability that n or more emergencies per day will occur and the $F_E(n)$ is the respective Cumulative Distribution Function.

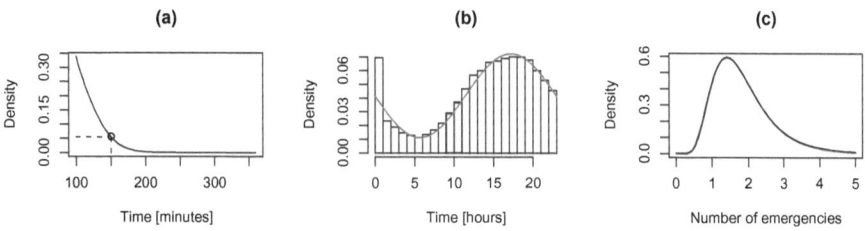

Fig. 2. Blockage probability as the composition of three elements (a) Probability of 'None returns', (b) Emergencies in time of the day, (c) Number of emergencies per day

By combining these three independent elements, the final formula for the blockage can be obtained (Formula 5).

$$P_B = P_N(t) \times P_D(t) \times P_E(n) \tag{5}$$

4 Experimental Verification

In this section we evaluate the model of blockages prediction empirically.

The data for the experiment were selected from IDRS and they represent the potential blockage from the Warsaw Fire Station $No.2$. The set consists of 615 situations when all (in this case three) of the units were out. The challenge was to predict the actual blockage.

The incidents in the set were labeled "B" for blockage and "N" for non-blockage. The collection of data contained 122 blockages and 493 non-blockages. The set was divided into two subsets: the training set included 95 blockages and 368 non-blockages, and the test set included 27 blockages and 125 non-blockages.

The first step of the experiment was to determine which value of P_B (formula (5)) separates best between blockages and non-blockages.

The calibration was performed on the training set. Figure 3a) outlines the density distributions functions of P_B for blockages and non-blockages. According to the picture, there is no evident point which separates the two curves. Therefore ROC curve is used to determine the border value (Figure 3b). Based on the AUC function we obtain the value of $P_B = 0,0138$ which is a trade-off between the benefit and the cost.

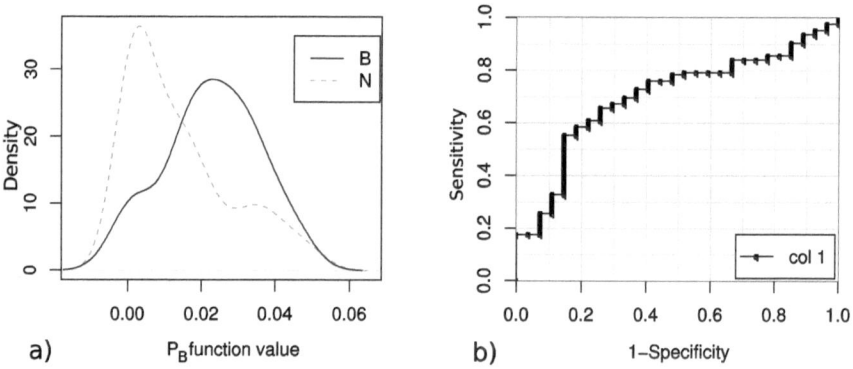

a) P_B function value   b)   1–Specificity

Fig. 3. The density distributions of (B)lockage and (N)on-blockage and ROC curve for the training set

Following are the results of validating the model against the test set.

Table 1. Confusion matrix of the prediction model

	Real Positive	Real Negative
Classified Positive	22	52
Classified Negative	5	73

The calculated measures validating the model: recall=0.814, precision=0.297, accuracy=0.625.

5   Conclusions

A method of estimating the blockage probability was presented in the article. The evaluation was done in the probability domain, by combining three elements: a) the probability that none of the unit will return within the given time interval (when all units are out), b) the probability that at the given part of the day a new emergency will occur and c) the probability that a number of emergencies will occur in the given day. As it is usually the case with such models, there is a possibility to extend the model by including additional parameters (e.g. weather conditions, commanders experience [6]), but the chosen three factors seem reasonable in authors' opinion and were considerably easily available from EWID database.

The proposed method is considerably easy to be implemented on the computer systems. Although the clustering needs to be periodically renewed and this is a computation intensive task, this needs to be done just occasionally, perhaps once a month. However, the actual calculating of the blockage probability (the supposed everyday usage) is trivial and can be done within seconds. It means that the algorithm can be implemented on workstations at the control room to support decision making when it comes to managing the blockage or relocating the units. It might be more purposeful to have such a system running at the city/district level, where the operators are capable of complementing the missing resources from another units. Additionally, the probability of the blockage can be calculated for the time of the travel of such a complementing unit – perhaps this probability will be low enough for the complementation to be unnecessary.

References

1. Bargiela, A., Pedrycz, W.: Granular computing: an introduction, vol. 717. Springer (2003)
2. Skowron, A., Wasilewski, P.: An Introduction to Perception Based Computing. In: Kim, T.-H., Lee, Y.-H., Kang, B.-H., Ślęzak, D. (eds.) FGIT 2010. LNCS, vol. 6485, pp. 12–25. Springer, Heidelberg (2010)
3. Skowron, A., Wasilewski, P.: Information systems in modeling interactive computations on granules. Theoretical Computer Science (2011)
4. Krasuski, A., Ślęzak, D., Kreński, K., Łazowy, S.: A Framework for Granular Knowledge Discovery – A Case Study of Incident Data Reporting System. In: Proceedings of the 16th East-European Conference on Advances in Databases and Information Systems (ADBIS 2012), Poznan, Poland, September 18-21. AISC. Springer, Heidelberg (2012)
5. Manning, C., Raghavan, P., Schutze, H.: Introduction to information retrieval. Cambridge University Press, Cambridge (2008)
6. Krasuski, A., Kreński, K., Łazowy, S.: A Method for Estimating the Efficiency of Commanding in the State Fire Service of Poland. Fire Technology Submitted for printing

Hybrid: A New Multigranulation Rough Set Approach

Xibei Yang[1,2], Cheng Yan[1], Cai Chen[1], and Jingyu Yang[2]

[1] School of Computer Science and Engineering,
Jiangsu University of Science and Technology,
Zhenjiang, Jiangsu, 212003, P.R. China
yangxibei@hotmail.com
[2] School of Computer Science and Technology,
Nanjing University of Science and Technology,
Nanjing, Jiangsu, 210094, P.R. China

Abstract. Multigranulation is a new developing approach, which can be used for the constructing approximations of target concept. Optimistic multigranulation rough set is consistent to the disjunctive explanation of multigranulation structure while pessimistic multigranulation rough set is consistent to the conjunctive explanation of multigranulation structure. To study the multigranulation structure with both disjunctive and conjunctive explanations, the concept of hybrid multigranulation structure is firstly proposed. The hybrid multigranulation rough set is also proposed. The hybrid multigranulation structure and the corresponding rough set are generalizations of traditional multigranulation structures and rough sets, respectively.

Keywords: multigranulation structure, multigranulation rough set, hybrid multigranulation structure, hybrid multigranulation rough set.

1 Introduction

As one of the important mathematical tools for granular computing, rough set [1] has been demonstrated to be useful in pattern recognition, medical diagnose, knowledge discovering, data mining and so on.

However, Pawlak's rough set does not reflect the depth, width and universality of granular computing in essence [2] since it cannot be used to reflect the multiview in granular computing. To solve such problem, Qian et al. [3–5] proposed the concept of multigranulation rough sets, which are constructed on the basis of a family of the binary relations. In Qian et al.'s multigranulation approach, there are two basic models: one is the optimistic multigranulation rough set, the second is the pessimistic multigranulation rough set. Through a further investigation, we can observe that optimistic multigranulation rough set is consistent to the disjunction explanation of the partitions in multigranulation structure, while the pessimistic multigranulation is consistent to the conjunction explanation of the partitions in multigranulation structure. In recent years,

T. Li et al. (Eds.): RSKT 2012, LNAI 7414, pp. 422–429, 2012.
© Springer-Verlag Berlin Heidelberg 2012

multigranulation rough set progressing rapidly. For example, Yang et al. and Xu et al. generalized multigranulation rough sets into fuzzy environment in Ref. [8] and Ref. [6], respectively. Yang et al. [7] also proposed multigranulation rough sets in incomplete information system.

Nevertheless, it should be noticed that neither optimistic nor pessimistic multigranulation rough sets hold the partitions in multigranulation structure with both disjunction and conjunction. To solve such problem, the concept of hybrid multigranulation structure is firstly proposed. In our hybrid multigranulation structure, the relationships among partitions are both disjunction and conjunction. Based on the hybrid multigranulation structure, the model of hybrid multigranulation rough set is also proposed. Some basic properties about hybrid multigranulation rough set are then addressed.

2   Multigranulation Rough Sets

2.1   Optimistic Multigranulation Rough Set

Suppose that U is the universe of discourse, $\boldsymbol{R} = \{R_1, R_2, \cdots, R_m\}$ is a set of the equivalence relations, then $(U, \boldsymbol{R})$ is referred to as a knowledge base. In Pawlak's rough set theory, the indiscernibility relation is the intersection of some equivalence relations in knowledge base and then indiscernibility relation is also an equivalence relation.

Different from Pawlak's rough set theory, multigranulation rough set approach is constructed on the basis of a family of the granulation structures. Since the equivalence relations are considered in this paper, the granulation structures are then partitions on the universe of discourse. In the following, to simply our discussions, the equivalence relations in knowledge base will be employed to construct multigranulation rough sets.

Let $(U, \boldsymbol{R})$ be a knowledge base, a family of the partitions generated from the equivalence relations in $\boldsymbol{R}$ are denoted by $U/R_1, U/R_2, \cdots, U/R_m$. If the logical relationships among these partitions are disjunctions, then the obtained multigranulation structure is referred to as the disjunctive multigranulation structure, it is denoted by

$$\bigvee_{i=1}^{m} U/R_i.$$

In $\bigvee_{i=1}^{m} U/R_i$, it is assumed that different two granulation structures are connected through the connective OR, such connective is also being used in an inclusive sense.

Definition 1. *Let $(U, \boldsymbol{R})$ be a knowledge base, then $\forall X \subseteq U$, the optimistic multigranulation lower and upper approximations are denoted by $\underline{\bigvee_{i=1}^{m} R_i}(X)$ and $\overline{\bigvee_{i=1}^{m} R_i}(X)$, respectively,*

$$\underline{\bigvee_{i=1}^{m} R_i}(X) = \{x \in U : \exists R_i \in \boldsymbol{R} \ s.t. \ [x]_{R_i} \subseteq X\}; \tag{1}$$

$$\overline{\bigvee_{i=1}^{m} R_i}(X) = \sim \bigvee_{i=1}^{m} \underline{R_i}(\sim X); \qquad (2)$$

$[x]_{R_i}$ is the equivalence class of x in terms of the equivalence relation R_i, $\sim X$ is the complement of set X.

$[\underline{\bigvee_{i=1}^{m} R_i}(X), \overline{\bigvee_{i=1}^{m} R_i}(X)]$ is referred to as the optimistic multigranulation rough set of X. By Definition 1, we can see that if an object has one of the equivalence classes, which is included by the target concept, then such object is in optimistic multigranulation lower approximation. Such explanation is consistent to the logical connective in disjunctive multigranulation structure $\bigvee_{i=1}^{m} U/R_i$.

2.2   Pessimistic Multigranulation Rough Set

Let $(U, \boldsymbol{R})$ be a knowledge base, a family of the partitions generated from the equivalence relations in $\boldsymbol{R}$ are denoted by $U/R_1, U/R_2, \cdots, U/R_m$. If the logical relationships among these partitions are conjunctions, then the obtained multigranulation structure is referred to as the conjunctive multigranulation structure, it is denoted by

$$\bigwedge_{i=1}^{m} U/R_i.$$

In $\bigwedge_{i=1}^{m} U/R_i$, it is assumed that different two granulation structures are connected through the connective *AND*.

Definition 2. *Let* $(U, \boldsymbol{R})$ *be a knowledge base, then* $\forall X \subseteq U$, *the pessimistic multigranulation lower and upper approximations are denoted by* $\underline{\bigwedge_{i=1}^{m} R_i}(X)$ *and* $\overline{\bigwedge_{i=1}^{m} R_i}(X)$, *respectively,*

$$\underline{\bigwedge_{i=1}^{m} R_i}(X) = \{x \in U : \forall R_i \in \boldsymbol{R} \ s.t. \ [x]_{R_i} \subseteq X\}; \qquad (3)$$

$$\overline{\bigwedge_{i=1}^{m} R_i}(X) = \sim \underline{\bigwedge_{i=1}^{m} R_i}(\sim X). \qquad (4)$$

$[\underline{\bigwedge_{i=1}^{m} R_i}(X), \overline{\bigwedge_{i=1}^{m} R_i}(X)]$ is referred to as the pessimistic multigranulation rough set of X. By Definition 2, we can see that if for an object, all of the equivalence classes are included by the target concept, then such object is in pessimistic multigranulation lower approximation. Similar to optimistic case, such explanation is consistent to the logical connective in conjunctive multigranulation structure $\bigwedge_{i=1}^{m} U/R_i$.

3   Hybrid Multigranulation Approach to Rough Set

In the multigranulation structures we mentioned in the above section, the relationships among granulation structures are either disjunction or conjunction. To study the multigranulation structures with both disjunction and conjunction explanations, the hybrid multigranulation approach will be proposed in this section. Firstly, let us define the concept of hybrid multigranulation structures.

Definition 3. *Let U be the universe of discourse,*

1. *if $\bigvee_{i=1}^{m_1} U/R_{1i}, \bigvee_{i=1}^{m_2} U/R_{2i}, \cdots, \bigvee_{i=1}^{m_J} U/R_{Ji}$ are J different disjunctive multi-granulation structures, then $\bigwedge_{j=1}^{J}(\bigvee_{i=1}^{m_j} U/R_{ji})$ is referred to as a hybrid multigranulation structure;*
2. *if $\bigwedge_{i=1}^{m_1} U/R_{1i}, \bigwedge_{i=1}^{m_2} U/R_{2i}, \cdots, \bigwedge_{i=1}^{m_J} U/R_{Ji}$ are J different conjunctive multi-granulation structures, then $\bigvee_{j=1}^{J}(\bigwedge_{i=1}^{m_j} U/R_{ji})$ is referred to as a hybrid multigranulation structure.*

By Definition 3, we can see that $\bigwedge_{j=1}^{J}(\bigvee_{i=1}^{m_j} U/R_{ji})$ is the conjunction of J different disjunctive multigranulation structures; while $\bigvee_{j=1}^{J}(\bigwedge_{i=1}^{m_j} U/R_{ji})$ is the disjunction of J different conjunctive multigranulation structures. If $J = 1$, then $\bigwedge_{j=1}^{J}(\bigvee_{i=1}^{m_j} U/R_{ji})$ will degenerate into the disjunctive multigranulation structure $\bigvee_{i=1}^{m_1} U/R_{1i}$, $\bigvee_{j=1}^{J}(\bigwedge_{i=1}^{m_j} U/R_{ji})$ will degenerate into the conjunctive multigranulation structure $\bigwedge_{i=1}^{m_1} U/R_{1i}$. From this point of view, hybrid multigranulation structure is a generalization of both disjunctive and conjunctive multigranulation structures. Moreover, if for each $j \in \{1, 2, \cdots, J\}$, $m_j = 1$, then $\bigwedge_{j=1}^{J}(\bigvee_{i=1}^{m_j} U/R_{ji})$ will degenerate into the conjunctive multigranulation structure $\bigwedge_{j=1}^{J} U/R_{j1}$, $\bigvee_{j=1}^{J}(\bigwedge_{i=1}^{m_j} U/R_{ji})$ will degenerate into the disjunctive multigranulation structure $\bigvee_{j=1}^{J} U/R_{j1}$. For such reason, hybrid multigranulation structure is still a generalization of both disjunctive and conjunctive multigranulation structures.

By the basic properties of disjunction and conjunction, it is not difficult to observe that $\bigwedge_{j=1}^{J}(\bigvee_{i=1}^{m_j} U/R_{ji})$ and $\bigvee_{j=1}^{J}(\bigwedge_{i=1}^{m_j} U/R_{ji})$ can be transformed into each other. For instance, given a hybrid multigranulation structure such that $(\bigwedge_{i=1}^{2} U/R_{1i}) \bigvee (\bigwedge_{i=1}^{3} U/R_{2i})$, then we can obtain an equivalent hybrid multi-granulation structure such that $(U/R_{11} \bigvee U/R_{21}) \bigwedge (U/R_{11} \bigvee U/R_{22}) \bigwedge (U/R_{11} \bigvee U/R_{23}) \bigwedge (U/R_{12} \bigvee U/R_{21}) \bigwedge (U/R_{12} \bigvee U/R_{22}) \bigwedge (U/R_{12} \bigvee U/R_{23})$. Therefore, we only need to construct rough set through one type of the hybrid multigranulation structure. In the following, we will propose a hybrid multigranulation rough set based on $\bigvee_{j=1}^{J}(\bigwedge_{i=1}^{m_j} U/R_{ji})$.

Definition 4. *Let U be the universe of discourse, $\bigvee_{j=1}^{J}(\bigwedge_{i=1}^{m_j} U/R_{ji})$ is a hybrid multigranulation structure, $\forall X \subseteq U$, the hybrid multigranulation lower approximation and hybrid multigranulation upper approximation of X are denoted by $\underline{\bigvee_{j=1}^{J}(\bigwedge_{i=1}^{m_j} R_{ji})}(X)$ and $\overline{\bigvee_{j=1}^{J}(\bigwedge_{i=1}^{m_j} R_{ji})}(X)$, respectively,*

$$\overline{\bigvee_{j=1}^{J}(\bigwedge_{i=1}^{m_j} R_{ji})}(X) = \{x \in U : \exists j \in \{1, \cdots, J\}, \forall i \in \{1, \cdots, m_j\}, [x]_{R_{ji}} \subseteq X\}; (5)$$

$$\overline{\bigvee_{j=1}^{J}(\bigwedge_{i=1}^{m_j} R_{ji})}(X) = \sim \underline{\bigvee_{j=1}^{J}(\bigwedge_{i=1}^{m_j} R_{ji})}(\sim X). \tag{6}$$

The pair $[\underline{\bigvee_{j=1}^{J}(\bigwedge_{i=1}^{m_j} R_{ji})}(X), \overline{\bigvee_{j=1}^{J}(\bigwedge_{i=1}^{m_j} R_{ji})}(X)]$ is referred to as a hybrid multigranulation rough set of X.

Theorem 1. *Let U be the universe of discourse, $\bigvee_{j=1}^{J}(\bigwedge_{i=1}^{m_j} U/R_{ji})$ is a hybrid multigranulation structure, then $\forall X \subseteq U$, we have*

$$\overline{\bigvee_{j=1}^{J}(\bigwedge_{i=1}^{m_j} R_{ji})}(X) = \{x \in U : \forall j \in \{1, \cdots, J\}, \exists i \in \{1, \cdots, m_j\}, [x]_{R_{ji}} \cap X \neq \emptyset\}.$$

Proof. $\forall x \in U$, by Definition 4, we have

$$x \in \overline{\bigvee_{j=1}^{J}(\bigwedge_{i=1}^{m_j} R_{ji})}(X) \Leftrightarrow x \notin \underline{\bigvee_{j=1}^{J}(\bigwedge_{i=1}^{m_j} R_{ji})}(\sim X)$$

$$\Leftrightarrow \forall j \in \{1, \cdots, J\}, \exists i \in \{1, \cdots, m_j\}, [x]_{R_{ji}} \not\subseteq (\sim X)$$

$$\Leftrightarrow \forall j \in \{1, \cdots, J\}, \exists i \in \{1, \cdots, m_j\}, [x]_{R_{ji}} \cap X \neq \emptyset.$$

$\square$

Theorem 2. *Let U be the universe of discourse, $\bigvee_{j=1}^{J}(\bigwedge_{i=1}^{m_j} U/R_{ji})$ is a hybrid multigranulation structure, then the hybrid multigranulation rough set has the following properties:*

1. $\underline{\bigvee_{j=1}^{J}(\bigwedge_{i=1}^{m_j} R_{ji})}(X) \subseteq X \subseteq \overline{\bigvee_{j=1}^{J}(\bigwedge_{i=1}^{m_j} R_{ji})}(X);$

2. $\underline{\bigvee_{j=1}^{J}(\bigwedge_{i=1}^{m_j} R_{ji})}(\emptyset) = \overline{\bigvee_{j=1}^{J}(\bigwedge_{i=1}^{m_j} R_{ji})}(\emptyset) = \emptyset,$

   $\underline{\bigvee_{j=1}^{J}(\bigwedge_{i=1}^{m_j} R_{ji})}(U) = \overline{\bigvee_{j=1}^{J}(\bigwedge_{i=1}^{m_j} R_{ji})}(U) = U;$

3. $\underline{\bigvee_{j=1}^{J}(\bigwedge_{i=1}^{m_j} R_{ji})}(\sim X) = \sim \overline{\bigvee_{j=1}^{J}(\bigwedge_{i=1}^{m_j} R_{ji})}(X),$

   $\overline{\bigvee_{j=1}^{J}(\bigwedge_{i=1}^{m_j} R_{ji})}(\sim X) = \sim \underline{\bigvee_{j=1}^{J}(\bigwedge_{i=1}^{m_j} R_{ji})}(X);$

4. $X \subseteq Y \Rightarrow \overline{\bigvee_{j=1}^{J}(\bigwedge_{i=1}^{m_j} R_{ji})}(X) \subseteq \overline{\bigvee_{j=1}^{J}(\bigwedge_{i=1}^{m_j} R_{ji})}(Y),$

   $X \subseteq Y \Rightarrow \underline{\bigvee_{j=1}^{J}(\bigwedge_{i=1}^{m_j} R_{ji})}(X) \subseteq \underline{\bigvee_{j=1}^{J}(\bigwedge_{i=1}^{m_j} R_{ji})}(Y);$

5. $\overline{\bigvee_{j=1}^{J}(\bigwedge_{i=1}^{m_j} R_{ji})}\left(\overline{\bigvee_{j=1}^{J}(\bigwedge_{i=1}^{m_j} R_{ji})}(X)\right) = \overline{\bigvee_{j=1}^{J}(\bigwedge_{i=1}^{m_j} R_{ji})}(X),$

   $\underline{\bigvee_{j=1}^{J}(\bigwedge_{i=1}^{m_j} R_{ji})}\left(\underline{\bigvee_{j=1}^{J}(\bigwedge_{i=1}^{m_j} R_{ji})}(X)\right) = \underline{\bigvee_{j=1}^{J}(\bigwedge_{i=1}^{m_j} R_{ji})}(X).$

Proof. 1. $\forall x \in \bigvee_{j=1}^{J}(\bigwedge_{i=1}^{m_j} R_{ji})(X)$, by Definition 4, there must be $j \in \{1,\cdots,J\}$ and $\forall i \in \{1,\cdots,m_j\}$ such that $[x]_{R_{ji}} \subseteq X$. Since R_{ji} is the equivalence relation, then $x \in [x]_{R_{ji}}$, it follows that $x \in X$, i.e. $\bigvee_{j=1}^{J}(\bigwedge_{i=1}^{m_j} R_{ji})(X) \subseteq X$. Similarity, it is not difficult to prove that $X \subseteq \overline{\bigvee_{j=1}^{J}(\bigwedge_{i=1}^{m_j} R_{ji})}(X)$.

2. By the result of 1 we have $\bigvee_{j=1}^{J}(\bigwedge_{i=1}^{m_j} R_{ji})(\emptyset) \subseteq \emptyset$. Therefore, it must be proven that $\bigvee_{j=1}^{J}(\bigwedge_{i=1}^{m_j} R_{ji})(\emptyset) \supseteq \emptyset$. $\forall x \notin \bigvee_{j=1}^{J}(\bigwedge_{i=1}^{m_j} R_{ji})(\emptyset)$, by Definition 4, we know that $\forall j \in \{1,\cdots,J\}$, there must be $i \in \{1,\cdots,m_j\}$ such that $[x]_{R_{ji}} \not\subseteq \emptyset$, from which we can conclude that $x \in U$ since $[x]_{ji} \neq \emptyset$, i.e. $\bigvee_{j=1}^{J}(\bigwedge_{i=1}^{m_j} R_{ji})(\emptyset) \supseteq \emptyset$.

Similarity, it is not difficult to prove that $\overline{\bigvee_{j=1}^{J}(\bigwedge_{i=1}^{m_j} R_{ji})}(\emptyset) = \emptyset$ and $\bigvee_{j=1}^{J}(\bigwedge_{i=1}^{m_j} R_{ji})(U) = \overline{\bigvee_{j=1}^{J}(\bigwedge_{i=1}^{m_j} R_{ji})}(U) = U$.

3. It can be derived directly from Definition 4 and Theorem 1.

4. $\forall x \in \bigvee_{j=1}^{J}(\bigwedge_{i=1}^{m_j} R_{ji})(X)$, by Definition 4, there must be $j \in \{1,\cdots,J\}$ and $\forall i \in \{1,\cdots,m_j\}$ such that $[x]_{R_{ji}} \subseteq X$. Since $X \subseteq Y$, then we have $x \in \bigvee_{j=1}^{J}(\bigwedge_{i=1}^{m_j} R_{ji})(Y)$, from which we can conclude that $\bigvee_{j=1}^{J}(\bigwedge_{i=1}^{m_j} R_{ji})(X) \subseteq \bigvee_{j=1}^{J}(\bigwedge_{i=1}^{m_j} R_{ji})(Y)$. Similarity, it is not difficult to prove the upper approximation case, i.e. $\overline{\bigvee_{j=1}^{J}(\bigwedge_{i=1}^{m_j} R_{ji})}(X) \subseteq \overline{\bigvee_{j=1}^{J}(\bigwedge_{i=1}^{m_j} R_{ji})}(Y)$.

5. By the result of 1 we have $\bigvee_{j=1}^{J}(\bigwedge_{i=1}^{m_j} R_{ji})(X) \subseteq X$, then by the result of 4 we obtain that $\bigvee_{j=1}^{J}(\bigwedge_{i=1}^{m_j} R_{ji})\left(\bigvee_{j=1}^{J}(\bigwedge_{i=1}^{m_j} R_{ji})(X)\right) \subseteq \bigvee_{j=1}^{J}(\bigwedge_{i=1}^{m_j} R_{ji})(X)$. Therefore, it must be proved that $\bigvee_{j=1}^{J}(\bigwedge_{i=1}^{m_j} R_{ji})\left(\bigvee_{j=1}^{J}(\bigwedge_{i=1}^{m_j} R_{ji})(X)\right) \supseteq \bigvee_{j=1}^{J}(\bigwedge_{i=1}^{m_j} R_{ji})(X)$. $\forall x \in \bigvee_{j=1}^{J}(\bigwedge_{i=1}^{m_j} R_{ji})(X)$, by Definition 4, there must be $j \in \{1,\cdots,J\}$ and $\forall i \in \{1,\cdots,m_j\}$ such that $[x]_{R_{ji}} \subseteq X$. $\forall y \in [x]_{R_{ji}}$, since $[x]_{R_{ji}}$ is the equivalence class, then we have $[y]_{R_{ji}} \subseteq X$, i.e. $y \in \bigvee_{j=1}^{J}(\bigwedge_{i=1}^{m_j} R_{ji})(X)$, from which we know that $[x]_{R_{ji}} \subseteq \bigvee_{j=1}^{J}(\bigwedge_{i=1}^{m_j} R_{ji})(X)$, i.e. $x \in \bigvee_{j=1}^{J}(\bigwedge_{i=1}^{m_j} R_{ji})\left(\bigvee_{j=1}^{J}(\bigwedge_{i=1}^{m_j} R_{ji})(X)\right)$. Similarity, it is not difficult to prove that $\overline{\bigvee_{j=1}^{J}(\bigwedge_{i=1}^{m_j} R_{ji})}\left(\overline{\bigvee_{j=1}^{J}(\bigwedge_{i=1}^{m_j} R_{ji})}(X)\right) = \overline{\bigvee_{j=1}^{J}(\bigwedge_{i=1}^{m_j} R_{ji})}(X)$.

$\square$

Theorem 2 shows the basic properties about hybrid multigranulation rough set. For example, 1 says that hybrid multigranulation lower and upper approximations satisfy the contraction and extension, respectively; 2 says that hybrid multigranulation rough set satisfies the normality and conormality; 3 shows the dual properties of hybrid multigranulation lower and upper approximations; 4 shows the monotone of hybrid multigranulation lower and upper approximations w.r.t. the variety of target concept; 5 expresses the invariant of hybrid multigranulation lower and upper approximations.

Theorem 3. *Let U be the universe of discourse, $\bigvee_{j=1}^{J}(\bigwedge_{i=1}^{m_j} U/R_{ji})$ is a hybrid multigranulation structure, then we have*

1. $\overline{\bigvee_{j=1}^{J}(\bigwedge_{i=1}^{m_j} R_{ji})}(X) = \bigcup_{j=1}^{J} \overline{\bigwedge_{i=1}^{m_j} R_{ji}}(X);$
2. $\underline{\bigvee_{j=1}^{J}(\bigwedge_{i=1}^{m_j} R_{ji})}(X) = \bigcap_{j=1}^{J} \underline{\bigwedge_{i=1}^{m_j} R_{ji}}(X).$

Proof. We only prove 1. $\forall x \in U$,

$$x \in \overline{\bigvee_{j=1}^{J}(\bigwedge_{i=1}^{m_j} R_{ji})}(X) \Leftrightarrow \exists j \in \{1, \cdots, J\}, \forall i \in \{1, \cdots, m_j\}, [x]_{R_{ji}} \subseteq X$$

$$\Leftrightarrow \exists j \in \{1, \cdots, J\}, x \in \overline{\bigwedge_{i=1}^{m_j} R_{ji}}(X)$$

$$\Leftrightarrow x \in \bigcup_{j=1}^{J} \overline{\bigwedge_{i=1}^{m_j} R_{ji}}(X).$$

□

Theorem 3 shows the relationships between hybrid multigranulation rough set and pessimistic multigranulation rough set.

Theorem 4. *Let U be the universe of discourse, $\bigvee_{j=1}^{J}(\bigwedge_{i=1}^{m_j} U/R_{ji})$ is a hybrid multigranulation structure,*

1. *if $J = 1$, then*
   $$\underline{\bigvee_{j=1}^{J}(\bigwedge_{i=1}^{m_j} R_{ji})}(X) = \underline{\bigwedge_{i=1}^{m_j} R_{ji}}(X), \overline{\bigvee_{j=1}^{J}(\bigwedge_{i=1}^{m_j} R_{ji})}(X) = \overline{\bigwedge_{i=1}^{m_j} R_{ji}}(X);$$
2. *if $\forall j \in \{1, \cdots, J\}$, $m_j = 1$, then*
   $$\underline{\bigvee_{j=1}^{J}(\bigwedge_{i=1}^{m_j} R_{ji})}(X) = \underline{\bigvee_{j=1}^{J} R_{j1}}(X), \overline{\bigvee_{j=1}^{J}(\bigwedge_{i=1}^{m_j} R_{ji})}(X) = \overline{\bigvee_{j=1}^{J} R_{j1}}(X).$$

Proof. It can be derived directly from Definition 1, Definition 2 and Definition 4.

□

Theorem 4 tells us that the proposed hybrid multigranulation rough set is a generalization of both optimistic and pessimistic multigranulation rough sets, optimistic and pessimistic multigranulation rough sets are special cases of hybrid multigranulation rough set.

4   Conclusions

The multigranulation rough sets approach reflects the multiview of granular computing. In this paper, by considering both disjunction and conjunction relationships among partitions in multigranulation structure, the hybrid multigranulation structure is proposed and then the corresponding hybrid multigranulation rough set is explored. It is proven that the proposed hybrid multigranulation rough set is a generalization of Qian et al.'s multigranulation rough set.

In our further researching, the hierarchy of hybrid multigranulation structures will be investigated. Moreover, the knowledge discovering approach through hybrid multigranulation rough set will also be explored.

Acknowledgment. This work is supported by Natural Science Foundation of China (No. 61100116), Natural Science Foundation of Jiangsu Province of China (No. BK2011492), Natural Science Foundation of Jiangsu Higher Education Institutions of China (No. 11KJB520004), Postdoctoral Science Foundation of China (No. 20100481149), Postdoctoral Science Foundation of Jiangsu Province of China (No. 1101137C).

References

1. Pawlak, Z.: Rough sets–theoretical aspects of reasoning about data. Kluwer Academic Publishers (1992)
2. Yao, Y.Y.: Triarchic theory of granular computing. In: Zhang, Y.P., Luo, B., Yao, Y.Y., et al. (eds.) Quotient Space Theory and Granular Computing, Theory and Practice of Structured Problem Solving, pp. 115–143. Science Press, Beijing (2010) (in Chinese)
3. Qian, Y.H., Liang, J.Y., Dang, C.Y.: Incomplete multigranulation rough set. IEEE Transactions on Systems, Man and Cybernetics, Part A 20, 420–431 (2010)
4. Qian, Y.H., Liang, J.Y., Yao, Y.Y., Dang, C.Y.: MGRS: a multi–granulation rough set. Information Sciences 180, 949–970 (2010)
5. Qian, Y.H., Liang, J.Y., Wei, W.: Pessimistic rough decision. In: Second International Workshop on Rough Sets Theory, Zhoushan, P.R. China, October 19-21, pp. 440–449 (2010)
6. Xu, W.H., Wang, Q.R., Zhang, X.T.: Multi–granulation fuzzy rough sets in a fuzzy tolerance approximation space. International Journal of Fuzzy Systems 13, 246–259 (2011)
7. Yang, X.B., Song, X.N., Chen, Z.H., Yang, J.Y.: On multigranulation rough sets in incomplete information system. International Journal of Machine Learning and Cybernetics, doi:10.1007/s13042-011-0054-8
8. Yang, X.B., Song, X.N., Dou, H.L., Yang, J.Y.: Multi–granulation rough set: from crisp to fuzzy case. Annals of Fuzzy Mathematics and Informatics 1, 55–70 (2011)

Inclusion Degrees of Graded Ill-Known Sets

Masahiro Inuiguchi

Graduate School of Engineering Science, Osaka University
1-3 Machikaneyama, Toyonaka, Osaka 560-8531, Japan
inuiguti@sys.es.osaka-u.ac.jp
http://www-inulab.sys.es.osaka-u.ac.jp/

Abstract. The potential usefulness of graded ill-known sets in decision making problems is recently observed. In this paper, to develop the application of graded ill-known sets to decision making under uncertainty, we investigate the inclusion degrees of graded ill-known sets. Several kinds of inclusion degrees are defined. Although those degrees are defined on the power set, it is shown that, in some useful cases, they can be calculated easily in the universe. A simple example is given to demonstrate the usage of the inclusion degrees.

Keywords: Ill-known set, possibility, necessity, inclusion degree, implication function, conjunction function.

1 Introduction

Ill-known sets and graded ill-known sets were proposed in 1988 by Dubois and Prade [1]. However, ill-known sets and graded ill-known sets had not gotten a lot of attention. As generalized models [2–4] of fuzzy sets and rough sets or alternative models [5, 6] of uncertainties are getting popular in recent years, ill-known sets and graded ill-known sets can be worthy of attention as a model of higher order fuzzy sets. Recently, studies on ill-known sets and graded ill-known sets can be found in the literature [4, 7, 8].

Ill-known sets and graded ill-known sets are crisp sets whose members are not known exactly. They are useful to treat set-valued variables. However, in decision making problems under uncertainty, the uncertain variables are usually single-valued. Considering the applications of graded ill-known sets to decision making, it is desirable to consider a way to treat single-valued uncertain variables by graded ill-known sets. Assuming the situation where the set of possible realizations of uncertain variable is not known exactly, the author [8] proposed to represent the set of possible realizations by a graded ill-known set. The function value calculations of ill-known sets of quantities have investigated in [8].

As in robust optimization [9] and possibilistic programming [10], the inclusion of all possible realizations in a satisfactory region is imposed on the solution for robust decisions. In this paper, continuing the author's studies [7, 8] on application of graded ill-known sets to decision making under uncertainty, we consider degrees of inclusion between graded ill-known sets. Because graded ill-known sets are characterized by a possibility distribution on the power set, the

T. Li et al. (Eds.): RSKT 2012, LNAI 7414, pp. 430–439, 2012.

treatment of graded ill-known sets are basically complex. We investigate the calculations of inclusion degrees of graded ill-known sets. We show that the inclusion degrees are calculated on the universe in some useful cases.

This paper is organized as follows. In Section 2, we briefly review graded ill-known sets and possibility and necessity measure calculations. We extend the inclusion relations to a relation between graded ill-known sets in Section 3. We investigate the calculations of inclusion degrees. In Section 4, a simple example is given to demonstrate a possible application.

2 Graded Ill-Known Sets

Let X be a universe. Let $\boldsymbol{A}$ be a crisp set which is ill-known, i.e., there exists at least one element $x \in X$, for which it is not known whether x belongs to $\boldsymbol{A}$ or not. To represent such an ill-known set, collecting possible realizations of $\boldsymbol{A}$, we obtain the following family:

$$\mathcal{A} = \{A_1, A_2, \ldots, A_n\}, \tag{1}$$

where A_i is a crisp set such that $\boldsymbol{A} = A_i$ can occur. We note that bold italic letters such as $\boldsymbol{A}$ show ill-known sets which are set-valued variables while script letters such as $\mathcal{A}$ and italic capital letters such as A_i show well-defined family and set, respectively.

Given $\mathcal{A}$, we obtain a set A^- of elements which is a certain member of $\boldsymbol{A}$ and a set A^+ of elements which is a possible member of $\boldsymbol{A}$ can be defined by

$$A^- = \bigcap \mathcal{A} = \bigcap_{i=1,\ldots,n} A_i, \quad A^+ = \bigcup \mathcal{A} = \bigcup_{i=1,\ldots,n} A_i. \tag{2}$$

We call A^- and A^+ "the lower approximation" of $\boldsymbol{A}$ and "the upper approximation" of $\boldsymbol{A}$, respectively.

In the real world, we may know only certain members and certain non-members. In other words, we know the lower approximation A^- as a set of certain members and the upper approximation A^+ as a complementary set of certain non-members. Given A^- and A^+ (or equivalently, the complement of A^+), we obtain a family $\mathcal{A}$ of possible realizations of $\boldsymbol{A}$ as

$$\mathcal{A} = \{A_i \mid A^- \subseteq A_i \subseteq A^+\}. \tag{3}$$

We note that A^- and A^+ are recovered by applying (2) to the family $\mathcal{A}$ induced from A^- and A^+ by (3). On the other hand, $\mathcal{A}$ cannot be always recovered by applying (3) to A^- and A^+ defined by (2).

If all A_i's of (1) are not regarded as equally possible, we may assign a possibility degree $\pi_{\mathcal{A}}(A_i)$ to each A_i so that

$$\max_{i=1,\ldots,n} \pi_{\mathcal{A}}(A_i) = 1. \tag{4}$$

A possibility distribution $\pi_{\mathcal{A}} : 2^X \to [0,1]$ can be seen as the membership function of fuzzy set $\mathcal{A}$ in 2^X. Graded ill-known set $\mathbf{A}$ is characterized by $\pi_{\mathcal{A}}$ of fuzzy set $\mathcal{A}$ in 2^X. Then we identify $\mathbf{A}$ with $\mathcal{A}$ in what follows. The ill-known set having such a possibility distribution is called "a graded ill-known set".

In this case, the lower approximation A^- and the upper approximation A^+ are defined by fuzzy sets with the following membership functions:

$$\mu_{A^-}(x) = \inf_{x \notin A_i} n(\pi_{\mathcal{A}}(A_i)), \quad \mu_{A^+}(x) = \sup_{x \in A_i} \varphi(\pi_{\mathcal{A}}(A_i)), \tag{5}$$

where $n : [0,1] \to [0,1]$ and $\varphi : [0,1] \to [0,1]$ are non-increasing and non-decreasing functions such that $n(0) = \varphi(1) = 1$ and $n(1) = \varphi(0) = 0$.

We have the following property (see [7]):

$$\forall x \in X, \ \mu_{A^-}(x) > 0 \text{ implies } \mu_{A^+}(x) = 1. \tag{6}$$

The specification of possibility distribution $\pi_{\mathcal{A}}$ may need a lot of information. However, as is often the case in real world applications, we know only the lower approximation A^- and the upper approximation A^+ as fuzzy sets satisfying (6). To have a consistent possibility distribution $\pi_{\mathcal{A}}$ for any A^- and A^+, we need to assume that n and φ are surjective, i.e.,

$$\{n(s) \mid s \in [0,1]\} = [0,1], \quad \{\varphi(s) \mid s \in [0,1]\} = [0,1]. \tag{7}$$

Although (7) is assumed, there are many possibility distributions $\pi_{\mathcal{A}}$ having given A^- and A^+ as their lower and upper approximations. We adopt the following maximal possibility distribution $\pi_{\mathcal{A}}^*(A_i)$:

$$\pi_{\mathcal{A}}^*(A_i) = \min \left(\inf_{x \notin A_i} n^*(\mu_{A^-}(x)), \inf_{x \in A_i} \varphi^*(\mu_{A^+}(x)) \right), \tag{8}$$

where we define $\inf \emptyset = 1$ and

$$n^*(a) = \sup\{s \in [0,1] \mid n(s) \geq a\}, \quad \varphi^*(a) = \sup\{s \in [0,1] \mid \varphi(s) \leq a\}. \tag{9}$$

We use the maximal possibility distribution $\pi_{\mathcal{A}}^*(A_i)$ as the possibility distribution corresponding to the given fuzzy sets A^- and A^+ unless the other information is available.

In what follows, we assume that n and φ are bijective so that we have $n^* = n^{-1}$ and $\varphi^* = \varphi^{-1}$, where n^{-1} and φ^{-1} are inverse functions of n and φ.

Calculations of possibility and necessity degrees of graded ill-known sets have been investigated in [1, 7]. The possibility degree of $\mathcal{B}$ under $\mathcal{A}$, $\Pi(\mathcal{B}|\mathcal{A})$ and the necessity degree of $\mathcal{B}$ under $\mathcal{A}$, $N(\mathcal{B}|\mathcal{A})$ are defined by

$$\Pi(\mathcal{B}|\mathcal{A}) = \sup_{C \in 2^X} T(\pi_{\mathcal{A}}(C), \pi_{\mathcal{B}}(C)), \tag{10}$$

$$N(\mathcal{B}|\mathcal{A}) = \inf_{C \in 2^X} I(\pi_{\mathcal{A}}(C), \pi_{\mathcal{B}}(C)), \tag{11}$$

where T and I are conjunction and implication functions, respectively. An implication function $I : [0,1] \times [0,1] \to [0,1]$ satisfies

(I1) $I(0,0) = I(0,1) = I(1,1) = 1$ and $I(1,0) = 0$,   (boundary condition)

(I2) $I(a,b) \leq I(c,d)$, $0 \leq c \leq a \leq 1$, $0 \leq b \leq d \leq 1$.   (monotonicity)

A conjunction function $T : [0,1] \times [0,1] \to [0,1]$ satisfies

(T1) $T(0,0) = T(0,1) = T(1,0) = 0$ and $T(1,1) = 1$,   (boundary condition)

(T2) $T(a,b) \leq T(c,d)$, $0 \leq a \leq c \leq 1$, $0 \leq b \leq d \leq 1$.   (monotonicity)

When graded ill-known sets are not specified by their lower and upper approximations, we have the following equations on calculations of possibility and necessity degrees of graded ill-known sets (see Inuiguchi [7]):

When I is upper semi-continuous,

$$N(\mathcal{B}|\mathcal{A}) = \min\left(\inf_{x \in X} I\left(\varphi^{-1}(\mu_{A^+}(x)), \varphi^{-1}(\mu_{B^+}(x)) \right), \right.$$

$$\left. \inf_{x \in X} I\left(n^{-1}(\mu_{A^-}(x)), n^{-1}(\mu_{B^-}(x)) \right) \right). \quad (12)$$

When $T = \min$,

$$\Pi(\mathcal{B}|\mathcal{A}) = \min\left(\inf_{x \in X} \max\left(n^{-1}(\mu_{B^-}(x)), \varphi^{-1}(\mu_{A^+}(x)) \right), \right.$$

$$\left. \inf_{x \in X} \max\left(n^{-1}(\mu_{A^-}(x)), \varphi^{-1}(\mu_{B^+}(x)) \right) \right). \quad (13)$$

3 Inclusion Degrees of Graded Ill-Known Sets

In robust optimization [9] and possibilistic programming [10], the inclusion of all possible realizations in a satisfactory region is imposed on the solution. The inclusion relation is important to formulate robust decision problems. Then we consider the inclusion relation between two graded ill-knwon sets $\mathcal{A}$ and $\mathcal{B}$ toward the applications of graded ill-knwon sets to robust decision making. Considering that a graded ill-known set can be seen as a fuzzy set in the power set, we utilize the extension procedure [11] of a relation between elements to that relation between fuzzy sets. To extend the inclusion relation, we introduce the following assumption:

(A) I is upper semi-continuous and T is lower semi-continuous.

Following the extension procedure [11], we first extend the inclusion relation to that relation between a graded ill-known set $\mathcal{A}$ and a usual set B. We obtain the following extended relations:

$$POS(B \subseteq \mathcal{A}) = \Pi(B^{\subseteq} \mid \mathcal{A}) = \sup_{A \supseteq B} T(\pi_{\mathcal{A}}(A), 1), \quad (14)$$

$$NES(B \subseteq \mathcal{A}) = N(B^{\subseteq} \mid \mathcal{A}) = \inf_{A \not\supseteq B} I(\pi_{\mathcal{A}}(A), 0), \quad (15)$$

$$POS(\mathcal{A} \subseteq B) = \Pi(B^{\supseteq} \mid \mathcal{A}) = \sup_{A \subseteq B} T(\pi_{\mathcal{A}}(A), 1), \quad (16)$$

$$NES(\mathcal{A} \subseteq B) = N(B^{\supseteq} \mid \mathcal{A}) = \inf_{A \not\subseteq B} I(\pi_{\mathcal{A}}(A), 0), \quad (17)$$

where we define $B^{\subseteq} = \{C \subseteq X \mid B \subseteq C\}$ and $B^{\supseteq} = \{C \subseteq X \mid C \subseteq B\}$.

The following theorems show that $NES(B \subseteq \mathcal{A})$ and $NES(\mathcal{A} \subseteq B)$ are calculated by membership functions of the lower and upper approximations of $\mathcal{A}$ without any additional assumption while $POS(B \subseteq \mathcal{A})$ and $POS(\mathcal{A} \subseteq B)$ are calculated by membership functions of upper and lower approximations of $\mathcal{A}$ with additional assumptions.

Theorem 1. *If assumption (A) holds, we have*

$$NES(B \subseteq \mathcal{A}) = \inf_{x \in B} I(n^{-1}(\mu_{A-}(x)), 0), \tag{18}$$

$$NES(\mathcal{A} \subseteq B) = \inf_{x \notin B} I(\varphi^{-1}(\mu_{A+}(x)), 0). \tag{19}$$

Proof. We prove only (18) because (19) can be proved in the same way.

$$NES(B \subseteq \mathcal{A}) = \inf_{A \not\supseteq B} I(\pi_{\mathcal{A}}(A), 0) = \inf_{\substack{x, A \\ x \notin A, \ x \in B}} I(\pi_{\mathcal{A}}(A), 0)$$

$$= \inf_{x \in B} I\left(n^{-1}\left(\inf_{x \notin A} n(\pi_{\mathcal{A}}(A))\right), 0\right) = \inf_{x \in B} I(n^{-1}(\mu_{A-}(x)), 0).$$

$\square$

Theorem 2. *If assumption (A) holds, we have*

$$POS(B \subseteq \mathcal{A}) \leq \inf_{x \in B} T(\varphi^{-1}(\mu_{A+}(x)), 1), \tag{20}$$

$$POS(\mathcal{A} \subseteq B) \leq \inf_{x \notin B} T(n^{-1}(\mu_{A-}(x)), 1). \tag{21}$$

However, if $\pi_{\mathcal{A}}$ satisfies

for any $\mathcal{Q} \subseteq 2^X$; $\pi_{\mathcal{A}}(A) \geq h$, $\forall A \in \mathcal{Q}$ implies $\pi_{\mathcal{A}}(\bigcup\{A \mid A \in \mathcal{Q}\}) \geq h$, (22)

we have the equality of (20) for any $B \subseteq X$. Moreover, if $\pi_{\mathcal{A}}$ satisfies

for any $\mathcal{Q} \subseteq 2^X$; $\pi_{\mathcal{A}}(A) \geq h$, $\forall A \in \mathcal{Q}$ implies $\pi_{\mathcal{A}}(\bigcap\{A \mid A \in \mathcal{Q}\}) \geq h$, (23)

we have the equality of (21) for any $B \subseteq X$.

Proof. We prove only (20) because (21) can be proved in the same way. For any $h \in [0, 1)$, we have

$$POS(B \subseteq \mathcal{A}) = \sup_{A \supseteq B} T(\pi_{\mathcal{A}}(A), 1) \geq h$$

$$\Leftrightarrow \forall \varepsilon > 0, \ \exists A \supseteq B, T(\pi_{\mathcal{A}}(A), 1) \geq h - \varepsilon$$

$$\Rightarrow \forall \varepsilon > 0, \ \forall x \in B \ \exists A \ni x, T(\pi_{\mathcal{A}}(A), 1) \geq h - \varepsilon$$

$$\Leftrightarrow \inf_{x \in B} \sup_{A \ni x} T(\pi_{\mathcal{A}}(A), 1) \geq h \Leftrightarrow \inf_{x \in B} T\left(\varphi^{-1}\left(\sup_{A \ni x} \varphi(\pi_{\mathcal{A}}(A))\right)\right) \geq h$$

$$\Leftrightarrow \inf_{x \in B} T(\varphi^{-1}(\mu_{A+}(x)), 1) \geq h.$$

Then we have (20). When (22) is satisfied, '$\Rightarrow$' in the third line of the equation above can be changed to '$\Leftrightarrow$'. Then we have the equality of (20). $\square$

When $\mathcal{A}$ is specified by its lower and upper approximation, i.e., $\mathcal{A}$ is characterized by $\pi_{\mathcal{A}}^*$ of (8), (22) and (23) are satisfied. Then we have the equalities in (20) and (21) in this case. Moreover this fact can be proved also by using (13) when $T = \min$.

Theorems 1 and 2 show the possibilities to calculate $POS(B \subseteq \mathcal{A})$, $NES(B \subseteq \mathcal{A})$, $POS(\mathcal{A} \subseteq B)$ and $NES(\mathcal{A} \subseteq B)$ on the universe while the original definitions require the calculations on the power set. We note that the calculations of $NES(B \subseteq \mathcal{A})$ and $NES(\mathcal{A} \subseteq B)$ on the universe do not request the assumption that $\mathcal{A}$ is specified by its lower and upper approximations.

Now let us extend the inclusion relation to the relation between graded ill-known sets. To this end, using $POS(B \subseteq \mathcal{A})$, $NES(B \subseteq \mathcal{A})$, $POS(\mathcal{A} \subseteq B)$ and $NES(\mathcal{A} \supseteq B)$, we define ill-known sets $\mathcal{A}_{\overline{\Pi}}^{\supseteq}$, $\mathcal{A}_{\overline{N}}^{\supseteq}$, $\mathcal{A}_{\overline{\Pi}}^{\subseteq}$ and $\mathcal{A}_{\overline{N}}^{\subseteq}$ associated with $\mathcal{A}$ by the following possibility distributions:

$$\pi_{\mathcal{A}_{\overline{\Pi}}^{\supseteq}}(B) = POS(B \subseteq \mathcal{A}), \qquad \pi_{\mathcal{A}_{\overline{N}}^{\supseteq}}(B) = NES(B \subseteq \mathcal{A}), \qquad (24)$$

$$\pi_{\mathcal{A}_{\overline{\Pi}}^{\subseteq}}(B) = POS(\mathcal{A} \subseteq B), \qquad \pi_{\mathcal{A}_{\overline{N}}^{\subseteq}}(B) = NES(\mathcal{A} \subseteq B). \qquad (25)$$

Finally, we obtain the following eight extended inclusion relations between graded ill-known sets $\mathcal{A}$ and $\mathcal{B}$:

$$POS(\mathcal{A} \subseteq \mathcal{B}) = \Pi(\mathcal{B}_{\overline{\Pi}}^{\supseteq} \mid \mathcal{A}) = \sup_{A \subseteq X} T\left(\pi_{\mathcal{A}}(A), \sup_{B \supseteq A} T(\pi_{\mathcal{B}}(B), 1)\right), \quad (26)$$

$$NEPO(\mathcal{A} \subseteq \mathcal{B}) = N(\mathcal{B}_{\overline{\Pi}}^{\supseteq} \mid \mathcal{A}) = \inf_{A \subseteq X} I\left(\pi_{\mathcal{A}}(A), \sup_{B \supseteq A} T(\pi_{\mathcal{B}}(B), 1)\right), \quad (27)$$

$$PONE(\mathcal{A} \subseteq \mathcal{B}) = \Pi(\mathcal{B}_{\overline{N}}^{\supseteq} \mid \mathcal{A}) = \sup_{A \subseteq X} T\left(\pi_{\mathcal{A}}(A), \inf_{B \not\supseteq A} I(\pi_{\mathcal{B}}(B), 0)\right), \quad (28)$$

$$NES(\mathcal{A} \subseteq \mathcal{B}) = N(\mathcal{B}_{\overline{N}}^{\supseteq} \mid \mathcal{A}) = \inf_{A \subseteq X} I\left(\pi_{\mathcal{A}}(A), \inf_{B \not\supseteq A} I(\pi_{\mathcal{B}}(B), 0)\right), \quad (29)$$

$$POS(\mathcal{B} \supseteq \mathcal{A}) = \Pi(\mathcal{A}_{\overline{\Pi}}^{\subseteq} \mid \mathcal{B}) = \sup_{B \subseteq X} T\left(\pi_{\mathcal{B}}(B), \sup_{A \subseteq B} T(\pi_{\mathcal{A}}(A), 1)\right), \quad (30)$$

$$NEPO(\mathcal{B} \supseteq \mathcal{A}) = N(\mathcal{A}_{\overline{\Pi}}^{\subseteq} \mid \mathcal{B}) = \inf_{B \subseteq X} I\left(\pi_{\mathcal{B}}(B), \sup_{A \subseteq B} T(\pi_{\mathcal{A}}(A), 1)\right), \quad (31)$$

$$PONE(\mathcal{B} \supseteq \mathcal{A}) = \Pi(\mathcal{A}_{\overline{N}}^{\subseteq} \mid \mathcal{B}) = \sup_{B \subseteq X} T\left(\pi_{\mathcal{B}}(B), \inf_{A \not\subseteq B} I(\pi_{\mathcal{A}}(A), 0)\right), \quad (32)$$

$$NES(\mathcal{B} \supseteq \mathcal{A}) = N(\mathcal{A}_{\overline{N}}^{\subseteq} \mid \mathcal{B}) = \inf_{B \subseteq X} I\left(\pi_{\mathcal{B}}(B), \inf_{A \not\subseteq B} I(\pi_{\mathcal{A}}(A), 0)\right). \quad (33)$$

When T is commutative and associative, we have $POS(\mathcal{A} \subseteq \mathcal{B}) = POS(\mathcal{B} \supseteq \mathcal{A})$. Moreover, when I satisfies $I(a, I(b, c)) = I(b, I(a, c))$, we have $NES(\mathcal{A} \subseteq \mathcal{B}) = NES(\mathcal{B} \supseteq \mathcal{A})$. Many famous conjunctive and implication functions such as triangular norm and S-implication satisfy those conditions. Then it is often that we only have six extensions. Moreover, when $I(1, a) \leq T(1, a)$ for any $a \in [0, 1]$, we obtain

$$POS(\mathcal{A} \subseteq \mathcal{B}) \geq NEPO(\mathcal{A} \subseteq \mathcal{B}) \geq NES(\mathcal{A} \subseteq \mathcal{B}), \tag{34}$$

$$POS(\mathcal{A} \subseteq \mathcal{B}) \geq PONE(\mathcal{A} \subseteq \mathcal{B}) \geq NES(\mathcal{A} \subseteq \mathcal{B}), \tag{35}$$

$$POS(\mathcal{B} \supseteq \mathcal{A}) \geq NEPO(\mathcal{B} \supseteq \mathcal{A}) \geq NES(\mathcal{B} \supseteq \mathcal{A}), \tag{36}$$

$$POS(\mathcal{A} \subseteq \mathcal{B}) \geq PONE(\mathcal{B} \supseteq \mathcal{A}) \geq NES(\mathcal{B} \supseteq \mathcal{A}). \tag{37}$$

When $I(a, T(b, c)) \geq T(b, I(a, c))$ for any $a, b \in [0, 1]$ and $c \in \{0, 1\}$, we have

$$NEPO(\mathcal{A} \subseteq \mathcal{B}) \geq PONE(\mathcal{B} \supseteq \mathcal{A}), \quad NEPO(\mathcal{B} \supseteq \mathcal{A}) \geq PONE(\mathcal{A} \subseteq \mathcal{B}). \tag{38}$$

When T is a t-norm and I is an S-implication, we have (34) to (38) as well as $POS(\mathcal{A} \subseteq \mathcal{B}) = POS(\mathcal{B} \supseteq \mathcal{A})$ and $NES(\mathcal{A} \subseteq \mathcal{B}) = NES(\mathcal{B} \supseteq \mathcal{A})$. In this case, we may understand that $POS(\mathcal{A} \subseteq \mathcal{B})$ and $NES(\mathcal{A} \subseteq \mathcal{B})$ are the weakest and strongest indices, respectively. Moreover, $NEPO(\mathcal{A} \subseteq \mathcal{B})$ and $NEPO(\mathcal{B} \supseteq \mathcal{A})$ (resp. $PONE(\mathcal{A} \subseteq \mathcal{B})$ and $PONE(\mathcal{B} \supseteq \mathcal{A})$) are second weakest (resp. strongest) indices but there is no weak-strong relation between them.

If assumption (A) holds, we have the following theorems about the calculations of the inclusion degrees by using lower and upper approximations.

Theorem 3. *If assumption (A) holds, we have*

$$NES(\mathcal{A} \subseteq \mathcal{B}) = \inf_{x \in X} I(\varphi^{-1}(\mu_{A^+}(x)), I(n^{-1}(\mu_{B^-}(x)), 0)), \tag{39}$$

$$NES(\mathcal{B} \supseteq \mathcal{A}) = \inf_{x \in X} I(n^{-1}(\mu_{B^-}(x)), I(\varphi^{-1}(\mu_{A^+}(x)), 0)), \tag{40}$$

$$NEPO(\mathcal{A} \subseteq \mathcal{B}) \leq \inf_{x \in X} I(\varphi^{-1}(\mu_{A^+}(x)), T(\varphi^{-1}(\mu_{B^+}(x)), 1)), \tag{41}$$

$$NEPO(\mathcal{B} \supseteq \mathcal{A}) \leq \inf_{x \in X} I(n^{-1}(\mu_{B^-}(x)), T(n^{-1}(\mu_{A^-}(x)), 1)). \tag{42}$$

We have the equality of (41) for any $\mathcal{A}$ when (22) corresponding to $\mathcal{B}$ is satisfied. Moreover, we have the equality of (42) for any $\mathcal{B}$ when (23) is satisfied.

Proof. The proof is omitted due to the page restriction. However, it can be proved in a similar way to the proof of Theorem 1. □

From Theorem 3, we have $NES(\mathcal{A} \subseteq \mathcal{B}) = NES(\mathcal{B} \supseteq \mathcal{A})$ when I satisfies $I(a, b) = I(\bar{n}(b), \bar{n}(a))$ and $\bar{n}(a) = \bar{n}^{-1}(a) = I(a, 0)$ for any $a, b \in [0, 1]$.

Theorem 4. *If assumption (A) holds, we have*

$$PONE(\mathcal{A} \subseteq \mathcal{B}) \leq \sup\{T(h_1, h_2) \mid (A^-)_{n(h_1)} \subseteq [B^-]_{n(I^*(h_2, 0))}\}, \tag{43}$$

$$PONE(\mathcal{B} \supseteq \mathcal{A}) \leq \sup\{T(h_1, h_2) \mid (A^+)_{\varphi(I^*(h_2, 0))} \subseteq [B^+]_{\varphi(h_1)}\}, \tag{44}$$

$$POS(\mathcal{A} \subseteq \mathcal{B}) \leq \sup\{T(h_1, h_2) \mid (A^-)_{n(h_1)} \subseteq [B^+]_{\varphi(T^*(h_2, 1))}\}, \tag{45}$$

$$POS(\mathcal{B} \supseteq \mathcal{A}) \leq \sup\{T(h_1, h_2) \mid (A^-)_{n(T^*(h_2, 1))} \subseteq [B^+]_{\varphi(h_1)}\}, \tag{46}$$

where $(A^-)_h$ and $(A^+)_h$ are strong h-level sets of fuzzy sets A^- and A^+ while $[B^-]_h$ and $[B^+]_h$ are h-level sets of fuzzy sets B^- and B^+, i.e., $(A^-)_h = \{x \in$

$X \mid \mu_{A^-}(x) > h\}$, $(A^+)_h = \{x \in X \mid \mu_{A^+}(x) > h\}$, $[B^-]_h = \{x \in X \mid \mu_{B^-}(x) \geq h\}$ and $[B^+]_h = \{x \in X \mid \mu_{B^+}(x) \geq h\}$. $I^*(\cdot, 0) : [0,1] \to [0,1]$ and $T^*(\cdot, 1) : [0,1] \to [0,1]$ are defined by

$$I^*(h, 0) = \sup\{q \in [0,1] \mid I(q, 0) \geq h\}, \quad T^*(h, 1) = \inf\{q \in [0,1] \mid T(q, 1) \geq h\}. \tag{47}$$

When $\mathcal{A}$ is specified by lower and upper approximations, (43) holds with equality. Similarly, when $\mathcal{B}$ is specified by lower and upper approximations, (44) holds with equality. When $\mathcal{A}$ is specified by lower and upper approximations and (22) holds for $\mathcal{B}$, (45) holds with equality. When $\mathcal{B}$ is specified by lower and upper approximations and (23) holds, (46) holds with equality.

Proof. We prove only (43). The others can be proved similarly. By definition and Theorem 1, we have

$$PONE(\mathcal{A} \subseteq \mathcal{B}) = \sup_{C \subseteq X} T\left(\pi_{\mathcal{A}}(C), \inf_{B \not\supseteq C} I(\pi_{\mathcal{B}}(B), 0)\right)$$

$$\leq \sup_{C \subseteq X} T\left(\min\left(\inf_{x \notin C} n^{-1}(\mu_{A^-}(x)), \inf_{x \in C} \varphi^{-1}(\mu_{A^+}(x))\right), \inf_{x \in C} I(n^{-1}(\mu_{B^-}(x)), 0)\right).$$

We have $\inf_{x \notin C} n^{-1}(\mu_{A^-}(x)) \geq h_1 \Leftrightarrow (A^-)_{n(h_1)} \subseteq C$, $\inf_{x \in C} \varphi^{-1}(\mu_{A^+}(x)) \geq h_1 \Leftrightarrow C \subseteq [A^+]_{\varphi(h_1)}$ and $\inf_{x \in C} I(n^{-1}(\mu_{B^-}(x)), 0) \geq h_2 \Leftrightarrow C \subseteq [B^-]_{n(I^*(h_2, 0))}$. Then, because $\mu_{A^-}(x) = 1$ implies $\mu_{A^+}(x) > 0$, we obtain

$$PONE(\mathcal{A} \subseteq \mathcal{B})$$
$$\leq \sup\left\{T(h_1, h_2) \mid (A^-)_{n(h_1)} \subseteq C \subseteq [A^+]_{\varphi(h_1)}, C \subseteq [B^-]_{n(I^*(h_2, 0))}\right\}$$
$$= \sup\{T(h_1, h_2) \mid (A^-)_{n(h_1)} \subseteq [B^-]_{n(I^*(h_2, 0))}\}.$$

When $\mathcal{A}$ is specified by the lower and upper approximations A^- and A^+, it has a possibility distribution π^* defined by (8). Then the inequality of the proof can be replaced with equality. Then we have (43) with equality. $\square$

Theorem 4 shows that the upper bounds of $PONE(\mathcal{A} \subseteq \mathcal{B})$, $PONE(\mathcal{B} \supseteq \mathcal{A})$, $POS(\mathcal{A} \subseteq \mathcal{B})$ and $POS(\mathcal{B} \supseteq \mathcal{A})$ can be calculated by the lower and upper approximations of $\mathcal{A}$ and $\mathcal{B}$ and these upper bounds become exact values in special cases. However, the calculations of these upper bounds are still complex because they are obtained as optimal values of optimization problems. However, when $T = \min$, those values can be calculated easily as shown in the following corollary.

Corollary 1. When $T = \min$, the right-hand sides of (43) to (46) become simpler, i.e., if assumption (A) holds, we have

$$PONE(\mathcal{A} \subseteq \mathcal{B}) \leq \inf_{x \in X} \max\left(n^{-1}(\mu_{A^-}(x)), I(n^{-1}(\mu_{B^-}(x)), 0)\right), \tag{48}$$

$$PONE(\mathcal{B} \supseteq \mathcal{A}) \leq \inf_{x \in X} \max\left(I(\varphi^{-1}(\mu_{A^+}(x)), 0), \varphi^{-1}(\mu_{B^+}(x))\right), \tag{49}$$

$$POS(\mathcal{A} \subseteq \mathcal{B}) \leq \inf_{x \in X} \max\left(n^{-1}(\mu_{A^-}(x)), \varphi^{-1}(\mu_{B^+}(x))\right), \tag{50}$$

$$POS(\mathcal{B} \supseteq \mathcal{A}) \leq \inf_{x \in X} \max\left(n^{-1}(\mu_{A^-}(x)), \varphi^{-1}(\mu_{B^+}(x))\right). \tag{51}$$

Table 1. Profit estimations and index values

project	probable	possible	POS	$NEPO$	$PONE$	NES
P_1	$(550, 570, 70, 30)$	$(470, 600, 30, 30)$	0.882353	0.538462	0.277778	0
P_2	$(520, 540, 30, 70)$	$(465, 605, 15, 20)$	0.923077	0.565217	0.142857	0

The conditions when the equalities of those equations are same as in Theorem 4.

Proof. Using Theorem 4, those can be proved straightforwardly by the following equivalence: $\inf_{x \in X} \max(\eta(x), \nu(x)) \geq h$ is equivalent to $\{x \in X \mid \eta(x) < h\} \subseteq \{x \in X \mid \nu(x) \geq h\}$ for any functions $\eta : X \to [0, 1]$ and $\nu : X \to [0, 1]$.   □

From Theorems 3 and 4, we can calculate all inclusion indices simply and correctly by lower and upper approximations when both graded ill-known sets are specified by their lower and upper approximations. In general cases, except $NES(\mathcal{A} \subseteq \mathcal{B})$ and $NES(\mathcal{B} \supseteq \mathcal{A})$, we cannot always calculate the inclusion degrees simply by the lower and upper approximations of $\mathcal{A}$ and $\mathcal{B}$. However the upper bounds of inclusion degrees are calculated simply by the lower and upper approximations of $\mathcal{A}$ and $\mathcal{B}$.

4   A Simple Example

In order to illustrate the potential applications of the proposed inclusion degrees in decision making, we consider a simple project selection problem. We assume there are two projects P_1 and P_2. The very probable profits and possible profits of projects are estimated by an expert as trapezoidal fuzzy numbers A_i^- and A_i^+ ($i = 1, 2$) shown in 'probable' and 'possible' columns of Table 1, respectively. A trapezoidal fuzzy numbers expressed by $M = (m^{\mathrm{L}}, m^{\mathrm{R}}, \alpha^{\mathrm{L}}, \alpha^{\mathrm{R}})$ is characterized by the following membership function:

$$\mu_M(x) = \begin{cases} 0, & \text{if } x < m^{\mathrm{L}} \text{ or } x > m^{\mathrm{R}}, \\ 1 - \dfrac{m^{\mathrm{L}} - x}{\alpha^{\mathrm{L}}}, & \text{if } x \in [m^{\mathrm{L}} - \alpha^{\mathrm{L}}, m^{\mathrm{L}}), \\ 1 - \dfrac{x - m^{\mathrm{R}}}{\alpha^{\mathrm{R}}}, & \text{if } x \in (m^{\mathrm{R}}, m^{\mathrm{R}} + \alpha^{\mathrm{R}}], \\ 1, & \text{if } x \in [m^{\mathrm{L}}, m^{\mathrm{R}}]. \end{cases} \tag{52}$$

The ill-known profit range $\mathcal{A}_i$ is defined by lower approximation A_i^- and upper approximation A_i^+. The possibility distribution $\pi_{\mathcal{A}_i}^*$ is obtained from A_i^- and A_i^+ ($i = 1, 2$). The satisfactory profits are given also by graded ill-known set $\mathcal{B}$ specified by its lower and upper approximations B^- and B^+. Then all inclusion indices are calculated exactly by lower and upper approximations. B^- and B^+ are characterized by the following two membership functions, respectively:

$$\mu_{B^-}(x) = \min\left(1, \max\left(0, \frac{x - 400}{100}\right)\right), \tag{53}$$

$$\mu_{B^+}(x) = \min\left(1, \max\left(0, \frac{x - 500}{110}\right)\right). \tag{54}$$

Let $\varphi(x) = x$, $n(x) = 1 - x$, $T = \min$ and $I(a, b) = \max(1 - a, b)$ (Dienes implication). In this special case, we have $POS(\mathcal{A}_i \subseteq \mathcal{B}) = POS(\mathcal{B} \supseteq \mathcal{A}_i)$, $NES(\mathcal{A}_i \subseteq \mathcal{B}) = NES(\mathcal{B} \supseteq \mathcal{A}_i)$, $NEPO(\mathcal{A}_i \subseteq \mathcal{B}) = PONE(\mathcal{B} \supseteq \mathcal{A}_i)$ and $PONE(\mathcal{A}_i \subseteq \mathcal{B}) = NEPO(\mathcal{B} \supseteq \mathcal{A}_i)$.

The values of $POS(\mathcal{A}_i \subseteq \mathcal{B})$, $NEPO(\mathcal{A}_i \subseteq \mathcal{B})$, $PONE(\mathcal{A}_i \subseteq \mathcal{B})$ and $NES(\mathcal{A}_i \subseteq \mathcal{B})$ are shown in POS, $NEPO$, $PONE$ and NES columns of Table 1, respectively. From Table 1, we observe the following two facts (I) and (II). (I) P_1 is better than P_2 in NEPO and PONE. Then by the degree that an estimated profit range of the project is included in a possible satisfactory range and by the degree that for each estimated profit range of the project there is a possible satisfactory range including it. project P_1 is better. However, (II) P_2 is better than P_1 in POS. Then by the degree that an estimated profit range of the project is included in all possible satisfactory ranges, project P_2 is better. If the analyst takes care of the certain satisfaction (or strong satisfaction), he/she can recommend P_2. On the contrary, if the analyst take care of the certainty to obtain somewhat satisfaction, he/she can recommend P_1.

Ackowledgement. This work was supported by KAKENHI No.23510169.

References

1. Dubois, D., Prade, H.: Incomplete Conjunctive Information. Comput. Math. Applic. 15(10), 797–810 (1988)
2. Atanassov, K.T.: Intuitionistic Fuzzy Sets. Fuzz. Sets Syst. 20(1), 87–96 (1986)
3. Dubois, D., Prade, H.: Rough Fuzzy Sets and Fuzzy Rough Sets. Int. J. General Syst. 17(2-3), 191–209 (1990)
4. Dubois, D., Prade, H.: Gradualness, Uncertainty and Bipolarity: Making Sense of Fuzzy Sets. Fuzzy Sets and Systems 192, 3–24 (2012)
5. Dubois, D., Prade, H.: A Set-Theoretic View of Belief Functions: Logical Operations and Approximations by Fuzzy Sets. Int. J. General Syst. 12(3), 193–226 (1986)
6. Augustin, T., Hable, R.: On the Impact of Robust Statistics on Imprecise Probability Models: A Review. Structural Safety 32(6), 358–365 (2010)
7. Inuiguchi, M.: Rough Representations of Ill-Known Sets and Their Manipulations in Low Dimensional Space. In: Skowron, A., Suraj, Z. (eds.) Rough Sets and Intelligent Systems - Professor Zdzisław Pawlak in Memoriam. ISRL, vol. 42, pp. 309–331. Springer, Heidelberg (2012)
8. Inuiguchi, M.: Ill-known Set Approach to Disjunctive Variables: Calculations of Graded Ill-known Intervals. In: Proceedings of IPMU 2012 (2012)
9. Ben-Tal, A., Nemirovski, A.: Robust Optimization – Methodology and Applications. Math. Program., Ser. B 92, 453–480 (2002)
10. Inuiguchi, M.: Robust Optimization by Fuzzy Linear Programming. In: Ermoliev, Y., Makowski, M., Marti, K. (eds.) Managing Safety of Heterogeneous Systems: Decisions under Uncertainties and Risks. LNEMS, vol. 658, pp. 219–239. Springer, Heidelberg (2012)
11. Inuiguchi, M., Ichihashi, H., Kume, Y.: Some Properties of Extended Fuzzy Preference Relations Using Modalities. Inform. Sci. 61(3), 187–209 (1992)

Information Granularity and Granular Structure in Decision Making

Baoli Wang[1,2], Jiye Liang[2], and Yuhua Qian[2]

[1] Department of Applied Mathematics, Yuncheng University,
Yuncheng 044000, Shanxi, China
[2] Key Laboratory of Computational Intelligence and
Chinese Information Processing of Ministry of Education,
School of Computer and Information Technology of Shanxi University,
Taiyuan 030006, Shanxi, China
pollycomputer@163.com,
{ljy,jinchengqyh}@sxu.edu.cn

Abstract. Multiple criteria decision making (MCDM) has received increasing attentions in both engineering and economic fields. Weights of the criteria directly affect decision results in MCDM, so it is important for us to acquire the appropriate weights of the criteria. In some decision making problems, experts always express their preference by multiplicative preference relation and fuzzy preference relation. In this paper, an objective method based on information granularity is proposed for acquiring weights of the criteria in MCDM. Moreover, we prove that the essence of a consistence preference relation is a partial relation, and analyze the corresponding partial granular structure of the alternative set according to the given partial relation.

Keywords: Information granularity, Granular structure, Multiplicative preference relation, Fuzzy preference relation, Partial relation.

1 Introduction

Decision making is a key issue of the decision theory. One of the most important decision making problems is the multi-criteria decision making problem (MCDM), which is characterized by the ranking of objects according to a set of criteria with pre-defined preference-ordered decision classes. It is widely used in credit approval, stock risk estimation, university ranking, etc[1–5]. In multi-criteria decision making problems, different weighting systems decide different results, so it is important for us to search for a rational weighting method. And weighting methods are classified into subjective method, objective method and the integrated method. Analytic hierarchy Process (AHP) introduced by Satty[6] is a very important approach to support the decision making. Using AHP, the decision maker(s) must compare all pairs of criteria and decision alternatives using a ratio scale to form some judgment matrixes, which are indeed the multiplicative preference relations. Fuzzy preference relation, the other common used

T. Li et al. (Eds.): RSKT 2012, LNAI 7414, pp. 440–449, 2012.
© Springer-Verlag Berlin Heidelberg 2012

preference format proposed by Basu[7], is also widely used in group decision making problems[8, 9].

Granular computing is an emerging field of study on human-centered, knowledge intensive problem solving using multiple levels of granularity [10, 11]. Granule, granulation and granularity are regarded as the three primitive notions of granular computing. A granule is a clump of objects drawn together by indistinguishability, similarity, and proximity of functionality [12, 13]. And granulation of an object leads to a collection of granules. The granularity is the measurement of the granulation degree of objects[14]. A triarchic theory of granular computing is proposed by Yao based on the three perspectives on philosophy, methodology and information processing paradigm. In methodological perspective, the granular computing is a structured problem solving method [15]. Of course, the granular computing can be used in the decision making problems, and many researchers have paid their attention to this field [16–19].

Granular computing has been used in decision problem in many fields. Granular reciprocal matrix was proposed by Pedrycz and Song in group decision making problems, and the flexibility offered by the level of granularity is used to increase the level of consensus within the group [16]. Different decision makers may provide multi-granular linguistic information in multi-criteria group decision making problems, so Herrera-Viedma et al. defined the measurements of consensus to help gain the more rational decision results[17], and paper[18] provided a way to use multi-granular linguistic model for management decision making in performance appraisal. In another study, Zheng et al. used granule sets to develop the bi-level decision models[19].

In this paper, we propose a special objective weights based on information granularity in multi-criteria decision problems. Moreover, we analyze the implied preference structure in the two preference relations: multiplicative preference relation and fuzzy preference relation. We prove that a consistent preference relation is indeed a partial relation under the given condition. And the corresponding partial relation induces a partial granular structure. This paper is organized as follows. Section 2 presents an objective weights acquisition method based on information granularity. Section 3 concludes that a consistent preference relation is truly a partial relation, and it can induce a partial granular structure. Section 4 concludes the paper and discusses the future research.

2   Weights Acquisition Based on Information Granularity

Multi-criteria decision making problems(MCDM)could be described by means of the following sets: $U = \{u_1, u_2, \cdots, u_n\}(n \geq 2)$ be a discrete set of n feasible alternatives; $A = \{a_1, a_2, \cdots, a_m\}$ be a finite set of attributes, and $a_i(u_j) = v_{ij}$ denote the value of u_j in the ith attribute; $\omega = (\omega_1, \omega_2, \cdots, \omega_m)$ be a weight vector of attributes, where $\omega_i \geq 0(i = 1, 2, \cdots, m), \sum_{i=1}^{m} \omega_i = 1$.

Several weighting methods have been summarized in reference[21]. In the following, we give a new weights acquisition method based on information granularity and we suppose all the attributes of the discrete numeric or linguistic values.

Given $a_i \in A$, let $ind(a_i) = \{(u_j, u_k)|a_i(u_j) = a_i(u_k)\}$, apparently, $ind(a_i)$ is the equivalence relation on U, and $U/ind(a_i)$ is the partition of the alternative set U, shortly and conveniently denoted by $U/a_i = \{[u_1]_{a_i}, [u_2]_{a_i}, \cdots, [u_n]_{a_i}\}$.

Definition 1. [20] *Let $U/a_i = \{[u_1]_{a_i}, [u_2]_{a_i}, \cdots, [u_n]_{a_i}\}$, then the information granularity $GR(a_i)$ of the attribute a_i is defined as*

$$GR(a_i) = \frac{1}{n} \sum_{j=1}^{n} \frac{|[u_j]_{a_i}|}{n}. \tag{1}$$

Where $|[u_j]_{a_i}|$ is the cardinality of the equivalence class of u_j .

Property 1. $\frac{1}{n} \leq GR(a_i) \leq 1$

Proof. $1 \leq |[u_j]_{a_i}| \leq n$, so $\frac{1}{n} \leq \frac{1}{n} \sum_{j=1}^{n} \frac{|[u_j]_{a_i}|}{n} \leq 1$.

There are two special cases, the first case is that every alternative has all the same value under the attribute a_i , the other case is that each alternative has unique value under the attribute a_i . In the former case, the attribute a_i contribute nothing to the decision making process, so we can set less weight even 0 for it. While in the other case, the a_i can distinguish the alternatives from each other, so we can set more weight for it. Just like information entropy, the information granulation $GR(a_i)$ can depict the distinguish ability of the attribute a_i . So we propose the weighting method based on information granularity in the following.

Definition 2. *In the MCDM problem given above, the distinguish importance of the attribute a_i is defined as*

$$DI(a_i) = 1 - GR(a_i). \tag{2}$$

Definition 3. *In the MCDM problem given above, the weight of the attribute a_i is defined as*

$$\omega_i = \frac{DI(a_i)}{\sum_{j=1}^{m} DI(a_j)}. \tag{3}$$

Apparently, $\omega_i \geq 0(i = 1, 2, \cdots, m), \sum_{i=1}^{m} \omega_i = 1$.

Example 1. An evaluation information system is given in table 1, we compute the weights of the criteria by the method given in this section.

$U/a_1 = \{\{u_1, u_3, u_4, u_8, u_9\}, \{u_2, u_7\}, \{u_5, u_6, u_{10}\}\}$

$U/a_2 = \{\{u_1, u_2, u_3, u_4\}, \{u_5, u_6, u_7, u_8, u_9\}, \{u_{10}\}\}$

$U/a_3 = \{\{u_1, u_8\}, \{u_2, u_5, u_{10}\}, \{u_3, u_6\}, \{u_4, u_7, u_9\}\}$

$U/a_4 = \{\{u_1\}, \{u_2, u_3, u_4, u_5, u_6, u_7, u_8, u_9, u_{10}\}\}$

$GR(a_1) = \frac{1}{10} \times (\frac{5}{10} + \frac{2}{10} + \frac{5}{10} + \frac{5}{10} + \frac{3}{10} + \frac{3}{10} + \frac{2}{10} + \frac{5}{10} + \frac{5}{10} + \frac{3}{10}) = \frac{38}{100}$

$GR(a_2) = \frac{1}{10} \times (\frac{4}{10} + \frac{4}{10} + \frac{4}{10} + \frac{4}{10} + \frac{5}{10} + \frac{5}{10} + \frac{5}{10} + \frac{5}{10} + \frac{5}{10} + \frac{1}{10}) = \frac{42}{100}$

$GR(a_3) = \frac{1}{10} \times (\frac{2}{10} + \frac{3}{10} + \frac{2}{10} + \frac{3}{10} + \frac{3}{10} + \frac{2}{10} + \frac{3}{10} + \frac{2}{10} + \frac{3}{10} + \frac{3}{10}) = \frac{26}{100}$

$GR(a_4) = \frac{1}{10} \times (\frac{1}{10} + \frac{9}{10} + \frac{9}{10} + \frac{9}{10} + \frac{9}{10} + \frac{9}{10} + \frac{9}{10} + \frac{9}{10} + \frac{9}{10} + \frac{9}{10}) = \frac{82}{100}$

$$DI(a_1) = 1 - GR(a_1) = 1 - \frac{38}{100} = \frac{62}{100}$$

$$DI(a_2) = 1 - GR(a_2) = 1 - \frac{42}{100} = \frac{58}{100}$$

$$DI(a_3) = 1 - GR(a_3) = 1 - \frac{26}{100} = \frac{74}{100}$$

$$DI(a_4) = 1 - GR(a_3) = 1 - \frac{82}{100} = \frac{18}{100}$$

$$\omega_1 = \frac{DI(a_1)}{\sum\limits_{j=1}^{4} DI(a_j)} = \frac{\frac{62}{100}}{\frac{62}{100} + \frac{58}{100} + \frac{74}{100} + \frac{18}{100}} = 0.292$$

$$\omega_2 = 0.274, \omega_3 = 0.349, \omega_4 = 0.085$$

From the Table 1, we can see that the discernibility of a_3 is the largest in the criteria and ω_3 is the same. The values of the alternatives under a_4 are almost all the same, so the importance of a_4 is the least. We just present the method of weights acquisition of this problem, the aggregation of it is not talked about here.

Table 1. An evaluation information system in a MCDM problem

U	a_1	a_2	a_3	a_4
u_1	good	medium	good	good
u_2	poor	medium	very poor	medium
u_3	good	medium	poor	medium
u_4	good	medium	medium	medium
u_5	medium	poor	very poor	medium
u_6	medium	poor	poor	medium
u_7	poor	poor	medium	medium
u_8	good	poor	good	medium
u_9	good	poor	medium	medium
u_{10}	medium	good	very poor	medium

The weighting method given above is an objective method, it can be used in MCDM, when it is hard to get the subjective weights. And it can also be combined with the subjective weights in integrated methods. Sometimes, the weights determined by objective methods are inconsistent with the DM's subjective preferences. Contrariwise, the judgments of the decision makers occasionally absolutely depend on their knowledge or experience, and the error in weights to some extent is unavoidable. As we all know, none of the two approaches is perfect, and a integrated method might be the most appropriate for determining the weights of criteria [21].

3   The Partial Granular Structure in Preference Relation

3.1   Multiplicative Preference Relation and Fuzzy Preference Relation

In some group decision making problems, decision makers expressed their preference by means of preference relation defined over a finite and fixed set of alternatives. Let $U = \{u_1, u_2, \cdots, u_n\}(n \geq 2)$ be a set of the alternatives. In a preference relation a decision maker associates to every pair of alternatives a value that reflects some degree of preference of the first alternative over the second one. Many important decision models have been developed using main two kinds of preference relations: multiplicative preference relation and fuzzy preference relation.

Multiplicative Preference Relations[22, 23]: A multiplicative preference relation A on the alternative set U is represented by a matrix $A = (a_{ij})$, while a_{ij} is interpreted as u_i is a_{ij} times as good as u_j. Satty suggests measuring a_{ij} using a ratio scale, and precisely the 1-9 scale [6]: $a_{ij} = 1$ indicates indifference between u_i and u_j , $a_{ij} = 9$ indicates u_i is absolutely preference to u_j , and $a_{ij} \in \{2, 3, \cdots, 8\}$ indicates intermediate preference evaluations, while $a_{ij} \in \{\frac{1}{2}, \frac{1}{3}, \cdots, \frac{1}{8}\}$ indicates the u_j is intermediate preference to u_i . In this case, the preference A is usually assumed multiplicative reciprocal, i.e., $a_{ij} \cdot a_{ji} = 1, \ \forall i, j \in \{1, 2, \cdots, n\}$.

Definition 4. *A reciprocal multiplicative preference relation $A = (a_{ij})$ is consistent if*

$$a_{ij} \cdot a_{jk} = a_{ik}, \forall i, j, k = 1, 2, \cdots, n. \tag{4}$$

Fuzzy Preference Relations[23]: A fuzzy preference relation B on the alternative set U is a fuzzy set on the product set $U \times U$, that is characterized by a membership function $\mu_B : U \times U \to [0, 1]$.

When cardinality of U is small, the preference relation may be conveniently represented by the $n \times n$ matrix $B = (b_{ij})$ being $b_{ij} = \mu_B(u_i, u_j) \forall i, j = 1, 2, \cdots, n$. b_{ij} is interpreted as the preference degree of the alternative u_i over u_j : $b_{ij} = \frac{1}{2}$ indicates indifference between u_i and $u_j(u_i \sim u_j)$, $b_{ij} = 1$ indicates u_i is absolutely preference to u_j , and $b_{ij} > \frac{1}{2}$ indicates u_i is preference to $u_j(u_i \succ u_j)$. In this case, the preference matrix B is usually assumed addictive reciprocal, i.e.,$b_{ij} + b_{ji} = 1, \ \forall i, j \in 1, 2, \cdots, n$.

Definition 5. [24] *A fuzzy preference relation $B = (b_{ij})$ is consistent if the relation satisfy the additive transitivity condition: reciprocal multiplicative preference relation $A = (a_{ij})$ is consistent if*

$$b_{ij} + b_{jk} - b_{ik} = \frac{1}{2} \ or \ b_{ij} + b_{jk} + b_{ki} = \frac{3}{2}(\forall i, j, k \in \{1, 2, \cdots, n\}). \tag{5}$$

3.2   Partial Granular Structure

Partial Relation and Consistent Multiplicative Preference Relation

Definition 6. *Let $A = (a_{ij})_{n \times n}$ be a consistent multiplicative preference relation, $\forall \alpha \in [\frac{1}{9}, 9]$, the binary relation R^{A^α} on U is defined as*

$$R^{A^\alpha} = \{(u_i, u_j) | a_{ij} > \alpha \ or \ i = j\}. \tag{6}$$

Theorem 1. *If $\alpha \geq 1$, the binary relation R^{A^α} induced by the consistent multiplicative relation A is a partial relation.*

Proof. Reflexivity. According to the definition of R^{A^α}, $\forall \mu_i \in U$, $(u_i, u_i) \in R^{A^\alpha}$.

Anti-symmetry. If$(u_i, u_j) \in R^{A^\alpha}$ and $(u_j, u_i) \in R^{A^\alpha}$ hold at the same time, according to the definition of R^{A^α}, and the condition $\alpha \geq 1$, we have the compound proposition $(a_{ij} > 1)$ *or* $(i = j)$ *and* $(a_{ji} > 1)$ *or* $(i = j)$ is true. If $a_{ij} > 1$ is true, then $a_{ji} = 1/a_{ij} < 1$, for the$(a_{ij} > 1)$ *or* $(i = j)$ is true, so $i = j$; If $a_{ij} \leq 1$, for $(a_{ij} > 1)$ *or* $(i = j)$ is true, then $i = j$;. So we can conclude that if $(u_i, u_j) \in R^{A^\alpha}$ and $(u_j, u_i) \in R^{A^\alpha}$ then $u_i = u_j$.

Transitivity. $(u_i, u_j) \in R^{A^\alpha}$ and $(u_j, u_k) \in R^{A^\alpha}$. When $i = j$ or $j = k$, obviously, $(u_i, u_k) \in R^{A^\alpha}$ is true. When $i \neq j$ *and* $j \neq k$, according to the definition of R^{A^α} and the condition $\alpha \geq 1$, we have $a_{ij} > 1, a_{jk} > 1$, then $a_{ik} = a_{ij} \times a_{jk} > 1$, so $(u_i, u_k) \in R^{A^\alpha}$.

This completes the proof.

Partial Relation and Consistent Fuzzy Preference Relation

Definition 7. *Let $B = (b_{ij})_{n \times n}$ be a consistence fuzzy preference relation, $\forall \alpha \in [0,1]$, the binary relation R^{B^α} on U is defined as*

$$R^{B^\alpha} = \{(u_i, u_j) | b_{ij} > \alpha \ \ or \ \ i = j\}. \tag{7}$$

Theorem 2. *If $\alpha \geq 0.5$, the binary relation R^{B^α} induced by the consistent fuzzy preference relation B is a partial relation.*

Proof. Reflexivity. According to the definition of R^{B^α}, $\forall \mu_i \in U$, $(u_i, u_i) \in R^{B^\alpha}$.

Anti-symmetry. If$(u_i, u_j) \in R^{B^\alpha}$ and $(u_j, u_i) \in R^{B^\alpha}$ hold at the same time, according to the definition of R^{B^α}, and the condition $\alpha \geq 0.5$, we have the compound proposition $(b_{ij} > 0.5)$ *or* $(i = j)$ *and* $(b_{ji} > 0.5)$ *or* $(i = j)$ is true. If $b_{ij} > 0.5$ is true, then $b_{ji} = 1 - b_{ij} < 0.5$, for the$(b_{ij} > 0.5)$ *or* $(i = j)$ is true, so $i = j$; If $b_{ij} \leq 0.5$, for $(b_{ij} > 0.5)$ *or* $(i = j)$ is true, then $i = j$;. So we can conclude that if $(u_i, u_j) \in R^{B^\alpha}$ and $(u_j, u_i) \in R^{B^\alpha}$ then $u_i = u_j$.

Transitivity. $(u_i, u_j) \in R^{B^\alpha}$ and $(u_j, u_k) \in R^{B^\alpha}$. When $i = j$ or $j = k$, obviously, $(u_i, u_k) \in R^{B^\alpha}$ is true. When $i \neq j$ *and*$j \neq k$, according to the definition of R^{B^α} and the condition $\alpha \geq 0.5$, we have $b_{ij} > 0.5, b_{jk} > 0.5$, then $b_{ik} = b_{ij} + b_{jk} - 0.5 > 0.5$, so $(u_i, u_k) \in R^{B^\alpha}$.

This completes the proof.

Let $U = \{u_1, u_2, \cdots, u_n\}$ be a set of alternatives, and P be a partial relation on U, $(u_i, u_j) \in P$, shortly, denoted by $u_i \succeq_P u_j$, means u_i preference to u_j under the partial relation P. The granule of knowledge induced by partial relation P is the set of objects dominating u_i, i.e. $[u_i]^{\succeq_P} = \{u_j | u_j \succeq u_i\}$.

Definition 8. *Let $U = \{u_1, u_2, \cdots, u_n\}$ be a set of alternatives, and P be a partial relation on U, define $G_P(U)$ is the partial granular structure of U induced by P, and denoted as*

$$G_P(U) = \{[u_1]^{\succeq_P}, [u_2]^{\succeq_P}, \cdots, [u_n]^{\succeq_P}\} \tag{8}$$

An approach to sorting for objects in set-valued ordered information systems are given based on the partial granulation induced by a partial relation in [25]. In this paper, we only discuss the partial granular structure induced by the two kinds of preference relation, and the applications are the further researches. The example given in the following is to demonstrate the partial granular structure implied in the consistent multiplicative preference relation and fuzzy preference relation.

Example 2. Suppose that we have a set of four alternatives $\{u_1, u_2, u_3, u_4, u_5\}$ and a decision maker gives his or her consistent multiplicative preference relation as follows:

$$A = \begin{bmatrix} 1 & \frac{1}{2} & 3 & 4 & 1 \\ 2 & 1 & 6 & 8 & 2 \\ \frac{1}{3} & \frac{1}{6} & 1 & \frac{4}{3} & \frac{1}{3} \\ \frac{1}{4} & \frac{1}{8} & \frac{3}{4} & 1 & \frac{1}{4} \\ 1 & \frac{1}{2} & 3 & 4 & 1 \end{bmatrix},$$

Then, we get

$R^{A^1} = \{(u_1, u_1), (u_1, u_3), (u_1, u_4), (u_2, u_1), (u_2, u_2), (u_2, u_3), (u_2, u_4), (u_2, u_5),$
$(u_3, u_3), (u_3, u_4), (u_4, u_4), (u_5, u_3), (u_5, u_4), (u_5, u_5)\}$

$R^{A^3} = \{(u_1, u_1), (u_1, u_4), (u_2, u_2), (u_2, u_3), (u_2, u_4), (u_3, u_3), (u_4, u_4), (u_5, u_4),$
$(u_5, u_5)\}$

$R^{A^6} = \{(u_1, u_1), (u_2, u_2), (u_2, u_4), (u_3, u_3), (u_4, u_4), (u_5, u_5)\}$

It is easily to prove that $R^{A^1}, R^{A^3}, R^{A^8}$ are partial relations, and their corresponding partial granular structure are:

$$G_{R^{A^1}}(U) = \{[u_1]^{\succeq}_{R^{A^1}}, [u_2]^{\succeq}_{R^{A^1}}, [u_3]^{\succeq}_{R^{A^1}}, [u_4]^{\succeq}_{R^{A^1}}, [u_5]^{\succeq}_{R^{A^1}}\}$$
$$= \{\{u_2, u_1\}, \{u_2\}, \{u_1, u_2, u_3, u_5\}, \{u_1, u_2, u_3, u_4, u_5\}, \{u_2, u_5\}\},$$

$$G_{R^{A^3}}(U) = \{[u_1]^{\succeq}_{R^{A^3}}, [u_2]^{\succeq}_{R^{A^3}}, [u_3]^{\succeq}_{R^{A^3}}, [u_4]^{\succeq}_{R^{A^3}}, [u_5]^{\succeq}_{R^{A^3}}\}$$
$$= \{\{u_1\}, \{u_2\}, \{u_2, u_3\}, \{u_1, u_2, u_4, u_5\}, \{u_5\}\},$$

$$G_{R^{A^6}}(U) = \{[u_1]^{\succeq}_{R^{A^6}}, [u_2]^{\succeq}_{R^{A^6}}, [u_3]^{\succeq}_{R^{A^6}}, [u_4]^{\succeq}_{R^{A^6}}, [u_5]^{\succeq}_{R^{A^6}}\}$$
$$= \{\{u_1\}, \{u_2\}, \{u_3\}, \{u_2, u_4\}, \{u_5\}\}.$$

From the example, we can see the smaller value of $\alpha \geq 1$, the more preference information can draw from the corresponding granular structure. So we can get the finest preference order is:

$$u_2 \succ \begin{pmatrix} u_1 \\ u_5 \end{pmatrix} \succ u_3 \succ u_4.$$

Example 3. Suppose that we have a set of four alternatives $\{u_1, u_2, u_3, u_4\}$ and a decision maker gives his or her fuzzy preference relation as follows:

$$B = \begin{bmatrix} 0.5 & 0.55 & 0.7 & 0.95 \\ 0.45 & 0.5 & 0.65 & 0.9 \\ 0.3 & 0.35 & 0.5 & 0.75 \\ 0.05 & 0.1 & 0.25 & 0.5 \end{bmatrix},$$

It is easy to prove that B is a consistent fuzzy preference relation. According to the definition 7, we get

$$R^{B^{0.5}} = \{(u_1, u_1), (u_2, u_2), (u_3, u_3), (u_4, u_4), (u_1, u_2), (u_1, u_3),$$
$$(u_1, u_4), (u_2, u_3), (u_2, u_4), (u_3, u_4)\},$$

$$R^{B^{0.7}} = \{(u_1, u_1), (u_2, u_2), (u_3, u_3), (u_4, u_4), (u_1, u_4)(u_2, u_4), (u_3, u_4), \},$$
$$R^{B^{0.95}} = \{(u_1, u_1), (u_2, u_2), (u_3, u_3), (u_4, u_4)\}.$$

And it is easy to see that the three relations $R^{B^{0.5}}, R^{B^{0.7}}, R^{B^{0.95}}$ are all partial relations, their respective partial granular structure are as follows:

$$G_{R^{B^{0.5}}}(U) = \{[u_1]^{\succeq}_{R^{B^{0.5}}}, [u_2]^{\succeq}_{R^{B^{0.5}}}, [u_3]^{\succeq}_{R^{B^{0.5}}}, [u_4]^{\succeq}_{R^{B^{0.5}}}\}$$
$$= \{\{u_1\}, \{u_1, u_2\}, \{u_1, u_2, u_3\}, \{u_1, u_2, u_3, u_4\}\},$$

$$G_{R^{B^{0.7}}}(U) = \{[u_1]^{\succeq}_{R^{B^{0.7}}}, [u_2]^{\succeq}_{R^{B^{0.7}}}, [u_3]^{\succeq}_{R^{B^{0.7}}}, [u_4]^{\succeq}_{R^{B^{0.7}}}\}$$
$$= \{\{u_1\}, \{u_2\}, \{u_3\}, \{u_1, u_2, u_3, u_4\}\},$$

$$G_{R^{B^{0.95}}}(U) = \{[u_1]^{\succeq}_{R^{B^{0.95}}}, [u_2]^{\succeq}_{R^{B^{0.95}}}, [u_3]^{\succeq}_{R^{B^{0.95}}}, [u_4]^{\succeq}_{R^{B^{0.95}}}\}$$
$$= \{\{u_1\}, \{u_2\}, \{u_3\}, \{u_4\}\}.$$

From this example, we can see the smaller value of $\alpha \geq 0.5$, the more preference information can draw from the corresponding granular structure. So we can get the finest preference order is:

$$u_1 \succ u_2 \succ u_3 \succ u_4.$$

4   Conclusions

To recapitulate, weight acquiring method is very important in multiple criteria decision making, and constructing granular structure of the set of alternatives is also helpful to comprehend the judgment of a decision maker. In this paper, we have proposed a weights acquisition method based on information granularity, and it can be combined with subjective weighting method to decide the final weights of the criteria. Moreover, we have proved that the partial structure is implied in the two preference relations. And a partial granular structure can be induced by the special partial relation implied in the two preference relation. The interesting topics for further study would be to construct suitable granules to solve decision making problems.

Acknowledgments. This work was supported by Special prophase project for the National Key Basic Research and Development Program of China (973) (No. 2011CB311805), the Scientific and Technical Plan of Shanxi Province(2011032102701), and Colledge Science and Technology Development of Shanxi Province(20101123).

References

1. Braszczynski, J., Greco, S., Slowinski, R.: Multi-criteria Classification-a New Scheme for Application of Dominace Based Decision Rules. Eur. J. Oper. Res. 181, 1030–1044 (2007)
2. Xu, Z.S.: Deviation Measures of Linguistic Preference Relations in Group Decision Making. Omega-Int. J. Manage. S. 33, 249–254 (2005)
3. Hwang, C.L., et al.: Group Decision under Multi-Criterion. Spering, Berlin (1987)
4. Keeney, R.A.: A Group Prefernce Axiomatization with Cardinal Utility. Manage. Sci. 23, 140–145 (1976)
5. Hu, Q.H., Yu, D.R., Guo, M.Z.: Fuzzy Preference Based on Rough Sets. Inf. Sci. 180, 2003–2022 (2011)
6. Satty, T.L.: Introduction to a Modeling of Social Decision Process. Math. Comput. Simul. 25, 105–107 (1983)
7. Basu, K.: Fuzzy Revealed Preference Theory. J. Econ. Theory 32, 212–227 (1984)
8. Chiclana, F., Herrera-Viedma, E., Herrera, F., Alonso, S.: Some Induced Ordered Weighted Averaging Operators and Their Use for Solving Group Decision Making Problems Based on Fuzzy Preference Relations. Eur. J. Oper. Res. 182, 383–399 (2007)
9. Chen, H.Y., Zhou, L.G.: An Approach to Group Decision Making with Interval Fuzzy Preference Relations Based on Induced Generalized Continuous Ordered Weighted Averaging Operator. Expert Syst. Appl. 38, 13432–13440 (2011)
10. Yao, Y.Y.: Granular Computing. Computer Science (Ji Suan Ji Ke Xue) 31, 1–5 (2004)
11. Banerjee, M., Yao, Y.: A Categorial Basis for Granular Computing. In: An, A., Stefanowski, J., Ramanna, S., Butz, C.J., Pedrycz, W., Wang, G. (eds.) RSFDGrC 2007. LNCS (LNAI), vol. 4482, pp. 427–434. Springer, Heidelberg (2007)
12. Liang, J.Y., Qian, Y.H.: Information Granules and Entropy Theory. Sci. China., Ser. F 51, 1427–1444 (2008)
13. Lin, T.Y.: Granular Computing on Binary Relations I: Data Mining and Neighborhood Systems. In: Skowron, A., Polkowski, L. (eds.) Knowledge Discovery, pp. 107–121. Physica-Verlag, New York (1998)
14. Qian, Y.H., Liang, J.Y., Wu, W.Z., Dang, C.Y.: Information Granularity in Fuzzy Binary GrC Model. IEEE T. Fuzzy Syst. 19, 253–265 (2011)
15. Yao, Y.: The Art of Granular Computing. In: Kryszkiewicz, M., Peters, J.F., Rybiński, H., Skowron, A. (eds.) RSEISP 2007. LNCS (LNAI), vol. 4585, pp. 101–112. Springer, Heidelberg (2007)
16. Pedrycz, W., Song, M.L.: Analytic Hierarchy Process in Group Decision Making and its Optimization with an Allocation of Information Granularity. IEEE T. Fuzzy Syst. 19, 527–540 (2011)
17. Herrera-Viedma, E., Mata, F.S., Martínez, L., Chiclana, F., Pérez, L.G.: Measurements of Consensus in Multi-granular Linguistic Group Decision-Making. In: Torra, V., Narukawa, Y. (eds.) MDAI 2004. LNCS (LNAI), vol. 3131, pp. 194–204. Springer, Heidelberg (2004)

18. de Andrés, R., García-Lapresta, J.L., Martínez, L.: A Multi-granular Linguistic Model for Management Decision-making in Performance Appraisal. Soft Comput. 14, 21–34 (2010)
19. Zheng, Z., He, Q., Shi, Z.Z.: Granule Sets Based Bilevel Decision Model. In: Wang, G., Peters, J.F., Skowron, A., Yao, Y. (eds.) RSKT 2006. LNCS (LNAI), vol. 4062, pp. 530–537. Springer, Heidelberg (2006)
20. Liang, J.Y., Qian, Y.H.: Information Granules and Entropy Theory in Information Systems. Science in China Series F: Inf. Sci. 51, 1427–1444 (2008)
21. Jahan, A., Mustapha, F., Sapuan, S.M., Ismail, M.Y., Bahraminasab, M.: A Framework for Weighting of Criteria in Ranking Stage of Material Selection Process. Int. J. Adv. Manuf. Technol. 58, 411–420 (2012)
22. Xu, Z.S., Cai, X.Q.: Group Consensus Algorithms based on Preference Relations. Information Science 181, 150–162 (2011)
23. Herrera-Viedma, E., Herrera, F., Chiclana, F., Luque, M.: Some Issues on Consistency of Fuzzy Preference Relations. Eur. J. Oper. Res. 154, 98–109 (2004)
24. Tanino, T.: Fuzzy Preference Orderings in Group Decision Making. Fuzzy Sets and Systems 12, 117–131 (1984)
25. Qian, Y.H., Liang, J.Y., Song, P., Dang, C.Y.: On Dominance Relations in Disjunctive Set-valued Ordered Information Systems. Int. J. Inf. Techn. & Mak. 9, 9–33 (2010)

Semi-supervised Clustering Ensemble
Based on Collaborative Training⋆

Jinyuan Zhang, Yan Yang, Hongjun Wang, Amjad Mahmood, and Feifei Huang

School of Information Science & Technology, Southwest Jiaotong University,
Chengdu, 610031, P.R. China
Key Lab. of Cloud Computing and Intelligent Technology, Chengdu, Sichuan
Province, 610031, P.R. China
lingsuch@qq.com, {yyang,wanghongjun}@swjtu.edu.cn,
amjad.pu@gmail.com, feifei-huang521@163.com

Abstract. Recent researches on data clustering is increasingly focusing
on combining multiple data partitions as a way to improve the robustness
of clustering solutions. Most of them focused on crisp clustering combination. Semi-supervised clustering uses a small amount of labeled data to
aid and bias the clustering of unlabeled data. However, in this paper, we
offer a semi-supervised clustering ensemble model based on collaborative
training (SCET) and an unsupervised clustering ensemble mode based
on collaborative training (UCET). In the ensemble step of SCET, semi-
supervised learning is introduced. While in UCET, the knowledge used
in SCET is replaced by information extracted from the base-clusterings.
Then tri-training is used as consensus of clustering ensemble. The experiments on datasets from UCI machine learning repository indicate that
the model improves the accuracy of clustering.

Keywords: Semi-supervised clustering ensemble model, collaborative
training, semi-supervised learning, clustering ensemble.

1   Introduction

Clustering for unsupervised data exploration and analysis has been investigated
for decades in the statistics, data mining, and machine learning communities.
The general goal of clustering is to partition a given set of data points in a
multidimensional space into clusters such that the points within a cluster are
similar to one another but points in different clusters are dissimilar. In other
words, intra-cluster similarity is maximized while the inter-cluster similarity is
minimized.

Clustering ensembles combine multiple base clusterings of a set of objects into
a single consolidated clustering, often referred to as the consensus solution [1–
7]. We proposed semi-supervised clustering ensemble model using collaborative
training.

⋆ This work is partially supported by the National Science Foundation of China (Nos.
61170111, 61003142 and 61152001) and the Fundamental Research Funds for the
Central Universities (No. SWJTU11ZT08).

T. Li et al. (Eds.): RSKT 2012, LNAI 7414, pp. 450–455, 2012.

The rest of this paper is organized as follows: Section 2 presents the clustering ensemble, semi-supervised learning and collaborative training. Section 3 explains the proposed semi-supervised clustering ensemble model using tri-training mechanism as consensus function. Finally, section 4 covers the experiments and analysis of results along with conclusion.

2   Related Work

2.1   Clustering Ensemble

Strehl and Ghosh [1] introduced three efficient heuristics to solve the cluster ensemble problem (i.e. CSPA, HGPA and MCLA). All of these algorithms approach the problem by first transforming the set of clusterings into a hypergraph representation.

Yang et al. [2] presented a new ant based clustering combination algorithm, which imitates the cooperative behavior of multi ant colonies. Initially each ant colony takes different types of ant moving speeds to generate independent clustering results, and then these results were combined using a hypergraph. Finally an ant based graph partitioning algorithm was used to cluster the hypergraph again. Wang et al. [7] proposed Bayesian cluster ensembles, a mixed-membership generative model to obtain a consensus clustering by combining multiple base clustering results.

2.2   Semi-supervised Learning

Semi-supervised clustering is another approach to improve the quality of cluster by using knowledge of original data. Semi-supervised clustering is also known as constrained clustering or clustering with side information, this is the cousin of semi-supervised classification. The goal is clustering but there are some "labeled data" in the form of must-links (two points must in the same cluster) and cannot-links (two points cannot in the same cluster).

2.3   Collaborative Training

Collaborative training is a good Semi-supervised learning method to improve the accuracy and quality of learning result. One of the useful techniques in this area is co-training. Most of the works follow the general co-training algorithm proposed by Blum and Mitchell [9] as shown in Fig. 1.

Tri-training [10] is new flavor of co-training algorithm. It generates 3 base classifiers from the original labeled data set. These classifiers are then refined using unlabeled instances in the tri-training process, where in each round, an unlabeled instance is labeled for a classifier if the other two classifiers agree on the labeling under certain conditions. Tri-training neither requires sufficient and redundant views nor puts any constraints on any supervised algorithm. It has a broader capability than the co-training style algorithms.

3   Semi-supervised Clustering Ensemble Model Based on Collaborative Training

3.1   Base Clustering Components Generation

A number of algorithms are available for the generation of base clusterings. We use K-means repeatedly with different initialization values (seeds).

3.2   Using Tri-training as Consensus

Generally, tri-training is used in classification, but we use it in clustering ensemble step. There are two types of selection. One is unsupervised clustering ensemble, where the knowledge is extracted from the results of clustering when it has the highest confidence. Another is the semi-supervised ensemble, where the knowledge is randomly extracted from the labeled dataset with a ratio, which can be defined or variously changed.

3.3   Semi-supervised Clustering Ensemble Model Based on Collaborative Training

As described earlier, we propose a semi-supervised clustering ensemble model based on collaborative training (SCET). In the first step, the base partitions are generated using unsupervised clustering or semi-supervised clustering. In the second step of ensembling, tri-training is used as a consensus function. The model is shown in Fig. 2.

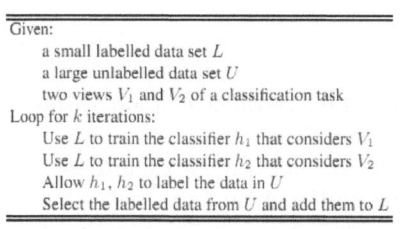

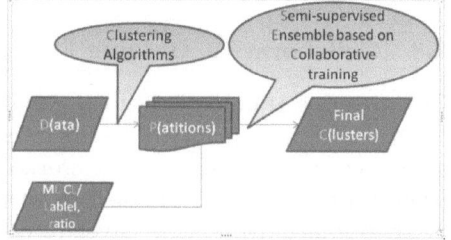

Fig. 1. General co-training algorithm   **Fig. 2.** Semi-supervised clustering ensemble model based on collaborative training

The model can also be described as Algorithm 1:

Algorithm 1. Semi-supervised clustering ensemble based on collaborative training

---

Input: $D(d_1, ..., d_n)$: dataset

      $L(l_1, ..., l_n)$: labels

      r: ratio of randomly selected labels to be the knowledge

Process:

    1. $P \leftarrow \Phi$

      for $i \in \{1, ..., n\}$ **do**

         $p_i \leftarrow K\text{-}means(d_i)$

         $P \leftarrow P \bigcup \{p_i\}$

      end for

    2. $PP \leftarrow$ randomly select $r * |D|$ instances from P,

      $LL \leftarrow$ corresponding labels from L.

      Generate *Labeled* using PP and LL.

    3. Generate *Unlabeled* using $P \bigcap \overline{PP}$.

    4. Final clusters: $C \leftarrow tri\text{-}training(Labeled, Unlabeled)$.

Output: C.

3.4 Unsupervised Clustering Ensemble Model Based on Collaborative Training

Meanwhile, an unsupervised clustering ensemble model based on collaborative training (UCET) is also proposed. The difference between UCET and SCET is that, in UCET, the knowledge used in SCET is replaced by the constraints extracted from the results of based clustering where each two instances have the highest confidence to be in the same cluster or in different clusters (i.e. $\forall i, j, number(p_i = p_j)$ is maximized $\|$ $number(p_i \neq p_j)$ is maximized).

4 Experiments

4.1 Dataset

The data sets used are gained at UCI Machine Learning Repository [11]. They are balance-scale (625, 4, 3), car (1728, 6, 4), contact-lenses (24, 4, 3), glass (214, 9, 7), haberman (306, 4, 2), hepatitis (155, 19, 2), iris (150, 4, 3), segment-challenge (2310, 19, 7) and soybean (683, 35, 19). The figures in the brackets indicate for each data set the number of instances, number of attributes and number of classes respectively.

4.2 Experiment Methodology

CSPA and K-means are chosen to compare with SCET and UCET. F1 and Accuracy are used as the validate measures for the comparison of the above algorithms.

The aim of these experiments is to compare K-means, MCLA, UCET and SCET. K-means are ran for 100 times with different initial values (seeds). Then average validate measures (i.e. F1 and Accuracy) of K-means and the validate measures of CSPA, UCET and SCET are compared to each other.

4.3   Results

F1 of the algorithms on each dataset is shown in Table 1 and Accuracy of the algorithms on each dataset is shown in Table 2.

Table 1. F1 of the algorithms on each dataset (max in bold type)

Alg \ Data	bal	car	cont	glass	hab	hep	iris	seg	soy
K-means	0.5890	0.4411	0.5235	0.4678	0.5498	0.7421	0.8803	0.6634	0.5444
CSPA	0.5883	0.3975	0.5521	0.3920	0.5601	0.7137	0.8800	0.6542	0.5007
UCET	0.6232	0.5406	0.6002	0.4570	0.5448	0.7603	**0.8853**	0.8130	0.6835
SCET(1%)	0.7721	0.7167	0.6151	0.5452	0.5508	0.7603	0.8800	0.8279	0.8986
SCET(6%)	0.8451	0.7560	0.6576	0.5432	0.5663	0.7603	**0.8853**	0.8371	0.9286
SCET(11%)	**0.8564**	**0.8220**	**0.7736**	**0.6177**	**0.6151**	0.7826	**0.8853**	**0.8428**	**0.9458**

Table 2. Accuracy of the algorithms on each dataset (max in bold type)

Alg \ Data	bal	car	cont	glass	hab	hep	iris	seg	soy
K-means	0.5314	0.3620	0.4792	0.3136	0.5042	0.7161	0.8803	0.6251	0.4831
CSPA	0.5296	0.3148	0.5417	0.2009	0.5327	0.6839	0.8800	0.6547	0.4392
UCET	0.5904	0.4948	0.6250	0.3178	0.5163	0.7355	**0.8867**	0.8093	0.6662
SCET(1%)	0.7136	0.6939	0.5833	0.3738	0.5229	0.7355	0.8800	0.8240	0.8990
SCET(6%)	0.8144	0.7390	0.6250	0.4439	0.5392	0.7355	**0.8867**	0.8353	0.9283
SCET(11%)	**0.8432**	**0.8119**	**0.7917**	**0.5140**	**0.5948**	0.7613	**0.8867**	**0.8407**	**0.9458**

1. It can be found, from the top 3 lines of Table 1 and Table 2, that UCET performs better than CSPA and K-means. UCET can be foundto be more stable than CSPA and K-means.
2. It can be found, from Table 1 and Table 2, that SCET performs best; since on all datasets, SCET(11%) gets the maximum value of F1 and Accuracy respectively. The decreasing order of performances (F1 and Accuracy) about each algorithms is SCET(11%) > SCET(6%) > SCET(1%) > UCET > CSPA > Base-cluster(K-means). Though on dataset Iris, the performances of the algorithms are nearly the same.
3. It can be found, from the last 3 lines in Table 1 and Table 2, that the more knowledge we use, the better performance shall we get. Since the decreasing order of performances is SCET(11%) > SCET(6%) > SCET(1%).

In addition, CSPA need a co-matrix, this require much space. However, in SCET and UCET, the co-matrix is not needed. So the SCET and UCET are more compatible than hypergraph approaches.

In summary, our approach performs good in the comparisons. It is due to not only the SCET is combination of the base clustering, but also guided by using semi-supervised knowledge and the knowledge is extended with the unlabeled part. UCET is another approach to be compatible for the condition where no semi-supervised information (i.e. labels or must-link / cannot-link) is given.

5 Conclusion

In this paper, a semi-supervised clustering ensemble model based on collaborative training (SCET) and an unsupervised model (UCET) are proposed. In the model, the semi-supervised learning is introduced, which uses a small amount of labeled data to aid and bias the clustering of unlabeled data. Then tri-training is regarded as consensus for clutering combination. Experiments indicate that the model improves the accuracy of clustering.

One of the future work is to improve the model to be more compatible in more conditions. Another future work is introduce the model into parallel computing or cloud computing.

References

1. Strehl, A., Gosh, J.: Cluster Ensembles - A Knowledge Reuse Framework for Combining Multiple Partitions. J. Mach. Learn. Res., 583–617 (2002)
2. Yang, Y., Kamel, M., Jin, F.: Topic Discovery from Document Using Ant-Based Clustering Combination. In: Zhang, Y., Tanaka, K., Yu, J.X., Wang, S., Li, M. (eds.) APWeb 2005. LNCS, vol. 3399, pp. 100–108. Springer, Heidelberg (2005)
3. Luo, H., Wei, H.: Clustering Ensemble Algorithm Based on Mathematical Morphology. Computer Science, 214–218 (2010) (in Chinese)
4. Ayad, H.G., Kamel, M.S.: On voting-based consensus of cluster ensembles. Pattern Recogn. 43, 1943–1953 (2010)
5. Vega-Pons, S., Ruiz-Shulcloper, J.: A survey of clustering ensemble algorithms. International Journal for Pattern Recognition and Artifitial Intelligence 25, 337–372 (2011)
6. Zhang, Y., Li, T.: Consensus Clustering + Meta Clustering = Multiple Consensus Clustering. In: Proceedings of the Twenty-Fourth International Florida Artificial Intelligence Research Society Conference, America (2011)
7. Wang, H.J., Li, Z.S., Qi, J.H., Cheng, Y., Zhou, P., Zhou, W.: Semi-supervised Cluster Ensemble Model Based on Bayesian Network. Journal of Software 21, 2814–2825 (2010) (in Chinese)
8. Du, J., Ling, C.X., Zhou, Z.H.: When Does Cotraining Work in Real Data? IEEE T. Knowl. Data En. 23, 788–799 (2011)
9. Blum, A., Mitchell, T.: Combining Labeled and Unlabeled Data with Co-Training, pp. 92–100 (1998)
10. Zhou, Z., Li, M.: Tri-training: exploiting unlabeled data using three classifiers. IEEE T. Knowl. Data En. 17, 1529–1541 (2005)
11. Frank, A., Asuncion, A.: UCI Machine Learning Repository. University of California, School of Information and Computer Science, Irvine (2010), http://archive.ics.uci.edu/ml/

Soft Set Approaches to Decision Making Problems

Keyun Qin[1], Jilin Yang[2], and Xiaohong Zhang[3]

[1] College of Mathematics, Southwest Jiaotong University, Chengdu, 610031, China
[2] College of Fundamental Education, Sichuan Normal University, Chengdu, 610068, China
[3] Department of Mathematics, College of Arts and Sciences, Shanghai Maritime University, Shanghai, 201306, China
{keyunqin,zxhonghz}@263.net, yjl524@163.com

Abstract. There are mainly two approaches applying soft set theory to decision making problems, namely, choice value based approach and comparison score based approach. Based on the analysis of these approaches, the main objective of this paper is to propose some improved algorithms which require relatively fewer calculations compared with the existing algorithms. In addition, it has been pointed out that the choice value based approach and comparison score based approach are equivalent for crisp soft set based decision making problems.

Keywords: Soft set, Fuzzy soft set, Choice value, Comparison score.

1 Introduction

Molodtsov [8] initiated the theory of soft sets as a mathematical tool for dealing with uncertain, fuzzy, not clearly defined objects. The soft set theory is different from traditional tools for dealing with uncertainties, such as probability theory, fuzzy set theory [14] and rough set theory [9], is that it is free from the inadequacy of the parametrization tools of these theories.

With the establishment of soft set theory, its application has boomed in recent years [1–4, 6, 10–13, 15]. Generally, there are two different approaches applying soft set theory to decision making problems. One is based on choice value and the other is based on comparison score. The choice value based approach is firstly proposed by Maji et al. [6] and is designed for soft set based decision making problems. The choice value of an object is the number of parameters possessed by the object. The object with the maximum choice value will be selected as the optimal decision. This approach has been generalized to dealing with fuzzy soft sets and interval-valued fuzzy soft sets [5, 13]. The comparison score based approach is proposed by Roy et al. [11] to dealing with fuzzy soft set based decision making problems. In this approach, we compare the membership values of two objects with respect to a common attribute to determine which one relatively possesses that attribute. This idea is implemented by introducing the notions of comparison table and comparison score of an object.

T. Li et al. (Eds.): RSKT 2012, LNAI 7414, pp. 456–464, 2012.
© Springer-Verlag Berlin Heidelberg 2012

We noticed that the weighted-average approach to incomplete soft set and comparison score based approach require many calculations, especially when the decision problem is associated with a large parameter set or a great number of objects. Thus sometimes it is not very efficient to select the optimal object by using these approaches. In addition, choice value based approach and comparison score based approach have not been connected closely among each other. Based on the analysis of these approaches, the objective of this paper is to improve the related algorithms and reduce the computational complexities. Concretely,

(1) An improved algorithm for comparison score based approach is proposed. It requires relatively fewer calculations due to the fact that the comparison table need not to be constructed.

(2) An improved algorithm is proposed for dealing with soft sets under incomplete information in decision making problems.

(3) It has been pointed out that, in crisp soft set based decision making problems, the choice value based approach and comparison score based approach are equivalent for selecting the optimal alternatives.

2   Soft Sets and Fuzzy Soft Sets

This section presents a review of some fundamental notions of soft sets and fuzzy soft set. We refer to [7, 8] for details.

Let U be an initial universe set and E the set of all possible parameters under consideration related to the objects in U. Usually, parameters are attributes, characteristics, or properties of objects in U. The pair (U, E) will be called a soft universe. Let $P(U)$ be the power set of U (i.e., the set of all subsets of U). Molodtsov defined the notion of a soft set in the following way:

Definition 1. *[8] A pair (F, A) is called a soft set over U, where $A \subseteq E$ and F is a mapping given by $F : A \rightarrow P(U)$.*

In other words, a soft set over U is a parameterized family of subsets of U, which gives an approximate description of the objects in U. For $e \in A$, $F(e)$ may be considered as the set of $e-$approximate elements of the soft set (F, A).

In [7], Maji et al. introduced the concept of fuzzy soft set by combining soft set and fuzzy set.

Definition 2. *[7] A pair (F, A) is called a fuzzy soft set over U, where $A \subseteq E$, F is a mapping given by $F : A \rightarrow F(U)$, and $F(U)$ is the set of all fuzzy sets on U.*

In this definition, fuzzy sets on the universe U are used as substitutes for crisp subsets of U. Hence it is easy to see that every soft set may be considered as a fuzzy soft set.

Example 1. Suppose that there are six houses in the universe U given by $U = \{h_1, h_2, h_3, h_4, h_5, h_6\}$ and $E = \{e_1, e_2, e_3, e_4, e_5\}$ is the set of parameters. Where e_1, e_2, e_3, e_4, e_5 stand for the parameter 'very costly', 'costly', 'cheap', 'beautiful' and 'in the green surroundings' respectively.

Consider the fuzzy soft set (F, A) which describes the 'cost of the houses', where $A = \{e_1, e_2, e_3\}$. Suppose that:

$$F(e_1) = \{(h_1, 0.7), (h_2, 1), (h_3, 0.8), (h_4, 1), (h_5, 0.9), (h_6, 0.3)\},$$
$$F(e_2) = \{(h_1, 0.8), (h_2, 1), (h_3, 1), (h_4, 1), (h_5, 0.9), (h_6, 0.5)\},$$
$$F(e_3) = \{(h_1, 0.3), (h_2, 0.2), (h_3, 0.2), (h_4, 0.1), (h_5, 0.3), (h_6, 0.7)\}.$$

Then the fuzzy soft set (F, A) is a parameterized family $\{F(e_i); 1 \leq i \leq 3\}$ of fuzzy sets on U and give us an approximate descriptions of the costs of houses.

3   Analysis of the Existing Approaches

3.1   Choice Value Based Approach

In [6], Maji et al. proposed a method for the application of soft set theory in decision making problems. We call the method choice value based approach.

A soft set (F, A) can be expressed as a binary table. If $h_i \in F(e_j)$ then $h_{ij} = 1$, otherwise $h_{ij} = 0$, where h_{ij} are the entries in the table. The choice value c_i of an object h_i is computed by $c_i = \sum_j h_{ij}$. Hence, the choice value of an object is just the number of parameters possessed by the object. In choice value based approach, the object with the maximum choice value is selected as the optimal decision. This approach may be summarized as the following algorithm.

Algorithm 1 [6]:
1. Input the (resultant) soft set (F, A) and place it in tabular form.
2. Compute the choice values c_i for each object h_i, where $c_i = \sum_j h_{ij}$.
3. The decision is h_i if $c_i = max_j c_j$.
4. If i has more than one value then any one of h_i may be chosen.

In this approach, the criterion for decision making is the comparison of choice values. For decision making problems using soft sets, the choice value of an object precisely represents the number of (good) attributes possessed by the object. Hence it is reasonable to select the object with the maximum choice value as the optimal alternative.

In real decision making problems, the choice parameters may not be of equal importance. To cope with such problems, we can impose different weights to different decision parameters. In addition, it has been generalized to deal with fuzzy soft set [5]. In this case, the choice value will be computed by: $c_i = \sum_j F(e_j)(h_i)$, where $F(e_j)(h_i)$ is the membership value of h_i with respect to fuzzy set $F(e_j)$. Table 1 and Table 2 are examples of weighted soft set and fuzzy soft set with choice values, respectively.

Feng et al. [2] pointed out that, for fuzzy soft set based decision making problems, the choice value should be exactly interpreted as 'fuzzy choice value'

Table 1. A weighted soft set (F, A) with choice values

	$e_1/0.3$	$e_2/0.2$	$e_3/0.1$	$e_4/0.3$	$e_5/0.1$	Choice value
h_1	0	1	1	1	1	0.7
h_2	1	0	1	1	0	0.7
h_3	0	0	1	1	1	0.5
h_4	1	0	1	1	0	0.7
h_5	1	0	1	0	0	0.4
h_6	1	1	0	1	1	0.8

Table 2. A fuzzy soft set (F, A) with choice values

	e_1	e_2	e_3	e_4	e_5	Choice value
h_1	0.1	0.5	0.3	0.4	0.3	1.6
h_2	0.3	0.5	0.2	0.3	0.6	1.9
h_3	0.1	0.7	0.4	0.5	0.1	1.8
h_4	0.7	0.2	0.2	0.2	0.3	1.6
h_5	0.2	0.6	0.3	0.2	0.3	1.6
h_6	0.9	0.2	0.1	0.1	0.8	2.1

and the direct addition of all the membership values with respect to different attributes is not always reasonable. In a certain sense this is just like, for instance, the addition of height and weight. Furthermore, an adjustable approach to fuzzy soft set based decision making problem was presented by means of level soft sets. We noticed that level soft sets of a fuzzy soft set are classical soft sets. Theoretically speaking, the adjustable approach presented in [2] is just the choice value based approach. The advantage is that we can select the corresponding 'level' based on the practical situations.

3.2   Comparison Score Based Approach

Roy and Maji [11] proposed comparison score based approach to solving fuzzy soft set based decision making problems. In this approach, rather than using the concept of choice values designed for crisp soft sets, we can compare the membership values of two objects with respect to a common attribute to determine which one relatively possesses that attribute. This idea is implemented by introducing the concept of a comparison table and a measure called the comparison score of an object.

The comparison table of a fuzzy soft set (F, A) is a square table in which rows and columns both are labelled by the objects $h_1, h_2, \cdots, h_n$ of the universe, and the entries c_{ij} indicate the number of parameters for which the membership value of h_i exceeds or equal to the membership value of h_j. That is, $c_{ij} = |\{e \in A; F(e)(h_i) \geq F(e)(h_j)\}|$. The row-sum r_i of an object h_i and the column-sum t_j of an object h_j are computed by

$$r_i = \sum_{j=1}^{n} c_{ij}, t_j = \sum_{i=1}^{n} c_{ij},$$

and the final comparison score of h_i is $s_i = r_i - t_i$. The object with the maximum comparison score will be selected as the optimal alternative.

Comparison score based approach can be summarized as the following algorithm.

Algorithm 2 [11]
1. Input the resultant fuzzy soft set (F, A) and place it in tabular form.
2. Construct the comparison table of the fuzzy soft set (F, A) and compute r_i and t_i for h_i, $\forall i$.
3. Compute the comparison score s_i of h_i, $\forall i$.
4. The decision is h_i if $s_i = max_j s_j$.
5. If i has more than one value then any one of h_i may be chosen.

Example 2. We consider fuzzy soft set (F, A) given in Table 2. The comparison table and the comparison score table of (F, A) are given in Table 3 and Table 4, respectively.

Table 3. Comparison table of fuzzy soft set (F, A)

	h_1	h_2	h_3	h_4	h_5	h_6
h_1	5	3	2	4	3	3
h_2	3	5	2	4	3	3
h_3	4	3	5	3	3	3
h_4	2	2	2	5	3	3
h_5	4	2	2	4	5	3
h_6	2	2	2	3	2	5

Table 4. Comparison score table of fuzzy soft set (F, A)

	Row-sum r_i	Column-sum t_i	Comparison score s_i
h_1	20	20	0
h_2	20	17	3
h_3	21	15	6
h_4	17	23	-6
h_5	20	19	1
h_6	16	20	-4

From Table 4, h_3 has the maximum score $s_3 = 6$. Hence, the optimal decision is to select h_3.

It is worth noticing that, by Table 2 and Table 4, the choice value based approach and comparison score based approach are not equivalent in general.

In comparison score based approach, we need to construct the comparison table. Because the comparison score is the difference of the related row-sum and column-sum, we need not take the condition $F(e)(h_i) = F(e)(h_j)$ into account. That is, the entries of comparison table can be revised as $c'_{ij} = |\{e \in E; F(e)(h_i) > F(e)(h_j)\}|$. The tabular form of fuzzy soft set (F, A) can be represented by a $n \times m$ matrix

$$H = (h_{ij})_{n \times m} = \begin{pmatrix} h_{11} & h_{12} & \cdots & h_{1m} \\ h_{21} & h_{22} & \cdots & h_{2m} \\ \cdots & \cdots & \cdots & \cdots \\ h_{n1} & h_{n2} & \cdots & h_{nm} \end{pmatrix}$$

where n is the number of objects in universe, and m is the number of parameters. We consider the matrix:

$$H_i = \begin{pmatrix} h_{i1} & h_{i2} & \cdots & h_{im} \\ h_{i1} & h_{i2} & \cdots & h_{im} \\ \cdots & \cdots & \cdots & \cdots \\ h_{i1} & h_{i2} & \cdots & h_{im} \end{pmatrix} - \begin{pmatrix} h_{11} & h_{12} & \cdots & h_{1m} \\ h_{21} & h_{22} & \cdots & h_{2m} \\ \cdots & \cdots & \cdots & \cdots \\ h_{n1} & h_{n2} & \cdots & h_{nm} \end{pmatrix}$$

$$= \begin{pmatrix} h_{i1} - h_{11} & h_{i2} - h_{12} & \cdots & h_{im} - h_{1m} \\ h_{i1} - h_{21} & h_{i2} - h_{22} & \cdots & h_{im} - h_{2m} \\ \cdots & \cdots & & \cdots \\ h_{i1} - h_{n1} & h_{i2} - h_{n2} & \cdots & h_{im} - h_{nm} \end{pmatrix}.$$

Assume that r'_i is the number of positive entries in H_i, whereas t'_i is the number of negative entries in H_i. It follows that $r'_i - t'_i$ is just the comparison score of h_i. Let $H_i^{(s)} = (b_{jk})_{n \times m}$ be a matrix where $b_{jk} = 1$ if $h_{ik} - h_{jk} > 0$, $b_{jk} = 0$ if $h_{ik} - h_{jk} = 0$, and $b_{jk} = -1$ if $h_{ik} - h_{jk} < 0$. We call $H_i^{(s)}$ the sign matrix of object h_i. Hence the sum of all entries in $H_i^{(s)}$ is the comparison score of h_i.

In view of above observations, Algorithm 2 can be revised to the following Algorithm 2' for fuzzy soft set based decision making problems:

Algorithm 2'
1. Input the resultant fuzzy soft set (F, A) and place it in tabular form.
2. Compute the sign matrix $H_i^{(s)}$ of object h_i, $\forall i$.
3. Compute the comparison score s_i of h_i, $\forall i$, where

$$s_i = (1, 1, \cdots, 1)_{1 \times n} H_i^{(s)} \begin{pmatrix} 1 \\ 1 \\ \cdots \\ 1 \end{pmatrix}_{m \times 1}. \tag{1}$$

4. The decision is h_i if $s_i = max_j s_j$.
5. If i has more than one value then any one of h_i may be chosen.

In Algorithm 2', we need not construct the comparison table. Hence Algorithm 2' requires relatively fewer calculations and it may be considered as an improvement of Algorithm 2.

In (1), if H_i is used as substitute for $H_i^{(s)}$, then we have:

Proposition 1. *Let (F, A) be a fuzzy soft set. Then*

$$n \times c_i - \triangle = (1, 1, \cdots, 1)_{1 \times n} H_i \begin{pmatrix} 1 \\ 1 \\ \cdots \\ 1 \end{pmatrix}_{m \times 1} \qquad (2)$$

Where n is the number of objects in the universe, c_i is the fuzzy choice value of h_i, and $\triangle = \sum_{i=1}^{n} \sum_{j=1}^{m} h_{ij}$ is the sum of all entries in tabular form of (F, A).

It is easy to see that $H_i = H_i^{(s)}$ for crisp soft set. Hence we have the following corollary.

Corollary 1. *Let (F, A) be a crisp soft set. Then $s_i = n \times c_i - \triangle$.*

This corollary shows the connections between choice value and comparison score for crisp soft set. In addition, in crisp soft set based decision making problems, the choice value based approach and comparison score based approach are equivalent for selecting the optimal alternatives.

3.3 Approaches for Incomplete Information

In [15], Zou and Xiao proposed an approach by using weighted-average to dealing with soft sets under incomplete information in decision making problems. The approach can be described as follows. Here some modifications on notations and technical terms have been made to fit the context of our discussion.

Let (F, E) be an incomplete soft set and $e \in E$. An object h belongs to $F(e)$, does not belong to $F(e)$ and h is uncertain with respect to $F(e)$ are denoted by $F(e)(h) = 1$, $F(e)(h) = 0$ and $F(e)(h) = *$, respectively. Let E_* be the set of parameters with incomplete information. Suppose p_e and q_e stand for the probabilities that an object h with $F(e)(h) = *$ belongs to and does not belong to $F(e)$, respectively, and defined by $p_e = \frac{n_1}{n_1 + n_0}, q_e = \frac{n_0}{n_1 + n_0}$, where n_1 and n_0 stand for the number of objects that belongs to and does not belong to $F(e)$, respectively. Let $E_*^h = \{e \in E_*; F(e)(h) = *\} = \{e_1, \cdots, e_m\}$ be the set of parameters having incomplete information with respect to h and $c = |\{e; F(e)(h) = 1\}|$. It follows that there are $m + 1$ possible choice values for h, namely, $c, c + 1, \cdots, c + m$. We denote by d_h the decision value of h. Hence the probability of $d_h = c$ is $\prod_{e \in E_*^h} q_e$, the probability of $d_h = c + 1$ is $p_{e_1} q_{e_2} \cdots q_{e_m} + q_{e_1} p_{e_2} \cdots q_{e_m} + \cdots + q_{e_1} q_{e_2} \cdots p_{e_m} = \sum_{i=1}^{m} p_{e_i} \prod_{1 \le j \le m, j \ne i} q_{e_j}$, and so on. Consequently $d_h = \sum_{x=0}^{m} (c + x) k_x$, and the weight k_x is computed by

$$k_x = \begin{cases} \prod_{e \in E_*^h} q_e, & \text{if } x = 0, \\ \prod_{e \in E_*^h} p_e, & \text{if } x = m, \\ \sum_{C_m^x} ((\prod_{e \in E_{*1}^h} p_e)(\prod_{e \in E_{*0}^h} q_e)), & \text{if } 0 < x < m. \end{cases} \qquad (3)$$

where $E_{*1}^h, E_{*0}^h \subseteq E_*^h$, E_{*1}^h and E_{*0}^h are the sets of parameters with the value of 1 and 0, respectively.

This approach can be summarized as the following algorithm.

Algorithm 3

1. Input the incomplete soft set (F, E) over the universe U and place it in tabular form, where the missing entries are represented by $*$.

2. Let E_* be the set of parameters with incomplete information. Calculate the probabilities p_e and q_e by the following formulas:

$p_e = \frac{n_1}{n_1 + n_0}, q_e = \frac{n_0}{n_1 + n_0}, e \in E_*$,

where $n_1 = |\{h \in U; F(e)(h) = 1\}|$, $n_0 = |\{h \in U; F(e)(h) = 0\}|$.

3. For each object h, let $E_*^h = \{e \in E_*; F(e)(h) = *\} = \{e_1, \cdots, e_m\}$ be the set of parameters having incomplete information with respect to h and $c = |\{e; F(e)(h) = 1\}|$. Compute the decision value d_h of h by weighted-average:

$d_h = \sum_{x=0}^{m}(c + x)k_x$.

4. The decision is h_i if $d_{h_i} = max_j d_{h_j}$.

5. If i has more than one value then any one of h_i may be chosen.

For decision value, we have the following proposition.

Proposition 2. *Let (F, E) be an incomplete soft set and h an object. Then*

$d_h = c + \sum_{e \in E_*^h} p_e$,

where $c = |\{e; F(e)(h) = 1\}|$, $E_*^h = \{e \in E; F(e)(h) = *\}$.

By this Proposition, the step 3 in Algorithm 3 can be simplified as:

$3'$. For each object h, computing the decision value d_h of h by weighted-average:

$d_h = c + \sum_{e \in E_*^h} p_e$,

where $c = |\{e; F(e)(h) = 1\}|$, $E_*^h = \{e \in E; F(e)(h) = *\}$.

This approach will improve Algorithm 3. In the revised algorithm, the decision values are computed directly based on the probabilities p_e and the probabilities of the possible choice values need not to be computed.

4   Conclusions

There are mainly two different approaches applying soft set theory to decision making problems. One is based on choice value and the other is based on comparison score. This paper is devoted to the analysis of these approaches and the related algorithms. Some improved algorithms are presented and the computational complexities of the existing algorithms are reduced. The connections between choice value based approach and comparison score based approach are also examined. Based on these results, we can further probe the practical applications of soft set theory in decision making problems.

Acknowledgements. This work has been supported by the National Natural Science Foundation of China (Grant No. 61175044), the Program of Education Office of Sichuan Province (Grant No. 11ZB068) and the Fundamental Research Funds for the Central Universities of China (Grant No. SWJTU11ZT29).

References

1. Chen, D.G., Tsang, E.C.C., Yeung, D.S., Wang, X.Z.: The paremeterization reduction of soft sets and its applications. Comput. Math. Appl. 49, 757–763 (2005)
2. Feng, F., Jun, Y.B., Liu, X.Y., Li, L.F.: An adjustable approach to fuzzy soft set based decision making. J. Comput. Appl. Math. 234, 10–20 (2010)
3. Feng, F., Li, Y.M., Violeta, L.F.: Application of level soft sets in decision making based on interval-valued fuzzy soft sets. Comput. Math. Appl. 60, 1756–1767 (2010)
4. Kong, Z., Gao, L.Q., Wang, L.F., Li, S.: The normal paremeter reduction of soft sets and its algorithm. Comput. Math. Appl. 56, 3029–3037 (2008)
5. Kong, Z., Gao, L.Q., Wang, L.F.: Comment on a fuzzy soft set theoretic approach to decision making problems. J. Comput. Appl. Math. 223, 540–542 (2009)
6. Maji, P.K., Biswas, R., Roy, A.R.: Fuzzy soft sets. J. Fuzzy Math. 9(3), 589–602 (2001)
7. Maji, P.K., Roy, A.R., Biswas, R.: An application of soft sets in a decision making problem. Comput. Math. Appl. 44, 1077–1083 (2002)
8. Molodtsov, D.: Soft set theory-First results. Comput. Math. Appl. 37, 19–31 (1999)
9. Pawlak, Z.: Rough sets. Int. J. Comput. Inf. Sci. 11, 341–356 (1982)
10. Qin, K.Y., Hong, Z.Y.: On soft equality. J. Comput. Appl. Math. 234, 1347–1355 (2010)
11. Roy, A.R., Maji, P.K.: A fuzzy soft set theoretic approach to decision making problems. J. Comput. Appl. Math. 203, 412–418 (2007)
12. Xiao, Z., Gong, K., Zou, Y.: A combined forecasting approach based on fuzzy soft sets. J. Comput. Appl. Math. 228, 326–333 (2009)
13. Yang, X.B., Lin, T.Y., Yang, J.Y., Li, Y., Yu, D.J.: Combination of interval-valued fuzzy set and soft set. Comput. Math. Appl. 58, 521–527 (2009)
14. Zadeh, L.A.: Fuzzy sets. Information and Control 8, 338–353 (1965)
15. Zou, Y., Xiao, Z.: Data analysis approaches of soft sets under incomplete information. Knowledge-Based Systems 21, 941–945 (2008)

The Models of Variable Precision Multigranulation Rough Sets

Huili Dou[1], Xibei Yang[1,2,*], Jiyue Fan[1], and Suping Xu[1]

[1] School of Computer Science and Engineering,
Jiangsu University of Science and Technology, Zhenjiang,
Jiangsu, 212003, P.R. China
[2] School of Computer Science and Technology,
Nanjing University of Science and Technology,
Nanjing, Jiangsu, 210094, P.R. China
douhuili@163.com, yangxibei@hotmail.com

Abstract. Multigranulation rough set is a new expansion of the classical rough set since the former use a family of the binary relations instead of a single binary relation for the constructing of approximations. In this paper, the basic idea of the variable precision rough set is introduced into the multigranulation environment and then the concept of the variable precision multigranulation rough set is proposed. Similar to the classical multigranulation rough set, the optimistic and pessimistic variable precision multigranulation rough sets are defined, respectively. Not only the properties of the variable precision multigranulation rough sets are investigated, but also the relationships among the variable precision rough set, multigranulation rough set and the variable precision multigranulation rough set are examined.

Keywords: Multigranulation rough set, variable precision optimistic multigranulation rough set, variable precision pessimistic multigranulation rough set, variable precision rough set.

1 Introduction

Rough set [1], proposed by Pawlak, is a powerful tool, which can be used to deal with the inconsistency problems by separation of certain and doubtful knowledge extracted from the exemplary decisions.

In Pawlak's rough set model, the classification analysis must be completely correct or certain. In practice, however, it seems that admitting some level of uncertainty in the classification process may lead to a deeper understanding and a better utilization of properties of the data being analyzed. To overcome such limitation in traditional rough set model, partial classification was taken into account by Ziarko in Ref. [2, 3]. Ziarko's approach is referred to as the Variable Precision Rough Set (VPRS) model.

---

* Corresponding author.

T. Li et al. (Eds.): RSKT 2012, LNAI 7414, pp. 465–473, 2012.
© Springer-Verlag Berlin Heidelberg 2012

Moreover, it should be noticed that in Ref. [4–8], Qian et al. argued that we often need to describe concurrently a target concept through multi binary relations (e.g. equivalence relation, tolerance relation, reflexive relation and neighborhood relation) on the universe according to a user's requirements or targets of problem solving. Therefore, they proposed the concept of Multigranulation Rough Set (MGRS) model. The first multigranulation rough set model was proposed by Qian et al. in Ref. [4]. Following such work, Qian classified his multigranulation rough set theory into two parts: one is the optimistic multigranulation rough set [5, 6] and the other is pessimistic multigranulation rough set [7]. In recent years, multigranulation rough set progressing rapidly. For example, Yang et al. and Xu et al. generalized multigranulation rough sets into fuzzy environment in Ref. [11] and Ref. [9], respectively. Yang et al. [10] also proposed multigranulation rough sets in incomplete information system.

The purpose of this paper is to further generalize Ziarko's variable precision rough set and Qian's multigranulation rough set. From this point of view, we will propose the concept of the variable precision multigranulation rough set. The variable precision multigranulation rough set is a generalization of the variable precision rough set since it uses a family of the binary relations instead of a single binary relation; on the other hand, the variable precision multigranulation rough set is also a generalization of the multigranulation rough set since it takes the partial classification into account under the frame of multigranulation rough set.

2   VPRS and MGRS

2.1   Variable Precision Rough Set

Given the universe U, $\forall X, Y \subseteq U$, we say that the set X is included in the set Y with an admissible error β if and only if:

$$X \subseteq^{\beta} Y \Leftrightarrow e(X, Y) \leq \beta \tag{1}$$

where

$$e(X, Y) = 1 - \frac{|X \bigcap Y|}{|X|}. \tag{2}$$

The quantity $e(X, Y)$ is referred to as the inclusion error of X in Y. The value of β should be limited: $0 \leq \beta < 0.5$.

Given an information system I in which $A \subseteq AT$, the variable precision lower and upper approximations in terms of $IND(A)$ are defined as

$$\underline{A}_{\beta}(X) = \{x \in U : e([x]_A, X) \leq \beta\}; \tag{3}$$

$$\overline{A}_{\beta}(X) = \{x \in U : e([x]_A, X) < 1 - \beta\}. \tag{4}$$

The pair $[\underline{P}^{\beta}(X), \overline{P}^{\beta}(X)]$ is referred to as the variable precision rough set of X with respect to the set of the attributes A.

2.2   Multigranulation Rough Set

The multigranulation rough set is different from Pawlak's rough set model because the former is constructed on the basis of a family of indiscernibility relations instead of single indiscernibility relation.

In Qian's multigranulation rough set theory, two different models have been defined. The first one is the optimistic multigranulation rough set and the other is pessimistic multigranulation rough set [7].

Optimistic Multigranulation Rough Set. In Qian's optimistic multigranulation rough set, the target is approximated through a family of the indiscernibility relations. In lower approximation, the word "optimistic" is used to express the idea that in multi independent granular structures, we need only at least one granular structure to satisfy with the inclusion condition between equivalence class and target. The upper approximation of optimistic multigranulation rough set is defined by the complement of the lower approximation. To know what is the multigranulation rough set, we need the following notions of the information system.

Formally, an information system can be considered as a pair $I = <U, AT>$, where U is a non–empty finite set of objects, it is called the universe; AT is a non–empty finite set of attributes, such that $\forall a \in AT$, V_a is the domain of attribute a.

$\forall x \in U$, let us denote by $a(x)$ the value that x holds on $a(a \in AT)$. For an information system I, one then can describe the relationship between objects through their attributes values. With respect to a subset of attributes such that $A \subseteq AT$, an indiscernibility relation $IND(A)$ may be defined as

$$IND(A) = \{(x, y) \in U^2 : a(x) = a(y), \forall a \in A\}. \tag{5}$$

The relation $IND(A)$ is reflexive, symmetric and transitive, then $IND(A)$ is an equivalence relation. Then, $[x]_A = \{y \in U : (x, y) \in IND(A)\}$ is the A–equivalence class, which contains x.

Definition 1. *[5] Let I be an information system in which $A_1, A_2, \cdots, A_m \subseteq AT$, then $\forall X \subseteq U$, the optimistic multigranulation lower and upper approximations are denoted by $\underline{\sum_{i=1}^{m} A_i}^O(X)$ and $\overline{\sum_{i=1}^{m} A_i}^O(X)$, respectively,*

$$\underline{\sum_{i=1}^{m} A_i}^O(X) = \{x \in U : [x]_{A_1} \subseteq X \vee [x]_{A_2} \subseteq X \vee \cdots \vee [x]_{A_m} \subseteq X\}; \tag{6}$$

$$\overline{\sum_{i=1}^{m} A_i}^O(X) = \sim \underline{\sum_{i=1}^{m} A_i}^O(\sim X); \tag{7}$$

where $[x]_{A_i}$ ($1 \leq i \leq m$) is the equivalence class of x in terms of set of attributes A_i, $\sim X$ is the complement of set X.

By the lower and upper approximations $\sum_{i=1}^{m} A_i{}^O(X)$ and $\overline{\sum_{i=1}^{m} A_i}{}^O(X)$, the optimistic multigranulation boundary region of X is

$$BN^O_{\sum_{i=1}^{m} A_i}(X) = \sum_{i=1}^{m} \overline{A_i}{}^O(X) - \sum_{i=1}^{m} \underline{A_i}{}^O(X). \tag{8}$$

Theorem 1. *Let I be an information system in which $A_1, A_2, \cdots, A_m \subseteq AT$, then $\forall X \subseteq U$, we have*

$$\sum_{i=1}^{m} \overline{A_i}{}^O(X) = \{x \in U : [x]_{A_1} \cap X \neq \emptyset \wedge [x]_{A_2} \cap X \neq \emptyset \wedge \cdots \wedge [x]_{A_m} \cap X \neq \emptyset\}. \tag{9}$$

By Theorem 1, we can see that though the optimistic multigranulation upper approximation is defined by the complement of the optimistic multigranulation lower approximation, it can also be considered as a set in which objects have non–empty intersection with the target in terms of each granular structure.

Pessimistic Multigranulation Rough Set. In Qian's pessimistic multigranulation rough set, the target is still approximated through a family of the indiscernibility relations. However, it is different from the optimistic case. In lower approximation, the word "pessimistic" is used to express the idea that in multi independent granular structures, we need all the granular structures to satisfy with the inclusion condition between equivalence class and target. The upper approximation of pessimistic multigranulation rough set is also defined by the complement of the pessimistic multigranulation lower approximation.

Definition 2. *[7] Let I be an information system in which $A_1, A_2, \cdots, A_m \subseteq AT$, then $\forall X \subseteq U$, the pessimistic multigranulation lower and upper approximations are denoted by $\sum_{i=1}^{m} \underline{A_i}{}^P(X)$ and $\overline{\sum_{i=1}^{m} A_i}{}^P(X)$, respectively,*

$$\sum_{i=1}^{m} \underline{A_i}{}^P(X) = \{x \in U : [x]_{A_1} \subseteq X \wedge [x]_{A_2} \subseteq X \wedge \cdots \wedge [x]_{A_m} \subseteq X\}; \tag{10}$$

$$\overline{\sum_{i=1}^{m} A_i}{}^P(X) = \sim \sum_{i=1}^{m} \underline{A_i}{}^P(\sim X). \tag{11}$$

By the lower and upper approximations $\sum_{i=1}^{m} \underline{A_i}{}^P(X)$ and $\overline{\sum_{i=1}^{m} A_i}{}^P(X)$, the pessimistic multigranulation boundary region of X is

$$BN^P_{\sum_{i=1}^{m} A_i}(X) = \sum_{i=1}^{m} \overline{A_i}{}^P(X) - \sum_{i=1}^{m} \underline{A_i}{}^P(X). \tag{12}$$

Theorem 2. *Let I be an information system in which $A_1, A_2, \cdots, A_m \subseteq AT$, then $\forall X \subseteq U$, we have*

$$\overline{\sum_{i=1}^{m} A_i}^{P}(X) = \{x \in U : [x]_{A_1} \cap X \neq \emptyset \vee [x]_{A_2} \cap X \neq \emptyset \vee \cdots \vee [x]_{A_m} \cap X \neq \emptyset\}. \quad (13)$$

Different from the upper approximation of optimistic multigranulation rough set, the upper approximation of pessimistic multigranulation rough set is represented as a set in which objects have non–empty intersection with the target in terms of at least one granular structure.

3 Variable Precision Multigranulation Rough Sets

3.1 Variable Precision Optimistic Multigranulation Rough Sets

Definition 3. *Let I be an information system in which $A_1, A_2, \cdots, A_m \subseteq AT$, then $\forall X \subseteq U$, the variable precision optimistic multigranulation lower and upper approximations are denoted by $\underline{\sum_{i=1}^{m} A_i}_{\beta}^{O}(X)$ and $\overline{\sum_{i=1}^{m} A_i}_{\beta}^{O}(X)$, respectively, where*

$$\underline{\sum_{i=1}^{m} A_i}_{\beta}^{O}(X) = \{x \in U : e([x]_{A_1}, X) \leq \beta \vee \cdots \vee e([x]_{A_m}, X) \leq \beta\}; \quad (14)$$

$$\overline{\sum_{i=1}^{m} A_i}_{\beta}^{O}(X) = \sim \underline{\sum_{i=1}^{m} A_i}_{\beta}^{O}(\sim X). \quad (15)$$

Theorem 3. *Let I be an information system in which $A_1, A_2, \cdots, A_m \subseteq AT$, then $\forall X \subseteq U$, we have*

$$\overline{\sum_{i=1}^{m} A_i}_{\beta}^{O}(X) = \{x \in U : e([x]_{A_1}, X) < 1 - \beta \wedge \cdots \wedge e([x]_{A_m}, X) < 1 - \beta\}. \quad (16)$$

Theorem 4. *Let I be an information system in which $A_1, A_2, \cdots, A_m \subseteq AT$, then $\forall X, Y \subseteq U$, we have*

1. $\underline{\sum_{i=1}^{m} A_i}_{\beta}^{O}(U) = \overline{\sum_{i=1}^{m} A_i}_{\beta}^{O}(U) = U$;
2. $\underline{\sum_{i=1}^{m} A_i}_{\beta}^{O}(\emptyset) = \overline{\sum_{i=1}^{m} A_i}_{\beta}^{O}(\emptyset) = \emptyset$;
3. $\beta_1 \geq \beta_2 \Rightarrow \underline{\sum_{i=1}^{m} A_i}_{\beta_1}^{O}(X) \supseteq \underline{\sum_{i=1}^{m} A_i}_{\beta_2}^{O}(X), \overline{\sum_{i=1}^{m} A_i}_{\beta_1}^{O}(X) \subseteq \overline{\sum_{i=1}^{m} A_i}_{\beta_2}^{O}(X)$;
4. (a) $\underline{\sum_{i=1}^{m} A_i}_{\beta}^{O}(X \cap Y) \subseteq \underline{\sum_{i=1}^{m} A_i}_{\beta}^{O}(X) \cap \underline{\sum_{i=1}^{m} A_i}_{\beta}^{O}(Y)$;
   (b) $\underline{\sum_{i=1}^{m} A_i}_{\beta}^{O}(X \cup Y) \supseteq \underline{\sum_{i=1}^{m} A_i}_{\beta}^{O}(X) \cup \underline{\sum_{i=1}^{m} A_i}_{\beta}^{O}(Y)$;
   (c) $\overline{\sum_{i=1}^{m} A_i}_{\beta}^{O}(X \cap Y) \subseteq \overline{\sum_{i=1}^{m} A_i}_{\beta}^{O}(X) \cap \overline{\sum_{i=1}^{m} A_i}_{\beta}^{O}(Y)$;
   (d) $\overline{\sum_{i=1}^{m} A_i}_{\beta}^{O}(X \cup Y) \supseteq \overline{\sum_{i=1}^{m} A_i}_{\beta}^{O}(X) \cup \overline{\sum_{i=1}^{m} A_i}_{\beta}^{O}(Y)$.

Proof. By Definition 3, it is not difficult to prove 1, 2 and 3.

$\forall x \in U,$

$$x \in \sum_{i=1}^{m} A_i \overset{O}{\underset{\beta}{}}(X \cap Y) \Rightarrow \exists i \in \{1, \cdots, m\}, e([x]_{A_i}, X \cap Y) \leq \beta$$

$$\Rightarrow \exists i \in \{1, \cdots, m\}, 1 - \frac{|[x]_{A_i} \cap X \cap Y|}{|[x]_{A_i}|} \leq \beta$$

$$\Rightarrow \exists i \in \{1, \cdots, m\}, 1 - \frac{|[x]_{A_i} \cap X|}{|[x]_{A_i}|} \leq \beta$$

$$\text{and } 1 - \frac{|[x]_{A_i} \cap Y|}{|[x]_{A_i}|} \leq \beta$$

$$\Rightarrow x \in \sum_{i=1}^{m} A_i \overset{O}{\underset{\beta}{}}(X) \text{ and } x \in \sum_{i=1}^{m} A_i \overset{O}{\underset{\beta}{}}(Y)$$

$$\Rightarrow x \in \sum_{i=1}^{m} A_i \overset{O}{\underset{\beta}{}}(X) \cap \sum_{i=1}^{m} A_i \overset{O}{\underset{\beta}{}}(Y).$$

Similarity, it is not difficult to prove other formulas.   □

Remark 1. It should be noticed that in variable precision optimistic multigranulation rough set model, $\underline{\sum_{i=1}^{m} A_i}_{\beta}^{O}(X) \subseteq \overline{\sum_{i=1}^{m} A_i}_{\beta}^{O}(X)$ does not always hold.

3.2   Variable Precision Pessimistic Multigranulation Rough Sets

Definition 4. *Let I be an information system in which $A_1, A_2, \cdots, A_m \subseteq AT$, then $\forall X \subseteq U$, the variable precision pessimistic multigranulation lower and upper approximations are denoted by $\underline{\sum_{i=1}^{m} A_i}_{\beta}^{P}(X)$ and $\overline{\sum_{i=1}^{m} A_i}_{\beta}^{P}(X)$, respectively, where*

$$\underline{\sum_{i=1}^{m} A_i}_{\beta}^{P}(X) = \{x \in U : e([x]_{A_1}, X) \leq \beta \wedge \cdots \wedge e([x]_{A_m}, X) \leq \beta\}; \quad (17)$$

$$\overline{\sum_{i=1}^{m} A_i}_{\beta}^{P}(X) = \sim \underline{\sum_{i=1}^{m} A_i}_{\beta}^{P}(\sim X). \quad (18)$$

Theorem 5. *Let I be an information system in which $A_1, A_2, \cdots, A_m \subseteq AT$, then $\forall X \subseteq U$, we have*

$$\overline{\sum_{i=1}^{m} A_i}_{\beta}^{P}(X) = \{x \in U : e([x]_{A_1}, X) < 1 - \beta \vee \cdots \vee e([x]_{A_m}, X) < 1 - \beta\}. \quad (19)$$

Theorem 6. *Let I be an information system in which $A_1, A_2, \cdots, A_m \subseteq AT$, then $\forall X, Y \subseteq U$, we have*

1. $\sum_{i=1}^{m} A_i{}_\beta^P(U) = \overline{\sum_{i=1}^{m} A_i}{}_\beta^P(U) = U$;

2. $\sum_{i=1}^{m} A_i{}_\beta^P(\emptyset) = \overline{\sum_{i=1}^{m} A_i}{}_\beta^P(\emptyset) = \emptyset$;

3. $\beta_1 \geq \beta_2 \Rightarrow \sum_{i=1}^{m} A_i{}_{\beta_1}^P(X) \supseteq \sum_{i=1}^{m} A_i{}_{\beta_2}^P(X), \overline{\sum_{i=1}^{m} A_i}{}_{\beta_1}^P(X) \subseteq \overline{\sum_{i=1}^{m} A_i}{}_{\beta_2}^P(X)$;

4. *(a)* $\sum_{i=1}^{m} A_i{}_\beta^P(X \cap Y) \subseteq \sum_{i=1}^{m} A_i{}_\beta^P(X) \cap \sum_{i=1}^{m} A_i{}_\beta^P(Y)$;

   (b) $\sum_{i=1}^{m} A_i{}_\beta^P(X \cup Y) \supseteq \sum_{i=1}^{m} A_i{}_\beta^P(X) \cup \sum_{i=1}^{m} A_i{}_\beta^P(Y)$;

   (c) $\overline{\sum_{i=1}^{m} A_i}{}_\beta^P(X \cap Y) \subseteq \overline{\sum_{i=1}^{m} A_i}{}_\beta^P(X) \cap \overline{\sum_{i=1}^{m} A_i}{}_\beta^P(Y)$;

   (d) $\overline{\sum_{i=1}^{m} A_i}{}_\beta^P(X \cup Y) \supseteq \overline{\sum_{i=1}^{m} A_i}{}_\beta^P(X) \cup \overline{\sum_{i=1}^{m} A_i}{}_\beta^P(Y)$.

Proof. By Definition 4, it is not difficult to prove this theorem. □

Remark 2. It should be noticed that in variable precision pessimistic multigranulation rough set model, $\sum_{i=1}^{m} A_i{}_\beta^P(X) \subseteq \overline{\sum_{i=1}^{m} A_i}{}_\beta^P(X)$ does not always hold.

4   Relationships among Several Models

Theorem 7. *Let I be an information system in which $A_1, A_2, \cdots, A_m \subseteq AT$, then $\forall X \subseteq U$, we have*

1. $\sum_{i=1}^{m} A_i{}_\beta^P(X) \subseteq \sum_{i=1}^{m} A_i{}_\beta^O(X)$;

2. $\overline{\sum_{i=1}^{m} A_i}{}_\beta^O(X) \subseteq \overline{\sum_{i=1}^{m} A_i}{}_\beta^P(X)$.

Proof. $\forall x \in \sum_{i=1}^{m} A_i{}_\beta^P(X)$, then we have $e([x]_{A_i}, X) \leq \beta$ for each $i = 1, 2, \cdots, m$. By the definition of the variable precision optimistic multigranulation lower approximation, $x \in \sum_{i=1}^{m} A_i{}_\beta^O(X)$ holds obviously, it follows that $\sum_{i=1}^{m} A_i{}_\beta^P(X) \subseteq \sum_{i=1}^{m} A_i{}_\beta^O(X)$.

Similarly, it is not difficult to prove $\overline{\sum_{i=1}^{m} A_i}{}_\beta^O(X) \subseteq \overline{\sum_{i=1}^{m} A_i}{}_\beta^P(X)$ through Theorem 3 and Theorem 5. □

Theorem 7 shows the relationships between variable precision optimistic and pessimistic multigranulation rough approximations. The details are: the variable precision pessimistic multigranulation lower approximation is included into the variable precision optimistic multigranulation lower approximation, the variable precision optimistic multigranulation upper approximation is included into the variable precision pessimistic multigranulation upper approximation.

Theorem 8. *Let I be an information system in which $A_1, A_2, \cdots, A_m \subseteq AT$, then $\forall X \subseteq U$, we have*

1. $\underline{\sum_{i=1}^{m} A_i}^{O}(X) \subseteq \underline{\sum_{i=1}^{m} A_i}_{\beta}^{O}(X)$;

2. $\overline{\sum_{i=1}^{m} A_i}^{O}(X) \supseteq \overline{\sum_{i=1}^{m} A_i}_{\beta}^{O}(X)$;

3. $\underline{\sum_{i=1}^{m} A_i}^{P}(X) \subseteq \underline{\sum_{i=1}^{m} A_i}_{\beta}^{P}(X)$;

4. $\overline{\sum_{i=1}^{m} A_i}^{P}(X) \supseteq \overline{\sum_{i=1}^{m} A_i}_{\beta}^{P}(X)$.

Proof. $\forall x \in \underline{\sum_{i=1}^{m} A_i}^{O}(X)$, there $\exists i = 1, 2, \cdots, m$ such that $[x]_{A_i} \subseteq X$, i.e. $e([x]_{A_i}, X) = 0$, it follows that $e([x]_{A_i}, X) \leq \beta$. Then by the definition of the variable precision optimistic multigranulation lower approximation, $x \in \underline{\sum_{i=1}^{m} A_i}_{\beta}^{O}(X)$, i.e. $\underline{\sum_{i=1}^{m} A_i}^{O}(X) \subseteq \underline{\sum_{i=1}^{m} A_i}_{\beta}^{O}(X)$.

Similarly, it is not difficult to prove other formulas.   □

Theorem 8 shows the relationships between the classical multigranulation rough sets and the variable precision multigranulation rough sets. The details are: the optimistic multigranulation lower approximation is included into the variable precision optimistic multigranulation lower approximation, the variable precision optimistic multigranulation upper approximation is included into the optimistic multigranulation upper approximation, the pessimistic multigranulation lower approximation is included into the variable precision pessimistic multigranulation lower approximation, the variable precision pessimistic multigranulation upper approximation is included into the pessimistic multigranulation upper approximation.

Theorem 9. *Let I be an information system in which $A_1, A_2, \cdots, A_m \subseteq AT$, then $\forall X \subseteq U$, we have*

1. $\underline{A_i}_{\beta}(X) \subseteq \underline{\sum_{i=1}^{m} A_i}_{\beta}^{O}(X), \forall i = 1, 2, \cdots, m$;

2. $\overline{\sum_{i=1}^{m} A_i}_{\beta}^{O}(X) \subseteq \overline{A_i}_{\beta}(X), \forall i = 1, 2, \cdots, m$;

3. $\underline{\sum_{i=1}^{m} A_i}_{\beta}^{P}(X) \subseteq \underline{A_i}_{\beta}(X), \forall i = 1, 2, \cdots, m$;

4. $\overline{A_i}_{\beta}(X) \subseteq \overline{\sum_{i=1}^{m} A_i}_{\beta}^{P}(X), \forall i = 1, 2, \cdots, m$.

Proof. $\forall i = 1, 2, \cdots, m$ and $\forall x \in \underline{A_i}_{\beta}(X)$, then we have $e([x]_{A_i}, X) \leq \beta$. By the definition of the variable precision optimistic multigranulation lower approximation, $x \in \underline{\sum_{i=1}^{m} A_i}_{\beta}^{O}(X)$ holds obviously, it follows that $\underline{A_i}_{\beta}(X) \subseteq \underline{\sum_{i=1}^{m} A_i}_{\beta}^{O}(X)$.

Similarly, it is not difficult to prove other formulas.   □

Theorem 9 shows the relationship between the classical variable precision rough set and the variable precision multigranulation rough sets. The details are: the classical variable precision lower approximation is include into the variable precision optimistic multigranulation lower approximation, the variable precision optimistic multigranulation upper approximation is include into the classical variable precision upper approximation, the variable precision pessimistic multigranulation lower approximation is included into the classical variable precision lower approximation, the classical variable precision upper approximation is included into the variable precision pessimistic multigranulation upper approximation.

5 Conclusions

In this paper, the concept of the variable precision multigranulation rough set is proposed. Such model is the combination of the variable precision rough set and multigranulation rough set. Different from Ziarko's variable precision rough set, we used a family of the indiscernibility relations instead of a single indiscernibility relation for constructing rough approximation. Moreover, different from Qian's multigranulation rough set, we presented the inclusion error into the multigranulation frame.

Furthermore, the reductions in variable precision multigranulation rough set is an interesting topic to be addressed.

Acknowledgment. This work is supported by Natural Science Foundation of China (No. 61100116), Natural Science Foundation of Jiangsu Province of China (No. BK2011492), Natural Science Foundation of Jiangsu Higher Education Institutions of China (No. 11KJB520004), Postdoctoral Science Foundation of China (No. 20100481149), Postdoctoral Science Foundation of Jiangsu Province of China (No. 1101137C).

References

1. Pawlak, Z.: Rough sets–theoretical aspects of reasoning about data. Kluwer Academic Publishers (1991)
2. Ziarko, W.: Variable precision rough set model. J. Compu. Syst. Sci. 46, 39–59 (1993)
3. Ziarko, W.: Probabilistic approach to rough sets. Int. J. Approx. Reason. 49, 272–284 (2008)
4. Qian, Y.H., Dang, C.Y., Liang, J.Y.: MGRS in incomplete information systems. In: 2007 IEEE International Conference on Granular Computing, pp. 163–168 (2007)
5. Qian, Y.H., Liang, J.Y., Yao, Y.Y., Dang, C.Y.: MGRS: A multi–granulation rough set. Inform. Sci. 180, 949–970 (2010)
6. Qian, Y.H., Liang, J.Y., Dang, C.Y.: Incomplete multigranulation rough set. IEEE T. Syst. Man Cy. B. 20, 420–431 (2010)
7. Qian, Y.H., Liang, J.Y., Wei, W.: Pessimistic rough decision. In: Second International Workshop on Rough Sets Theory, pp. 440–449 (2010)
8. Qian, Y.H., Liang, J.Y., Witold, P., Dang, C.Y.: Positive approximation: an accelerator for attribute reduction in rough set theory. Artif. Intell. 174, 597–618 (2010)
9. Xu, W.H., Wang, Q.R., Zhang, X.T.: Multi–granulation fuzzy rough sets in a fuzzy tolerance approximation space. International Journal of Fuzzy Systems 13, 246–259 (2011)
10. Yang, X.B., Song, X.N., Chen, Z.H., Yang, J.Y.: On multigranulation rough sets in incomplete information system. International Journal of Machine Learning and Cybernetics, doi:10.1007/s13042-011-0054-8
11. Yang, X.B., Song, X.N., Dou, H.L., Yang, J.Y.: Multi–granulation rough set: from crisp to fuzzy case. Annals of Fuzzy Mathematics and Informatics 1, 55–70 (2011)

Topological Properties of the Pessimistic Multigranulation Rough Set Model

Yanhong She* and Xiaoli He

College of Science, Xi'an Shiyou University, Xi'an 710065, China

Abstract. Multigranulation rough set theory is a newly proposed mathematical tool for dealing with inexact, uncertain or vague information. This paper concerns the topological properties of pessimistic multigranulation rough sets. It is shown that the collection of definable sets in pessimistic rough set model determines a clopen topology on the universe. Furthermore, it forms a Boolean algebra under the usual set-theoretic operations.

Keywords: Multigranulation rough set, topology, Boolean algebra.

1 Introduction

Rough set theory, initiated by Pawlak [1, 2], is a mathematical tool for the study of intelligent systems characterized by insufficient and incomplete information. By using the concepts of lower and upper approximations in rough set theory, knowledge hidden in information systems may be unraveled and expressed in the form of decision rules.

Pawlak's rough set is constructed on the basis of an approximation space (U, R) with U being a nonempty set (also called the universe of discourse) and R being the equivalence relation imposed upon U. However, the equivalence relation R is too restrictive in many practical applications, particularly in handling real-valued or symbolic attribute values. In view of this fact, several extensions of Pawlak's rough set model have been accomplished in the past years, which include arbitrary binary relation based rough set model $[3 - 8]$, covering based rough set model$[9 - 12]$, fuzzy rough set model$[13 - 16]$, et al. One important characteristic of these extensions is that a target concept is always characterized via the so called upper and lower approximations under a single granulation, i.e., the concept is depicted by available knowledge induced from a single relation on the universe. In view of granular computing, an equivalence relation (or a tolerance relation) on the universe can be regarded as a granulation, and a partition (or a covering) on the universe can be regarded as a granulation space. It can thus be concluded that the existing rough set models are based on a single granulation. However, as illustrated in [17], in some cases it is more reasonable to describe the target concept through multiple relations on the universe according to user requirements or targets of problem solving. To more widely apply the

---

* Corresponding author.

T. Li et al. (Eds.): RSKT 2012, LNAI 7414, pp. 474–481, 2012.

rough set theory in practical applications, Qian and Liang extended Pawlak's single-granulation rough set model to a multigranulation rough set model [17]. In Qian et al.'s multigranulation rough set theory, there are two different basic models: one is the optimistic multigranulation rough set and the other is the pessimistic multigranulation rough set. Presently, the multigranulation rough set theory progresses rapidly. For instance, Qian et al. extends the rough set model based on a tolerance relation to an incomplete rough set model based on multi-granulations [18], where set approximations are defined through using multiple tolerance relations on the universe. Yang et al. generalized the multigranulation rough sets into fuzzy environments in [19]. Abu-Donia [20] studied the rough approximations through multi knowledge base. Wu and Leung [21] investigated the multi-scale information system, which reflects the explanation of the same problem at different scales (levels of granulations). Khan and Banerjee [22] investigated the reasoning approach in multiple source approximation systems in which information arrives from multiple sources.

In this paper, we aim to consider the topological properties of multigranulation rough set model. It is worth pointing out that topological or lattice-theoretic approach to rough set theory has been extensively studied in the existing literatures [23 − 29], however, researches about these aspects are mainly confined to the single-granulation rough set model. How to study the mathematical structure of definable sets in the context of multigranulation rough set model is an unexplored but interesting topic to be addressed, more importantly, such a study will help to gain more insights into the mathematical structure of multigranulation rough set model and also hopefully develop methods for potential applications. As will be shown below, the collection of lower definable sets, upper definable sets and definable set in the pessimistic multigranulation rough set model coincide with each other, and they determine a clopen topology on the universe, furthermore, they form a Boolean algebra.

To facilitate our discussion, we first present some basic notions of the multigranulation rough set model in Section 2. Then in Section 3, the structure of the pessimistic multigranulation rough set model is examined in detail. This paper is completed with some concluding remarks, as stated in Section 4.

2   Review of Multigranulation Rough Set Model

For the elementary knowledge of Pawlak's rough set theory, we refer the readers to [1, 2] for details.

Definition 1 [17]. Let (U, AT, f) be a complete information system, $X \subseteq U$ and $\hat{P}_1, \hat{P}_2, \cdots, \hat{P}_m$ be m partitions induced by the attributes $P_1, P_2, \cdots, P_m$, respectively. Define

$$\underline{X}^o_{\sum_{i=1}^m \hat{P}_i} = \{x : [x]_{\hat{P}_1} \subseteq X, or\ [x]_{\hat{P}_2} \subseteq X, \cdots, or\ [x]_{\hat{P}_m} \subseteq X\}, \quad (1)$$

$$\overline{X}^o_{\sum_{i=1}^m \hat{P}_i} =\sim\ \sim \underline{X}^o_{\sum_{i=1}^m \hat{P}_i}. \quad (2)$$

Then we call $\underline{X}^o_{\Sigma^m_{i=1}\hat{P}_i}, \overline{X}^o_{\Sigma^m_{i=1}\hat{P}_i}$ the optimistic multi-lower approximation and the optimistic multi-upper approximation of X, respectively. Particularly, if $\underline{X}^o_{\Sigma^m_{i=1}\hat{P}_i} = X$, then we call X a lower definable set of optimistic multigranulation rough set model, if $X = \overline{X}^o_{\Sigma^m_{i=1}\hat{P}_i}$, then we call X a upper definable set of optimistic multigranulation rough set model and if $\underline{X}^o_{\Sigma^m_{i=1}\hat{P}_i} = X = \overline{X}^o_{\Sigma^m_{i=1}\hat{P}_i}$, then we call X a definable set of optimistic multigranulation rough set model.

Definition 2 [17]. Let (U, AT, f) be a complete information system, $X \subseteq U$ and $\hat{P}_1, \hat{P}_2, \cdots, \hat{P}_m$ be m partitions induced by the attributes $P_1, P_2, \cdots, P_m$, respectively. Define

$$\underline{X}^p_{\Sigma^m_{i=1}\hat{P}_i} = \{x : [x]_{\hat{P}_1} \subseteq X, and \ [x]_{\hat{P}_2} \subseteq X, \cdots, and \ [x]_{\hat{P}_m} \subseteq X\}, \tag{3}$$

$$\overline{X}^p_{\Sigma^m_{i=1}\hat{P}_i} = \sim \underline{\sim X}^p_{\Sigma^m_{i=1}\hat{P}_i}. \tag{4}$$

Then we call $\underline{X}^p_{\Sigma^m_{i=1}\hat{P}_i}, \overline{X}^p_{\Sigma^m_{i=1}\hat{P}_i}$ the pessimistic multi-lower approximation and the pessimistic multi-upper approximation of X, respectively. Particularly, if $\underline{X}^p_{\Sigma^m_{i=1}\hat{P}_i} = X$, then we call X a lower definable set of pessimistic multigranulation rough set model, if $X = \overline{X}^p_{\Sigma^m_{i=1}\hat{P}_i}$, then we call X a upper definable set of pessimistic multigranulation rough set model and if $\underline{X}^p_{\Sigma^m_{i=1}\hat{P}_i} = X = \overline{X}^o_{\Sigma^m_{i=1}\hat{P}_i}$, then we call X a definable set of pessimistic multigranulation rough set model.

Proposition 1 [17]. Let (U, AT, f) be a complete information system, $X \subseteq U$ and $\hat{P}_1, \hat{P}_2, \cdots, \hat{P}_m$ be m partitions induced by the attributes $P_1, P_2, \cdots, P_m$, respectively. Then the following properties hold: $\forall X, Y \subseteq U$,

(i) $\underline{\varnothing}^o_{\Sigma^m_{i=1}\hat{P}_i} = \overline{\varnothing}^o_{\Sigma^m_{i=1}\hat{P}_i} = \varnothing, \ \underline{U}^o_{\Sigma^m_{i=1}\hat{P}_i} = \overline{U}^o_{\Sigma^m_{i=1}\hat{P}_i} = U$,

(ii) $\underline{X}^o_{\Sigma^m_{i=1}\hat{P}_i} \subseteq X \subseteq \overline{X}^o_{\Sigma^m_{i=1}\hat{P}_i}$,

(iii) $\underline{\underline{X}^o_{\Sigma^m_{i=1}\hat{P}_i}}^o_{\Sigma^m_{i=1}\hat{P}_i} = \underline{X}^o_{\Sigma^m_{i=1}\hat{P}_i}, \ \overline{\overline{X}^o_{\Sigma^m_{i=1}\hat{P}_i}}^o_{\Sigma^m_{i=1}\hat{P}_i} = \overline{X}^o_{\Sigma^m_{i=1}\hat{P}_i}$,

(iv) $X \subseteq Y$ implies that $\underline{X}^o_{\Sigma^m_{i=1}\hat{P}_i} \subseteq \underline{Y}^o_{\Sigma^m_{i=1}\hat{P}_i}$,

(v) $\overline{\cap^n_{j=1}X_j}^o_{\Sigma^m_{i=1}\hat{P}_i} \subseteq \cap^n_{j=1}(\overline{X_j}^o_{\Sigma^m_{i=1}\hat{P}_i})$.

Proposition 2 [17]. Let (U, AT, f) be a complete information system, $X \subseteq U$ and $\hat{P}_1, \hat{P}_2, \cdots, \hat{P}_m$ be m partitions induced by the attributes $P_1, P_2, \cdots, P_m$, respectively. Then the following properties hold: $\forall X, Y \subseteq U$,

(i) $\underline{\varnothing}^p_{\Sigma^m_{i=1}\hat{P}_i} = \overline{\varnothing}^p_{\Sigma^m_{i=1}\hat{P}_i} = \varnothing, \ \underline{U}^p_{\Sigma^m_{i=1}\hat{P}_i} = \overline{U}^p_{\Sigma^m_{i=1}\hat{P}_i} = U$,

(ii) $\underline{X}^p_{\Sigma^m_{i=1}\hat{P}_i} \subseteq X \subseteq \overline{X}^p_{\Sigma^m_{i=1}\hat{P}_i}$,

(iii) $\overline{\overline{X}^p_{\Sigma^m_{i=1}\hat{P}_i}}^p_{\Sigma^m_{i=1}\hat{P}_i} = \overline{X}^p_{\Sigma^m_{i=1}\hat{P}_i}, \ \overline{\overline{X}^p_{\Sigma^m_{i=1}\hat{P}_i}}^p_{\Sigma^m_{i=1}\hat{P}_i} = \overline{X}^p_{\Sigma^m_{i=1}\hat{P}_i}$,

(iv) $X \subseteq Y$ implies that $\underline{X}^p_{\Sigma^m_{i=1}\hat{P}_i} \subseteq \underline{Y}^p_{\Sigma^m_{i=1}\hat{P}_i}$,

(v) $\overline{\cap^n_{j=1}X_j}^p_{\Sigma^m_{i=1}\hat{P}_i} \subseteq \cap^n_{j=1}(\overline{X_j}^p_{\Sigma^m_{i=1}\hat{P}_i})$.

The main objective of this paper is to investigate the structure of multigranulation rough set model, to this end, some basic notions and their properties in lattice theory and topology are reviewed below.

A nonempty subset τ of 2^U is referred to a topology [30] on U if it satisfies the properties: (i) $\emptyset, U \in \tau$, (ii) τ is closed w.r.t. arbitrary set-theoretic union, i.e., $\forall X_j \in \tau(j \in J), \cup_{j \in J} X_j \in \tau$, (iii) τ is closed w.r.t. finite set-theoretic intersection, i.e., $\forall X_j \in \tau(j = 1, 2, \cdots, n), \cap_{j=1}^n X_j \in \tau$. A topology τ on U satisfying the condition that $\forall X \subseteq U, X \in \tau \Leftrightarrow X^c \in \tau$ is called a clopen topology. A topological space is a T_0 space if and only if for each pair x and y of distinct points, there is a neighborhood of one point to which the other does not belong. A topological space is a T_1 space if and only if for each pair x and y of distinct points, there is a neighborhood of each point to which the other does not belong to. A topological space is a Hausdorff space whenever x and y are disjoint points of the space there exists disjoint neighborhoods of x and y. A space is normal if and only if for each disjoint pair of closed sets, A and B, there are disjoint open sets V and W such that $A \subseteq V$ and $B \subseteq W$. Similarly, a topological space is regular if and only if for each point x and each closed set A, if $x \notin A$, then there exists disjoint open sets V and W such that $x \in V$ and $A \subseteq W$. A lattice L is called a Boolean algebra[31] if (i) L is distributive, (ii) L is bounded, (iii) each $a \in L$ has a complement $a' \in L$.

The following proposition concerning the structure of definable sets in Pawlak's rough set model can be obtained immediately.

Proposition 3. The collection of definable sets in Pawlak's single-granulation rough set model determines a clopen topology on the universe, moreover, it forms a Boolean algebra under the usual set-theoretic operations.

3 Topological Properties of the Pessimistic Multigranulation Rough Set Model

In this section, we will consider some fundamental properties of the pessimistic multigranulation rough set model, more specifically, we will focus on the structure of three different kinds of definable sets, they are

$$H_1^p = \{X : \underline{X}_{\Sigma_{i=1}^m \hat{P}_i}^p = X\}, \tag{5}$$

$$H_2^p = \{X : \overline{X}_{\Sigma_{i=1}^m \hat{P}_i}^p = X\} \tag{6}$$

and

$$H_3^p = \{X : \underline{X}_{\Sigma_{i=1}^m \hat{P}_i}^p = X = \overline{X}_{\Sigma_{i=1}^m \hat{P}_i}^p\}. \tag{7}$$

Proposition 4. (i) $\emptyset, U \in H_1^p$,
    (ii) $X_j \in H_1^p (j = 1, \cdots, n)$ implies that $\cup_{j=1}^n X_j \in H_1^p$,
    (iii) $X_j \in H_1^p (j = 1, \cdots, n)$ implies that $\cap_{j=1}^n X_j \in H_1^p$,
    (iv) $X \in H_1^p$ if and only if $X^c \in H_2^p$,
    (v) $X \in H_1^p$ if and only if $X^c \in H_1^p$.

Proof. (i) It follows immediately from Prop.2(i).

(ii) It suffices to show that $\underline{(\cup_{j=1}^{n}X_j)^p}_{\Sigma_{i=1}^{m}\hat{P}_i} = \cup_{j=1}^{n}X_j$. By Prop.2(ii), we have $\underline{(\cup_{j=1}^{n}X_j)^p}_{\Sigma_{i=1}^{m}\hat{P}_i} \subseteq \cup_{j=1}^{n}X_j$. To show $\cup_{j=1}^{n}X_j \subseteq \underline{(\cup_{j=1}^{n}X_j)^p}_{\Sigma_{i=1}^{m}\hat{P}_i}$, it suffices to show that $\forall j \in \{1, \cdots, n\}, X_j \subseteq \underline{(\cup_{j=1}^{n}X_j)^p}_{\Sigma_{i=1}^{m}\hat{P}_i}$. Choose arbitrarily an element $x \in X_j$, we have from $X_j = \underline{X_j^p}_{\Sigma_{i=1}^{m}\hat{P}_i}$ that $[x]_{\hat{P}_i} \subseteq X_j$,consequently, $[x]_{\hat{P}_i} \subseteq \cup_{j=1}^{n}X_j$ for all $\hat{P}_i \in \{\hat{P}_1, \cdots, \hat{P}_m\}$, which shows that $x \in \underline{(\cup_{j=1}^{n}X_j)^p}_{\Sigma_{i=1}^{m}\hat{P}_i}$. And hence, we have $\underline{(\cup_{j=1}^{n}X_j)^p}_{\Sigma_{i=1}^{m}\hat{P}_i} = \cup_{j=1}^{n}X_j$.

(iii) It suffices to show that $\underline{(\cap_{j=1}^{n}X_j)^p}_{\Sigma_{i=1}^{m}\hat{P}_i} = \cap_{j=1}^{n}X_j$. By Prop. 2(ii), we have $\underline{(\cap_{j=1}^{n}X_j)^p}_{\Sigma_{i=1}^{m}\hat{P}_i} \subseteq \cap_{j=1}^{n}X_j$. To show $\cap_{j=1}^{n}X_j \subseteq \underline{(\cap_{j=1}^{n}X_j)^p}_{\Sigma_{i=1}^{m}\hat{P}_i}$, we choose arbitrarily an element $x \in \cap_{j=1}^{n}X_j$, then will show that $x \in \underline{(\cap_{j=1}^{n}X_j)^p}_{\Sigma_{i=1}^{m}\hat{P}_i}$. By Def. 2, it suffices to show that $[x]_{\hat{P}_i} \subseteq X_j$ for all $\hat{P}_i \in \{\hat{P}_1, \cdots, \hat{P}_m\}$ and $X_j \in \{X_1, \cdots, X_n\}$, which follows immediately from $x \in X_j$ and $\underline{X_j^p}_{\Sigma_{i=1}^{m}\hat{P}_i} = X_j$.

(iv) It follows immediately from (4), (5) and (6).

(v) It suffices to show that $\overline{X^{cp}}_{\Sigma_{i=1}^{m}\hat{P}_i} = X^c$. By Prop. 2 (ii), we have $\overline{X^{cp}}_{\Sigma_{i=1}^{m}\hat{P}_i} \subseteq X^c$. For the converse inclusion, choose arbitrarily $x \in X^c$, we will show that $x \in \overline{X^{cp}}_{\Sigma_{i=1}^{m}\hat{P}_i}$ below. Indeed, as a preliminary result, we can show that $[x]_{\hat{P}_i} \cap X = \emptyset$ for all $\hat{P}_i \in \{\hat{P}_1, \cdots, \hat{P}_m\}$. Suppose, on the contrary, that there exists $y \in [x]_{\hat{P}_i} \cap X$, then by $X = \underline{X^p}_{\Sigma_{i=1}^{m}\hat{P}_i}$, we have that $[y]_{\hat{P}_i} \subseteq X$. Moreover, since $y \in [x]_{\hat{P}_i}$, we further have $[x]_{\hat{P}_i} = [y]_{\hat{P}_i} \subseteq X$, which, however, contradicts with $x \in X^c$. And hence, $\forall x \in X^c$, $x \in \overline{X^{cp}}_{\Sigma_{i=1}^{m}\hat{P}_i}$, which, together with $\overline{X^{cp}}_{\Sigma_{i=1}^{m}\hat{P}_i} \subseteq X^c$, yields the desired result.

Proposition 5. $H_1^p = H_2^p = H_3^p$

Proof. It follows from Prop. 4(iv) and (v).

In view of Prop. 5, the following discussion pertains primarily to H_1^p.

Proposition 6. H_1^p forms a clopen topology on U.

Proof. It follows immediately from Proposition 4(i), (ii), (iii) and (v).

Proposition 7. X is definable in pessimistic multigranulation rough set model if and only if X is definable in each approximation space $(U, \hat{P}_i), i = 1, \cdots, m$.

Proof. " $\Rightarrow$ " Let $(U, \hat{P}_i)$ be an arbitrarily given approximation space, to show X is definable in $(U, \hat{P}_i)$, it suffices to show that $X = \cup_{x \in U}[x]_{\hat{P}_i}$. Indeed, $\forall x \in U$, it follows from (3) that $[x]_{\hat{P}_j} \subseteq X$ for all $\hat{P}_j \in \{\hat{P}_1, \cdots, \hat{P}_m\}$, which, of course, holds for the given P_i, i.e., $[x]_{\hat{P}_i} \subseteq X$. Due to the arbitrariness of x, we thus conclude that $X \subseteq \cup_{x \in U}[x]_{\hat{P}_i} \subseteq X$, i.e., $X = \cup_{x \in U}[x]_{\hat{P}_i}$.

" $\Leftarrow$ " If X is definable in each approximation space $(U, \hat{P}_i), i = 1, \cdots, m$, then $\forall x \in X, [x]_{\hat{P}_i} \subseteq X$, by (3), we thus have $x \in \underline{X}^p_{\Sigma^m_{i=1} \hat{P}_i}$, and hence, $X \subseteq \underline{X}^p_{\Sigma^m_{i=1} \hat{P}_i}$, which, together with Prop.2(ii), yields the desired result.

Proposition 8. H^p_1 is a T_0 space if and only if $\forall x \in U, \hat{P}_i \in \{\hat{P}_1, \cdots, \hat{P}_m\}$, $[x]_{\hat{P}_i} = \{x\}$.

Proof. " $\Rightarrow$ " Suppose, on the contrary, that there exist $x \in U$ and $\hat{P}_i \in \{\hat{P}_1, \cdots, \hat{P}_m\}$, such that $[x]_{\hat{P}_i} \neq \{x\}$, then there exists $y \in U$ such that $y \neq x$ and $y \in [x]_{\hat{P}_i}$. Let X be an arbitrarily given open set in H^p_1 containing x, it follows immediately from (3) that $[x]_{\hat{P}_i} \subseteq X$, on account of $y \in [x]_{\hat{P}_i}$, we further have $y \in X$. Due to the arbitrariness of X, we conclude that there exists no open set containing x such that $y \notin X$. Similarly, we can show that there does not exist an open set containing y such that $x \notin X$. This contradicts with the fact that H^p_1 is a T_0 space. And hence, $\forall x \in U, \hat{P}_i \in \{\hat{P}_1, \cdots, \hat{P}_m\}, [x]_{\hat{P}_i} = \{x\}$.

" $\Leftarrow$ " If $\forall x \in U, \hat{P}_i \in \{\hat{P}_1, \cdots, \hat{P}_m\}, [x]_{\hat{P}_i} = \{x\}$, then by (3), it can be easily show that $\{x\} \in H^p_1$ for all $x \in U$, which directly leads to the desired result.

The following two propositions can be obtained in a similar manner.

Proposition 9. H^p_1 is a T_1 space if and only if $\forall x \in U, \hat{P}_i \in \{\hat{P}_1, \cdots, \hat{P}_m\}$, $[x]_{\hat{P}_i} = \{x\}$.

Proposition 10. H^p_1 is a Hausdorff space if and only if $\forall x \in U$, $\hat{P}_i \in \{\hat{P}_1, \cdots, \hat{P}_m\}, [x]_{\hat{P}_i} = \{x\}$.

Proposition 11. H^p_1 is a regular topological space.

Proof. Let $x \in U$ and X be a closed set satisfying $x \notin X$, we have Prop.6 that $X \in H^p_1$ and $X^c \in H^p_1$, then it can be observed immediately that X and X^c are the disjoint open sets such that $x \in X^c$ and $X \subseteq X$, which shows that H^p_1 is a regular topological space.

Proposition 12. H^p_1 is a normal topological space.

Proof. It can be shown in a similar manner as that of Prop.11.

Proposition 13. H^p_1, equipped with the set-theoretic inclusion order, forms a Boolean algebra.

Proof. It follows immediately from Prop. 4.

Example 1. Table 1 depicts a complete information system containing some information about an emporium investment project. Locus, investment and population density are the attributes of the system. Let $U = \{e_1, e_2, e_3, e_4, e_5, e_6, e_7, e_8\}$ and L and P denote the attributes Locus and Population density, respectively, then $V_L = \{good, common, bad\}$ and $V_P = \{big, media, small\}$. The partitions induced from L and P are as follows: $\hat{L} = \{\{e_1, e_7\}, \{e_2, e_3, e_4, e_5, e_6\}, \{e_8\}\}$ and $\hat{P} = \{\{e_1, e_7, e_8\}, \{e_3, e_4, e_5\}, \{e_2, e_6\}\}$. Then by computing, one can obtain that $H^p_1 = H^p_2 = H^p_3 = \{\emptyset, \{e_1, e_7, e_8\}, \{e_2, e_3, e_4, e_5, e_6\}, U\}$. Observe immediately that H^p_1 (or equivalently, H^p_2, H^p_3) determines a clopen topology on U,

Table 1. A complete target information system about emporium investment project

Project	Locus	Population density
e1	Common	Big
e2	Bad	Medium
e3	Bad	Small
e4	Bad	Small
e5	Bad	Small
e6	Bad	Medium
e7	Common	Big
e8	Good	Big

such a topology is also regular and normal. Moreover, since $[e_1]_{\check{P}} = \{e_1, e_7, e_8\}$, it follows immediately from Prop. 8-10 that (H_1^p, U) (or equivalently, (H_2^p, U), (H_3^p, U)) is not a T_0 space, not T_1 space and also not a Hausdorff space.

4   Concluding Remarks

In this paper, we consider the topological property of the collection of definable sets in the pessimistic multigranulation rough set model. We prove that the collections of three different kinds of definable sets coincide with each other, and they determine a clopen topology on the universe, furthermore, they form a Boolean algebra under the usual set-theoretic operations. We hope that the study of the mathematical structure of multigranulation rough set theory may bring the multigranulation rough set model to a new horizon of applications in the real world.

Acknowledgments. The authors greatly appreciate the financial support of the NSF Project (No. 61103133) of China, Scientific Research Program Funded by Shaanxi Provincial Education Department(No. 11JK0473) and Research Program for Young Scholars, Xi'an Shiyou University (No. YS29030909).

References

1. Pawlak, Z.: Rough sets. Int. J. Comput. Inf. Sci. 11, 341–356 (1982)
2. Pawlak, Z.: Rough Sets-Theoretical Aspects to Reasoning about Data. Kluwer Academic Publisher, Boston (1991)
3. Skowron, A., Stepaniuk, J.: Tolerance Approximation Spaces. Fundam. Inform. 27, 245–253 (1996)
4. Slowinski, R., Vanderpooten, D.: A Generalized Definition of Rough Approximations Based on Similarity. IEEE Trans. Knowl. Data En. 12, 331–336 (1990)
5. Yao, Y.Y.: Relational Interpretations of Neighborhood Operators and Rough Set Approximation Operators. Inf. Sci. 111, 239–259 (1998)
6. Yao, Y.Y.: Constructive and Algebraic Methods of Theory of Rough Sets. Inf. Sci. 109, 21–47 (1998)

7. Yao, Y.Y.: Two Views of The Theory of Rough Sets in Finite Universes. Int. J. Approx. Reason. 15, 291–317 (1999)
8. Zhu, W.: Generalized Rough Sets Based on Relations. Inf. Sci. 177, 4997–5011 (2007)
9. Zhu, P.: Covering Rough Sets Based on Neighborhoods: An Approach without Using Neighborhoods. Int. J. Approx. Reason. 52, 461–472 (2011)
10. Zhu, W., Wang, F.Y.: Reduction and Axiomization of Covering Generalized Rough Sets. Inf. Sci. 152, 217–230 (2003)
11. Zhu, W.: Topological Approaches to Covering Rough Sets. Inf. Sci. 177, 1499–1508 (2007)
12. Zhu, W., Wang, F.Y.: On Three Types of Covering Rough Sets. IEEE Trans. Knowl. Data En. 19, 1131–1144 (2007)
13. Dubois, D., Prade, H.: Rough Fuzzy Sets and Fuzzy Rough Sets. Int. J. Gen. Syst. 17, 191–209 (1990)
14. Mi, J.S., Zhang, W.X.: An Axiomatic Characterization of A Fuzzy Generalization of Rough Sets. Inf. Sci. 160, 235–249 (2004)
15. Pei, D.: A Generalized Model of Fuzzy Fough Sets. Int. J. Gen. Syst. 34, 603–613 (2005)
16. Wu, W.Z., Mi, J.S., Zhang, W.X.: Generalized Fuzzy Rough Sets. Inf. Sci. 151, 263–282 (2003)
17. Qian, Y.H., Liang, J.Y., Yao, Y.Y., Dang, C.Y.: MGRS: A Multigranulation Rough Set. Inf. Sci. 180, 949–970 (2010)
18. Qian, Y.H., Liang, J.Y., Dang, C.Y.: Incomplete Multigranulation Rough Set. IEEE Trans. Syst., Man, Cybern. A, Syst., Humans. 40, 420–431 (2010)
19. Yang, X.B., Song, X.N., Dou, H.L., Yang, J.Y.: Multigranulation Rough Set: from Crisp to Fuzzy Case. Ann. Fuzzy Math. Inform. 1, 55–70 (2011)
20. Abu-Donia, H.M.: Multi-knowledge Based Rough Approximations and Applications. Knowl.-Based Syst. 26, 20–29 (2012)
21. Wu, W.Z., Leung, Y.: Theory and Applications of Granular Labelled Partitions in Multi-scale Decision Tables. Inf. Sci. 181, 3878–3897 (2011)
22. Khan, M.A., Banerjee, M.: Formal Reasoning with Rough Sets in Multiple-source Approximation Systems. Int. J. Approx. Reason. 49, 466–477 (2008)
23. Cattaneo, G.: An Investigation about Rough Set Theory: Some Foundational and Mathematical Aspects. Fundam. Inform. 108, 197–221 (2011)
24. Iwinski, T.B.: Algebraic Approach to Rough Sets. Bull. Pol. Ac. Math. 35, 673–683 (1987)
25. Kondo, M.: On the Structure of Generalized Rough Sets. Inf. Sci. 176, 589–600 (2006)
26. Liu, G.L., Zhu, W.: The Algebraic Structures of Generalized Rough Set Theory. Inf. Sci. 178, 4105–4113 (2008)
27. Pagliani, P., Chakraborty, M.: A Geometry of Approximation Rough Set Theory: Logic, Algebra and Topology of Conceptual Patterns. Springer, New York (2008)
28. Qin, K.Y., Pei, Z.: On The Topological Properties of Fuzzy Rough Sets. Fuzzy Sets and Systems 151, 601–613 (2005)
29. Yang, L.Y., Xu, L.S.: Topological Properties of Generalized Approximation Spaces. Inf. Sci. 181, 3570–3580 (2011)
30. Kelley, J.: General Topology. Van Nostrand Co., Princeton (1955)
31. Davey, B.A., Priestley, H.A.: Introduction to Lattices and Order. Cambridge University Press, London (1990)

Axiomatic Granular Approach
to Knowledge Correspondences

A. Mani*

Department of Pure Mathematics
University of Calcutta
9/1B, Jatin Bagchi Road
Kolkata-700029, India
a.mani.cms@gmail.com
http://www.logicamani.co.cc

Abstract. An axiomatic approach towards granulation in general rough set theory (RST) was introduced by the present author in [4] and extended in [7] over general rough Y-systems (RYS). In the present brief paper a restricted first order version is formulated and granular correspondences between simpler RYS are considered. These correspondences are also relevant from the perspective of knowledge interpretation of rough sets, where we may find admissible concepts of a knowledge being a sub-object of another. Proofs will appear separately.

Keywords: Axiomatic Theory of Granules, Granular Rough Semantics, First Order Theory, Rough Y-Systems, RYS, Sub Natural Correspondences, Contamination Problem, Theories of Knowledge.

1   Introduction

Rough approximations defined by operators may be used to approximate concepts and these may built upon to generate more complex concepts. We view this process as a *rough evolution* of concepts. When the approximations are defined through explicit granules, then we speak of *granular rough evolutions of concepts*. Given two different domains with distinct granular rough evolutions of concepts, we seek to characterize natural concepts of correspondence between the two. The formal approach introduced by the present author in [7] is restricted to set theoretic settings in a first order formalism, a correspondence of interest is identified and a few results are proved in the present research paper. The eventual goal is to clarify the nature of correspondences between two agents or human reasoners from a rough perspective subject to the entire rough evolution being not necessarily known beforehand, at least some of the corresponding concepts being known and all concepts having knowable granular evolution.

In Pawlak's concept of knowledge in classical RST, if S is a set of attributes and P an indiscernibility relation on it, then sets of the form A^l and A^u represent

---

* The present author would like to thank Prof Mihir Chakraborty for discussions and the referees for useful remarks.

T. Li et al. (Eds.): RSKT 2012, LNAI 7414, pp. 482–487, 2012.

clear and definite concepts and related semantics may encode knowledge. In the extension of knowledge interpretation to tolerance spaces TAS of [4] by the present author, any non degenerate use of choice operations over granules in the construction of *definite* objects (in any of the rough senses) can be seen to result in clear concepts or beliefs with ontology or knowledge. More generally, in RYS, concepts are handled as *rough sets*. This extended knowledge interpretation provides the most difficult example of a RYS [7].

Measures for consistency between two knowledges have been introduced in [3]. In [5], these have been generalised from a granular paradigm with stress on the difference between the rough (Meta-R) and classical (Meta-C) semantic domains. Meta-R is the meta level corresponding to the observer or agent experiencing the vagueness or reasoning in vague terms (but without rough inclusion functions and degrees of inclusion), while Meta-C is the more usual higher order classical meta level at which the semantics is formulated. It should be noted that rough membership functions and similar measures are defined at Meta-C, but they do not exist at Meta-R. In [6], algebraic semantics for these contexts are developed by the present author. In all these there is no provision for handling correspondences of concepts across contexts with different kinds of granular evolution of knowledge. A formal setting is necessary for handling this.

2   General Rough Y-Systems – A First Order Approach

This is a simplified approach to RYS (relative [7]) that avoids the operator ι and is focussed on a general set-theoretic perspective sufficient for the purposes of this paper. RYS maybe seen as a generalisation of *abstract approximation spaces* [2] and other general structures. It includes most relation-based RST, cover-based RST and more. These structures are provided with enough structure so that a Meta-C and at least one Meta-R of roughly equivalent objects along with admissible operations and predicates are associable.

Definition 1. *The language of RYS $\mathfrak{L}$ will consist of symbols for binary predicates $\mathbf{P}, =, \overset{e}{=}$, binary partial operations $\oplus, \cdot$, unary operations $(l_i)_1^n, (u_i)_1^n, \sim$ and 0-place operations 1 . Existence equations ($t \overset{e}{=} s$) will simply be pairs of terms and will be treated as atomic formulas (see [1]). The axioms of RYS $\mathcal{A}$ will be the basic equality axioms and the following:*

* *A1: $(\forall x)\mathbf{P}xx$; A2: $(\forall x, y)(\mathbf{P}xy, \mathbf{P}yx \longrightarrow x = y)$,*
* *For each i, A3: $(\forall x, y)(\mathbf{P}xy \longrightarrow \mathbf{P}(l_i x)(l_i y), \mathbf{P}(u_i x)(u_i y))$,*
* *A4: $(\forall x)\mathbf{P}(l_i x)x, \mathbf{P}(x)(u_i x))$; A5: $(\forall x)(\mathbf{P}(u_i x)(l_i x) \longrightarrow x = l_i x = u_i x)$.*

The partial operations $\oplus, \odot, \ominus$ shall satisfy and the derived operations $\mathbf{O}, \mathbb{P}$, $\mathbb{X}, \mathbb{O}$ will be assumed to be defined uniquely as follows:

Overlap: $\mathbf{O}xy \leftrightarrow (\exists z)\mathbf{P}zx \wedge \mathbf{P}zy$, **Proper Part:** $\mathbb{P}xy \leftrightarrow \mathbf{P}xy \wedge \neg\mathbf{P}yx$,
Overcross: $\mathbb{X}xy \leftrightarrow \mathbf{O}xy \wedge \neg\mathbf{P}xy$, **Proper Overlap:** $\mathbb{O}xy \leftrightarrow \mathbb{X}xy \wedge \mathbb{X}yx$,
wDifference1: $(\forall x, y, z)(x \ominus y = z \to (\forall w)(\mathbf{P}wz \leftrightarrow (\mathbf{P}wx \wedge \neg\mathbf{O}wy)))$

wDifference2: $(\forall x, y, z, a)(x \ominus y \stackrel{e}{=} x \ominus y, \Phi(x,y,z), \Phi(x,y,a) \to x \ominus y = z = a)$, where $\Phi(x,y,z)$ stands for $(\forall w)(\mathbf{P}wz \leftrightarrow (\mathbf{P}wx \wedge \neg \mathbf{O}wy))$.

Sum1: $(\forall x, y, z)(x \oplus y = z \to (\forall w)(\mathbf{O}wz \leftrightarrow (\mathbf{O}wx \vee \mathbf{O}wy)))$

Sum2: $(\forall x, y, z, a)(x \oplus y \stackrel{e}{=} x \oplus y, \Omega(x,y,z), \Omega(x,y,a) \to x \oplus y = z = a)$, where $\Omega(x,y,z)$ stands for $(\forall w)(\mathbf{O}wz \leftrightarrow (\mathbf{O}wx \vee \mathbf{O}wy))$

Product1: $(\forall x, y, z)(x \odot y = z \to (\forall w)(\mathbf{P}wz \leftrightarrow (\mathbf{P}wx \wedge \mathbf{P}wy)))$

Product2: $(\forall x, y, z, a)(x \odot y \stackrel{e}{=} x \odot y, \Pi(x,y,z), \Pi(x,y,a) \to x \odot y = z = a)$, where $\Pi(x,y,z)$ stands for $(\forall w)(\mathbf{P}wz \leftrightarrow (\mathbf{P}wx \wedge \mathbf{P}wy))$

wAssociativity $x \oplus (y \oplus z) \stackrel{\omega *}{=} (x \oplus y) \oplus z$ and similarly for product. For two terms s, t, $t \stackrel{\omega *}{=} s$ is an abbreviation for $(s \stackrel{e}{=} s \to s \stackrel{e}{=} t) \wedge (t \stackrel{e}{=} t \to s \stackrel{e}{=} t)$.

wCommutativity $x \oplus y \stackrel{\omega *}{=} y \oplus x$; $x \odot y \stackrel{\omega *}{=} y \odot x$

Definition 2. *If all the operations are total, then a model of the theory of RYS $\langle \mathfrak{L}, \mathcal{A} \rangle$ will be an $\mathfrak{L}$-structure of the form $\langle \underline{S}, \mathbf{P}, (l_i)_1^n, (u_i)_1^n, \oplus, \odot, \ominus, 1 \rangle$ that satisfies the axioms $\mathcal{A}$. In this case the existence equality will coincide with the usual equality and the the operations $\oplus, \odot$ would be defined by the axioms.*

If partial operations are present, then the formulation of a 2-valued model or a model will be more involved [7,1]. The intended concept of a *rough set* in a RYS is as a collection of some sense definite elements of the form $\{a_1, a_2, \ldots a_n, b_1, b_2, \ldots b_r\}$ subject to a_is being 'part of' of some of the b_js and obtainable from the multiple approximation operators $\{l_i, u_i\}$. $\mathbf{P}xy$ can be read as 'x is a part of y' and is intended to generalise the inclusion of the classical case. The elements of S may be approximable and/or exact objects (or collections thereof). The justification for using a non-transitive parthood relation can be traced to various situations in which restrictive criteria operating on inclusion of attributes happen (see [4,7] for examples). A fragment of the axioms of set-theory compatible mereology in [9] are assumed. This mereology is not ontologically compatible with Lesniewskian mereology and reformulations in the original language of ontology are needed for *comparisons*. In the Lesniewski-inspired mereology of Polkowski and Skowron [8], the basic functor "part to a degree r", definitely assumes knowledge of the degree r and is responsible for the strong collectivizing properties of the class operator. The reason for avoiding such an approach, explained in [7], primarily relate to the contamination problem that lays stress on the difference between the awareness at a rough semantic domain and the classical semantic domain.

Note that the definitions of $\odot, \oplus, \ominus$ require two statements as they are intended to be interpreted as partial operations. Existence equalities of the form $s \stackrel{e}{=} s$ are intended to be interpreted as *s exists*. $\oplus$ is intended to be interpreted as a general aggregation operation, $\odot$ as a general commonality-extracting operation and $\ominus$ is a reserve operation for accommodating differences or complementation via 1. $\ominus$ may become completely redundant in applications to concept approximation in human learning contexts. In a model of RYS, the definition of sum, product and difference operations do not follow in general from the conditions on $\mathbf{P}$ in the above. But do so from the assumptions of closed extensional mereology ([7]). Note that a minimalistic concept of *admissible granulation* was defined in [7] as those granulations satisfying the conditions WRA, LFU, LS.

Definition 3. *By the theory of* Classical RST-RYS, *we will mean a theory* $\mathfrak{Th}$ *of RYS in which* $\oplus, \odot, \ominus$ *correspond respectively to set* $\cup, \cap, \setminus$ *respectively. S is a power-set of some set A, 1 = A, and the additional granular axioms RA, ACG, MER, FU, NO, PS, ST, I (see [7]) hold. Idempotence (I) is* $\forall i, (\forall x \in \mathcal{G})x^{u_i} = x^{u_i u_i}, x^{l_i} = x^{l_i l_i}$.

Theorem 1. *The theory of classical RST-RYS is well defined, is not categorical or* κ*-categorical (*κ *being a cardinal) and is consistent.*

The different types of granulations used in TAS of the form $\langle S, T \rangle$, in the literature are considered in [7]. Here we will be concerned with $\mathcal{T} = \{[x] ; [x] = \{y : (x, y) \in T\}\}$ and $\mathcal{B}^* = \{\cap \beta(x) : x \in S\}$, where $\beta(x)$ is the set *blocks*(maximal squares) containing the singleton set $\{x\}$ (For a subset H, $\beta(H)$ will be the set of all blocks containing H). If $\mathcal{G}$ is one of the above, for $A \subseteq S$, we will associate the approximations $A^{l\mathcal{G}} = \bigcup\{H : H \subseteq A, H \in \mathcal{G}\}$ and $A^{u\mathcal{G}} = \bigcup\{H : H \cap A \neq \emptyset, H \in \mathcal{G}\}$. For an arbitrary binary reflexive relation T on S, elements of $\mathcal{T}$ are also referred to as *successor neighbourhoods*. In what follows, new results are stated without proofs for simpler cases involving partitions, system of blocks and successor neighbourhoods within the formal framework of RYS.

Theorem 2. *In the TAS context, with* $\mathcal{T}$ *being the set of granules and restricting to the approximations* $l\mathcal{T}, u\mathcal{T}$, *all of RA, MER, ST and weakenings thereof hold, but others do not hold in general.*

3   Sub-Natural Correspondences (SNC) across RYS

A map from a RYS S_1 to another S_2 will be referred to as a *correspondence*. It will be called a *morphism* if and only if it preserves the operations $\oplus$ and $\odot$. We will also speak of $\oplus$-morphisms and $\odot$-morphisms if the correspondence preserves just one of the partial/total operations. Correspondences are of interest for the relativised representation problems introduced in [7] and our present focus on the knowledge interpretation of general rough sets. Explicit use of higher order choice functions acting on granules will not be permitted as they would require an extension of the language in the natural scheme of things. By Sub-Natural Correspondences (SNC), we seek to capture simpler correspondences that associate granules with elements representable by granules and do not commit the context to Galois connections.

Definition 4. *Let If* S_1 *and* S_2 *are two RYS with granulations,* $\mathcal{G}_1$ *and* $\mathcal{G}_2$ *respectively, consisting of successor neighbourhoods or neighbourhoods. An injective correspondence* $\varphi : S_1 \longmapsto S_2$ *will be said to be a* SNC *if and only if there is a term function t in the signature of* S_2 *such that*

$$(\forall [x] \in \mathcal{G}_1)(\exists y_1, \ldots y_n \in \mathcal{G}_2) \varphi([x]) = t(y_1, \ldots, y_n) \text{ and}$$

y_i *is generated by* $\varphi(\{x\})$ *for each i (*$\{x\}$ *is a singleton).*

Note that the base sets of the RYS need not be power-sets under the assumptions and these may be semi-algebras of sets. As intersection of distinct equivalence classes are empty, we have

Theorem 3. *If φ is a SNC and both $\mathcal{G}_1$ and $\mathcal{G}_2$ are partitions, then the non-trivial cases are equivalent to one of the following:*

* B1: $(\forall \{x\} \in S_1)\, \varphi([x]) = [\varphi(\{x\})]$. B2: $(\forall \{x\} \in S_1)\, \varphi([x]) = \sim [\varphi(\{x\})]$.
* B3: $(\forall \{x\} \in S_1)\, \varphi([x]) = \bigcup_{y \in [x]} [\varphi(\{y\})]$. B4: $(\forall \{x\} \in S_1)\, \varphi([x]) = \sim (\bigcup_{y \in [x]} [\varphi(\{y\})])$.

Theorem 4. *If φ is a SNC and both $\mathcal{G}_1$ and $\mathcal{G}_2$ are partitions and $(\forall \{x\} \in S_1)\, \varphi([x]) = [\varphi(\{x\})]$, then the image of the restriction of φ to $\mathcal{G}_1$ is a subset of $\mathcal{G}_2$ and*

$$(\forall x \in S_1)\, \varphi(x^l) \subseteq (\varphi(x))^l \text{ and } (\forall x \in S_1)\, \varphi(x^u) \subseteq (\varphi(x))^u \text{ hold.}$$

Theorem 5. *If φ is a SNC that is a $\oplus$-morphism and both $\mathcal{G}_1$ and $\mathcal{G}_2$ are partitions and $(\forall \{x\} \in S_1)\, \varphi([x]) = \sim [\varphi(\{x\})]$ and further if S_1 and S_2 are derived from approximation spaces, then these must be trivial.*

Theorem 6. *If φ is a SNC that is a morphism and both $\mathcal{G}_1$ and $\mathcal{G}_2$ are partitions and $(\forall \{x\} \in S_1)\, \varphi([x]) = [\varphi(\{x\})]$, then*

$$(\forall x \in S_1)\, \varphi(x^l) = (\varphi(x))^l \text{ and } (\forall x \in S_1)\, \varphi(x^u) = (\varphi(x))^u.$$

The condition $(\forall \{x\} \in S_1)\, \varphi([x]) = \bigcup_{y \in [x]} [\varphi(\{y\})]$ is extremely interesting in that without additional restrictions on φ, S_2 ends up with objects that behave like S_1 granules. But even if we assume that φ is a morphism, we get few properties relating to lower and upper approximations of objects.

Theorem 7. *If φ is a SNC that is a morphism and both $\mathcal{G}_1$ and $\mathcal{G}_2$ are partitions and $(\forall \{x\} \in S_1)\, \varphi([x]) = \bigcup_{y \in [x]} [\varphi(\{y\})]$ and further if S_1 and S_2 are derived from approximation spaces, then the following hold:*

$$(\forall A \in S_1)\, \varphi(A^u) = \bigcup_{[x] \cap A \neq \emptyset} \bigcup_{y \in [x]} [\varphi(\{y\})] = \bigcap_{A \subseteq S_1 \setminus [x]} \varphi(S \setminus [x]).$$

Theorem 8. *If φ is a SNC, $\mathcal{G}_1$ is a partition and $\mathcal{G}_2$ is a system of blocks, then the non-trivial cases should be equivalent to one of the following ($\sim (\beta(x))$ is the set $\{\sim B : B \in \beta(x)\}$ and $\{x\} \in S_1$):*

* C1: $\varphi([x]) = \cup \beta(\varphi(\{x\}))$. C2: $\varphi([x]) = \sim (\cup \beta(\varphi(\{x\})))$.
* C3: $\varphi([x]) = \cap \beta(\varphi(\{x\}))$. C4: $\varphi([x]) = \sim (\cap \beta(\varphi(\{x\})))$.
* C5: $\varphi([x]) = \bigcup_{y \in [x]} \cup \beta(\varphi(\{x\}))$. C6: $\varphi([x]) = \sim (\bigcup_{y \in [x]} \cup \beta(\varphi(\{x\})))$.
* C7: $\varphi([x]) = \bigcap_{y \in [x]} \cup \beta(\varphi(\{x\}))$. C8: $\varphi([x]) = \sim (\bigcup_{y \in [x]} \cap \beta(\varphi(\{x\})))$.

Theorem 9. *If we take S_1 to be a classical RST-RYS and S_2 is a TAS-RYS with approximations lB^* and uB^* and φ is a SNC is a $\oplus$-morphism satisfying the first condition above, then*

$$\varphi(x^l) \subseteq (\varphi(x))^{lB^*} \text{ and } \varphi(x^u) \subseteq (\varphi(x))^{uB^*}.$$

If φ is a morphism, that preserves $\emptyset$ and 1, then equality holds, but the converse is not true in general.

Theorem 10. *If we take S_1 to be a classical RST-RYS and S_2 is a TAS-RYS with approximations lT and uT and φ is a SNC and a $\oplus$ - morphism satisfying for each singleton $\{x\} \in S_1$, $\varphi([\{x\}]) = [\varphi(x)]$, then*

$$\varphi(x^l) \subseteq (\varphi(x))^{lT} \text{ and } \varphi(x^u) \subseteq (\varphi(x))^{uT}$$

Concluding Remarks: In this research, the concept of sub natural correspondences of granulations is introduced in a restriction of the formal framework of RYS [7]. Further work is also motivated by applications in both the representation theory of granules and the problem of identifying sub-objects in knowledge interpretation of general RST.

References

1. Burmeister, P.: A Model-Theoretic Oriented Approach to Partial Algebras. Akademie-Verlag (1986, 2002)
2. Cattaneo, G., Ciucci, D.: Lattices with Interior and Closure Operators and Abstract Approximation Spaces. In: Peters, J.F., Skowron, A., Wolski, M., Chakraborty, M.K., Wu, W.-Z. (eds.) Transactions on Rough Sets X. LNCS, vol. 5656, pp. 67–116. Springer, Heidelberg (2009)
3. Chakraborty, M.K., Samanta, P.: Consistency-Degree Between Knowledges. In: Kryszkiewicz, M., Peters, J.F., Rybiński, H., Skowron, A. (eds.) RSEISP 2007. LNCS (LNAI), vol. 4585, pp. 133–141. Springer, Heidelberg (2007)
4. Mani, A.: Choice Inclusive General Rough Semantics. Information Sciences 181(6), 1097–1115 (2011), http://dx.doi.org/10.1016/j.ins.2010.11.016
5. Mani, A.: Dialectics of Counting and Measures of Rough Set Theory. In: IEEE Proceedings of NCESCT 2011, Pune, February 1-3, p. 17 (2011), Arxiv:1102.2558, http://arxiv.org/abs/1102.2558
6. Mani, A.: Towards Logics of Some Rough Perspectives of Knowledge. In: Suraj, Z., Skowron, A. (eds.) Intelligent Systems Reference Library Dedicated to the Memory of Prof. Pawlak, pp. 342–367. Springer (2011-2012)
7. Mani, A.: Dialectics of Counting and the Mathematics of Vagueness. In: Peters, J.F., Skowron, A. (eds.) Transactions on Rough Sets XV. LNCS, vol. 7255, pp. 122–180. Springer, Heidelberg (2012)
8. Polkowski, L., Skowron, A.: Rough Mereology: A New Paradigm for Approximate Reasoning. Internat. J. Appr. Reasoning 15(4), 333–365 (1996)
9. Varzi, A.: Parts, Wholes and Part-Whole Relations: The Prospects of Mereotopology. Data and Knowledge Engineering 20, 259–286 (1996)

Definable and Rough Sets
in Covering-Based Approximation Spaces

Arun Kumar and Mohua Banerjee

Department of Mathematics and Statistics,
Indian Institute of Technology, Kanpur 208016, India
{kmarun,mohua}@iitk.ac.in

Abstract. This article explores the notion of definable sets in approximation spaces based on coverings, and presents properties of algebraic structures obtained from the collection of all definable sets. A granule-based definition of lower and upper approximations of sets, is then investigated. A rough set is defined in the usual way, viz. as a pair of the lower and upper approximations of a set, and we investigate the algebraic structure of the collection of all rough sets.

Keywords: Granules, Approximations of sets.

1 Introduction

The essence of classical rough set theory proposed by Pawlak is to approximate a set with the help of granules, where a granule is an equivalence class in the domain. A union of granules is considered as a *definable set*. A set A is approximated using the lower and upper approximation operators, where the lower approximation is the largest definable set contained in A, while the upper approximation of A is the smallest definable set which contains A. In practice however, granules may not be disjoint, or may not arise from an equivalence relation. A lot of ground has been covered in the study of generalized approximation spaces, and we have a number of different notions of approximations of sets (cf. [1]).

It is interesting to see how a set may be approximated by granules in these general situations. Study in this regard has been done, for instance, by Yao [2–5]. In this paper, we focus on covering-based approximation spaces, and definitions of a definable set in these spaces. Two definitions of lower and upper approximations that are granule-based are investigated. The interest of this article lies finally, in the algebraic structures that may be formed out of definable and rough sets. In Sections 2 and 3, we present the basic definitions and properties. Section 4 studies the algebraic structure of the collection of all rough sets. Section 5 concludes the article.

T. Li et al. (Eds.): RSKT 2012, LNAI 7414, pp. 488–495, 2012.

2   Definable Sets and Their Algebraic Structure

Let U be a non-empty set, and consider a collection $\mathcal{C}$ of non-empty subsets of U, whose union is U. As we know, $\mathcal{C}$ is called a *cover* for U, in topology. Our interest is in the covering-based approximation space $(U, \mathcal{C})$.

Let $x \in U$ and define

$$n(x) := \cap \{C \in \mathcal{C} : x \in C\}.$$

We consider $n(x)$ as a *granule* containing x, in the approximation space $(U, \mathcal{C})$. $n(x)$ has been used to define approximations of sets in [6–9]. Note that $n(x)$ gives rise to a quasi-order R on U: for $x, y \in U$, xRy if and only if $y \in n(x)$. R is reflexive and transitive, and for each $x \in U$, $n(x) = R(x)$, where $R(x) := \{y \in U : xRy\}$. We should mention here that Yao has considered $R(x)$ as a granule in [2–5].

Every quasi-order R on a set corresponds to an Alexandrov topology τ_R on that set, for which the set $\{R(x) : x \in U\}$ is the smallest base (cf. [10]). So, for every covering-based approximation space $(U, \mathcal{C})$, we get an Alexandrov topology. Let us denote it as $\tau_{\mathcal{C}}$. Its smallest base is then given by the collection $\{n(x) : x \in U\}$. We note that τ_R is just the same as $\tau_{\mathcal{C}}$, but the different notations are used in the sequel according as we refer to R or $\mathcal{C}$.

Definition 1. *A subset D of U is said to be* definable *if and only if D can be written as a union of granules, i.e. there exists a set $X(\subseteq U)$ such that*

$$D = \bigcup_{x \in X} n(x).$$

As an immediate consequence, $\emptyset$ becomes a definable set. Notice that D is an open set of the topology $\tau_{\mathcal{C}}$.

Let $\mathcal{D} := \{D \subseteq U : D \text{ is definable}\}$. It is clear that $\mathcal{D} = \tau_{\mathcal{C}}$.

Observation 1. Every Alexandrov topology is a complete lattice (a sublattice of $\mathcal{P}(U)$, the power set of U), and so $\mathcal{D}$ is a complete lattice.

More explicitly: $\mathcal{D}$ is closed with respect to arbitrary union. For the case of arbitrary intersection, observe that $n(x) \cap n(y) = \cup\{n(z) : z \in n(x) \cap n(y)\}$, for any $x, y \in U$.

Let us look a little deeper into the algebraic structure of $\mathcal{D}$. Every topological space can be turned into a Heyting algebra [11], and we get

Proposition 1. $(\mathcal{D}, \cap, \cup, \Rightarrow)$ *is a* Heyting algebra, *where for any $D_1, D_2 \in \mathcal{D}$, $D_1 \Rightarrow D_2 := \cup\{D \in \mathcal{D} : D \subseteq D_1^c \cup D_2\}$. In particular, $\mathcal{D}$ is a pseudo-complemented lattice, where for $D \in \mathcal{D}$, the pseudo-complement of D is given by $\neg D := \cup\{n(x) : n(x) \cap D = \emptyset\}$. Here D_1^c is the set theoretic complement of D_1.*

Observation 2. As a special case, if $\mathcal{C}$ is a partition of U, the following are immediate.

1. Definable sets become unions of equivalence classes – this is in consonance with the classical definition of definable sets.

2. $\{n(x) : x \in U\}$ is the set of all equivalence classes which forms the smallest base for a 0-dimensional topology.
3. $\mathcal{D}$ forms a Boolean algebra.

3   Approximations of Sets

We now turn to definitions of lower and upper approximation operators L, U respectively in a covering-based approximation space $(U, \mathcal{C})$.

Definition 2. *For any* $X \subseteq U$, $\mathsf{L}, \mathsf{U} : \mathcal{P}(\mathsf{U}) \to \mathcal{P}(\mathsf{U})$ *are such that*
$$\mathsf{L}(X) := \bigcup \{D \in \mathcal{D} : D \subseteq X\},$$
$$\mathsf{U}(X) := \bigcap \{D \in \mathcal{D} : X \subseteq D\}.$$

So, for $X \subseteq U$, $\mathsf{L}(X)$ and $\mathsf{U}(X)$ are definable.

Observation 3

1. Definition 2 has the following equivalent forms. For any $X \subseteq U$,
$$\mathsf{L}(X) = \bigcup \{n(x) : n(x) \subseteq X\}$$
$$= \{x \in U : n(x) \subseteq X\}.$$
$$\mathsf{U}(X) = \bigcup \{n(x) : x \in X\}.$$
2. In terms of open sets of the topology $\mathcal{D}$, for any $X \subseteq U$, $\mathsf{L}(X)$ is the largest open set contained in X and $\mathsf{U}(X)$ the smallest open set, which contains X.
3. If $\mathcal{C}$ is a partition of U, the operators L, U are just the classical lower and upper approximation operators respectively on $\mathcal{P}(U)$.

Note 1. In the above equivalent form, L as a lower approximation operator is considered in [12], and U as an upper approximation operator is considered in [12, 8]. The difference is that we would like the pair (L, U) to define a rough set for us (as given formally in the sequel, cf. Definition 3).

In Proposition 2 below, we compile a list of properties of L and U.

Proposition 2. *For any* $X, Y \subseteq U$ *and* $D \in \mathcal{D}$, L *and* U *satisfy the following properties.*

1. $\mathsf{L}(U) = U$ *and* $\mathsf{L}(\emptyset) = \emptyset$.
2. $\mathsf{U}(U) = U$ *and* $\mathsf{U}(\emptyset) = \emptyset$.
3. $\mathsf{L}(X) \subseteq X \subseteq \mathsf{U}(X)$.
4. $\mathsf{L}(D) = D$, $\mathsf{U}(D) = D$.
5. $\mathsf{LL}(X) = \mathsf{L}(X)$ *and* $\mathsf{LU}(X) = \mathsf{U}(X)$.
6. $\mathsf{UU}(X) = \mathsf{U}(X)$ *and* $\mathsf{UL}(X) = \mathsf{L}(X)$.
7. $\mathsf{L}(X \cap Y) = \mathsf{L}(X) \cap \mathsf{L}(Y)$ *and* $\mathsf{U}(X \cup Y) = \mathsf{U}(X) \cup \mathsf{U}(Y)$.
8. $\mathsf{L}(X) \cup \mathsf{L}(Y) \subseteq \mathsf{L}(X \cup Y)$ *and* $\mathsf{U}(X \cap Y) \subseteq \mathsf{U}(X) \cap \mathsf{U}(Y)$.
9. $\mathsf{L}(X) = X$ *if and only if* $\mathsf{U}(X) = X$.

Let us illustrate the definitions and properties through an example.

Example 1. Let $U := \{a, b, c, d\}$ and $\mathcal{C} := \{\{a, b, c\}, \{b\}, \{c\}, \{c, d\}\}$ be a covering of U. Then $n(a) = \{a, b, c\}$, $n(b) = \{b\}$, $n(c) = \{c\}$ and $n(d) = \{c, d\}$.

For $X \subseteq U$, $\mathsf{L}(X)$, $\mathsf{U}(X)$ is then given by Table 1. We abuse standard notations slightly for simplicity: a set $\{x, y, z\}$ will be written as xyz in the table.

Table 1.

X	$L(X)$	$U(X)$	X	$L(X)$	$U(X)$
$\emptyset$	$\emptyset$	$\emptyset$	ac	c	abc
a	$\emptyset$	abc	ad	$\emptyset$	U
b	b	b	bd	b	bcd
c	c	c	abc	abc	abc
d	$\emptyset$	cd	bcd	bcd	bcd
ab	b	abc	abd	b	U
bc	bc	bc	acd	c	U
cd	cd	cd	U	U	U

Observation 4

1. Consider $X := \{b\}$ in Example 1, then $U(X) = \{b\}$, but $\{b\} \neq \{a,b\} = (L(\{a,c,d\}))^c = (L(\{b\}^c))^c$, hence (L, U) are not dual operators. However, it can be easily verified that they are adjoint to each other.
2. From Proposition 2, we observe that L and U are interior and closure operators respectively on $\mathcal{P}(U)$, but the fact that they are not dual shows that they induce two different topologies on U. With L, $\mathcal{D}$ is the set of all open sets in the Alexandrov topology τ_R, while with U, $\mathcal{D}$ is the set of all closed sets in the Alexandrov topology $\tau_{R^{-1}}$. (τ_R and $\tau_{R^{-1}}$ are the Alexandrov topologies corresponding to the quasi-orders R and R^{-1} respectively, as mentioned in Section 2.) Hence for any $X \subseteq U$, $U(X) = (L^{-1}(X^c))^c$, where L^{-1} is the interior operator in the Alexandrov topology $\tau_{R^{-1}}$.
3. Following [13], if we consider $\mathcal{T}_R(U) = \{L(X) : X \subseteq U\}$ and $\mathcal{T}^R(U) = \{U(X) : X \subseteq U\}$, then, by Proposition 2, $\mathcal{T}_R(U) = \mathcal{T}^R(U) = \mathcal{D}$. In [13], it is also observed that $\mathcal{T}_R(U)$ and $\mathcal{T}^R(U)$ form Heyting algebras.

3.1   Comparison with Other Neighborhood Based Operators

In this section, we give a comparison between L, U, and other operators in literature that are defined using $n(x)$. The operators in 1-4 are from [12], while those in 5 are from [8].

1. $\quad L_1(X) := \{x \in U : n(x) \subseteq X\}$,
   $\quad U_1(X) := \{x \in U : n(x) \cap X \neq \emptyset\}$.
2. $\quad L_2(X) := \{x \in U : \exists u(u \in n(x) \wedge n(u) \subseteq X)\}$,
   $\quad U_2(X) := \{x \in U : \forall u(u \in n(x) \rightarrow n(u) \cap X \neq \emptyset)\}$.
3. $\quad L_3(X) := \{x \in U : \forall u(x \in n(u) \rightarrow n(u) \subseteq X)\}$,
   $\quad U_3(X) := \cup\{n(x) : n(x) \cap X \neq \emptyset\}$.
4. $\quad L_4(X) := \{x \in U : \forall u(x \in n(u) \rightarrow u \in X)\}$,
   $\quad U_4(X) := \cup\{n(x) : x \in X\}$.

$(L_1, U_1) - (L_4, U_4)$ are dual operator pairs. There is a non-dual operator pair defined using $n(x)$ also.

5. $L_5(X) := \cup\{C \in \mathcal{C} : C \subseteq X\}$,
    $U_5(X) := L_5(X) \cup (\cup\{n(x) : x \in (X \setminus L_5(X))\})$
    $\qquad = \cup\{n(x) : x \in X\}$.

Proposition 3. *Let $(U, \mathcal{C})$ be a covering-based approximation space and $X \subseteq U$.*

1. $L_1 = L$.
2. $L \subseteq L_2$.
3. $L_3(X) \subseteq L(X)$ *and* $U(X) \subseteq U_3(X)$.
4. $U_4 = U$.
5. $L_5(X) \subseteq L(X)$ *and* $U_5 = U$.

We show through the following example that U_1, U_2 and L_4 are not comparable with U and L respectively.

Example 2. Let $U := \{a, b, c, d\}$ and $\mathcal{C} := \{\{a, b, c\}, \{b, c\}, \{c\}, \{c, d\}\}$. Then $n(a) = \{a, b, c\}$, $n(b) = \{b, c\}$, $n(c) = \{c\}$, $n(d) = \{c, d\}$.

Let $X := \{a, b, c\}$. Then $U(\{a, b, c\}) = \{a, b, c\}$ and $U_1(\{a, b, c\}) = U = U_2(\{a, b, c\})$. So, here $U(X) \subseteq U_1(X) = U_2(X)$. But if we consider $X := \{a, d\}$, then $U(\{a, d\}) = U$, while $U_1(\{a, d\}) = \{a, d\}$, and $U_2(\{a, d\}) = \emptyset$. Hence, in this case, $U_1(X) \subseteq U(X)$, as well as $U_2(X) \subseteq U(X)$. So U and U_1, U_2 are not comparable.

Again consider $X := \{a, b, c\}$. Then $L(\{a, b, c\}) = \{a, b, c\}$ and $L_4(\{a, b, c\}) = \{a, b\}$. So, $L_4(X) \subseteq L(X)$. But taking $X := \{a, d\}$ as before, we find $L(\{a, d\}) = \emptyset$, while $L_4(\{a, d\}) = \{a, d\}$. So, $L(X) \subseteq L_4(X)$, in this case. Hence L and L_4 are not comparable either.

Observation 5. Thus (L, U) is comparable only with (L_3, U_3) and (L_5, U_5), and is better than either as a pair of approximations.

4   Rough Sets and Their Algebraic Structure

Let us now define a rough set in a covering-based approximation space $(U, \mathcal{C})$.

Definition 3. *For each $X \subseteq U$, the pair $(L(X), U(X))$ is called a* rough set.
Let $\mathcal{RS} := \{(L(X), U(X)) : X \subseteq U\}$, i.e. $\mathcal{RS}$ is the collection of all rough sets in $(U, \mathcal{C})$.

We first look at a slightly more general picture. A pair (D_1, D_2), where $D_1 \subseteq D_2$ and $D_1, D_2 \in \mathcal{D}$, may be considered to be a generalized version of a rough set [14]. Suppose $\mathcal{R} := \{(D_1, D_2) : D_1 \subseteq D_2 \text{ and } D_1, D_2 \in \mathcal{D}\}$. We have seen earlier that $\mathcal{D}$ is a complete sublattice of $\mathcal{P}(U)$. This lattice structure can also be induced in $\mathcal{R}$. We have

Proposition 4. $(\mathcal{R}, \cup, \cap, (\emptyset, \emptyset), (U, U))$ *is a complete lattice, where, for (D_1, D_2), $(D_1', D_2') \in \mathcal{R}$,*

$$(D_1, D_2) \cup (D_1', D_2') := (D_1 \cup D_1', D_2 \cup D_2'),$$
$$(D_1, D_2) \cap (D_1', D_2') := (D_1 \cap D_1', D_2 \cap D_2').$$

As $\mathcal{R}$ is a complete sublattice of $\mathcal{P}(U) \times \mathcal{P}(U)$, it is a completely distributive lattice. Hence we have,

Proposition 5. $(\mathcal{R}, \cup, \cap, \Rightarrow, (\emptyset, \emptyset), (U, U))$ *is a Heyting algebra, with*
$(D_1, D_2) \Rightarrow (D'_1, D'_2) := \cup\{(D, D') \in \mathcal{R} : (D_1, D_2) \cap (D, D') \subseteq (D'_1, D'_2)\},$
for $(D_1, D_2), (D'_1, D'_2) \in \mathcal{R}.$
In particular, $\mathcal{R}$ *is a pseudo-complemented lattice, where the pseudo-complement of* $(D_1, D_2) \in \mathcal{R}$ *is given by*
$$\neg_s(D_1, D_2) := (\neg D_2, \neg D_2),$$
where $\neg D_2$ *is the pseudo-complement of* D_2 *in* $\mathcal{D}.$

Now, we already have $\mathcal{RS} \subseteq \mathcal{R}$. When $\mathcal{C}$ is a partition, it is observed in [14] that $\mathcal{R}$ and $\mathcal{RS}$ are equivalent if and only if each equivalence class of $(U, \mathcal{C})$ contains at least two elements. In other words, for any pair (D_1, D_2) of definable sets in $\mathcal{R}$, there is $X(\subseteq U)$ such that the lower approximation of X, $\mathsf{L}(X) = D_1$ and the upper approximation of X, $\mathsf{U}(X) = D_2$, if and only if $|D_2 \setminus D_1| \neq 1$. On investigating the situation here, we find that the same condition works here as well.

Theorem 1. $\mathcal{R} = \mathcal{RS}$ *if and only if whenever* $(D_1, D_2) \in \mathcal{R}$, $|D_2 \setminus D_1| \neq 1$.

However, we have the following.

Observation 6. The condition of the above theorem implies that
$$|n(x)| \geq 2, \text{ for all } x \in U. \tag{*}$$
But the converse is not true, as we show in Example 4 below. We note that the two statements are equivalent when $\mathcal{C}$ is a partition of U.

Proof of the theorem. First let us assume $\mathcal{R} = \mathcal{RS}$.
Suppose the condition of the theorem does not hold. Then there exist $D_1, D_2 \in \mathcal{D}$ such that $D_1 \subseteq D_2$, and $|D_2 \setminus D_1| = 1$. Let $D_1 = \cup_{y \in Z_1} n(y)$ and $D_2 = \cup_{z \in Z_2} n(z)$, where $Z_1, Z_2 \subseteq U$. Now, $D_1 \subseteq D_2$, and $|D_2 \setminus D_1| = 1$ imply that there is $x \in D_2$ such that $x \notin D_1$. $x \in D_2$ means $x \in n(z)$ for some $z \in Z_2$. If $x \neq z$, then $z \in D_1$, and so $n(z) \subseteq D_1$. This implies that $x \in D_1$, a contradiction to our assumption. Hence if $x \in n(z) \subseteq D_2$, $z = x$. Now suppose there is $X \subseteq U$ such that $\mathsf{L}(X) = D_1$ and $\mathsf{U}(X) = D_2$. $\mathsf{U}(X) = D_2$ implies $x \in X$: otherwise, $x \in \mathsf{U}(X) = \cup\{n(z) : z \in X\} = D_2$ implies there is $z \in X \subseteq D_2$ such that $x \in n(z)$, a contradiction. Hence $x \in X$. We also have $\mathsf{L}(X) = D_1$, hence $\mathsf{L}(X) = D_1 \subseteq X \subseteq \mathsf{U}(X) = D_2$ reduces to $\mathsf{L}(X) = D_1 = X = \mathsf{U}(X) = D_2$, a contradiction to the assumption that $|D_2 \setminus D_1| = 1$.

Conversely, let us assume that the condition of the theorem holds, and let $(D_1, D_2) \in \mathcal{R}$. Let $D_2 = \cup_{y \in W} n(y)$, where $W \subseteq U$ is such that, for all $y, z \in W$ with $y \neq z$, neither of $n(y) \subseteq n(z)$ or $n(z) \subseteq n(y)$ holds. Note that we can always find such $W \subseteq U$. Let $Z := \{y \in W : y \notin D_1\}$. Now, let us consider $X := D_1 \cup Z$. We claim that $\mathsf{L}(X) = D_1$ and $\mathsf{U}(X) = D_2$.

Let us first prove that $\mathsf{L}(X) = D_1$. As $D_1 \subseteq X$, $\mathsf{L}(D_1) = D_1 \subseteq \mathsf{L}(X)$. Now, let $x \in \mathsf{L}(X)$. Then $n(x) \subseteq X$.
Case 1: If $n(x) \subseteq D_1$, we have nothing to prove.
Case 2: Assume $n(x) \not\subseteq D_1$. Then we have two subcases.
Subcase 1: If $n(x) \subseteq Z$, then $x = y$ for some $y \in Z$. So using (*) of Observation 6, $|n(y)| \geq 2$, which implies that there exists some $z \neq y$ such that $z \in n(y)$.

This implies $n(z) \subseteq n(y)$, which is a contradiction. Hence this subcase is ruled out.

Subcase 2: Now, assume there are two disjoint non-empty sets A and B such that $A \cup B = n(x)$, and $A \subseteq D_1$, $B \subseteq Z$. Clearly $x \notin A$, otherwise $x \in A \subseteq D_1$ would imply $n(x) \subseteq D_1$, a contradiction to our assumption. Hence $x \in B \subseteq Z$ and so $x = y$ for some $y \in Z$. As D_1 and Z are disjoint, $n(x) \setminus D_1 = B \subseteq Z$. By the condition of the theorem, $B = n(x) \setminus D_1 = n(x) \setminus (D_1 \cap n(x))$, has at least two elements. Hence there is $z \in B$ and $z \neq x$. This implies that $z \in n(x)$, and hence $n(z) \subseteq n(x)$, which is a contradiction to the fact that $z, x \in Z$.

Hence case 2 is ruled out and we have $n(x) \subseteq D_1$. Thus $\mathsf{L}(X) = D_1$.

Let us prove that $\mathsf{U}(X) = D_2$. We already have $X \subseteq D_2$, which implies $\mathsf{U}(X) \subseteq \mathsf{U}(D_2) = D_2$. Now let $x \in D_2$. Then there exists a y where $y \in W$, such that $x \in n(y)$. If $y \in D_1$, then $x \in n(y) \subseteq D_1$. If $y \notin D_1$, then $y \in Z$. This implies that $n(y) \subseteq \mathsf{U}(Z) \subseteq \mathsf{U}(X)$, and hence $x \in \mathsf{U}(X)$. So, $\mathsf{U}(X) = D_2$.

Corollary 1. *If, whenever $(D_1, D_2) \in \mathcal{R}$, $|D_2 \setminus D_1| \neq 1$, then $\mathcal{RS}$ is a complete sublattice of $\mathcal{P}(U) \times \mathcal{P}(U)$. Moreover, $\mathcal{RS}$ is a Heyting algebra.*

Let us illustrate the above theorem through an example.

Example 3. Let us recall Example 1. It is clear that the condition of the theorem is violated: $(\emptyset, n(b)) = (\emptyset, b) \in \mathcal{R}$. So there is no $X \subseteq U$, such that $\mathsf{L}(X) = \emptyset$ and $\mathsf{U}(X) = \{b\}$.

We can make another interesting point through this example. Let us consider $X := \{a, b\}$ and $Y := \{c, d\}$. Then $\mathsf{L}(X) \cup \mathsf{L}(Y) = \{b, c, d\}$, $\mathsf{U}(X) \cup \mathsf{U}(Y) = \{a, b, c, d\}$, and $(\mathsf{L}(X) \cup \mathsf{L}(Y), \mathsf{U}(X) \cup \mathsf{U}(Y)) \in \mathcal{R}$. This pair also violates the condition of the theorem, and so there does not exist any $Z(\subseteq U)$ such that $\mathsf{L}(X) \cup \mathsf{L}(Y) = \mathsf{L}(Z)$ and $\mathsf{U}(X) \cup \mathsf{U}(Y) = \mathsf{U}(Z)$. Note that this is contrary to the case in classical rough set theory, when $\mathcal{C}$ is a partition of U (cf. [15, 14]).

We give the following example in the context of Observation 6.

Example 4. Let $U := \{a, b, c, d, e\}$ and $\mathcal{C} := \{\{a, b, c\}, \{b, c\}, \{d, e\}\}$. Then $n(a) = \{a, b, c\}$, $n(b) = \{b, c\}$, $n(c) = \{b, c\}$, $n(d) = \{d, e\}$ and $n(e) = \{d, e\}$. (*) in Observation 6 is satisfied by $(U, \mathcal{C})$, but it does not satisfy the condition of the theorem: consider the pair $(n(b), n(a))$ in $\mathcal{R}$.

Now, let us define a binary relation on $\mathcal{P}(U)$, corresponding to the classical notion of *rough equality*.

Definition 4. *X is roughly equal to Y in the covering-based approximation space $(U, \mathcal{C})$, denoted $X \approx Y$, if and only if $\mathsf{L}(X) = \mathsf{L}(Y)$ and $\mathsf{U}(X) = \mathsf{U}(Y)$.*

$\approx$ is an equivalence relation on $\mathcal{P}(U)$, and it is clear that the quotient set $\mathcal{P}(U)/\approx$ and $\mathcal{RS}$ are order isomorphic. Further, recall that in classical rough set theory, equivalence classes of $\mathcal{P}(U)/\approx$ are called rough sets.

Corollary 2. *Under the condition given in Corollary 1, $\mathcal{P}(U)/\approx$ becomes a Heyting algebra.*

5   Conclusions

In this article, we have studied definable and rough sets in covering-based approximation spaces. In particular, the algebraic structures that are formed by them are investigated. It would be worthwhile to check whether these structures can be extended, and study corresponding logical systems. One could also check if the construction from $\mathcal{D}$ to $\mathcal{R}$ may be abstracted in the lines of the work in [15].

Acknowledgements. We thank the referees for their valuable comments.

References

1. Samanta, P., Chakraborty, M.K.: Generalized Rough Sets and Implication Lattices. In: Peters, J.F., Skowron, A., Sakai, H., Chakraborty, M.K., Slezak, D., Hassanien, A.E., Zhu, W. (eds.) Transactions on Rough Sets XIV. LNCS, vol. 6600, pp. 183–201. Springer, Heidelberg (2011)
2. Yao, Y.Y.: Relational interpretations of neighborhood operators and rough set approximation operators. Information Sciences 111, 239–259 (1998)
3. Yao, Y.Y.: Granular computing using neighborhood system. In: Roy, R., Furuhashi, T., Chawdhry, P.K. (eds.) Advances in Soft Computing: Engineering Design and Manufacturing, pp. 539–553. Springer, New York (1999)
4. Yao, Y.Y.: Rough sets, neighborhood system and granular computing. In: Proceedings of IEEE Canadian Conf. on Electrical and Computer Engineering, pp. 1553–1558. IEEE press (1999)
5. Yao, Y.Y.: Information granulation and rough set approximation. Int. J. Intelligent Systems 16(1), 87–104 (2001)
6. Li, T.J.: Rough Approximation Operators in Covering Approximation Spaces. In: Greco, S., Hata, Y., Hirano, S., Inuiguchi, M., Miyamoto, S., Nguyen, H.S., Słowiński, R. (eds.) RSCTC 2006. LNCS (LNAI), vol. 4259, pp. 174–182. Springer, Heidelberg (2006)
7. Pagliani, P.: Pretopologies and dynamic spaces. Fundamenta Informaticae 59(2-3), 221–239 (2004)
8. Zhu, W.: Topological approaches to covering rough sets. Information Sciences 177, 1499–1508 (2007)
9. Zhu, W., Wang, F.Y.: Relationship among three types of covering rough sets. In: Proceedings of IEEE Int. Conf. on Granular Computing, pp. 43–48 (2006)
10. Järvinen, J., Radeleczki, S., Veres, L.: Rough sets determined by quasiorders. order 26, 337–355 (2009)
11. Rasiowa, H.: An Algebraic Approach to Non-classical Logics. North-Holland (1974)
12. Qin, K., Gao, Y., Pei, Z.: On Covering Rough Sets. In: Yao, J., Lingras, P., Wu, W.-Z., Szczuka, M.S., Cercone, N.J., Ślęzak, D. (eds.) RSKT 2007. LNCS (LNAI), vol. 4481, pp. 34–41. Springer, Heidelberg (2007)
13. Järvinen, J., Pagliani, P., Radeleczki, S.: Information completeness in Nelson algebras of rough sets induced by quasiorders (2012), http://arxiv.org/pdf/1203.2136
14. Banerjee, M., Chakraborty, M.K.: Algebras from rough sets. In: Pal, S.K., Polkowski, L., Skowron, A. (eds.) Rough-neuro Computing: Techniques for Computing with Words, pp. 157–184. Springer, Berlin (2004)
15. Banerjee, M., Chakraborty, M.: Rough sets through algebraic logic. Fundamenta Informaticae 28(3-4), 211–221 (1996)

Intuitionistic Fuzzy Topologies in Crisp Approximation Spaces

You-Hong Xu and Wei-Zhi Wu

School of Mathematics, Physics and Information Science,
Zhejiang Ocean University, Zhoushan, Zhejiang, 316004, P.R. China
{xyh,wuwz}@zjou.edu.cn

Abstract. A rough intuitionistic fuzzy set is the result of approximation of an intuitionistic fuzzy set with respect to a crisp approximation space. In this paper, we investigate topological structures of rough intuitionistic fuzzy sets. We examine that a reflexive crisp rough approximation space can induce an intuitionistic fuzzy Alexandrov space. It is proved that the lower and upper rough intuitionistic fuzzy approximation operators are, respectively, an intuitionistic fuzzy interior operator and an intuitionistic fuzzy closure operator if and only if the binary relation in the crisp approximation space is reflexive and transitive. Finally, we show that a similarity crisp approximation space can produce an intuitionistic fuzzy clopen topological space.

Keywords: Approximation operators, Intuitionistic fuzzy sets, Intuitionistic fuzzy topologies, Rough intuitionistic fuzzy sets, Rough sets.

1 Introduction

Topology is a branch of mathematics, whose concepts exist not only in almost all branches of mathematics, but also in many real life applications. One direction for the study of rough set theory [12] is to investigate relationships between rough approximation operators and topological structures of rough sets. The relationships between crisp rough sets and crisp topological spaces were studied by many authors (see e.g. [5,8,10,14,15,20,22]). Some authors also investigated topological structures of rough sets in the fuzzy environment [3,9,13,17,18,21]. One of the main results is that a fuzzy reflexive and transitive relation on a universe of discourse can yield a fuzzy topology on the same universe.

In the present paper, we investigate topological structure of rough intuitionistic fuzzy sets. In the next section, we review basic concepts related to intuitionistic fuzzy (IF) sets. In Section 3, we define rough IF approximation operators and present their properties. In Section 4, we introduce some notions and theoretical results of IF topological spaces. In Section 5, we examine relationships between rough IF sets and IF topological spaces. We then conclude the paper with a summary in Section 6.

T. Li et al. (Eds.): RSKT 2012, LNAI 7414, pp. 496–503, 2012.

2   Basic Notions of Intuitionistic Fuzzy Sets

Throughout this paper, U will be a nonempty set called the universe of discourse which may be infinite. The class of all subsets (fuzzy subsets, respectively) of U will be denoted by $\mathcal{P}(U)$ (by $\mathcal{F}(U)$, respectively). For $y \in U$, 1_y will denote the fuzzy singleton with value 1 at y and 0 elsewhere; 1_M will denote the characteristic function of a crisp set $M \subseteq U$. We will use the symbols $\vee$ and $\wedge$ to denote the supremum and the infimum, respectively.

Definition 1. (Lattice $(L^*, \leq_{L^*})$) *Denote*
$\quad L^* = \{(x_1, x_2) \in [0, 1] \times [0, 1] \mid x_1 + x_2 \leq 1\}.$
A relation $\leq_{L^*}$ on L^* is defined as follows: $\forall (x_1, x_2), (y_1, y_2) \in L^*,$
$\quad (x_1, x_2) \leq_{L^*} (y_1, y_2) \iff x_1 \leq y_1 \text{ and } x_2 \geq y_2.$
The relation $\leq_{L^*}$ is a partial ordering on L^* and the pair $(L^*, \leq_{L^*})$ is a complete lattice with the smallest element $0_{L^*} = (0, 1)$ and the greatest element $1_{L^*} = (1, 0)$ [7]. The meet operator $\wedge$ and join operator $\vee$ on $(L^*, \leq_{L^*})$ which are linked to the ordering $\leq_{L^*}$ are, respectively, defined as follows: $\forall (x_1, x_2), (y_1, y_2) \in L^*,$
$\quad (x_1, x_2) \wedge (y_1, y_2) = (\min(x_1, y_1), \max(x_2, y_2)),$
$\quad (x_1, x_2) \vee (y_1, y_2) = (\max(x_1, y_1), \min(x_2, y_2)).$
Meanwhile, an order relation $\geq_{L^*}$ on L^* is defined as follows: $\forall x = (x_1, x_2), y = (y_1, y_2) \in L^*,$
$\quad (y_1, y_2) \geq_{L^*} (x_1, x_2) \iff (x_1, x_2) \leq_{L^*} (y_1, y_2),$
and
$\quad x = y \iff x \leq_{L^*} y \text{ and } y \leq_{L^*} x.$

Definition 2. [2] *Let a set U be fixed. An intuitionistic fuzzy set A in U is an object having the form*
$\quad A = \{\langle x, \mu_A(x), \gamma_A(x) \rangle \mid x \in U\},$
where the functions $\mu_A : U \to [0, 1]$ and $\gamma_A : U \to [0, 1]$ define the degree of membership and the degree of non-membership of the element $x \in U$ to A, respectively. The function μ_A and γ_A satisfy: $\mu_A(x) + \gamma_A(x) \leq 1$ for all $x \in U$. The family of all IF subsets in U is denoted by $\mathcal{IF}(U)$. The complement of an IF set A is denoted by $\sim A = \{\langle x, \gamma_A(x), \mu_A(x) \rangle \mid x \in U\}$.

Formally, an IF set A associates two fuzzy sets $\mu_A : U \to [0, 1]$ and $\gamma_A : U \to [0, 1]$ and can be represented as $A = (\mu_A, \gamma_A)$. Obviously, any fuzzy set $A = \mu_A = \{\langle x, \mu_A(x) \rangle \mid x \in U\}$ may be identified with the IF set in the form $A = \{\langle x, \mu_A(x), 1 - \mu_A(x) \rangle \mid x \in U\}$.

We now introduce operations on $\mathcal{IF}(U)$ as follows [2]: $\forall A, B \in \mathcal{IF}(U),$

- $A \subseteq B$ if and only if (iff) $\mu_A(x) \leq \mu_B(x)$ and $\gamma_A(x) \geq \gamma_B(x)$ $\forall x \in U$,
- $A \supseteq B$ iff $B \subseteq A$,
- $A = B$ iff $A \subseteq B$ and $B \subseteq A$,
- $A \cap B = \{\langle x, \min(\mu_A(x), \mu_B(x)), \max(\gamma_A(x), \gamma_B(x)) \rangle \mid x \in U\}$,
- $A \cup B = \{\langle x, \max(\mu_A(x), \mu_B(x)), \min(\gamma_A(x), \gamma_B(x)) \rangle \mid x \in U\}$.

Here we define some special IF sets: a constant IF set $\widehat{(\alpha, \beta)} = \{\langle x, \alpha, \beta \rangle \mid x \in U\}$, where $(\alpha, \beta) \in L^*$; the IF universe set is $U = 1_U = \widehat{(1, 0)} = \{\langle x, 1, 0 \rangle \mid x \in U\}$ and the IF empty set is $\emptyset = 0_U = \widehat{(0, 1)} = \{\langle x, 0, 1 \rangle \mid x \in U\}$.

For any $y \in U$, IF sets 1_y and $1_{U-\{y\}}$ are, respectively, defined by: $\forall x \in U$,

$$\mu_{1_y}(x) = \begin{cases} 1, & \text{if } x = y, \\ 0, & \text{if } x \neq y. \end{cases} \qquad \gamma_{1_y}(x) = \begin{cases} 0, & \text{if } x = y, \\ 1, & \text{if } x \neq y. \end{cases}$$

$$\mu_{1_{U-\{y\}}}(x) = \begin{cases} 0, & \text{if } x = y, \\ 1, & \text{if } x \neq y. \end{cases} \qquad \gamma_{1_{U-\{y\}}}(x) = \begin{cases} 1, & \text{if } x = y, \\ 0, & \text{if } x \neq y. \end{cases}$$

3   Rough Intuitionistic Fuzzy Sets

Definition 3. *Let U be a nonempty universe of discourse which may be infinite. A subset $R \in \mathcal{P}(U \times U)$ is referred to as a (crisp) binary relation on U. The relation R is referred to as reflexive if for all $x \in U$, $(x, x) \in R$; R is referred to as symmetric if for all $x, y \in U$, $(x, y) \in R$ implies $(y, x) \in R$; R is referred to as transitive if for all $x, y, z \in U$, $(x, y) \in R$ and $(y, z) \in R$ imply $(x, z) \in R$; R is referred to as a similarity relation if R is reflexive and symmetric; R is referred to as a preorder if R is reflexive and transitive; and R is referred to as an equivalence relation if R is reflexive, symmetric and transitive.*

For an arbitrary crisp relation R on U, we can define a set-valued mapping $R_s : U \to \mathcal{P}(U)$ by:

$$R_s(x) = \{y \in U | (x, y) \in R\}, \quad x \in U. \tag{1}$$

$R_s(x)$ is called the successor neighborhood of x with respect to (w.r.t.) R [19].

An rough IF set is the approximation of an IF set w.r.t. a crisp approximation space.

Definition 4. *Let (U, R) be a crisp approximation space. For $A \in \mathcal{IF}(U)$, the upper and lower approximations of A w.r.t. (U, R), denoted by $\overline{RI}(A)$ and $\underline{RI}(A)$, respectively, are defined as follows:*

$$\overline{RI}(A) = \{\langle x, \mu_{\overline{RI}(A)}(x), \gamma_{\overline{RI}(A)}(x)\rangle \mid x \in U\},$$
$$\underline{RI}(A) = \{\langle x, \mu_{\underline{RI}(A)}(x), \gamma_{\underline{RI}(A)}(x)\rangle \mid x \in U\},$$

where

$$\mu_{\overline{RI}(A)}(x) = \bigvee_{y \in R_s(x)} \mu_A(y), \qquad \gamma_{\overline{RI}(A)}(x) = \bigwedge_{y \in R_s(x)} \gamma_A(y),$$
$$\mu_{\underline{RI}(A)}(x) = \bigwedge_{y \in R_s(x)} \mu_A(y), \qquad \gamma_{\underline{RI}(A)}(x) = \bigvee_{y \in R_s(x)} \gamma_A(y).$$

It is easy to verify that $\overline{RI}(A)$ and $\underline{RI}(A)$ are two IF sets in U, thus IF mappings $\overline{RI}, \underline{RI} : \mathcal{IF}(U) \to \mathcal{IF}(U)$ are referred to as the upper and lower rough IF approximation operators, respectively, and the pair $(\underline{RI}(A), \overline{RI}(A))$ is called the rough IF set of A w.r.t. the approximation space (U, R).

The following Theorem 1 presents some basic properties of rough IF approximation operators [16].

Theorem 1. *Let (U, R) be a crisp approximation space, then the upper and lower rough IF approximation operators defined in Definition 4 satisfy the following properties: $\forall A, B, A_j \in \mathcal{IF}(U), j \in J, J$ is an index set, $\forall (\alpha, \beta) \in L^*$,*

(IL1) $\underline{RI}(\sim A) =\sim \overline{RI}(A)$,

(IU1) $\overline{RI}(\sim A) =\sim \underline{RI}(A)$;

(IL2) $\underline{RI}(A \cup \widehat{(\alpha, \beta)}) = \underline{RI}(A) \cup \widehat{(\alpha, \beta)}$,

(IU2) $\overline{RI}(A \cap \widehat{(\alpha, \beta)}) = \overline{RI}(A) \cap \widehat{(\alpha, \beta)}$;

(IL2) ' $\underline{RI}(U) = U$,

(IU2) ' $\overline{RI}(\emptyset) = \emptyset$;

(IL3) $\underline{RI}(\bigcap\limits_{j \in J} A_j) = \bigcap\limits_{j \in J} \underline{RI}(A_j)$,

(IU3) $\overline{RI}(\bigcup\limits_{j \in J} A_j) = \bigcup\limits_{j \in J} \overline{RI}(A_j)$;

(IL4) $\underline{RI}(A \cup B) \supseteq \underline{RI}(A) \cup \underline{RI}(B)$,

(IU4) $\overline{RI}(A \cap B) \subseteq \overline{RI}(A) \cap \overline{RI}(B)$;

(IL5) $A \subseteq B \Longrightarrow \underline{RI}(A) \subseteq \underline{RI}(B)$,

(IU5) $A \subseteq B \Longrightarrow \overline{RI}(A) \subseteq \overline{RI}(B)$.

In the case of connections between special types of crisp relations and properties of rough IF approximation operators, we have the following

Theorem 2. *Let R be an arbitrary crisp binary relation on U, and $\underline{RI}, \overline{RI} : \mathcal{IF}(U) \to \mathcal{IF}(U)$ the lower and upper rough IF approximation operators. Then*

(1) R *is reflexive*

$\Longleftrightarrow$ (ILR) $\underline{RI}(A) \subseteq A \quad \forall A \in \mathcal{IF}(U)$,

$\Longleftrightarrow$ (IUR) $A \subseteq \overline{RI}(A) \quad \forall A \in \mathcal{IF}(U)$.

(2) R *is symmetric*

$\Longleftrightarrow$ (ILS) $\overline{RI}(\underline{RI}(A)) \subseteq A \quad \forall A \in \mathcal{IF}(U)$,

$\Longleftrightarrow$ (IUS) $A \subseteq \underline{RI}(\overline{RI}(A)) \quad \forall A \in \mathcal{IF}(U)$,

$\Longleftrightarrow$ (ILS)' $\mu_{\underline{RI}(1_{U-\{x\}})}(y) = \mu_{\underline{RI}(1_{U-\{y\}})}(x) \quad \forall (x, y) \in U \times U$,

$\Longleftrightarrow$ (IUS)' $\mu_{\overline{RI}(1_x)}(y) = \mu_{\overline{RI}(1_y)}(x) \quad \forall (x, y) \in U \times U$,

$\Longleftrightarrow$ (ILS)'' $\gamma_{\underline{RI}(1_{U-\{x\}})}(y) = \gamma_{\underline{RI}(1_{U-\{y\}})}(x) \quad \forall (x, y) \in U \times U$,

$\Longleftrightarrow$ (IUS)'' $\gamma_{\overline{RI}(1_x)}(y) = \gamma_{\overline{RI}(1_y)}(x) \quad \forall (x, y) \in U \times U$.

(3) R *is transitive*

$\Longleftrightarrow$ (ILT) $\underline{RI}(A) \subseteq \underline{RI}(\underline{RI}(A)) \quad \forall A \in \mathcal{IF}(U)$,

$\Longleftrightarrow$ (IUT) $\overline{RI}(\overline{RI}(A)) \subseteq \overline{RI}(A) \quad \forall A \in \mathcal{IF}(U)$.

4   Basic Concepts of Intuitionistic Fuzzy Topological Spaces

In this section we introduce basic concepts related to IF topological spaces.

Definition 5. [6] *An IF topology in the sense of Lowen* [11] *on a nonempty set* U *is a family* τ *of IF sets in* U *satisfying the following axioms:*

(T_1) $\widehat{(\alpha, \beta)} \in \tau$ *for all* $(\alpha, \beta) \in L^*$,
(T_2) $G_1 \cap G_2 \in \tau$ *for any* $G_1, G_2 \in \tau$,
(T_3) $\bigcup_{i \in J} G_i \in \tau$ *for a family* $\{G_i | i \in J\} \subseteq \tau$, *where* J *is an index set.*

In this case the pair (U, τ) *is called an IF topological space and each IF set* A *in* τ *is referred to as an IF open set in* (U, τ). *The complement of an IF open set in the IF topological space* (U, τ) *is called an IF closed set in* (U, τ).

It should be pointed out that if axiom T_1 in Definition 5 is replaced by axiom

(T_1') $\emptyset \in \tau$ *and* $U \in \tau$,

then τ is an IF topology in the sense of Chang [4]. Clearly, an IF topology in the sense of Lowen must be an IF topology in the sense of Chang. Throughout this paper, we always consider the IF topology in the sense of Lowen.

Now we define IF closure and interior operations in an IF topological space.

Definition 6. *Let* (U, τ) *be an IF topological space and* $A \in \mathcal{IF}(U)$. *Then the IF interior and IF closure of* A *are, respectively, defined as follows:*

$int(A) = \cup \{G \mid G$ *is an IF open set and* $G \subseteq A\}$,
$cl(A) = \cap \{K \mid K$ *is an IF closed set and* $A \subseteq K\}$,

and int and $cl : \mathcal{IF}(U) \to \mathcal{IF}(U)$ *are, respectively, called the IF interior operator and the IF closure operator of* τ, *and sometimes in order to distinguish, we denote them by* int_τ *and* cl_τ.

It can be shown that $cl(A)$ is an IF closed set and $int(A)$ is an IF open set in (U, τ), and

(a) A is an IF open set in (U, τ) iff $int(A) = A$,
(b) A is an IF closed set in (U, τ) iff $cl(A) = A$.

Moreover,

$$cl(\sim A) = \sim int(A) \quad \forall A \in \mathcal{IF}(U), \tag{2}$$

$$int(\sim A) = \sim cl(A) \quad \forall A \in \mathcal{IF}(U). \tag{3}$$

It can be verified that the IF closure operator satisfies following properties:

(Cl1) $cl(\widehat{(\alpha, \beta)}) = \widehat{(\alpha, \beta)} \; \forall (\alpha, \beta) \in L^*$.
(Cl2) $cl(A \cup B) = cl(A) \cup cl(B) \; \forall A, B \in \mathcal{IF}(U)$.
(Cl3) $cl(cl(A)) = cl(A) \; \forall A \in \mathcal{IF}(U)$.
(Cl4) $A \subseteq cl(A) \; \forall A \in \mathcal{IF}(U)$.

Properties (Cl1)–(Cl4) are called the IF closure axioms. Similarly, the IF interior operator satisfies following properties:

(Int1) $int(\widehat{(\alpha, \beta)}) = \widehat{(\alpha, \beta)} \; \forall (\alpha, \beta) \in L^*$.
(Int2) $int(A \cap B) = int(A) \cap int(B) \; \forall A, B \in \mathcal{IF}(U)$.
(Int3) $int(int(A)) = int(A) \; \forall A \in \mathcal{IF}(U)$.
(Int4) $int(A) \subseteq A \; \forall A \in \mathcal{IF}(U)$.

Definition 7. *A mapping* $cl : \mathcal{IF}(U) \to \mathcal{IF}(U)$ *is referred to as a fuzzy closure operator if it satisfies axioms* (Cl1)-(Cl4).

Similarly, the fuzzy interior operator can be defined by corresponding axioms.

Definition 8. *A mapping $int : \mathcal{IF}(U) \to \mathcal{IF}(U)$ is referred to as a fuzzy interior operator if it satisfies axioms* (Int1)-(Int4).

It is easy to show that an IF interior operator int defined in Definition 8 determines an IF topology

$$\tau_{int} = \{A \in \mathcal{IF}(U) | int(A) = A\}. \tag{4}$$

So, the IF open sets are the fixed points of int. Dually, from an IF closure operator defined in Definition 7, we can obtain an IF topology on U by setting

$$\tau_{cl} = \{A \in \mathcal{IF}(U) | cl(\sim A) =\sim A\}. \tag{5}$$

The results are summarized as the following

Theorem 3. (1) *If an operator $int : \mathcal{IF}(U) \to \mathcal{IF}(U)$ satisfies axioms* (Int1) - (Int4), *then τ_{int} defined as Eq. (4) is an IF topology on U and*

$$int_{\tau_{int}} = int. \tag{6}$$

(2) *If an operator $cl : \mathcal{IF}(U) \to \mathcal{IF}(U)$ satisfies axioms* (Cl1)-(Cl4), *then τ_{cl} defined as Eq. (5) is an IF topology on U and*

$$cl_{\tau_{cl}} = cl. \tag{7}$$

Similar to the crisp Alexandrov topology [1] and crisp clopen topology [8], we now introduce the concepts of an IF Alexandrov topology and an IF clopen topology.

Definition 9. *An IF topology τ on U is called an IF Alexandrov topology if the intersection of arbitrarily many IF open sets is still open, or equivalently, the union of arbitrarily many IF closed sets is still closed. An IF topological space (U, τ) is said to be an IF Alexandrov space if τ is an IF Alexandrov topology on U. An IF topology τ on U is called an IF clopen topology if, for every $A \in \mathcal{IF}(U)$, A is IF open in (U, τ) if and only if A is IF closed in (U, τ). An IF topological space (U, τ) is said to be an IF clopen space if τ is an IF clopen topology on U.*

5   Intuitionistic Fuzzy Topologies of Rough Intuitionistic Fuzzy Sets

In this section we discuss the relationship between IF topological spaces and rough IF sets. Throughout this section we always assume that U is a nonempty universe of discourse, R a crisp binary relation on U, and $\underline{RI}$ and $\overline{RI}$ the rough IF approximation operators defined in Definition 4.

Denote

$$\tau_R = \{A \in \mathcal{IF}(U) | \underline{RI}(A) = A\}. \tag{8}$$

The next theorem shows that any reflexive binary relation determines an IF Alexandrov topology.

Theorem 4. *If R is a reflexive crisp binary relation on U, then τ_R defined by Eq. (8) is an IF Alexandrov topology on U.*

Theorem 5 below shows that the lower and upper rough IF approximation operators induced from a reflexive and transitive relation are, respectively, an IF interior operator and an IF closure operator.

Theorem 5. *Assume that R is a crisp binary relation on U. Then the following statements are equivalent:*

(1) *R is a preorder, i.e., R is a reflexive and transitive relation;*

(2) *the upper rough IF approximation operator $\overline{RI} : \mathcal{IF}(U) \to \mathcal{IF}(U)$ is an IF closure operator;*

(3) *the lower rough IF approximation operator $\underline{RI} : \mathcal{F}(U) \to \mathcal{F}(U)$ is an IF interior operator.*

Lemma 1. *If R is a symmetric crisp binary relation on U, then for all $A, B \in \mathcal{IF}(U)$,*

$$\overline{RI}(A) \subseteq B \Longleftrightarrow A \subseteq \underline{RI}(B). \tag{9}$$

Theorem 6. *Let R be a similarity crisp binary relation on U, and $\underline{RI}$ and $\overline{RI}$ the rough IF approximation operators defined in Definition 4. Then $\underline{RI}$ and $\overline{RI}$ satisfy property (Clop): for $A \in \mathcal{IF}(U)$,*

$$(Clop) \quad \underline{RI}(A) = A \Longleftrightarrow A = \overline{RI}(A) \Longleftrightarrow \underline{RI}(\sim A) =\sim A \Longleftrightarrow\sim A = \overline{RI}(\sim A).$$

The next theorem shows that an IF topological space induced from a reflexive and symmetric crisp approximation space is an IF clopen topological space.

Theorem 7. *Let R be a similarity crisp binary relation on U, and $\underline{RI}$ and $\overline{RI}$ the rough IF approximation operators defined in Definition 4. Then τ_R defined as Eq. (8) is an IF clopen topology on U.*

6 Conclusion

In this paper we have studied the topological structure of rough IF sets. We have shown that a reflexive crisp relation can induce an IF Alexandrov space. We have also examined that the lower and upper rough IF approximation operators are, respectively, an IF interior operator and an IF closure operator if and only if the binary relation in the crisp approximation space is reflexive and transitive. We have further presented that an IF topological space induced from a reflexive and symmetric crisp approximation space is an IF clopen topological space.

Acknowledgement. This work was supported by grants from the National Natural Science Foundation of China (Nos. 61075120, 11071284, and 61173181) and the Zhejiang Provincial Natural Science Foundation of China (No. LZ12F03002).

References

1. Arenas, F.G.: Alexandroff Spaces. Acta Mathematica Universitatis Comenianae 68, 17–25 (1999)
2. Atanassov, K.: Intuitionistic Fuzzy Sets. Physica-Verlag, Heidelberg (1999)
3. Boixader, D., Jacas, J., Recasens, J.: Upper and Lower Approximations of Fuzzy Sets. International Journal of General Systems 29, 555–568 (2000)
4. Chang, C.L.: Fuzzy Topological Spaces. Journal of Mathematical Analysis and Applications 24, 182–189 (1968)
5. Chuchro, M.: On Rough Sets in Topological Boolean Algebras. In: Ziarko, W. (ed.) Rough Sets, Fuzzy Sets and Knowledge Discovery, pp. 157–160. Springer, Berlin (1994)
6. Coker, D.: An Introduction of Intuitionistic Fuzzy Topological spaces. Fuzzy Sets and Systems 88, 81–89 (1997)
7. Cornelis, C., Deschrijver, G., Kerre, E.E.: Implication in Intuitionistic Fuzzy and Interval-valued Fuzzy Set Theory: Construction, Classification, Application. International Journal of Approximate Reasoning 35, 55–95 (2004)
8. Kondo, M.: On the Structure of Generalized Rough Sets. Information Sciences 176, 589–600 (2006)
9. Lai, H., Zhang, D.: Fuzzy Preorder and Fuzzy Topology. Fuzzy Sets and Systems 157, 1865–1885 (2006)
10. Lashin, E.F., Kozae, A.M., Khadra, A.A.A., et al.: Rough Set Theory for Topological Spaces. International Journal of Approximate Reasoning 40, 35–43 (2005)
11. Lowen, R.: Fuzzy Topological Spaces and Fuzzy Compactness. Journal of Mathematical Analysis and Applications 56, 621–633 (1976)
12. Pawlak, Z.: Rough Sets: Theoretical Aspects of Reasoning about Data. Kluwer Academic Publishers, Boston (1991)
13. Qin, K.Y., Pei, Z.: On the Topological Properties of Fuzzy Rough Sets. Fuzzy Sets and Systems 151, 601–613 (2005)
14. Qin, K.Y., Yang, J., Pei, Z.: Generalized Rough Sets Based on Reflexive and Transitive Relations. Information Sciences 178, 4138–4141 (2008)
15. Wiweger, R.: On Topological Rough Sets. Bulletin of Polish Academy of Sciences: Mathematics 37, 89–93 (1989)
16. Wu, W.-Z., Xu, Y.-H.: Rough Approximations of Intuitionistic Fuzzy Sets in Crisp Approximation Spaces. In: Proceedings of Seventh International Conference on Fuzzy Systems and Knowledge Discovery (FSKD 2010), vol. 1, pp. 309–313 (2010)
17. Wu, W.-Z., Yang, Y.-F., Xu, Y.-H.: Some Fuzzy Topologies Induced by Rough Fuzzy Sets. In: Yao, J., Ramanna, S., Wang, G., Suraj, Z. (eds.) RSKT 2011. LNCS, vol. 6954, pp. 156–165. Springer, Heidelberg (2011)
18. Wu, W.-Z., Zhou, L.: On Intuitionistic Fuzzy Topologies Based on Intuitionistic Fuzzy Reflexive and Transitive Relations. Soft Computing 15(6), 1183–1194 (2011)
19. Yao, Y.Y.: Constructive and Algebraic Methods of the Theory of Rough Sets. Journal of Information Sciences 109, 21–47 (1998)
20. Zhang, H.-P., Yao, O.Y., Wang, Z.D.: Note on Generlaized Rough Sets Based on Reflexive and Transitive Relations. Information Sciences 179, 471–473 (2009)
21. Zhou, L., Wu, W.-Z., Zhang, W.-X.: On Intuitionistic Fuzzy Rough Sets and Their Topological Structures. International Journal of General Systems 38, 589–616 (2009)
22. Zhu, W.: Topological Approaches to Covering Rough Sets. Information Sciences 177, 1499–1508 (2007)

Oppositions in Rough Set Theory

Davide Ciucci[1,2,*], Didier Dubois[1], and Henri Prade[1]

[1] IRIT, Université Paul Sabatier
118 route de Narbonne, 31062 Toulouse cedex 9, France
[2] DISCo - Università di Milano – Bicocca
Viale Sarca 336 – U14, 20126 Milano, Italia

Abstract. The role of opposition in rough set theory is laid bare. There are two sources which generate oppositions in rough sets: approximations and relations. In the former case, we outline a hexagon and a cube of oppositions. In the second case, we define a classical square of oppositions and also a tetrahedron when considering the standpoint of two agents.

1 Introduction

Starting from the Greek philosophy, in particular Aristotle, oppositions have been organized in a so-called *square of opposition* where each vertex represents a different statement involving two entities X and Y: "Every X is Y" (A); "Some X is not Y" (O); "Every X is not Y" (E) and "Some X is Y" (I). Clearly, (A) and (I) are in opposition to (O) and (E) (and vice-versa), (A) implies (I) and (E) implies (O). (A) and (E) can be false together but not true, and for (I) and (O) it is the converse. This organization of oppositions have been discussed by several authors until the Middle Ages, then it progressively lost its importance in logic. Only since the 1950s, we can observe a renewed interest on this topic. In particular, in the last years, several authors tried to generalize the notion of square to describe more complex situations and to apply the square and other geometrical organizations of oppositions to several fields [2,3]. We just mention the link with Belnap and paraconsistent logic, fixed-point calculus, possibility theory, formal concept analysis [1]. Here, we lay bare the links between several forms of opposition and rough sets: cube and hexagon of opposition naturally arise by lower and upper approximations and square and tetrahedron of oppositions are generated by relations. The aim is to give a new theoretical approach to rough sets with possible new links (and differences) to other paradigms such as formal concept analysis and possibility theory. New ideas about approximations with two relations/agents are also proposed.

2 Preliminary Notions

We now give the introductory notions about the geometrical organization of oppositions and rough set theory.

---

* Supported by FP7-Marie Curie Action (IEF) n.276158.

T. Li et al. (Eds.): RSKT 2012, LNAI 7414, pp. 504–513, 2012.

2.1   Oppositions and Geometrical Organization

The Aristotelian square of opposition described in the introduction can be generalized in two directions: either by negation of X and Y, or by considering (U), the disjunction of (A) and (E), and (Y) the conjunction of (I) and (O).

In the first case we obtain a cube of oppositions, where, besides the typical square statements we have: "Every (not X) is (not Y)" (a); "Some (not X) is Y" (o); "Every (not X) is Y" (e) and "Some (not X) is (not Y)" (i).

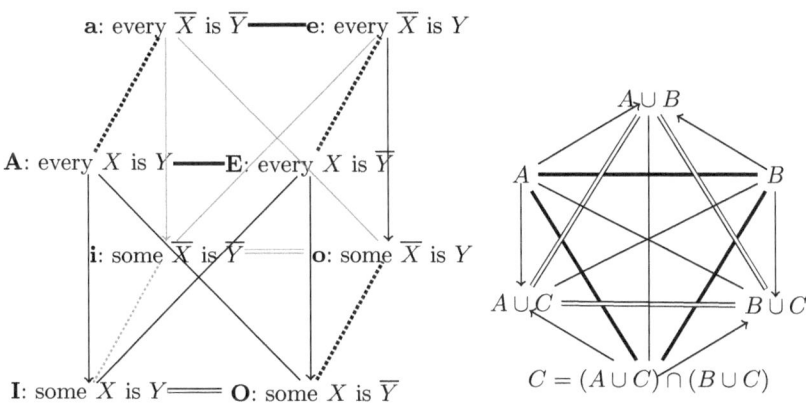

Fig. 1. Cube and hexagon of opposition

So, in the cube we have two squares of opposition: one in the front and the other in the back facet. In figure 1(left), the thick segment relates the contraries, the double thin undirected segments the sub-contraries, the diagonal non-directed segments the contradictories, and the vertical uni-directed segments point to subalterns (and express entailments if the set of X's and the set of $\overline{X}$'s are not empty).

Example 2.1. Let X=apple and Y=red. Then, we have **A:** every apple is red; **E:** every apple is not red; **I:** some apple is red; **O:** some apple is not red; **a:** anything that is not an apple is not red; **e:** anything that is not an apple is red; **i:** something that is not an apple is not red; **o:** something that is not an apple is red.

In the second case, we obtain the hexagon of opposition proposed by Blanché as the starting point in his analysis of oppositions [4]. Hexagons naturally arise in different context, for instance in comparative relations in mathematics: (A): = " > ", (E): = " < ", (U): = " ≠ ", (I): = " ≥ ", (O): = " ≤ ", and (Y): = " = " or with tastes: (A) = "I like it", (E): = "I dislike it", (U): = "I am not indifferent", (I): = "I do not dislike it", (O): = " I do not like it", and (Y): = "I am indifferent". Basically, we encounter an hexagon when we have a partition in three mutually exclusive situations [9]. Indeed, once considered a 3-partition of a universe of objects $Obj = A \cup B \cup C$, we obtain the hexagon of figure 1(right). In this figure, thick segments link contraries and the double segments link sub-contraries.

As a further step, we can consider a 4-partition, i.e. a partition of a universe in four mutually exclusive subsets. In this case we obtain a tetrahedron of oppositions. An important example of tetrahedron is obtained by all the combinations of two Boolean statements p, q. Figure 2 gives a rich view of the relations among the binary connectives. We notice indeed that:

- any connective that appears between two others on the same segment (which may be an edge of the tetrahedron, or a segment linking a vertex to the middle of an opposite edge) is implied by each of the two connectives at the extremities;
- contradiction and tautology can both be associated with the center of the tetrahedron. Indeed, a segment linking the middle of two opposite edges, or a segment linking a vertex to the center of the opposite facet, is associated with a pair of connectives (at its extremities) whose conjunction yields the contradiction, and their disjunction yields the tautology.
- the position of a connective on a segment reflects the number of bits to switch for moving from it to the connectives that are at its extremities.

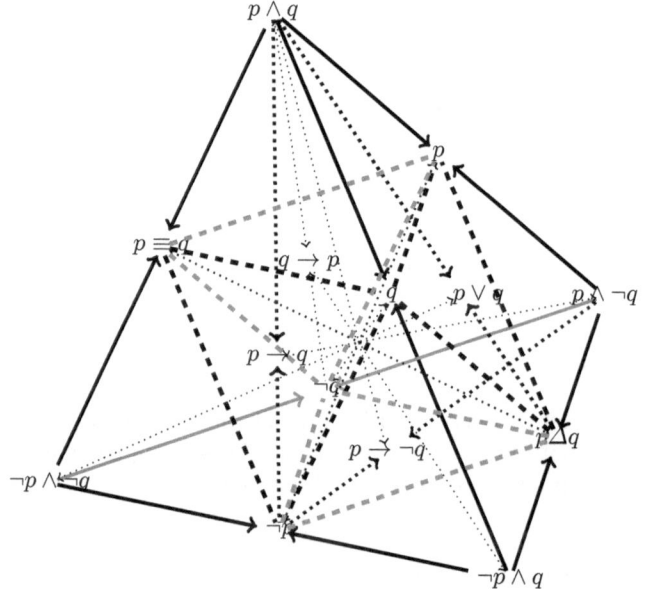

Fig. 2. The tetrahedron of the 16 binary connectives

2.2   Rough Sets

In rough set theory, knowledge about objects is represented in terms of observables (attributes) collected in an Information Table (or System) [13,14].

Definition 2.1. *An* Information Table *is a structure* $\langle Obj, A, \mathbf{val}, F \rangle$ *where:*

- *the universe Obj is a non empty set of* objects;
- *A is a non empty set of* condition attributes;
- **val**, *the set of all* possible values *that can be observed for all attributes;*
- *F (the* information map*) is a mapping* $F : Obj \times A \to \mathbf{val}$ *which associates to any pair object–attribute, the value* $F(x, a)$ *assumed by a for the object x.*

In standard rough set theory we define an indiscernibility relation with respect to a set of attributes $B \subseteq A$ as $x\mathcal{I}_B y$ iff $\forall a \in B,\ F(x, a) = F(y, a)$. This is an equivalence relation, which partitions X into equivalence classes $[x]_B$, our *granules* of information. Due to a lack of knowledge we are not able to distinguish objects inside the granules, thus, not all subsets of Obj can be precisely characterized in terms of the available attributes B. However, any set $X \subseteq Obj$ can be approximated by a lower and an upper approximation, respectively defined as:

$$L_B(X) = \{x : [x]_B \subseteq X\} \quad U_B(X) = \{x : [x]_B \cap X \neq \emptyset\} \tag{1}$$

By set complementation of Obj we can also define the *exterior region* $E_B(X) = (U_B(X))^c$, that is the objects surely not belonging to X. Finally, the difference between upper and lower approximations is named *boundary* and denoted as $Bnd(X) = U(X) \backslash L(X)$. An important (with respect to the present work) property of approximations is the duality between lower and upper approximations: $L(X^c) = U^c(X)$ which permits to define the lower approximation given the upper one and vice versa.

Several generalizations of this approach are known in literature, which can concern the indiscernibility relation (hence, we have for instance similarity rough sets [16,17]), the subsethood relation (see the Variable Precision Rough Sets [10]) or the data under investigation which can be described by fuzzy sets instead of classical sets [8]. We will give more details in the following sections when needed.

3   Opposition from Approximations

The first obvious way to define oppositions in the rough set framework is to consider approximations. Indeed, we have that the lower approximation and the exterior region are disjoint sets (also, orthopairs [7]). So, given a subset of objects $X \subseteq Obj$, we have a tri-partition of the universe $Obj = L(X) \cup Bnd(X) \cup E(X)$ and consequently a hexagon of oppositions.

We note that in the upper part of the hegaxon we have the sets representing certain knowledge on the universe, namely $L(X), E(X)$ and their union. On the other hand on the other half of the hegaxon we have uncertain knowledge: $U(X), U(X^c)$ and at the bottom the boundary region.

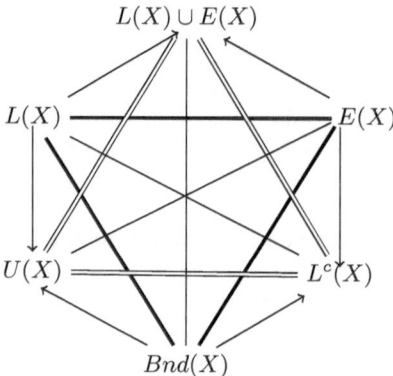

Fig. 3. Hexagon induced by Pawlak approximations

In more generalized models of rough sets, we do not have the duality between lower and upper approximations [6]. This happens, for instance, in Probabilistic Rough Sets [18] or Fuzzy Rough sets [8,15]. When considering opposition in this generalized framework we have a splitting in two parts of the hexagon. In figure 4 the cube of opposition resulting from this splitting is represented, where the upper and lower part are omitted. We also note that the square on the diagonals, i.e., those with vertices $(L(X), L(X^c), L^c(X), L^c(X^c))$ and $(U(X), U(X^c), U^c(X), U^c(X^c))$ are examples of Piaget group[1][9].

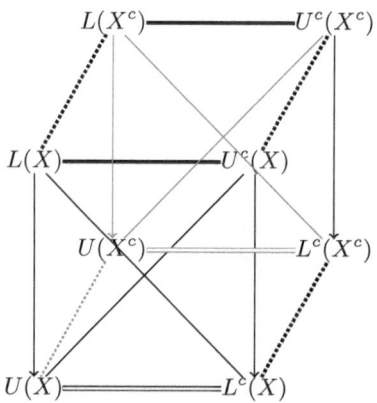

Fig. 4. Cube of opposition induced by generalized approximations

---

[1] Given a logical formula $\phi = f(p, q, r, \ldots)$, the Piaget group is the set of four transformations: identity $I(\phi) = \phi$; negation $N(\phi) = \neg\phi$; reciprocation $R(\phi) = f(\neg p, \neg q, \neg r, \ldots)$ and correlation $C(\phi) = \neg f(\neg p, \neg q, \neg r, \ldots)$.

4   Opposition from Relations

Looking at the definition of the indiscernibility relation xR_Ey iff $\forall a \in B \subseteq A$, $F(x,a) = F(y,a)$, it can be easily seen that there other kinds of relations that can be generated, changing $\forall$ with $\exists$ and $=$ with $\neq$. Thus, given a subset of attributes $B \subseteq A$, we can define four relations:

$$xR_Ey \quad \text{iff} \quad \forall a \in B \subseteq A, \; F(x,a) = F(y,a) \quad \text{Indiscernibility}$$
$$xR_Sy \quad \text{iff} \quad \exists a \in B \subseteq A, \; F(x,a) = F(y,a) \quad \text{Partial Identity}$$
$$xR_Dy \quad \text{iff} \quad \exists a \in B \subseteq A, \; F(x,a) \neq F(y,a) \quad \text{Discernibility}$$
$$xR_Py \quad \text{iff} \quad \forall a \in B \subseteq A, \; F(x,a) \neq F(y,a) \quad \text{Complete Difference}$$

The discernibility relation and complete difference are the negation, respectively, of the equivalence relation and partial identity. So, it can be easily seen that these four relations form a classical square of oppositions (figure 5, left) of the AEIO kind. Moreover, this situation can be generalized considering any similarity (i.e., reflexive and symmetric) relation [16,17] instead of partial identity.

Examples of a similarity relation are: two objects are similar if they have a fixed percentage of equal attributes or two objects are similar if the distance between their value of some numerical attribute is less than a fixed threshold. Then, instead of the complete difference we consider the negation of the similarity: xR_Py iff *not* xR_Sy, which is a preclusivity relation, i.e., a relation which is anti-reflexive and symmetric. If we characterize each relation by its required properties, we obtain the square of opposition on the right in figure 5, where r=reflexivity, s=symmetry, t=transitivity and i=irreflexivity.

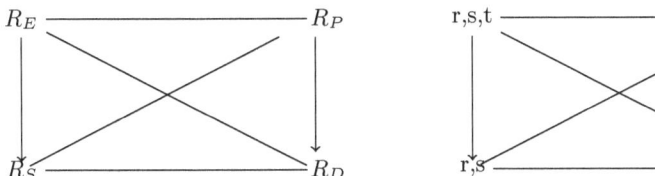

Fig. 5. Square induced by the 4 relations

Example 4.1. Let us consider a medical information table where patients are classified according to five attributes: Temperature, Pressure, Headache, Cold and Muscle-Pain. Then, two objects are equivalent if they have all the attributes equal; they are discernible if they differ for at least one attribute and we can define similarity as having three (out of five) attributes equal and consequently preclusivity is having at least three attributes different.

As far as approximations are concerned, we notice that based on a similarity relation, lower and upper approximations can be defined as in equations 1 (other possibilites have been investigated in literature, see for instance [12,19]). On the other hand, given a preclusivity relation R_p, and a set of objects X, we can build the set of objects which are different from all the objects in X: $X^{\#} = \{y \in Obj, \forall x \in X \, (x R_p y)\}$. Then, we define the upper approximation of a set as $U(X) = X^{\#\#}$ and the lower approximation is defined by duality as $L(X) = (U(X^c))^c$ [5]. Discernibility relation is a special case of preclusivity, so we can proceed in the same manner. So, also with these approximations, we obtain the hexagon of Figure 3.

It is worth noticing that the approximations obtained by the discernibility relation are equal to the approximations obtained by the equivalence relations computed on the same set of attributes. That is, given a set of attributes B and a set of objects X we get $L_{R_E}^B(X) = L_{R_D}^B(X)$ and the same for the upper approximation. So, with respect to approximations, it is the same to ask for equivalence or discernibility of objects. On the other hand, in the case of similarity and preclusivity, given one relation we can define the other by negation but the approximations differ. In general, it holds $L_{R_S}^B(X) \subseteq L_{R_P}^B(X) \subseteq X \subseteq U_{R_P}^B(X) \subseteq U_{R_S}^B(X)$.

4.1   Tetrahedron from Two Relations

Let us consider two relations R_1 and R_2 of any of the four types outlined above: equivalence, similarity, discernibility and preclusivity. If we want to aggregate them, we have 14 different combinations (including R_1, R_2 and their negations) using the classical logic connectives. So, for instance (R_1 AND R_2) requires that two objects are simultaneously related according to R_1 and to R_2. Clearly, these combinations form a tetrahedron of oppositions as described in section 2.1. Contradiction and tautology, which can be found in the middle of the tetrahedron, can here be interpreted as the fact that no objects are in the aggregated relation or that all objects are related.

Let us consider the rough set standpoint and address the problem of computing the lower and upper approximations generated by the aggregation of the two relations $R_1 \odot R_2$. First of all, let us note that the effect of NOT is to turn a similarity relation into a preclusivity one (resp., equivalence into discernibility) and vice versa. Then, generally speaking, it is not possible to use directly the relation $R_1 \odot R_2$ to compute approximations since it can be of none of the above four kinds.

Example 4.2. Let us consider the information table of example 4.1. Then, we define $R_1=$ equivalence relation on the attributes {Temperature, Cold} and $R_2=$ preclusivity relation consisting in having at least two (out of five) different attributes. Clearly, we are able to compute the approximations with respect to R_1 and R_2 but not with respect to the relation R_1 AND R_2 = having equal values for Temperature, Cold AND at least two different attributes.

So, the corresponding lower and upper approximations can only be obtained by computing separately the approximations obtained by the two relations and then aggregating them by the set operation corresponding to $\odot$. That is, we can give the following definitions

$$L_{R_1 \text{AND} R_2}(X) := L_{R_1}(X) \cap L_{R_2}(X) \quad U_{R_1 \text{AND} R_2}(X) := U_{R_1}(X) \cap U_{R_2}(X)$$
$$L_{R_1 \text{OR} R_2}(X) := L_{R_1}(X) \cup L_{R_2}(X) \quad U_{R_1 \text{OR} R_2}(X) := U_{R_1}(X) \cup U_{R_2}(X)$$

and similarly, XOR is turned into union minus intersection, $\equiv$ in intersection plus the complement of the union.

Example 4.3. Let us consider the two relations of example 4.2 and the following information table.

Table 1. Medical information table

Patient	Temperature	Pressure	Headache	Cold	Muscle Pain
p_1	high	normal	yes	yes	yes
p_2	high	high	no	yes	yes
p_3	normal	low	yes	no	no
p_4	very high	normal	yes	yes	no

Let us define the set of patients $X = \{p_2, p_3\}$. Then, we have $L_{R_1}(X) = \{p_3\}$, $U_{R_1}(X) = \{p_1, p_2, p_3\}$, $L_{R_2}(X) = U_{R_2}(X) = \{p_2, p_3\}$. So, $L_{R_1 \text{AND} R_2}(X) = \{p_3\}$, $U_{R_1 \text{AND} R_2}(X) = \{p_2, p_3\}$, $L_{R_1 \text{OR} R_2}(X) = \{p_2, p_3\}$ and $U_{R_1 \text{OR} R_2}(X) = \{p_1, p_2, p_3\}$.

In this way we obtain a new tetrahedron where instead of the aggregation of relation we have the aggregation of approximations. Contradiction and tautology correspond to the emptyset and the whole universe, respectively.

Finally, we can give two interpretations to R_1, R_2 and their aggregation:

1. the relations represent the standpoint of two different agents with their own knowledge about the same phenomenon. Then, by aggregation of the two relations, we obtain an aggregation of their points of view. Thus, we are in a situation similar to [11] with the difference that our relations are not necessarily equivalence ones. In analogy to the work by Khan and Banerjee, the approximations obtained by AND of the two relations can be named *strong* approximations and the one obtain by the OR as *weak* approximations.
2. R_1 and R_2 correspond to a single agent having different requirements on different attributes. For instance, consider again the medical information table. In this framework, an agent can ask for instance that two patients are in relation if they have same value for the attributes $B_1 = \{$Headache, Muscle-Pain, Cold$\}$ and similar values for $B_2 = \{$Temperature, Pressure$\}$. This corresponds to search for the pair $xR_E^{B_1}y$ AND $xR_S^{B_2}y$, R_S needing to be properly defined.

5   Conclusions

We have analyzed rough set theory in the light of the theory of oppositions. Several kinds of opposition and their geometrical organizations can be laid bare.

New scenarios are also highlighted on the aggregation of two relations and on the possibility that a single agent expresses different requirements on the same data. One can also suppose to further generalize this approach to n relations. In this case the number of possible connectives is 2^{2^n} (including contradiction and tautology). The interpretation in rough sets is a clear extension of what is said above: n agents or n different requirements from a single agent. On the contrary, the geometrical organization of oppositions in this setting becomes problematic, since the number of connectives would become intractable even for small values of n.

Finally, we note that the hexagon and the cube of Figure 3 and 4 have hexagon and cube counterparts in FCA (and possibility theory) in [9]. But as explained in [9] they are *not* regular hexagon and cube (indeed, some links have a different meaning) while in the rough set case the hexagon and the cube are regular ones (in the sense of Blanché). This reveals that the structure of opposition in rough sets and FCA are different, which offers a new departure point for comparing the two theories.

References

1. Béziau, J.Y., Gan-Krzywoszyńska, K. (eds.): The Square of Opposition (2010), http://www.square-of-opposition.org/Square2010-handbook.pdf
2. Béziau, J.Y., Payette, G.: Preface. Special issue on the square of opposition. Logica Universalis 2, 1 (2008)
3. Béziau, J.Y., Payette, G. (eds.): Handbook of the Second World Congress on the Square of Opposition. Peter Lang, Pieterlen (2012)
4. Blanché, R.: Sur l'opposition des concepts. Theoria 19, 89–130 (1953)
5. Cattaneo, G.: Generalized rough sets (preclusivity fuzzy-intuitionistic BZ lattices). Studia Logica 58, 47–77 (1997)
6. Ciucci, D.: Approximation algebra and framework. Fundamenta Informaticae 94(2), 147–161 (2009)
7. Ciucci, D.: Orthopairs: A simple and widely usedway to model uncertainty. Fundamenta Informaticae 108(3-4), 287–304 (2011)
8. Dubois, D., Prade, H.: Putting rough sets and fuzzy sets together. In: Słowinski, R. (ed.) Intelligent Decision Support – Handbook of Applications and Advances of the Rough Sets Theory, pp. 203–232. Kluwer Academic Publishers, Dordrecht (1992)
9. Dubois, D., Prade, H.: From Blanché hexagonal organization of concepts to formal concept analysis and possibility theory. Logica Universalis, 1–21 (in press, 2012), http://dx.doi.org/10.1007/s11787-011-0039-0
10. Katzberg, J., Ziarko, W.: Variable precision extension of rough sets. Fundamenta Informaticae 27, 155–168 (1996)
11. Khan, M. A., Banerjee, M.: A Study of Multiple-Source Approximation Systems. In: Peters, J.F., Skowron, A., Słowiński, R., Lingras, P., Miao, D., Tsumoto, S. (eds.) Transactions on Rough Sets XII. LNCS, vol. 6190, pp. 46–75. Springer, Heidelberg (2010)
12. Lin, T.Y., Huang, K.J., Liu, Q., Chen, W.: Rough sets, neighborhood systems and approximation. In: Proceedings of the Fifth International Symposium on Methodologies of Intelligent Systems, Selected Papers, pp. 130–141 (1990)

13. Pawlak, Z.: Information systems - theoretical foundations. Information Systems 6, 205–218 (1981)
14. Pawlak, Z., Skowron, A.: Rudiments of rough sets. Information Sciences 177, 3–27 (2007)
15. Radzikowska, A., Kerre, E.: A comparative study of fuzzy rough sets. Fuzzy Sets and Systems 126, 137–155 (2002)
16. Skowron, A., Stepaniuk, J.: Tolerance approximation spaces. Fundamenta Informaticae 27, 245–253 (1996)
17. Słowinski, R., Vanderpooten, D.: Similarity relation as a basis for rough approximations. In: Wang, P. (ed.) Advances in Machine Intelligence and Soft-Computing, vol. IV, pp. 17–33. Duke University Press, Durham (1997)
18. Yao, J.T., Yao, Y., Ziarko, W.: Probabilistic rough sets: Approximations, decision-makings, and applications. Int. J. Approx. Reasoning 49(2), 253–254 (2008)
19. Yao, Y.: Relational interpretations of neighborhood operators and rough set approximation operators. Information Sciences 111, 239–259 (1998)

Partial First-order Logic
with Approximative Functors
Based on Properties

Tamás Mihálydeák

Department of Computer Science,
Faculty of Informatics, University of Debrecen
Kassai út 26, H-4028 Debrecen, Hungary
mihalydeak.tamas@inf.unideb.hu

Abstract. In the present paper a logically exact way is presented in order to define approximative functors on object level in the partial first-order logic relying on approximation spaces. By the help of defined approximative functors one can determine what kind of approximations has to be taken into consideration in the evaluating process of a formula. The representations of concepts (properties) of our available knowledge can be used to approximate not only any concept (property) but any relation. In the last section lower and upper characteristic matrixes are introduced. These can be very useful in different applications.

Keywords: Approximation of sets, rough set, partial logic, partial semantics.

1 Introduction

In recent years, a number of theoretical attempts have appeared in order to approximate sets. For example, rough set theory was originally proposed by Pawlak (see in [1], [2]), its different generalizations (see, e.g. in [3]) and granular computing[1] play a crucial role in computer sciences. Rough set theory provides a powerful foundation to reveal and discover important structures and patterns in data, and to classify complex objects.[2]

In most cases, we have a family of base sets — as subsets of a universe of discourse. In philosophy these sets represent our available knowledge, we consider them as the extensions of our available concepts/properties, and their members are the instances of these concepts/properties. The primary goal of different systems of set approximation is to "approximate/learn/express an unknown concept/property (represented by an arbitrary subset of the universe)" ([8] p. 520).

---

[1] Rough set theory has served as a "pattern" of granular computing developments, see, e.g. in [4], [5], [6], [7], [8].

[2] An overview of some research trends on rough set foundations and rough set–based methods can be found in [12].

T. Li et al. (Eds.): RSKT 2012, LNAI 7414, pp. 514–523, 2012.

Rough set theory can be considered as the foundation of various kinds of deductive reasoning. Particularly, various kinds of logics based on the rough set approach have been investigated, rough set methodology contributed essentially to modal logics, many valued logic, intuitionistic logic (see in [12]). There are many papers about the logical features of different systems of rough sets (or, in general, set approximation). A summary of this research can be found in [13].

In this paper we carry on our logical investigations about the possibility of using different systems of set approximation in logical semantics. At RST2011 we presented a very general framework of set approximation and showed that the semantics of a partial first-order logic could rely on general partial approximation spaces.[3] The common set theoretical framework proved a useful tool to compare the results and consequences of different approximations from the logical point of view. The theoretical results appeared on the meta level of our logical system.

In the present paper we show that

1. there is a logically exact way to define approximative functors on object level in order to determine what kind of approximation has to be taken into consideration in the evaluating process of a formula;
2. the representations of concepts (properties) of our available knowledge can be used to approximate not only any concept (property) but any relation.

After giving the language of our logical system, a general partial approximation space is generated by the help of an interpretation of our language in order to give semantic rules. Finally central semantic notions are defined in order to give some fundamental laws.

2   Tool-based Partial First-order Logic (TbPFoL) with Approximative Functors

2.1   Language of TbPFoL with Approximative Functors

If we want to represent approximative functors in object language we need a specific first-order language. Its main reason is that in standard first-order language there is no predicate functors, i.e. functors whose inputs and outputs are predicates. There are two different types of approximative functors: the first ones produce predicates from predicates, and the second ones produce formulae from formulae. We can treat the first ones as primitives, the second ones can be introduced by contextual definitions. Informally these functors tell us what kind of approximations (lower or upper) has to be used in order to determinate the truth value of a given formula.

Definition 1. $L = \langle LC, Var, Con, Term, \mathcal{T}, Pred, Form \rangle$ *is a language of TbPFoL with approximative functors, if*

---

[3] http://rst.disco.unimib.it/RoughSetTheory/Slides2011_files/
8-Csajbok-Mihalydeak.pdf

1. $LC = \{\neg, \supset, +, \forall, \exists, ^\nabla, ^\triangle, (,)\}$, LC is the set of logical constants;[4]
2. $Var = \{x_i : i = 0, 1, 2, \ldots\}$, Var is the denumerable infinite set of individual variables;
3. $Con = \mathcal{N} \cup (\bigcup_{n=0}^{\infty} \mathcal{P}(n))$, where $\mathcal{N}$ is the set of name parameters, and $\mathcal{P}(n)$ is the set of n–argument predicate parameters. Con is the denumerable set of non–logical constants;
4. The sets LC, Var, $\mathcal{P}(n)$ $(n = 0, 1, 2, \ldots)$ are pairwise disjoint;
5. $Term = Var \cup \mathcal{N}$;
6. $\mathcal{T} \subseteq \mathcal{P}(1)$, $\mathcal{T} \neq \emptyset$ and the set $\mathcal{T}$ (the set of tools) is finite;
7. $Pred = \bigcup_{n=1}^{\infty} Pred(n)$ (the set of predicates) and the set of n–argument predicates $Pred(n)$ is given by the following inductive definition:
   (a) $\mathcal{P}(n) \subseteq Pred(n)(n = 1, 2, \ldots)$;
   (b) If $P \in Pred(n)$, then $P^\nabla, P^\triangle, \in Pred(n)$;
8. The set $Form$ (the set of formulae) is given by the following inductive definition:
   (a) $\mathcal{P}(0) \subseteq Form$;
   (b) If $P \in Pred(n)(n = 1, 2, \ldots)$ and $t_1, t_2, \ldots, t_n \in Term$, then $P(t_1, t_2, \ldots, t_n) \in Form$;
   (c) If $A, B \in Form$, then $\neg A, (A \supset B), +A \in Form$;
   (d) If $A \in Form$, $x \in Var$, then $\forall x A, \exists x A \in Form$;

Definition 2. *Approximative sentence functors* $\blacktriangledown$, $\blacktriangle$ *can be introduced in the following inductive contextual way:*

1. If $p \in \mathcal{P}(0)$, then $\blacktriangledown p =_{def} p$, $\blacktriangle p =_{def} p$
2. It $P \in Pred(n)$, and $t_1, t_2, \ldots, t_n \in Term$, then
   (a) $\blacktriangledown P(t_1, t_2, \ldots, t_n) =_{def} P^\nabla(t_1, t_2, \ldots, t_n)$
   (b) $\blacktriangle P(t_1, t_2, \ldots, t_n) =_{def} P^\triangle(t_1, t_2, \ldots, t_n)$.
3. If $A \in Form$, then
   (a) $\blacktriangle \neg A =_{def} \neg \blacktriangle A$, and $\blacktriangledown \neg A =_{def} \neg \blacktriangledown A$
   (b) $\blacktriangle + A =_{def} + \blacktriangle A$, and $\blacktriangledown + A =_{def} + \blacktriangledown A$.
4. If $A, B \in Form$, then $\blacktriangle (A \supset B) =_{def} (\blacktriangle A \supset \blacktriangle B)$ and $\blacktriangledown (A \supset B) =_{def} (\blacktriangledown A \supset \blacktriangledown B)$.
5. If $A \in Form$ and $x \in Var$ then
   (a) $\blacktriangle \forall x A =_{def} \forall x \blacktriangle A$ and $\blacktriangle \exists x A =_{def} \exists x \blacktriangle A$.
   (b) $\blacktriangledown \forall x A =_{def} \forall x \blacktriangledown A$ and $\blacktriangledown \exists x A =_{def} \exists x \blacktriangledown A$.

2.2   Interpretations of TbPFol with Approximation Functors

Definition 3. *Let* $L = \langle LC, Var, Con, Term, \mathcal{T}, Pred, Form \rangle$ *be a language of TbPFoL with approximative functors. The ordered pair* $\langle U, \varrho \rangle$ *is a tool-based interpretation of* L, *if*

---

[4] $+$: It is true that $\ldots$, and the predicate functors $^\nabla, ^\triangle$ are the lower and upper approximative functors, respectively.

1. U is a nonempty set;
2. ϱ is a function such that $Dom(\varrho) = Con$ and
    (a) if $a \in \mathcal{N}$, then $\varrho(a) \in U$;
    (b) if $p \in \mathcal{P}(0)$, then $\varrho(p) \in \{0, 1\}$;
    (c) if $P \in \mathcal{P}(n)$ $(n = 1, 2, \ldots)$, then $\varrho(P) \in \{0, 1\}^{U^{(n)}}$;
    (d) if $T \in \mathcal{T}$, then $\varrho(T) \neq \emptyset$.

In order to give semantic rules we only need the notions of assignment and modified assignment:

Definition 4. *Function v is an assignment relying on the interpretation $\langle U, \varrho \rangle$ if $v : Var \to U$.*

Definition 5. *Let v be an assignment relying on the interpretation $\langle U, \varrho \rangle$, $x \in Var$ and $u \in U$. $v[x : u]$ is a modified assignment of v, if $v[x : u] : Var \to U$, $v[x : u](y) = v(y)$ if $x \neq y$, and $v[x : u](x) = u$.*

2.3   Generated Tool-based General Partial Approximation Spaces

If we have a tool-based interpretation of a language of TbPFoL, then the semantic values of tools (the members of set $\mathcal{T}$) determine a general partial approximation space with respect to the given interpretation. The generated approximation space is logically relevant in the sense, that it gives the lower and upper approximations of any predicate P to be taken into consideration in the definition of semantic rules.

Definition 6. *Let $L = \langle LC, Var, Con, Term, \mathcal{T}, Pred, Form \rangle$ be a language of TbPFoL with approximative functors and $\langle U, \varrho \rangle$ be a tool-based interpretation of L.*

The ordered 5–tuple $\mathcal{PAS}(\mathcal{T}) = \langle \mathcal{PR}(U), \mathfrak{B}(\mathcal{T}), \mathfrak{D}_{\mathfrak{B}(\mathcal{T})}, \mathsf{l}, \mathsf{u} \rangle$ is a logically relevant general partial approximation space generated by set $\mathcal{T}$ of tools with respect to the interpretation $\langle U, \varrho \rangle$ if

1. $\mathcal{PR}(U) = \bigcup_{n=1}^{\infty} U^{(n)}$, where $U^{(1)} = U$, $U^{(n)} = U \times U \times \cdots \times U$;
2. $\mathfrak{B}(\mathcal{T}) = \bigcup_{n=1}^{\infty} \mathfrak{B}_n(\mathcal{T})$ where $\mathfrak{B}_n(\mathcal{T}) = \{\varrho(T_1) \times \cdots \times \varrho(T_n) : T_1, \ldots, T_n \in \mathcal{T}\}$;
3. $\mathfrak{D}_{\mathfrak{B}(\mathcal{T})}$ is an extension of $\mathfrak{B}(\mathcal{T})$, i.e. $\mathfrak{B}(\mathcal{T}) \subseteq \mathfrak{D}_{\mathfrak{B}(\mathcal{T})}$, such that $\emptyset \in \mathfrak{D}_{\mathfrak{B}(\mathcal{T})}$;
4. the functions l, u forms an approximation pair $\langle \mathsf{l}, \mathsf{u} \rangle$, i.e.
    (a) $\mathsf{l}, \mathsf{u} : 2^{\mathcal{PR}(U)} \to 2^{\mathcal{PR}(U)}$ such that if $S \subseteq U^{(n)}$, then $\mathsf{l}(S), \mathsf{u}(S) \subseteq U^{(n)}$;
    (b) $\mathsf{l}(2^{\mathcal{PR}(U)}), \mathsf{u}(2^{\mathcal{PR}(U)}) \subseteq \mathfrak{D}_{\mathfrak{B}(\mathcal{T})}$ (definability of l, u);
    (c) the functions l and u are monotone, i.e. for all $S_1, S_2 \in 2^{\mathcal{PR}(U)}$ if $S_1 \subseteq S_2$ then $\mathsf{l}(S_1) \subseteq \mathsf{l}(S_2)$ and $\mathsf{u}(S_1) \subseteq \mathsf{u}(S_2)$;
    (d) $\mathsf{u}(\emptyset) = \emptyset$ (normality of u)
    (e) if $S \in \mathfrak{D}_{\mathfrak{B}(\mathcal{T})}$, then $\mathsf{l}(S) = S$ (granularity of $\mathfrak{D}_{\mathfrak{B}(\mathcal{T})}$, i.e. l is standard);
    (f) if $S \in 2^{\mathcal{PR}(U)}$, then $\mathsf{l}(S) \subseteq \mathsf{u}(S)$ (weak approximation property).

Remark 1. The semantic values of tools (given by the interpretation) generate the set $\mathfrak{B}(\mathcal{T})$. It contains those sets by which the semantic value of any predicate parameter is approximated.

The definitions of definable (totally definable) and crisp (totally crisp) sets can be given as usual in rough set theory.

Remark 2. In general case members of $\mathfrak{D}_{\mathfrak{B}(\mathcal{T})}$ are definable, but not necessarily totally definable and they are not crisp or totally crisp.

A general partial approximation space can be specified in the following different ways:

- giving some requirements for the base set;
- giving a special way how to get the set of definable sets;
- specifying the approximation pair.

Definition 7. *Let* $\mathcal{PAS}(\mathcal{T}) = \langle \mathcal{PR}(U), \mathfrak{B}(\mathcal{T}), \mathfrak{D}_{\mathfrak{B}(\mathcal{T})}, \mathsf{l}, \mathsf{u} \rangle$ *be a logically relevant general partial approximation space generated by set* $\mathcal{T}$ *of tools with respect to the interpretation* $\langle U, \varrho \rangle$.

- *Requirements from the base set point of view:*
    - $\mathcal{PAS}(\mathcal{T})$ *is total if* $\bigcup\{\varrho(T) : T \in \mathcal{T}\} = U$;
    - $\mathcal{PAS}(\mathcal{T})$ *is partial if it is not total, i.e.* $\bigcup\{\varrho(T) : T \in \mathcal{T}\} \neq U$;
    - $\mathcal{PAS}(\mathcal{T})$ *is single layered if* $\varrho(T) \not\subseteq \bigcup\{\varrho(T') : T' \in \mathcal{T}, T' \neq T\}$ *for all* $T \in \mathcal{T}$;
    - $\mathcal{PAS}(\mathcal{T})$ *is one–layered if* $\varrho(T_1) \cap \varrho(T_2) = \emptyset$ *for all* $T_1, T_2 \in \mathcal{T}$, *such that* $T_1 \neq T_2$;
    - $\mathcal{PAS}(\mathcal{T})$ *relies on Pawlakian base if* $\{\varrho(T) : T \in \mathcal{T}\}$ *is a partition of the set* U.
- *Requirement from the set* $\mathfrak{D}_{\mathfrak{B}}(\mathcal{T})$ *point of view*
    - $\mathcal{PAS}(\mathcal{T})$ *is a union type partial approximation space if* $\cup \mathfrak{B}' \in \mathfrak{D}_{\mathfrak{B}(\mathcal{T})}$ *for any* $\mathfrak{B}' \subseteq \mathfrak{B}_n(\mathcal{T}), n = 1, 2, \ldots$;
    - $\mathcal{PAS}(\mathcal{T})$ *is an intersection type partial approximation space if* $\cap \mathfrak{B}' \in \mathfrak{D}_{\mathfrak{B}(\mathcal{T})}$ *for any* $\mathfrak{B}' \subseteq \mathfrak{B}_n(\mathcal{T}), n = 1, 2, \ldots$;
- *Requirements from the approximation pair point of view:*
    - $\mathcal{PAS}(\mathcal{T})$ *is lower semi–strong, if* $\mathsf{l}(S) \subseteq S$ *for all* $S \in 2^{\mathcal{PR}(U)}$;
    - $\mathcal{PAS}(\mathcal{T})$ *is upper semi–strong, if* $S \subseteq \mathsf{u}(S)$ *for all* $S \in 2^{\mathcal{PR}(U)}$;
    - $\mathcal{PAS}(\mathcal{T})$ *is strong, if lower and upper semi–strong, i.e.* $\mathsf{l}(S) \subseteq S \subseteq \mathsf{u}(S)$ *for all* $S \in 2^{\mathcal{PR}(U)}$;
    - $\mathcal{PAS}(\mathcal{T})$ *fulfills the approximation hypothesis, if* $\langle 2^{\mathcal{PR}(U)}, \mathsf{u}, 2^{\mathcal{PR}(U)}, \mathsf{l} \rangle$ *is a Galois connection on* $\langle \mathcal{PR}(U), \subseteq \rangle$, *i.e.* $\mathsf{u}(S_1) \subseteq S_2$ *if and only if* $S_1 \subseteq \mathsf{l}(S_2)$ *for all* $S_1, S_2 \in 2^{\mathcal{PR}(U)}$.
    - $\mathcal{PAS}(\mathcal{T})$ *is a partial approximation space with Pawlakian approximation pair if for any set* $S, S \subseteq \mathcal{PR}(U)$
        1. $\mathsf{l}(S) = \bigcup\{B : B \in \mathfrak{B}(\mathcal{T}) \wedge B \subseteq S\}$;
        2. $\mathsf{u}(S) = \bigcup\{B : B \in \mathfrak{B}(\mathcal{T}) \wedge B \cap S \neq \emptyset\}$.

For the sake of simplicity general partial approximation spaces can be taken as weak ones.

Corollary 1. $\mathcal{PAS}(\mathcal{T})$ *relies on Pawlakian base iff* $\mathcal{PAS}(\mathcal{T})$ *is total and one-layered.*

2.4 Semantic Rules of TbPFoL with Approximative Functors

In the semantics of TbPFol the semantic value of an expression depends on
a given interpretation $Ip = \langle U, \varrho \rangle$, a given logically relevant general partial
approximation space $\mathcal{PAS}(\mathcal{T}) = \langle \mathcal{PR}(U), \mathfrak{B}(\mathcal{T}), \mathfrak{D}_{\mathfrak{B}(\mathcal{T})}, \mathsf{l}, \mathsf{u} \rangle$ generated by set
$\mathcal{T}$ of tools with respect to the interpretation $\langle U, \varrho \rangle$. For the sake of simplicity
we use a null entity to represent partiality of semantic rules. We use number 0
for falsity, number 1 for truth and number 2 for null entity. The semantic value
of an expression A with respect to $Ip = \langle U, \varrho \rangle$, $\mathcal{PAS}(\mathcal{T})$ and the assignment
v is denoted by $|A|_v^{Ip, \mathcal{PAS}(\mathcal{T})}$ or $|A|_v^{\langle U, \varrho \rangle, \mathcal{PAS}(\mathcal{T})}$. For the sake of simplicity the
superscripts are omitted.

Definition 8. *The semantic rules of TbPFoL with approximative functors with
respect to the interpretation $\langle U, \varrho \rangle$ and assignment v based on the logically rele-
vant general partial approximation space $\mathcal{PAS}(\mathcal{T}) = \langle \mathcal{PR}(U), \mathfrak{B}(\mathcal{T}), \mathfrak{D}_{\mathfrak{B}(\mathcal{T})}, \mathsf{l}, \mathsf{u} \rangle$
are the following:*

1. *If $a \in \mathcal{N}$, then $|a|_v = \varrho(a)$.*
2. *If $x \in Var$, then $|x|_v = v(x)$.*
3. *If $p \in \mathcal{P}(0)$, then $|p|_v = \varrho(p)$*
4. *If $T \in \mathcal{T}$, then $|T|_v = s$, where $s : U \to \{0, 1, 2\}$ is a function such that*
$$s(u) = \begin{cases} 1 \text{ if } u \in \varrho(T) \\ 0 \text{ if } u \in \mathsf{l}(U \setminus \varrho(T)) \\ 2 \text{ otherwise} \end{cases}$$
5. *If $P \in \mathcal{P}(n) \setminus \mathcal{T}$ $(n \neq 0)$, then $|P|_v = \varrho(P)$.*
6. *If $P \in Pred(n)$ $(n = 1, 2, \dots)$, then $|P^\nabla|_v = s$, where $s : U^{(n)} \to \{0, 1, 2\}$ is
a function such that*
$$s(u_1, u_2, \dots, u_n) = \begin{cases} 1 \text{ if } \langle u_1, u_2, \dots, u_n \rangle \in \mathsf{l}(|P|_v) \\ 0 \text{ if } \langle u_1, u_2, \dots, u_n \rangle \in \mathsf{l}(U^{(n)} \setminus \mathsf{u}(|P|_v)) \\ 2 \text{ otherwise} \end{cases}$$
7. *If $P \in Pred(n)$ $(n = 1, 2, \dots)$, then $|P^\triangle|_v = s$, where $s : U^{(n)} \to \{0, 1, 2\}$ is
a function such that*
$$s(u_1, u_2, \dots, u_n) = \begin{cases} 1 \text{ if } \langle u_1, u_2, \dots, u_n \rangle \in \mathsf{u}(|P|_v) \\ 0 \text{ if } \langle u_1, u_2, \dots, u_n \rangle \in \mathsf{l}(U^{(n)} \setminus \mathsf{u}(|P|_v)) \\ 2 \text{ otherwise} \end{cases}$$
8. *If $P \in Pred(n)$ $(n \neq 0)$, $t_1, \dots, t_n \in Term$, then
$|P(t_1, \dots, t_n)|_v = |P|_v(|t_1|_v, \dots, |t_n|_v)$*
9. *If $A \in Form$, then $| + A|_v = \begin{cases} 1 \text{ if } |A|_v = 1 \\ 0 \text{ otherwise} \end{cases}$*
10. *If $A \in Form$, then $|\neg A|_v = \begin{cases} 2 \quad\quad\; \text{if } |A|_v = 2 \\ 1 - |A|_v \text{ otherwise} \end{cases}$*
11. *If $A, B \in Form$, then*
$$|(A \supset B)|_v = \begin{cases} 0 \text{ if } |A|_v = 1, \text{ and } |B|_v = 0; \\ 2 \text{ if } |A|_v = 2, \text{ or } |B|_v = 2; \\ 1, \text{ otherwise} \end{cases}$$

12. If $A \in Form, x \in Var$, then
$$|\forall x A|_v = \begin{cases} 0 & \text{if there is an } u \in U\text{: } |A|_{v[x:u]} = 0; \\ 2 & \text{if for all } u \in U\text{: } |A|_{v[x:u]} = 2 \\ 1, & \text{otherwise} \end{cases}$$
$$|\exists x A|_v = \begin{cases} 1 & \text{if there is an } u \in U\text{: } |A|_{v[x:u]} = 1; \\ 2 & \text{if for all } u \in U\text{: } |A|_{v[x:u]} = 2 \\ 0, & \text{otherwise} \end{cases}$$

The semantic rules of classical logical constants as negation and implication are the conservative extensions of classical two–value ones. Conjunction, disjunction and (material) equivalence can be introduced by contextual definition: If $A, B \in Form$, then

- $(A \wedge B) =_{def} \neg(A \supset \neg B)$;
- $(A \vee B) =_{def} (\neg A \supset B)$;
- $(A \equiv B) =_{def} ((A \supset B) \wedge (B \supset A))$.

For example if P is a one–argument predicate parameter which is not a tool and $u \in U$, then

1. P^∇ is true/false at u if our tools evaluate P as certainly true/false at u;
2. P^∇ is undefined u if our tools are not enough to decide whether P is certainly true or certainly false at u);
3. $P^\triangle$ is true/false at u if our tools evaluate P as maybe true/certainly false at u;
4. $P^\triangle$ is undefined at u if our tools are not enough to decide whether P is maybe true or certainly false at u;

2.5   Theorems of Representations

In the following let $\langle T_1, T_2, \ldots, T_k \rangle$ be a fixed sequence of all tools (i. e. $\mathcal{T} = \{T_1, T_2, \ldots T_k\}$), $\langle U, \varrho \rangle$ be an interpretation, and $\mathcal{PAS}(\mathcal{T})$ be a logically relevant union type general partial approximation space with Pawlakian approximation pair generated by the set of tools $\mathcal{T}$ with respect to the interpretation $\langle U, \varrho \rangle$. Let $\mathsf{A} = \langle a_1, a_2, \ldots a_k \rangle$ be a sequence containing 0 and 1 and $T_\vee^\mathsf{A}(t) = T_{i_1}(t) \vee T_{i_2}(t) \vee \cdots \vee T_{i_l}(t)$ if $a_{i_1}, a_{i_2}, \ldots a_{i_l} = 1$ and $T_\vee^\mathsf{A}(t) =\downarrow$ if $a_i = 0$ for all i, where $\downarrow$ is a formula which is always false.

Theorem 1. *Theorem of representation*
   Let $P \in Pred(n)$. Then there are two $n \times k$ matrixes
$$\mathsf{A}_P^l = \begin{pmatrix} a_{11} & a_{12} & \ldots & a_{1k} \\ a_{21} & a_{22} & \ldots & a_{2k} \\ \vdots & \vdots & \ddots & \vdots \\ a_{n1} & a_{n2} & \ldots & a_{nk} \end{pmatrix} \quad \mathsf{B}_P^u = \begin{pmatrix} b_{11} & b_{12} & \ldots & b_{1k} \\ b_{21} & b_{22} & \ldots & b_{2k} \\ \vdots & \vdots & \ddots & \vdots \\ b_{n1} & b_{n2} & \ldots & b_{nk} \end{pmatrix}$$
   such that $a_{ij}, b_{ij} \in \{0, 1\}(i = 1, 2, \ldots, n, j = 1, 2, \ldots, k)$ and

– if $|\exists x_1 x_2 \ldots x_n P^\triangledown(x_1, x_2, \ldots, x_n)|_v = 1$, then
$|\forall x_1 x_2 \ldots x_n (P^\triangledown(x_1, x_2, \ldots, x_n) \equiv T_\triangledown^{A_1}(x_1) \wedge T_\triangledown^{A_2}(x_2) \wedge \cdots \wedge T_\triangledown^{A_n}(x_n))|_v = 1$,
where $A_i = \langle a_{i1}, a_{i2}, \ldots, a_{ik} \rangle$.
– if $|\exists x_1 x_2 \ldots x_n P^\triangle(x_1, x_2, \ldots, x_n)|_v = 1$, then
$|\forall x_1 x_2 \ldots x_n (P^\triangle(x_1, x_2, \ldots, x_n) \equiv T_\triangledown^{B_1}(x_1) \wedge T_\triangledown^{B_2}(x_2) \wedge \cdots \wedge T_\triangledown^{B_n}(x_n))|_v = 1$,
where $B_i = \langle b_{i1}, b_{i2}, \ldots, b_{ik} \rangle$.

There is no enough space to give the proof exactly. The main idea is the following: If the lower/upper approximation of an n–argument predicate is not empty, then the lower/upper approximation is a set of n–dimensional cubes. If the projection of lower/upper approximation to a fixed dimension is taken, then it can be given by the union of the semantic values of some tools. If projection to every dimension is taken, we can reconstruct the lower/upper approximation by the (1–argument) tools. The characteristic matrixes of a given n–argument predicate P show how the relation represented by P can be expressed by our available knowledge, i.e. properties represented by tools.

Definition 9. *The marixes* A_P^l, B_P^u *are the lower and upper characteristic matrixes of* P, *respectively.*

Corollary 2.   *1. If the lower approximation of predicate P is not empty, then there is a lower characteristic matrix of P.*
2. *If the upper approximation of predicate P is not empty, then there is an upper characteristic matrix of P.*
3. *If there is a lower characteristic matrix of P, then there is an upper characteristic matrix of P.*

The notions of characteristic matrixes can be introduced for any formula A which has at least one free variable by introducing a new predicate parameter.

Of course we can ask whether a characteristic matrix of a predicate P is unique.

Definition 10. *Let* $P \in Pred(n)$ *and* $\mathcal{PAS}(\mathcal{T}) = \langle \mathcal{PR}(U), \mathfrak{B}(\mathcal{T}), \mathfrak{D}_{\mathfrak{B}(\mathcal{T})}, l, u \rangle$.

1. *The predicate P is crisp with respect to the interpretation $\langle U, \varrho \rangle$ if*
$|\forall x_1 x_2 \ldots x_n (P^\triangledown(x_1, x_2, \ldots, x_n) \equiv P^\triangle(x_1, x_2, \ldots, x_n))|_v = 1$.
2. *The predicate P is definable with respect to the interpretation $\langle U, \varrho \rangle$ if*
$\{\langle x_1 x_2 \ldots x_n \rangle : |P(x_1, x_2, \ldots, x_n)|_v = 1\} \in \mathfrak{D}_{\mathfrak{B}(\mathcal{T})}$.

Theorem 2. *Let* $P \in Pred(n)$.

1. *If* $|\exists x_1 x_2 \ldots x_n P^\triangledown(x_1, x_2, \ldots, x_n)|_v = 1$, *i.e. the lower approximation of predicate P is not empty, then*
   (a) *if $\mathcal{PAS}(\mathcal{T})$ is single layered, then the lower characteristic matrix of P is unique;*
   (b) *if $\mathcal{PAS}(\mathcal{T})$ is one–layered, then the lower and upper characteristic matrixes of P are unique;*

(c) *if the predicate* P *is definable with respect to the interpretation* $\langle U, \varrho \rangle$, *then*

$$|\forall x_1 x_2 \ldots x_n (P(x_1, x_2, \ldots, x_n) \equiv T_\vee^{A_1}(x_1) \wedge T_\vee^{A_2}(x_2) \wedge \cdots \wedge T_\vee^{A_n}(x_n))|_v = 1$$

(d) *if the predicate* P *is crisp with respect to the interpretation* $\langle U, \varrho \rangle$, *and* $\mathcal{PAS}(\mathcal{T})$ *is one-layered, then the lower and upper characteristic matrixes of* P *are the same.*

2. *If* $|\exists x_1 x_2 \ldots x_n P^\Delta (x_1, x_2, \ldots, x_n)|_v = 1$, *i.e. the upper approximation of predicate* P *is not empty, and* $\mathcal{PAS}(\mathcal{T})$ *is one-layered, then the upper characteristic matrix of* P *is unique.*

The proof relies on some results in [14], where the authors introduced a very general version of the theory of set approximation. The partial approximative set theory can be considered as a generalization of the rough set theory.

3   Conclusion

If we want to investigate the different systems of set approximation from the logical point of view or we want to apply its results in inferences, we have to face a certain problem. What happens if in the semantics of first-order logic we use the approximations of sets as the semantic values of predicate parameters instead of sets given by their total interpretation? In this paper the semantic system of a partial first-order logic with approximative functors is presented. The semantics relies on a very general notion of approximation spaces generated by available knowledge (appearing in properties), and so it gives a very flexible common framework of different systems of set approximation. Approximative functors can appear in object language, and so the properties of set approximation can be given as logical laws. Presented semantic system can be considered as an alternative to logical systems which rely on a version of rough set theory and use Kripke–style possible world semantics. In the Sect. 2.5 the author shows, how a predicate can be represented by a given set of tools in a given interpretation. In [16] the authors introduce central logical notions (e.g. semantic representation, model, satisfiability, irrefutability, consequence relation) with respect to a given set of interpretations and assignments. Further features which are hold in a given interpretation or valid in a set of interpretations will also be addressed in a future work.

Acknowledgements. The work is supported by the TÁMOP 4.2.1./B-09/1/ KONV-2010-0007 project. The project is co-financed by the European Union and the European Social Fund.

The author is thankful to the anonymous referees for valuable suggestions.

References

1. Pawlak, Z.: Rough sets. International Journal of Information and Computer Science 11(5), 341–356 (1982)
2. Pawlak, Z.: Rough Sets: Theoretical Aspects of Reasoning about Data. Kluwer Academic Publishers, Dordrecht (1991)

3. Yao, Y.Y.: On Generalizing Rough Set Theory. In: Wang, G., Liu, Q., Yao, Y., Skowron, A. (eds.) RSFDGrC 2003. LNCS (LNAI), vol. 2639, pp. 44–51. Springer, Heidelberg (2003)

4. Yao, Y.Y.: Information granulation and rough set approximation. International Journal of Intelligent Systems 16(1), 87–104 (2001)

5. Skowron, A., Stepaniuk, J.: Information granules: Towards foundations of granular computing. International Journal of Intelligent Systems 16(1), 57–85 (2001)

6. Skowron, A., Świniarski, R., Synak, P.: Approximation Spaces and Information Granulation. In: Peters, J.F., Skowron, A. (eds.) Transactions on Rough Sets III. LNCS, vol. 3400, pp. 175–189. Springer, Heidelberg (2005)

7. Zadeh, L.A.: Granular Computing and Rough Set Theory. In: Kryszkiewicz, M., Peters, J.F., Rybiński, H., Skowron, A. (eds.) RSEISP 2007. LNCS (LNAI), vol. 4585, pp. 1–4. Springer, Heidelberg (2007)

8. Lin, T.Y.: Approximation Theories: Granular Computing vs Rough Sets. In: Chan, C.-C., Grzymala-Busse, J.W., Ziarko, W.P. (eds.) RSCTC 2008. LNCS (LNAI), vol. 5306, pp. 520–529. Springer, Heidelberg (2008)

9. Pawlak, Z., Polkowski, L., Skowron, A.: Rough sets: An approach to vagueness. In: Rivero, L.C., Doorn, J., Ferraggine, V. (eds.) Encyclopedia of Database Technologies and Applications, pp. 575–580. Idea Group Inc., Hershey (2005)

10. Zhu, P.: Covering rough sets based on neighborhoods: An approach without using neighborhoods. International Journal of Approximate Reasoning 52(3), 461–472 (2011)

11. Düntsch, I., Gediga, G.: Approximation Operators in Qualitative Data Analysis. In: de Swart, H., Orłowska, E., Schmidt, G., Roubens, M. (eds.) Theory and Applications of Relational Structures as Knowledge Instruments. LNCS, vol. 2929, pp. 214–230. Springer, Heidelberg (2003)

12. Pawlak, Z., Skowron, A.: Rudiments of rough sets. Information Sciences 177(1), 3–27 (2007)

13. Polkowski, L.: Rough Sets: Mathematical Foundations. AISC. Physica-Verlag, Heidelberg (2002)

14. Csajbók, Z., Mihálydeák, T.: Partial approximative set theory: A generalization of the rough set theory. International Journal of Computer Information System and Industrial Management Applications 4, 437–444 (2012)

15. Csajbók, Z.: Partial approximative set theory: A generalization of the rough set theory. In: Martin, T., Muda, A.K., Abraham, A., Prade, H., Laurent, A., Laurent, D., Sans, V. (eds.) Proceedings of the International Conference of Soft Computing and Pattern Recognition (SoCPaR 2010), Pontoise/Paris, France, December 7-10, pp. 51–56. IEEE (2010)

16. Mihálydeák, T., Csajbók, Z.: Tool-based partial first-order logic with approximative functors (forthcoming)

Author Index